Arid-land ecosystems: structure, functioning and management

Volume 2

THE INTERNATIONAL BIOLOGICAL PROGRAMME

The International Biological Programme was established by the International Council of Scientific Unions in 1964 as a counterpart of the International Geophysical Year. The subject of the IBP was defined as 'The Biological Basis of Productivity and Human Welfare', and the reason for its establishment was recognition that the rapidly increasing human population called for a better understanding of the environment as a basis for the rational management of natural resources. This could be achieved only on the basis of scientific knowledge, which in many fields of biology and in many parts of the world was felt to be inadequate. At the same time it was recognised that human activities were creating rapid and comprehensive changes in the environment. Thus, in terms of human welfare, the reason for the IBP lay in its promotion of basic knowledge relevant to the needs of man.

The IBP provided the first occasion on which biologists throughout the world were challenged to work together for a common cause. It involved an integrated and concerted examination of a wide range of problems. The Programme was co-ordinated through a series of seven sections representing the major subject areas of research. Four of these sections were concerned with the study of biological productivity on land, in freshwater, and in the seas, together with the processes of photosynthesis and nitrogen fixation. Three sections were concerned with adaptability of human populations, conservation and ecosystems and the use of biological resources.

After a decade of work, the Programme terminated in June 1974 and this series of volumes brings together, in the form of syntheses, the results of national and international activities.

INTERNATIONAL BIOLOGICAL PROGRAMME

Arid-land ecosystems: structure, functioning and management

Volume 2

EDITED BY

D. W. Goodall
R. A. Perry
with the assistance of
K. M. W. Howes
Division of Land Resources Management
CSIRO, Wembley, Western Australia

CAMBRIDGE UNIVERSITY PRESS
CAMBRIDGE
LONDON · NEW YORK · NEW ROCHELLE
MELBOURNE · SYDNEY

CAMBRIDGE UNIVERSITY PRESS
Cambridge, New York, Melbourne, Madrid, Cape Town, Singapore, São Paulo, Delhi

Cambridge University Press
The Edinburgh Building, Cambridge CB2 8RU, UK

Published in the United States of America by Cambridge University Press, New York

www.cambridge.org
Information on this title: www.cambridge.org/9780521105569

First published 1981
This digitally printed version 2009

A catalogue record for this publication is available from the British Library

ISBN 978-0-521-22988-3 hardback
ISBN 978-0-521-10556-9 paperback

Contents

List of Collaborators xiii
Preface xv
J. B. Cragg

Introduction 1
R. A. Perry

Part I. Composite and interactive processes 3

1 Introduction 5
F. H. Wagner
2 Soil–vegetation–atmosphere interactions 9
M. A. Ayyad
3 Plant–plant interactions 33
M. G. Barbour
4 Animal–animal interactions 51
F. H. Wagner & R. D. Graetz
5 Plant–animal interactions 85
R. D. Graetz
6 Effects of biotic components on abiotic components 105
K. E. Lee
7 Population dynamics 125
F. H. Wagner
8 Primary productivity 169
L. E. Rodin
9 Production by desert animals 199
F. B. Turner & R. M. Chew

Part II. Ecosystem dynamics 261

10 Introduction 263
J. A. MacMahon
11 Short-term water and energy flow in arid ecosystems 271
J. A. Ludwig & W. G. Whitford
12 Nutrient cycling in desert ecosystems 301
N. E. West

13 Short-term dynamics of minerals in arid ecosystems 325
P. Binet
14 Long-term dynamics in arid-land vegetation and ecosystems of North Africa 357
H. N. Le Houérou
15 The modelling of arid ecosystem dynamics 385
D. W. Goodall
16 Spatial effects in modelling of arid ecosystems 411
I. Noy-Meir
17 Simulation of plant production in arid regions: a hierarchical approach 433
H. van Keulen & C. T. de Wit
18 Understanding arid ecosystems: the challenge 447
I. Noy-Meir

Part III. Management of arid lands 451

19 Introduction 453
R. A. Perry
20 Management of arid-land resources for domestic livestock forage 455
C. M. McKell & B. E. Norton
21 Management of arid-land resources for dryland and irrigated crops 479
H. S. Mann
22 Recreation and tourism in arid lands 495
M. D. Sutton
23 Management of water resources in arid lands 519
J. L. Thames & J. N. Fischer
24 Some social aspects of managing arid ecosystems 549
M. D. Young
25 Synthesis 555
M. Evenari

Index 593

Table des matières

Liste des collaborateurs xiii
Préface xv
J. B. Cragg
Introduction 1
R. A. Perry

Part I. Processus complexus et interactions 3
1 Introduction 5
F. H. Wagner
2 Interactions sol–végétation–atmosphère 9
M. A. Ayyad
3 Interactions plante–plante 33
M. G. Barbour
4 Interactions animal–animal 51
F. H. Wagner & R. D. Graetz
5 Interactions plante–animal 85
R. D. Graetz
6 Effets des éléments biotiques sur les éléments abiotiques 105
K. E. Lee
7 Dynamique des populations 125
F. H. Wagner
8 Productivité primaire 169
L. E. Rodin
9 Animaux désertiques comme producteurs 199
F. B. Turner & R. M. Chew

Part II. Dynamique des écosystèmes 261
10 Introduction 263
J. A. MacMahon
11 Ecoulement de l'eau et d'énergie de nature transitoire dans les écosystèmes arides 271
J. A. Ludwig & W. G. Whitford
12 Cycles alimentaires dans les écosystèmes désertiques 301
N. E. West

13 Dynamique à court terme des substances minerales dans les écosystèmes arides 325
P. Binet
14 Dynamique à long terme de la végétation et écosystèmes arides de l'Afrique du Nord 357
H. N. Le Houérou
15 La modelisation de la dynamique de l'écosystème aride 385
D. W. Goodall
16 Effets spatiaux dans la modélisation des écosystèmes arides 411
I. Noy-Meir
17 Simulation de la production végétale dans les régions arides: une approche hiérarchi e 433
H. van Keulen & C. T. de Wit
18 Comprendre les écosystèmes arides: le défi 447
I. Noy-Meir

Part III. Gestion des pays arides 451
19 Introduction 453
R. A. Perry
20 Gestion en pays aride des ressources pour le fourrage du bétail domestique 455
C. M. McKell & B. E. Norton
21 Gestion des ressources en pays arides pour les cultures sêches et irriguées 479
H. S. Mann
22 Loisirs et tourisme en pays arides 495
M. D. Sutton
23 Gestion en pays arides des ressources en eau 519
J. L. Thames & J. N. Fischer
24 Aspects sociaux de la gestion des écosystèmes arides 549
M. D. Young
25 Synthèse 555
M. Evenari

Index 593

Contenido

Lista de Colaboradores xiii
Prólogo xv
J. B. Cragg
Introducción 1
R. A. Perry

Parte I. Procesos compuestos e interactivos 3
1 Introducción 5
F. H. Wagner
2 Interacciones suelo–vegetación–atmósfera 9
M. A. Ayyad
3 Interacciones planta–planta 33
M. G. Barbour
4 Interacciones animal–animal 51
F. H. Wagner & R. D. Graetz
5 Interacciones animal–planta 85
R. D. Graetz
6 Efectos de los componentes bióticos sobre los abióticos 105
K. E. Lee
7 Dinámica de poblaciones 125
F. H. Wagner
8 Productividad primaria 169
L. E. Rodin
9 Producción en animales de desierto 199
F. B. Turner & R. M. Chew

Parte II. Dinámica del ecosistema 261
10 Introducción 263
J. A. MacMahon
11 El flugo de agua y energía de periodo breve en los ecosistemas de tipo árido 271
J. A. Ludwig & W. G. Whitford
12 Ciclo de nutrientes en ecosistemas de desierto 301
N. E. West

13 Dinámica de periodo breve en los minerales de ecosistemas de tipo árido 325
P. Binet

14 Dinámica de largo periodo en la vegetación y ecosistemas de tierras áridas del Norte de África 357
H. N. Le Houérou

15 La concepción de modelos de dinámica de ecosistemas áridas 385
D. W. Goodall

16 Efectos espaciales en la confección de modelos de ecosistemas áridos 411
I. Noy-Meir

17 Simulación de la producción vegetal en regiones áridas: un enfoque jerárquico 433
H. van Keulen & C. T. de Wit

18 Comprender los ecosistemas de tipo árido: el problema 447
I. Noy-Meir

Parte III. Gestión de las tierras áridas 451

19 Introducción 453
R. A. Perry

20 Gestión de los recursos de las tierras áridas para el pasto del ganado doméstico 455
C. M. McKell & B. E. Norton

21 Gestión de los recursos de las tierras áridas para cultivos de secano y de regadío 479
H. S. Mann

22 Esparcimiento y turismo en las tierras áridas 495
M. D. Sutton

23 Gestión de los recursos hídricos de las tierras áridas 519
J. L. Thames & J. N. Fischer

24 Aspectos sociales de la gestión de las tierras áridas 549
M. D. Young

25 Síntesis 555
M. Evenari

Índice 593

Содержание

Список авторов xiii

Предисловие xv

J. B. Cragg

Введение 1

R. A. Perry

Часть I. Сложные и взаимодействующие процессы 3

1 Введение 5
F. H. Wagner

2 Взаимодействия между почвой, растителбностьбю и атмосферой 9
M. A. Ayyad

3 Взаимоотнощения между растениями 33
M. G. Barbour

4 Взаимоотношения между животными 51
F. G. Wagner & R. D. Graetz

5 Взаимоотношения между растениями и животными 85
R. D. Graetz

6 Воздействие биотических компонентов на абиотические 105
K. E. Lee

7 Динамика популяций 125
F. H. Wagner

8 Первичная продуктивность 169
L. E. Rodin

9 Продукция пустынных животных 199
F. B. Turner & R. M. Chew

Часть II. Динамика экосистем 261

10 Введение 263
J. A. MacMahon

11 кратковремжнное течение вопы и знергии в аридных зкосистемах 271
J. A. Ludwig & W. G. Whitford

12 Циклы питательных веществ в пустынных экосистемах 301
N. E. West

13 Динамика минеральных веществ в аридных экосисистемах в течение коротких периодов 325
P. Binet
14 Долговременная динамика аридной наземной растительности и экосистем Северной Африки 357
H. N. Le Houérou
15 Моделирование динамики аридных экосистем 385
D. W. Goodall
16 Пространственные эффекры при моделировании аридных экосистем 411
I. Noy-Meir
17 Моделирование продуцирования растений в аридных зонах: иерархический подход 433
H. van Keulen & C. T. de Wit
18 Попытка понимания аридных экосистем 447
I. Noy-Meir

Часть III. Управление аридными землями 451
19 Введение 453
R. A. Perry
20 Управление ресурсами аридных земель как кормовой базы домашнего скота 455
C. M. McKell & B. E. Norton
21 Управление ресурсами аридных земель для неорошаемого и орошаемого растениеводства 479
H. S. Mann
22 Рекреация и туризм в аридных земелях 495
M. D. Sutton
23 Управление водными ресурсами в аридных странах 519
J. L. Thames & J. N. Fischer
24 Социальные аспекты управления аридными экосистемами 549
M. D. Young
25 Синтез 555
M. Evenari

Указатель 593

List of Collaborators

M. A. Ayyad	Botany Department, Faculty of Science, University of Alexandria, Moharran Bay, Alexandria, Egypt
M. G. Barbour	Botany Department, University of California, Davis, California 95616, USA
P. Binet	Laboratoire de Physiologie végétale, Université de Caen, France
R. M. Chew	Department of Biological Sciences, University of Southern California, Los Angeles, California 9007, USA
C. T. de Wit	Department of Theoretical Production Ecology, Agricultural University, Wageningen, The Netherlands
M. Evenari	Department of Botany, Hebrew University, Jerusalem, Israel
J. N. Fischer	College of Agriculture, University of Arizona, Tucson, Arizona 85721, USA
D. W. Goodall	CSIRO, Division of Land Resources Management, Private Bag, P. O. Wembley, Western Australia 60014
R. D. Graetz	CSIRO, Division of Land Resources Management, Deniliquin, N.S.W., Australia
K. Lee	CSIRO, Division of Soils, Private Bag No. 2, Glen Osmond, South Australia
H. N. Le Houérou	P.O. Box 5689, Addis Ababa, Ethiopia, Africa
J. A. Ludwig	Department of Biology, New Mexico State University, Las Cruces, New Mexico, USA
C. M. McKell	Department of Range Science, Utah State University, Logan, Utah 84322, USA
J. A. MacMahon	Department of Biology and Ecology Center, Utah State University, Logan, Utah 84322, USA
H. S. Mann	Central Arid Zone Research Institute, Jodhpur, India
B. E. Norton	College of Natural Resources, Utah State University, Logan, Utah 84322, USA
I. Noy-Meir	Department of Botany, Hebrew University, Jerusalem, Israel
R. A. Perry	CSIRO, Division of Land Resources Management, P.O. Wembley, Western Australia 60014

List of collaborators

L. E. Rodin	Komarov Botanical Institute, USSR Academy of Science, Leningrad, USSR
M. D. Sutton	1909 Eardale Court, Alexandria, Virginia 22306, USA
J. L. Thames	College of Agriculture, University of Arizona, Tucson, Arizona 85721, USA
F. B. Turner	Laboratory of Nuclear Medicine and Radiation Biology, University of California, Los Angeles, California 90024, USA
H. van Keulen	Centre for Agrobiological Research, Wageningen, The Netherlands
F. H. Wagner	College of Natural Resources, Utah State University, Logan 84322, USA
W. G. Whitford	Department of Biology, New Mexico State University, Las Cruces, New Mexico, USA
M. D. Young	CSIRO, Division of Land Resources Management, Deniliquin, N.S.W., Australia

Preface

The anonymous author of a recent article, reflecting on UNESCO's many years of research on natural resources, began by quoting Francis Bacon: *Naturae enim non imperatur, nisi parendo*. These words, even in translation, remain linguistically impressive and important: 'We cannot command nature except by obeying her.' Within a few years of its establishment in 1946, UNESCO, acting as though it had taken Bacon's aphorism to heart, initiated a worldwide programme on arid lands. The numerous volumes and papers which have emerged from thirty years of study have indicated some of the ways in which these fragile ecosystems can be utilized and at the same time safeguarded against destruction. IBP's Arid Lands programme followed many of the pathways established by UNESCO and it is important to remember that IBP itself has had UNESCO's support from its inception.

Whilst arid lands defy precise definition, they cover a large part of the earth. A conservative estimate puts their area at about one-eighth of the land surface of the globe. If semi-arid areas are included, the total reaches one-third of the world's land surface with, perhaps, one-tenth of the world's population dependent on it for survival.

Characterized by long periods without precipitation and by extreme temperature gradients, arid lands impose considerable restraints on the organisms which inhabit them. Above all, their ecosystems, as recent years have demonstrated so clearly in the Sahel, are subject to severe and, in some cases, irreversible destruction. The Arid Lands Theme in IBP was shaped to establish a firm basis of integrated knowledge on which to base the future management of arid lands.

Studies on arid lands within the Terrestrial Productivity section of IBP were defined as Theme 3 in the following terms: 'Arid Lands present important problems to man's rational use of natural resources. In these areas vegetation is in sensitive balance with an irregular and limited water supply and, in general, consumers are either nomadic (or migratory) or are capable of high protein production during favourable periods. The balance of arid zone communities with the environment is so precarious that man has devastated large areas of the world by unwise management practices. The IBP aim is to further our understanding of the structure and functioning of these arid zone communities, so that we may be better able to predict the consequences of man's efforts to utilize the natural resources of these areas.'

In 1969, eleven National IBP Committees were prepared to contribute projects to the Theme. Some of the projects never materialized and only the USA mounted a major coordinated biome study. Nevertheless, world coverage was extensive and it was no easy matter to organize the synthesis of many different lines of enquiry. The Steering Committee which guided this final phase of IBP's arid lands investigations consisted of: Dr D. W. Goodall (USA – now in Australia), Professor M. Evenari (Israel), Professor M. Kassas (Egypt), Professor L. A. Rodin (USSR), Dr V. Roig (Argentina), Dr R. K. Gupta (India) and Dr R. A. Perry (Australia). The latter member acted as Convener. In 1972, the Committee met twice – in Leningrad in June and in Logan, USA, in September. A Symposium was to have been held in New Delhi in 1973 but the meeting had to be cancelled through lack of funds. However, the plan proposed for that meeting under the title *Arid Ecosystems and Management* represented an important step in the preparation of the synthesis volumes.

At Leningrad it was proposed to have three volumes: the first concerned with the description and structure of arid land ecosystems with some emphasis on ecosystem processes; the second was to be devoted to ecosystem dynamics both short-term and long-term; the final volume was to take as its theme the all-important subject of the management of arid lands and was to deal with the practical aspects of management and the utilization of optimization models. These overall plans, as happened with early proposals in other biome studies, foundered through lack of funds. Once the field investigations for IBP had been completed, some individuals had to take up other activities and were unable to contribute in a major way to the final synthesis.

At the Logan meeting the plan, as a consequence, had to be reduced in size and, in my opinion, the synthesis has gained rather than lost by the twin processes of attrition and compression. The two volumes which have now emerged present a succinct view of the extensive problems associated with arid lands. Of special interest is the approach to modelling which has been followed in this Theme. The very nature of arid lands lent substance to Goodall's view stated some years ago in a meeting at Montreal: 'The dynamics of most ecosystems constitute a succession of transients, and they are never in anything resembling a steady state.' Thus modelling in the Arid Lands Biome has included a number of sub-models built to answer specific questions, whereas elsewhere in IBP attempts were made to create large models capable of answering a multiplicity of questions. Wherever modelling has been discussed in IBP, the small-model vs large-model approach has been heatedly contested. I cannot disguise my own sympathies for Goodall's insistence on the necessity for concentrating efforts on sub-models and, of course, on their improvement and validation especially in relation to measuring productive capacity.

Again quoting Goodall: 'Prediction is the acid test of a set of hypotheses about the real world; the pure gold survives, the counterfeit is rejected and must be replaced.' His ideas influenced much of the thinking in the Arid Lands Biome studies. In all his modelling studies, Goodall never lost sight of the importance of fundamental biological knowledge for he considers that attempts '...to understand the dynamics of an ecosystem without taking biological individuality into account is doomed to failure.'

No programme is ever completed to the satisfaction of its participants. There have been many moments of despondence in the synthesis phase as gaps emerged in our knowledge of the basic components of arid land ecosystems. Information essential for the development of a broadly based general theory was lacking.

The absence of a firm data base cannot be attributed to an absence of funding or to a shortage of scientific personnel although these deficiencies did contribute in some measure. There is another reason for the 'information gap' – the piecemeal approach to many environmental problems. These two volumes on Arid Lands, together with other IBP Biome volumes now in press or in the final stages of preparation, emphasize the need for a holistic approach to the investigation of ecosystems and related environmental problems. In fact, the exposure of gaps in our present knowledge can be looked upon as one of the major successes of the programme! Topics which now require detailed study have been defined and some have already become parts of two major international operations: the desertification studies of UNEP (United Nations Environment Program) and UNESCO's *Man and the Biosphere*. The latter is combining the predominantly biological approaches of IBP with social and economic studies.

J. B. CRAGG
Killam Memorial Professor

Introduction

R. A. PERRY

Arid lands comprise a significant proportion of the land surface of the world and support about 300 million people. The history of man's use of arid lands is, in general, a sad record of progressive, man-induced deterioration of the natural resource base and of generally low and declining living standards for the human population. Developing ways to reverse this trend in resource degeneration and to raise the living standards of the increasing human population supported on these resources is an important challenge to mankind. One pre-requisite to meeting the challenge is a sound knowledge of the natural ecosystems. It is to this objective that the IBP arid-land studies reported in this, and a succeeding volume, are addressed.

Arid-land studies are part of only one of the themes of the International Biological Programme. Throughout the world the main contribution to this sub-theme, in terms of research specifically funded and identified with the IBP, was from the USA. However, a very considerable amount of research in other countries had similar objectives and so is relevant to the IBP although not identified specifically with it. These two volumes have drawn on information from all relevant research and thus include much more than the results of IBP studies.

The subject matter of the two volumes is organised into five major topics; two are included in the first volume and three in the second volume.

The first section of Volume 1 describes in detail the structure of the arid ecosystems in terms of climate, soils, geomorphology, hydrology and flora and fauna. Unfortunately South America is not covered because, despite very considerable and repeated attempts, the editors were unable to find an author willing to prepare the chapter. The areas which are described include North America, north Africa, south Africa, Australasia, southwest Asia and central Asia. In the second section of Volume 1 the processes which operates within, and control, the ecosystem are dealt with individually. The section is subdivided into four parts which deal with aspects of atmospheric soil, plant, and animal processes.

The interaction of the various components of the ecosystem is covered in the first section of Volume 2 and this is followed in the second section by an integration of the composite and interactive processes as they influence the short- and long-term dynamics of the whole system. Over the IBP period modelling has played an increasingly important part in the integrative views of ecosystem functioning and three chapters are devoted to this topic. The third section in Volume 2 deals with the impact of man on the arid areas and the various options available for management of the arid lands. A final chapter is included in Volume 2 to pull together the contributions to both volumes and to indicate areas where further research could improve our knowledge and management of the arid lands.

PART I

Composite and interactive processes

1. Introduction

F. H. WAGNER

Historically, the most commonly perceived biological characteristic of deserts has been the dearth of life and biological production. While long understood by nomadic inhabitants, the extreme temporal and spatial variability of desert ecosystems has only come into scientific prominence in the past few decades. The temporal variability results, of course, from the extreme moisture constraint under which the system exists, and from the fact that precipitation is relatively more variable in arid climates than in more mesic ones.

The most evident ecological manifestation of temporal variability in desert ecosystems is the great, relative variation in annual primary production. Annual, net primary production in forests may vary by less than a factor of one and a half between years, and in grasslands by a factor of two to four, but desert production commonly varies by a factor of five to ten between years with below and above average precipitation (Wagner, 1979). One explicit means of representing these differences is in response curves which depict the logarithm of annual production as functions of annual precipitation (Le Houérou & Hoste, 1977). The desert curve has a much steeper slope than the grassland and forest curves (Wagner, 1978).

A number of scientific and applied implications follow from the tight moisture relationship and high variability of desert primary production. The steep response curve implies that any factor producing slight change in the moisture regime of an area will elicit marked change in vegetation structure and function. In Chapter 2 Ayyad explores such effects induced by variations in topography, soil texture and salinity as well as in meteorologic variables themselves. And while most attention tends to be focused on the overriding influence of the physical environment on desert biotas, Lee explores the very real reciprocal effects of biota on the physical environment in Chapter 6.

While not explored in the chapters of Part 1, the steep moisture–response curve would seem to have important applied implications. On the negative side, any land-use practice which tended to reduce available moisture could be expected to elicit marked reduction in primary production and change in vegetation structure. The system could then be said to be fragile, if fragility is defined as the magnitude of response from a measured perturbation. Anthropogenic effects that could alter moisture regimes are changes in microclimate through vegetative removal for fuel, building material, and livestock forage; and changes in soil structure from livestock trampling or heavy recreational use. Rodin discusses the effects of such uses on primary production in Chapter 8. In this same connection, Otterman (1974) proposes the extremely interesting hypothesis that anthropogenic reduction in plant cover can expose high-albedo soils, lower surface temperature and atmospheric ascent, and ultimately reduce adiabatically produced rainfall.

On the positive side, any land-management measure that tended to enhance the moisture conditions of an area could be expected to elicit a marked increase in production and change in vegetation structure. The system could then be said to be resilient, if resilience is defined as the magnitude of response following release or reduction of a constraint. This is somewhat contrary to the traditional view of desert systems having great inertia and recovering slowly from perturbation. And yet, the steep moisture–response curve implies a high degree of responsiveness in the system with slight changes in moisture conditions. If recovery is slow and inertia high, it would seem to imply little or no mean change in the moisture conditions.

On the basis of contemporary ecological theory, the extreme variability of desert precipitation and primary production leads to several predictions about the structure of desert communities, the niche characteristics of constituent species, and the influences limiting their populations. Where resources are scarce and unpredictable, it is to the adaptive advantage of species to evolve broad niche characteristics and remain relatively unspecialized (MacArthur, 1969, 1971; Cody, 1974). Graetz explores these characteristics of several desert species in Chapter 5, while in Chapter 3 Barbour examines the evidence for competition in desert vegetation which would be an important force in the evolution of niche structure. Wagner and Graetz weigh similar considerations in Chapter 4 with regard to animals.

If niche spaces are wide in desert communities, those communities would be expected to be biotically simple according to contemporary theory. While Pianka examines several aspects of diversity in desert communities in Volume 1, Chapter 10, a great deal of interbiome comparison remains to be done to test this implication of niche theory. Is α-diversity in general lower in plant, arthropod, reptilian, mammalian, and avian communities of deserts than in more mesic systems? The question fairly cries out for an answer.

As Graetz points out in Chapter 5, it is also to the adaptive advantage of animals in an unpredictable environment to exploit resources rather conservatively. This implies that populations in desert systems should be rather strongly self-limited, or by combinations of self limitation and direct limitation of physical factors. As a corollary, desert populations would not be expected to be significantly limited by resource ceilings, interspecific competition, and predation. Wagner & Graetz in Chaper 4 and Wagner in Chaper 7 address these hypotheses in terms of population mechanisms and limiting influences.

Conservative resource use by animals should be reflected in energy-flow patterns. While research has not yet been sufficiently complete to elucidate energy-flow patterns of whole desert ecosystems in the detail needed to test the implications of conservative resource use, Turner & Chew examine energy flow in a number of desert animal species in meticulous detail in Chapter 9. When analyses of these kinds are available for the entire fauna, the resource-use patterns will become evident.

A final implication of desert variability lies in the demographic strategies of the constituent species. It is to the adaptive advantage of species in an unpredictable environment to be able to make prompt numerical responses to favorable conditions, according to contemporary theory. Such species would be expected to have higher potential rates of increase than related species in more mesic, less variable environments. Wagner explores this question in Chapter 7.

In conclusion, a great deal of ecological theory and many applied implications, relate to the variability of different ecosystems. Deserts are among the most variable, and therefore provide excellent ground for empirical test. The authors of Part 1 proceed with some of this testing.

References

Cody, M. L. (1974). *Competition and the structure of bird communities.* Monograph in Population Biology No. 7. Princeton University Press.

Le Houérou, H. N. & Hoste, C. H. (1977). Rangeland production and annual rainfall relations in the Mediterranean Basin and in the African Sahelo-Sudanian Zone. *Journal of Range Management*, **30**, 181–9.

MacArthur, R. H. (1969). Species packing and what interspecies competition minimizes. *Proceedings of the National Academy of Science, USA*, **64**, 1396–471.

MacArthur, R. H. (1971). Patterns of terrestrial bird communities. In: *Avian biology*, vol. 1 (ed. D. S. Farner & J. R. Kings), pp. 189–222. Academic Press, New York.

Otterman, J. (1974). Baring high-albedo soils by overgrazing: A hypothesized desertification mechanism. *Science*, **186**, 531–3.

Wagner, F. H. (1978). Desert ecosystem. In: *1978 McGraw-Hill Yearbook of Science and Technology*, pp. 137–9. McGraw-Hill, New York.

Wagner, F. H. (1979). Integrating and control mechanisms in arid and semiarid systems – considerations for impact assessment. In: *Proceedings of a symposium on biological evaluation of environmental impact.* 27th Annual AIBS Meeting, New Orleans. Council on Environmental Quality, Washington, DC, in press.

Manuscript received by the editors September 1978

2. Soil–vegetation–atmosphere interactions

M. A. AYYAD

Introduction

Variability of ecosystem structure and function is generally a product of interactions between its different components. With the austerity of environmental conditions in arid lands, these interactions incur high significance, so that slight irregularities in one component of the ecosystem are likely to entail remarkable spatial variations in others, creating distinct micro-habitats. The smallest incident may trigger a chain of temporal changes encompassing the whole ecosystem and induce far-reaching effects on the plants. A considerable number of ecological studies have dealt with different aspects of variability in desert ecosystems and have undoubtedly added to our comprehension of the nature of interactions between different components (see the reviews by Shreve, 1942, 1951; Beadle, 1948; Chapman, 1960; Zohary, 1962, 1972; Walter, 1964; Quézel, 1965; Hastings & Turner, 1965; Le Houérou, 1969; Evenari, Shanan & Tadmor, 1971; Emberger, 1971). But these interactions are often too complex to be fully explained through individual studies and one hopes that the integrative approach of simulation modelling will be helpful in this respect (Goodall, 1972, 1973*a*, *b*; Noy-Meir, 1974).

The following account is a review of studies dealing with the nature of interactions between soil, vegetation and atmosphere in arid ecosystems, and with local spatial variations and short-term temporal changes in vegetational attributes affected by the soil–atmosphere complex. The soil component is viewed here with regard to edaphic and physiographic characters, and the atmospheric component with regard to climatic and microclimatic elements. Vegetational variations with time and space are considered in terms of distributional patterns, productivity, life-form and phenology.

The nature of interactions in desert ecosystems

The inter-relationships between soil, vegetation and atmosphere in arid lands are so intricate that, in an ecological perspective, they can hardly be contemplated as separate entities. While physiographic and edaphic features of a desert ecosystem are frequently determined by climatic factors, microclimatic variations are produced by landform patterns. Whereas the soil–atmosphere complex controls the behaviour of plant species, these species usually contribute to the process of soil formation and may modify the microclimate.

Composite and interactive processes

Climatically-induced processes of weathering, erosion and deposition are continuously at work, dissecting desert landscapes into a variety of landforms and fragmenting the physical environment into a complex mosaic of micro-environments (Emberger & Lemée, 1962). The impact of rainfall is unmistakable. A decade or even a century may pass before a desert ecosystem experiences a heavy downpour, but when it does fall, it achieves a great deal of erosion and deposition due to the sparseness of vegetation which offers little or no protection to the soil. Major erosional forms now present in desert generally result from fluvial action (Hills, Ollier & Twidale, 1966). Some, such as wadis and their affluents (ancient water streams which dried up with increasing aridity), are undoubtedly relict features deriving from past periods of heavier rainfall, but many are attributable to occasional heavy rainfall of the present regime.

During the initial stages of land-mass denudation in arid land, numerous large depressions exist between ridges (Strahler, 1960). In the depressions, the extent of waterlogging depends on the relative proximity of the water table to the soil surface. However, even with a high water table, a depression will usually remain dry except during short rainy periods, due to excessive evaporation in the hot, dry climate. As denudation proceeds, depressions become progressively filled with alluvial material from adjoining ridges which are worn lower. This process of alluviation exerts a multitudinous effect on edaphic characters.

On ridges, the depth of soil will depend upon the intensity of rainfall, and the degree and position of slope. The lower position of a gentle slope will receive more soil than it loses, while the upper position of a steep slope will be almost permanently bare, since any products of erosion will have little chance to remain *in situ* for long.

In depressions, as run-off water slows down, eroded material is deposited in sequence of weight: coarser sediments at the borders and finer ones towards the centre. The concentration of different ions in these sediments will depend upon the chemical nature of the parent rock of bordering ridges.

At the beginning of the erosion cycle, a desert plateau has an intact rocky surface which affords little, if any, possibility for plant growth. As erosion proceeds, fissures and notches form where fine soil material may collect, furnishing favourable micro-habitats for a few plants. At a later stage, the rock surface may become veneered with residual rock which may be produced by weathering *in situ*, forming an erosion pavement. Alternatively it may be transported by run-off water, forming a gravel desert (Kassas & Imam, 1959; Kassas, 1966). Both forms are said to be mature after the removal of soft material by deflation or washing so that the coarse lag material at the surface becomes closely strewn, forming a 'desert armour'. Beside the impenetrability of these surfaces, geochemical processes lead to the formation of one or more subsurface layers of gypsum and other salts, which adds to their hostility for plant growth. Occasionally, wind- or water-borne soft material may form a surface sheet over these sterile surfaces where shallow-rooted plants may become established; otherwise, the plant cover will usually be confined to shallow drainage runnels.

Where the parent rock is disintegrated by weathering into finer soil material, and where the plant cover is scanty, wind becomes an extremely active agent of transportation, erosion and deposition. In arid lands, blowouts and sand dunes are the prominent landforms shaped by wind action. Blowouts are formed either in plains in the form of shallow depressions which become progressively enlarged by deflation, or on rock surfaces where the rock is being weathered or desiccated. Dunes accumulate in different forms; they may be crescentic, parabolic, longitudinal or transverse, but they are all characterized by the coarse loose nature of sand and present a variable environment because of their changing stability. When first formed, a dune is unstable, and relatively few plant species endowed with the ability of sand-binding and resistance to burial may survive. Later, with the progressive stabilization of the dune, these specialized species give way to ones less adapted to unstable situations.

Due to the paucity of rainfall, the high evaporation rate and the sparseness of vegetation in arid lands, salt accumulation close to the soil surface becomes a common phenomenon. This is especially obvious in areas where drainage is impeded by such factors as the high proportion of clay and silt, and the close proximity of water table to the soil surface.

Climatically induced variations in landform and soil features in arid lands lead to modifications in the 'average' climatic conditions. Of these modifications, the most crucial to plant life are those related to the amount of rainfall. In fact, the degree of moisture availability in a desert ecosystem is not a direct product of the amount of rainfall, but rather a result of the modification of that amount by the complex interactions of numerous physiographic, edaphic and microclimatic factors. Conceivably, the amount of available moisture varies remarkably according to topographic position; in plains and wadis, it will usually be several times greater than on the tops of hills and slopes (Tadmor, Orshan & Rawitz, 1962). Moisture availability on slopes is largely determined, directly or indirectly, by their orientation, angle and elevation (Kassas & Imam, 1957; Ayyad & Ammar, 1974). At subtropical latitudes, where the largest part of the arid zone is situated, differences in the amounts of radiation (and consequently, the amounts of evapotranspiration) received by slopes of opposite orientations may be smaller than those at higher latitudes (Ayyad, 1971). However, even these small differences may affect plant life in deserts, where moisture availability is a limiting factor. The more decisive role of slopes is, indeed, through their control of the amount of moisure received or lost by run-off. The upper positions of steep slopes would retain the least, and the lower positions of gentle slopes the largest amount of moisture.

Compared to other landform types, a wadi has the great merit of being a drainage system collecting water from an extensive catchment area and thus forms one of the most favourable habitats for plant growth. Except for the central part of its bed and the outer curve of its meander, where the torrent action of rainfall provides little chance for soil build-up, the wadi usually supports richer vegetation than other types of habitat (Kassas & Imam, 1954). The soil profile of a wadi bed is characterized by the presence of layers of fine material alternating with layers of coarse gravel. As the texture of these sediments is indicative of the transporting capacity of water bodies in the wadi, this alternation of layers indicates episodic variations in water resources.

Interacting with the effects of topographic peculiarities on the amount of moisture availability in desert ecosystems, are the effects of the nature of surface, depth and texture of the soil (Shreve, 1951; Hillel & Tadmor, 1962; Ayyad & Ammar, 1973, 1974). It is at the surface of the lithosphere where partitioning between infiltration, run-off, evapotranspiration and deep seepage is initiated. An intact rock surface retains almost no moisture, and except for certain lithophytes, affords no possibility for plant growth, while a fissured or notched rock surface may furnish suitable micro-habitats for chasmophytes and annuals where soft soil material and moisture accumulate. The amount of moisture retained by a continuous sheet of soft material depends primarily on its depth. A thin soil sheet is moistened during the rainy season, but becomes desiccated by the approach of the dry season and thus can only support a plant cover of shallow-rooted annuals. A deep soil retains and stores moisture for longer periods and offers a possibility for deep-rooted perennials to become established.

The role of texture in determining the degree of moisture availability in arid lands is no less important. Sandy and gravelly soils usually have more favourable moisture conditions and support denser and taller vegetation than silty and clayey soils (Shreve, 1942; Hillel & Tadmor, 1962; Noy-Meir, 1973). This may be attributed to the greater infiltration capacity of coarse soils which safeguards against excessive losses by run-off and evaporation, allowing for storage of greater amounts of water in deeply-seated layers.

The role played by vegetation in desert ecosystems varies with the degree of aridity. In extremely arid regions with very scanty vegetation, this role is insignificant, and soil development is essentially a geochemical process where calcareous, siliceous and gypsum crusts or subsurface pans are formed. As rainfall increases and the vegetation becomes more dense, plants assume an active role in modifying edaphic and climatic conditions. They influence infiltration through their effects on pore size, wettability, surface tension and viscosity (Fletcher, 1960; Kincaid, Gardner & Schreiber, 1964; Lyford & Qashu, 1969). Desert plants may also have an active role in stabilizing surface deposits. Some are capable of building mounds and hillocks (Batanouny & Batanouny, 1968, 1969) which form suitable micro-habitats for certain annuals, while others are instrumental in arresting the movement of large dunes, rendering them less mobile and more hospitable for other plants. Interception of rainfall by desert shrubs and trees may cause only minor evaporative losses due to their low cover, and may, through stem flow, induce patterns of soil wetting nearby (Slatyer, 1961; Qashu, Evans, Wheeler & Hanks, 1972), creating favourable micro-environments for smaller plants.

Spatial interactions

Variations in plant community composition

To the casual observer, spatial variations in desert vegetation seem to be associated with landform patterns: the assemblage of plants in wadis will be different from that on tops of hills, and common species on sand dunes will not be the same as those in depressions. In ecological surveys, this fact is well recognized and plant communities are generally related to landforms. The landform pattern of a desert ecosystem may be regarded as a simple expression of the totality of its physical environment.

It is true that, in studying areas with a more or less uniform topography, or in large-scale vegetational surveys, the patchiness of desert communities could be directly related to spatial variations in the mean rainfall (e.g. Zohary, 1947, 1962; Went, 1953; Long, 1954; Schoenenberger, Floret & Soler, 1966; Schoenenberger & Soler, 1967; Le Houérou, 1969; El-Ghonemy & Tadros, 1970; Emberger, 1971; Batanouny & Abou El-Souod, 1972). However, in more local vegetational patterns, or in areas with rapid topographic changes, spatial variations are more explicable in relation to landform patterns (Zohary, 1942, 1945, 1947; Evenari, 1951; Zohary & Feinburn, 1951; Davis, 1953; Kassas, 1953, 1956*a*, *b*, 1960; Long, 1955; Zohary & Orshan, 1956; Tadros & Atta, 1958*a*; Migahid & Ayyad, 1959; Migahid, Shafei, Abdel-Rahman & Hammouda, 1959; Tadros & El-Sharkawi, 1960; Box, 1961; Halwagy, 1961; Orshan & Zohary, 1963; Danin, Orshan & Zohary, 1964; Quézel, 1965; Beard, 1969; Girgis, 1970; Migahid, Batanouny & Zaki, 1971; Wiedemann, 1971; Brown, 1971; Obeid & Mahmoud, 1971; Barbour & Diaz, 1972; Saxena, 1972; Harniss & West, 1973; Ayyad & Hilmy, 1974). Through their influence on microclimatic and edaphic characters, landforms control moisture availability, salinity and soil stability in arid-land ecosystems. Of these prominent factors, moisture availability is responsible for the largest component of variability in the spatial distribution of species. Accordingly, the classification of desert vegetation into communities is commonly related to soil physical characters, nature of surface and topographic peculiarities which all act through modifying the amounts of available moisture.

In Egyptian deserts, for example, runnels of the limestone plateau form a variety of habitats supporting different communities (Kassas & Girgis, 1964): rill-lines across rocky slopes are dominated by *Fagonia kaherica*, precipitous cliffs by *Capparis spinosa*, stepped cliffs by *Limonium pruinosum*, stepped runnels by *Erodium glaucophyllum*, short shallow runnels by *Fagonia mollis*, long shallow runnels by *Iphiona mucronata*, and long deep runnels by *Zygophyllum coccineum*. In the gravel desert, gravel slopes support mostly ephemerals, runnels with gravel slopes are dominated by *Haloxylon salicornicum* and *Mesembryanthemum forskalie*, and the channels communities are dominated by *Haloxylon salicornicum*, *Panicum turgidum*, *Lasiurus hirsutus*, *Zilla spinosa* and *Artemisia monosperma* (Kassas & Imam, 1959). In the Negev Desert of Palestine, Tadmor *et al.* (1962) recognize communities of *Zygophyllum dumosum* on hill tops and slopes, of *H. articulatum* in loessial plains, and of *Retama retam* and *Thymelaea hirsuta* in gravelly wadis. In the Rajasthan Desert in India, plant communities of the desertic zone (mean annual rainfall less than 200 mm) are clearly associated with landforms. Thus, a *H. salicornicum* community is found in old flood plains, a *Calligonum* spp.–*Haloxylon* spp. community on sand dunes, a *Calligonum* spp.–*Acacia jacquemontii* community on concentric dune slopes, a *Calligonum* spp.–*Prosopis* spp.–*Capparis* spp. community on interdunal areas, and *Acacia senegal*–*Salvadora oleoides*–*Commiphora* spp. community on sandstone hills.* ,In the shadscale zone of south-eastern Utah, West & Ibrahim (1968) distinguished the following four distinctive units of landscape, soil type and vegetation.

(1) Level pediment remnants with coarse-textured and well-developed soil profiles derived from sandstone. The soil is non-saline down to 75 cm but saline at greater depths. This habitat supports a community of *Atriplex confertifolia* and *Hilaria jamesii*.
(2) Eroded pediment slopes with loamy and non-alkaline profiles. Soils are non-saline down to 38 cm. These slopes are dominated by *A. nuttallii* and *H. jamesii*.
(3) Lower Mancos shale badlands with fine-textured and non-alkaline profiles. Soils are non-saline down to 30 cm and are dominated by *A. nuttallii* and *Aster xylorhiza*.
(4) Alluvial basins where material from the other three habitats has been deposited over Mancos shale. Soils are heavy-textured, alkaline and saline throughout the profile. These basins support a community of *A. corrugata*.

In southern Texas, Box (1961) recognizes communities of mesquite (*Prosopis juliflora*) and chaparral (*Acacia amentacea*, *Condalia obtusifolia*, *C. obovata*, *Zanthoxylon fagara*, *Celtis pallida* and *Berberis trifoliata*) on heavy clays; communities of bunchgrass (*Andropogon scoparius* and *Elyonurus tripsacoides*) on very fine sand; and communities of prickly pear (*Opuntia lindheimeri*) on clay loam.

In a study of gradients in species composition of desert vegetation near

* Information was provided by Dr R. K. Gupta of the Soil Conservation Research Centre, Dehha Dun, India, to whom the author is grateful.

El Paso, Texas, Williams (1969) correlates the relative importance of species with physiographic variations in the arroyos (i.e. dry channels). Xeric sites are dominated by *Larrea divaricata*, accompanied by *Krameria parviflora*, *Coldenia canescens* and *Nama hispidum*. More favourable conditions, as in crevices or in sites with deeper soil where water is trapped by impervious layers, result in a greater complexity of species composition and a greater number and size of plants.

Using a multivariate analysis, Noy-Meir, Tadmor & Orshan (1970) classify the vegetation of the Avedat Desert of Palestine into communities of *Artemisia herba-alba* and *Helianthemum kahericum* on very stony slopes, *Zygophyllum dumosum* and/or *Hammada scoparia* in loessial plains, and *Artemisia herba-alba*, *Poa sinaica* and *Reaumuria negevensis* on moderately stony slopes, in runnels and partly in plains.

Regression analysis between catenae of the vegetation of semi-arid Australia and different environmental factors (Noy-Meir, 1973) indicates that the vegetational variation is produced by an interaction between rainfall and texture, with increased sandiness operating in the same direction as increased rainfall. In the western Mediterranean coastal desert of Egypt, the distribution of ridge vegetation as represented by an ordination of stands correlates significantly with a group of physiographic and edaphic factors which determine moisture availability in these stands (Ayyad & Ammar, 1974). The spatial patterns of species within the communities of *Z. dumosum* in the Negev Desert (Waisel, 1971), and the *Acacia-Capparis* semi-desert shrub in the Sudan (Greig-Smith & Chadwick, 1965) are also related to variations in the soil factors which affect soil moisture relations.

Beside the role played by topographic and edaphic factors in determining the effectiveness of precipitation in desert ecosystems, there may be a significant interaction between annual rainfall (or snowfall in cold deserts) and other climatic factors. The composition and structure of vegetation in arid lands will vary according to whether precipitation occurs in the cold, cool or warm season, or in two seasons. In arid and semi-arid Australia, Rogers (1972) demonstrates the interactive effect of rainfall and temperature on the distribution of lichens. Thus four groups of these lichens are arrayed in an order from wet to dry and from cool to hot habitats. The different patterns of solar energy induced by varying slope and exposure may also affect the moisture balance and consequently the distributional behaviour of species (Kassas & Imam, 1957). Variations in the moisture level may again occur due to changes in the amount of the accumulated snow produced by wind action. This is clearly demonstrated in Curlew Valley, Utah, where snow is blown from *Eurotia* stands to adjacent stands, causing less moisture recharge and salinity dilution in the soil under the pure *Eurotia* community (West & Caldwell, 1973); as a result, the larger *Atriplex* and *Artemisia tridentata*, which catch snow more efficiently, invade the pure *Eurotia* stands.

Next to moisture availability, salinity is the most prominent factor having major consequences on plant life in arid lands. In many areas, salinity controls the spatial distribution of species, and plant communities are generally recognized in association with its pattern of variation. In the meadows around the Great Salt Lake (Utah), *Salicornia rubra* and *S. utahensis* are found in sites with a soil-salt content as high as 2.5%, while those of *Artemisia tridentata* fail to occur in areas with a soil salinity higher than 0.4% (Shreve, 1942). In playas, the vegetational zones arranged in order of decreasing salinity are those of *S. rubra* and *S. utahensis*, *Suaeda erecta*, *Allenrolfea occidentalis*, *Distichlis stricta*, and *Suaeda torreyana* (Flowers & Evans, 1966).

Along the western coast of the Red Sea, zonation of the vegetation is affected by varying levels of salinity and other related factors (Kassas, 1957; Kassas & Zahran, 1962, 1965, 1967; Zahran, 1969). The shore-line zone is occupied by a mangrove growth of *Avicennia marina*, and the less moist zones by communities of *Arthrocnemon glaucum*, *Halacnemon strobilaceum*, *Cressa cretica*, *Suaeda fruticosa*, *Limonium pruinosum*, *Suaeda vermiculata*, *Z. album*, *Nitraria retusa* and *Tamarix mannifera*, in order of decreasing salinity. A similar zonation is recognized on the eastern coast of the Red Sea south of Jedda in Saudi Arabia (Vesey-Fitzgerald, 1955). The halophilous vegetation of the Mediterranean coast of Egypt is classified into several associations related to salinity and level of water table (Tadros, 1953; Tadros & Atta, 1958*a*). Of these, *Phragmitetum communis*, *Typheto–Scirpetum littoralis* and *Salicornietum herbaceae* occupy successive zones on lake shores. In relatively dry areas, *Halocnemetum strobilaceae*, *Salicornietum fruticosae*, *Arthrocnemetum–Limoniastretum monopetalae*, *Zygophylletum albae*, *Salsoletum tetrandrae*, and *Lycieto–Limoniastretum halimi*, occupy areas of successively decreasing salinity.

In so far as salinity adds an osmotic component to the matric potential of soil solution, inducing 'physiological dryness', it acts in the same direction as the paucity of rainfall, less moisture by run-off and/or excessive evaporation. The level of salinity, therefore, is one of the factors which determine the moisture availability, and the distribution of desert vegetation may thus be related to the magnitude of water stress caused by the scarcity of moisture and/or salinity (Fig. 2.1).

The idea expressed by many investigators that the zonation of halophilous vegetation is essentially related to soil salinity may be applicable only in explaining the macro-distribution of species. In fact the osmotic pressure as such is not the most decisive factor in determining the growth. Life limits of plants (Heimann, 1966), and the micro-distribution of halophytic species is more likely related to the relative concentration of different ions (Gates, Stoddart & Cook, 1956; Chapman, 1966; Ayyad & El-Ghareeb, 1972).

Notwithstanding the overwhelming effect of moisture availability and salinity on plant life in arid zones, other factors may, in some cases, share control. On sand dunes, an important factor which determines the

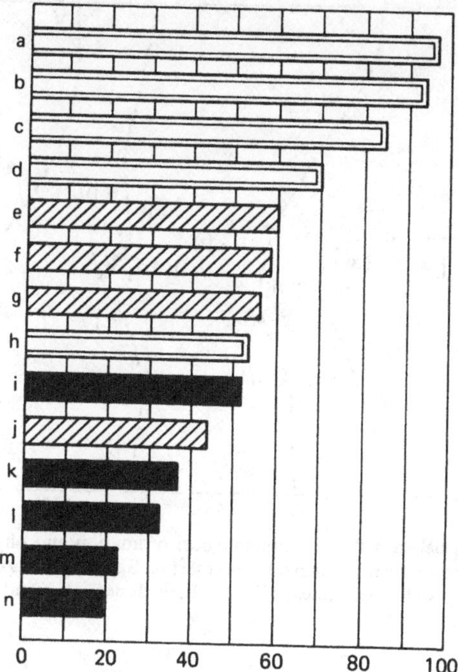

Fig. 2.1. Total soil moisture stress (plant community stress limit) for 14 plant communities in the Willow Creek Basin, Montana. Open bars indicate halophytes; bars with inclined hatches, non-halophytic northern desert shrubs; and black bars, types that are common in the intermountain region. (After Bronson, Miller & McQueen, 1970.) *a*, Nuttall saltbush slick; *b*, Nuttall saltbush hilltop; *c*, Nuttall saltbush semislick; *d*, greasewood strip; *e*, big sage–pricklypear; *f*, big sagebrush strip; *g*, big sagebrush; *h*, greasewood–western wheatgrass; *i*, blue grama; *j*, silver sagebrush–western wheatgrass; *k*, buckwheat; *l*, western wheatgrass; *m*, foxtail; *n*, mixed shrub.

distributional behaviour of species is the physical process of dune stabilization. Ayyad (1973) recognizes five physiographic categories representing successive stages of dune stabilization on the western Mediterranean coastal land of Egypt: active, partly stabilized and stabilized dunes, sand shadows and consolidated dunes. These categories correlate significantly with the phytosociological structure of the dune habitat as represented by a stand ordination (Fig. 2.2). Although *Ammophila arenaria* dominates the first three forms, its abundance becomes markedly lower from the active to the stabilized dunes, and the number of associated species increases from five to seventeen. On the other hand, the sand shadows are co-dominated by a group of shrubby deep-rooted species, such as *Crucianella maritima*, *Echinops spinosissimus* and *Thymelaea hirsuta*, and the consolidated dunes by a group of species with marked affinities towards skeletal soils.

The spatial distribution of species may also be affected by the geological formation; accordingly, communities may be associated with certain types of

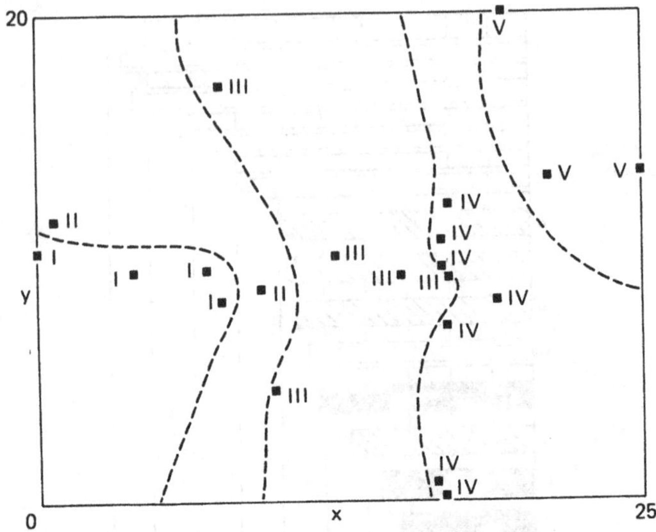

Fig. 2.2. The relation between the phytosociological ordination and physiographic categories of coastal dunes in the western Mediterranean desert of Egypt. (After Ayyad, 1973.) I, Active sand dunes; II, partly stabilized dunes; III, stabilized dunes; IV, sand shadows; V, shallow substrates of the coastal ridge.

parent rock. On the western foothills of the Judean Mountains of Palestine, for example, a *Poterieto-Avenetum* community co-dominated by *Poterium spinosum* and *Avena sterilis*, and a *Hyperrhenietum hirtae* community co-dominated by *Hyperrhenia hirta* and *Avena sterilis* occupy slopes of either 'Nari' and residual rendzina, or hard limestone and terra rossa, while a *Poterietum spinosi cretatrium* community dominated by *Poterium spinosum* is found on chalks with grey calcareous rendzina (Lativ, 1967).

Spatial variations in the soil–atmosphere complex of desert ecosystems, especially those related to moisture availability, usually induce parallel variations in plant biomass and productivity. In general, the productivity of plant communities is normally greater in regions of higher rainfall. For example, the peak of the annual productivity of an *Artemisietum* community in the Negev Desert, with a mean annual rainfall of 80 mm, is 156.7 g m^{-2}, while that for the same community in Algeria with a mean annual rainfall of 301 mm is 418.7 g m^{-2}. In pre-Saharan Tunisia, variations in productivity along a gradient of annual rainfall in arid and semi-arid Saharan regions are synthesized by Le Houérou (1969, 1972) and are illustrated in Fig. 2.3 for two soil types. Primary productivity obviously increases as the mean annual rainfall increases, especially where this is less than 100 mm.

On a more local scale, productivity varies with differences in physiographic and/or edaphic features that influence the moisture balance of arid ecosystems. This is clear in Fig. 2.3 which indicates that with the same mean annual

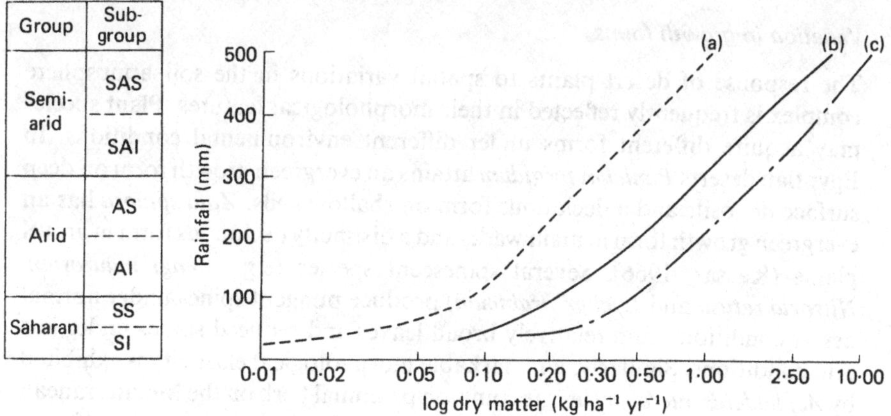

Fig. 2.3. A synthesis of actual knowledge about the productivity of pre-Saharan Tunisia. (After Le Houérou, 1969, 1972.) SAS, Mediterranean semi-arid (subgroup superior); SAI, Mediterranean semi-arid (subgroup inferior); AS, Mediterranean arid (subgroup superior); AI, Mediterranean arid (subgroup inferior); SS, Mediterranean Saharan (subgroup superior); SI, Mediterranean Saharan (subgroup inferior). *a*, Groups of low productivity (on skeletal soils); *b*, mean for all groups; *c*, groups of high productivity (on gravelly soils).

Table 2.1. *Productivity (air-dry matter) of the associations of the pre-Caspian lowlands*

Association	Soil type	Average stem biomass ($t\ ha^{-1}$)	Average root biomass ($t\ ha^{-1}$)
Artemisia pauciflora	Solonochachous solonetz	0.9	7.8
Agropyron desertorum–Pyrethrum achilleifolium	Light chestnuts	1.3	10.9
Herbaceous grass cover	Dark chernozem-like soils	2.8	17.5

rainfall in Pre-Saharan Tunisia, vegetational groups on gravelly soils are more productive than those on skeletal soils which are less capable of retaining moisture. In the Mohave Desert, productivity attains an annual maximum of 365.9 g m^{-2} for terrace vegetation, and 314.5 g m^{-2} for slope vegetation. In the Negev Desert, the peak of productivity in the communities of stony slopes (*Zygophylletum* and *Artemisietum*) ranges from 89.8 to 160.7 g m^{-2}, as compared to 322.8 g m^{-2} for the *Hammadetum* community in flood plains with deep loess soils (Evenari *et al.*, 1971). Similarly, notable differences are recorded by Bolshakov & Rode (1972) in the productivity of the associations dominating areas with different soil types in the pre-Caspian lowlands (Table 2.1).

Variation in growth forms

The response of desert plants to spatial variations in the soil–atmosphere complex is frequently reflected in their morphological features. Plant species may acquire different forms under different environmental conditions. In Egyptian deserts *Panicum turgidum* attains an evergreen growth form on deep surface deposits and a deciduous form on shallow soils. *Zilla spinosa* has an evergreen growth form in main wadis and a distinctly deciduous form in gravel plains (Kassas, 1966). Several spinescent species (e.g. *Alhagi maurorum*, *Nitraria retusa* and *Lycium arabicum*) produce pungent spines under normal desert conditions, and relatively broad leaves and reduced spines under less arid conditions. Similarly, a remarkable morphological elasticity is exhibited by *Asphodelus microcarpus*, the common perennial herb on the Mediterranean coastal land of the Middle East, in response to variations in moisture availability (Ayyad & Hilmy, 1972). The dimensions of its leaf, tuber and scape in sites of above average moisture conditions attain values at least five times greater than those in sites of below average moisture conditions.

The regulation of size under the influence of environmental factors is also a notable feature of desert ephemerals (Shreve, 1942). This remarkable morphological elasticity of desert plants has led to the recognition of a number of vegetation forms on the bases of structural features and mode of growth, which are valuable indicators of the soil–atmosphere complex, especially with regard to the water regime.

The following classification system of vegetation forms is suggested by Kassas (1970).

(*A*) Accidental vegetation form.
(*B*) Ephemeral vegetation: (1) succulent-ephemeral form, (2) ephemeral-grass form, (3) herbaceous-ephemeral form.
(*C*) Suffrutescent perennial vegetation: (1) succulent-half-shrub form, (2) perennial grassland form, (3) woody-perennial form.
(*D*) Frutescent perennial vegetation: (1) succulent-shrub form, (2) scrubland form.

In this order, these vegetation forms represent an improvement of soil moisture relationships, and embrace a range from accidental vegetation, where rainfall is not an annually recurring incident, to scrubland that is associated with sustained moisture resources. Each of these categories comprises numerous types of phyrocoenosis that differ according to local habitat conditions. Thus Kassas & Imam (1959) note that ephemeral communities in the Egyptian gravel deserts are represented on gravelly south facing slopes and runnels of Pliocene gravels, summer deciduous types represented by *Lasiurus hirsutus* community in runnels, the succulent types represented by

Haloxylon salicornicum community in large runnels, and the evergreen types represented by communities of *Panicum turgidum*, *Zilla spinosa* and *Artemisia monosperma* in main channels.

Temporal interactions

Annual variations

The ever-changing physical environment of desert ecosystems evokes temporal vegetational variations which are reflected in the abundance, productivity and phenology of species. One of the most distinct features of this environment that has far-reaching effects on plant growth is the seasonal and year-to-year erratic fluctuations in rainfall. These fluctuations control the appearance and the abundance of ephemerals. Accordingly, ephemerals can be divided into winter annuals which appear during winter and early spring in regions with winter rainfall (e.g. the Mediterranean belt of North Africa), and summer annuals which appear after the first heavy rain of summer in regions with summer rainfall (e.g. Sudanese deserts). In areas with two rainy seasons (e.g. some American deserts) the restriction of the two groups of ephemerals to their respective seasons is dependent on the optimum temperature for the germination of their seeds (Shreve, 1942). In fact, it is usually an interaction between the timing and intensity of rainfall and temperature which controls the germination and growth of ephemerals and the sprouting of ephemeroids (geophytes) and other perennials (Went, 1948, 1949, 1953, 1955; Tevis, 1958*a*, 1958*b*; Beatley, 1967).

In the Joshua Tree National Monument, California, a correlation of the appearance and abundance of ephemerals with rainfall and temperature (Juhren, Went & Philips, 1956) provides evidence that the temperatures which prevail in a brief period following each rain, together with the amount and duration of precipitation, are major factors that bring about germination, and that the percentage of germination is proportional to the amount of precipitation up to an optimum beyond which germination decreases. The intensity of the first winter rain and the accompanying temperature are of special significance. Patten & Smith (1973*a*) also note that, in the Sonoran Desert, ephemerals fail to develop when extremely high temperatures prevail, although the amount of moisture may be non-limiting. Likewise, observations on the germination and growth of plants in the Death Valley region (California) indicate that a rainfall followed by a minimum temperature of 30 °C results in no germination, while a minimum temperature of 15–16 °C causes germination of *Larrea* only. Full germination of winter annuals with no germination of *Larrea* takes place if the rain is followed by a minimum temperature of 8–10 °C (Went & Westergaard, 1949). In years of above average rainfall, the abundance of ephemerals increases, while in exceptionally dry years, the ephemeral growth is very much reduced, and many deciduous perennials and ephemeroids may fail to appear. Thus, Zohary (1962) remarks that, depending on the intensity of annual rainfall, *Poa* species in Palestine may not sprout at all, may produce basal leaves only, or develop flowering culms. In the semi-desert of southern New Mexico, the cover of *Bouteloua eriopoda* is generally reduced during dry years (Herbel, Ares & Wright, 1972).

In arid ecosystems, where water supply is a limiting factor, the rate of metabolic processes is primarily governed by the amount of available moisture and the efficiency with which plants regulate the expenditure of that moisture. Temporal variations in the productivity of desert plants, therefore, may generally be expected to correlate with fluctuations in precipitation. Thus in communities of a salt-desert shrub range in southern Utah, a large proportion of variation in the biomass of new growth could be accounted for by linear regression on precipitation, with correlation coefficients ranging from 0.84–0.93.* In western Pamirs, the production of two communities of *Artemisia vachanica* varied from 0.10–0.24 t ha^{-1} in years of average rainfall. During more humid years it varied from 0.1–0.37 t ha^{-1} in one community and from 0.20–0.47 t ha^{-1} in the other (Rousyaeva, 1972).

However, lower rainfall with an almost ideal distribution during the growing season may result in better yields than higher rainfall with recurring droughts in mid-season (Tadmor, Eyal & Benjamin, 1972). Remarkable year-to-year variations in the biomass of winter annuals were also recorded in southern Nevada by Beatley (1969), who noted that these variations were better correlated with the annual rainfall rather than with the length of the growing season or with the dominant shrub species.

Other climatic factors may interact with annual rainfall in determining the productivity of desert plants, especially air and soil temperature. In the northern Aral Sea area, the productivity of plant parts above the ground fluctuates considerably according to the hydrometeorological conditions (Kirichenko, 1972). The highest yearly increments are obtained in years with above average precipitation, with more days above 0 °C during the growing season, and with least vapour pressure deficit.

In the Sonoran Desert, Patten (1971, 1972*a*, *b*), and Patten & Smith (1973*b*) indicated that the growth rates of cacti responded to rainfall and temperature fluctuations. Late autumn and winter appeared to have the best environmental conditions for maximum productivity. In various regions of Uzbekistan, precipitation in 1968 with warm autumn and mild winter ensured the normal development of ephemerals, while in 1969 with higher precipitation, but with prolonged periods of frost, ephemeral growth was retarded (Burygin & Markova, 1972).

* Data were provided by Dr D. Wilkins of the US Desert Biome, Utah State University, to whom the author is grateful.

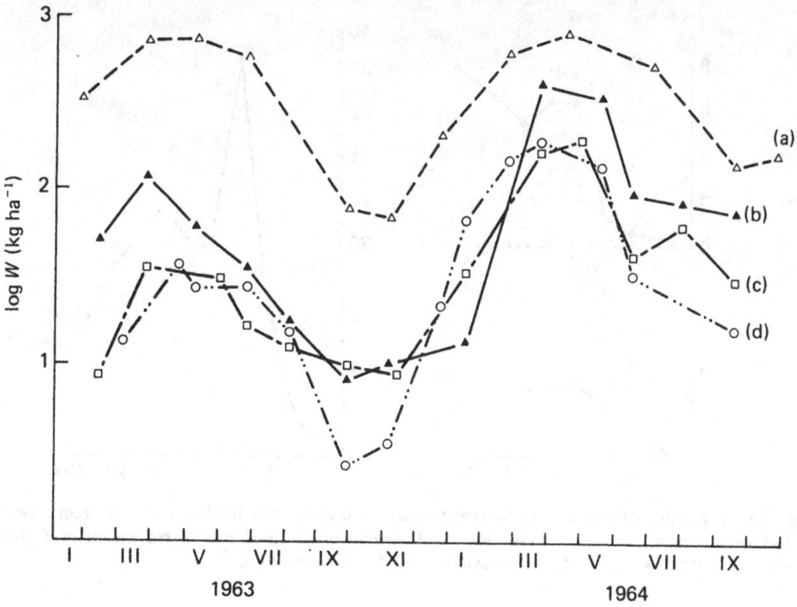

Fig. 2.4. Annual variation in the dry weight of green plant material in some desert communities of the Middle East. (After Orshan & Diskin, 1968). *a*, *Poterium spinosi*; *b*, *Anabasidetum articulatae arenarium*; *c*, *Artemisietum herbae-albae*; *d*, *Zygophylletum dumosi.*

Seasonal variations

The seasonal variation in the productivity of desert communities is also a function of climatic fluctuations. In deserts with a single rainy season, there is usually a short growth period followed by a prolonged period of considerable reduction in the amount of green plant material (Fig. 2.4). This is partly due to the disappearance of ephemerals and ephemeroids, and partly to the body reduction of shrubs (Orshan & Diskin, 1968). During the dry season, the arido-active plants (i.e. metabolically active during the dry season) continually reduce their shoot surface area, so that the amount of water needed for metabolic functions is reduced, and consequently the overall dry matter production is also reduced.

The pattern of seasonal variation in the productivity of desert communities is also related to the growth form of dominant species (Fig. 2.5). In an association dominated by the evergreen shrub *Artemisia herba-alba* in the Algerian desert, the percentage of new green matter during the winter of 1968 was insignificant, and in spite of adequate precipitation only a small amount was added during early spring as a result of low air and soil temperature (Botschantzev *et al.*, 1970). With the advent of warm weather in late spring, the percentage increased notably. During mid-summer, some leaves were shed

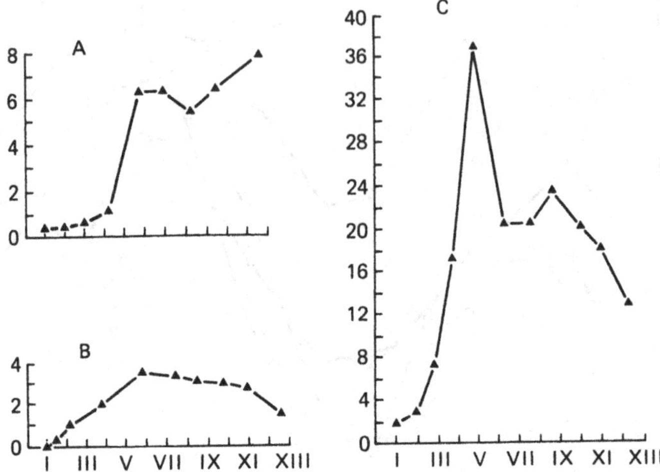

Fig. 2.5. Dynamics of dry matter increase of some associations in the Algerian desert. (After Botschantzev *et al.*, 1970.) A, Association of *Artemisia herba-alba*; B, association of *Lygeum spartum*; C, association of *Atriplex halimus* and *Suaeda pruinosa*.

and the percentage of new green matter dropped, then increased steadily during autumn until it reached a maximum at the beginning of winter. In the community of the perennial grass *Lygeum spartum*, active growth took place during the second half of winter and spring; at the beginning of summer, most shoots were dropped and the percentage of new dry matter declined. A third pattern was exhibited by the community co-dominated by the halophilous species *Atriplex halimus* and *Suaeda pruinosa*: after a rapid increase in productivity from a minimum in early winter to a maximum at the end of spring, productivity in this community decreased sharply during early summer. In mid-summer it increased slightly to another maximum after which it declined steadily throughout autumn.

The phenological behaviour of desert plants in relation to fluctuations in climatic conditions is, in fact, a morphological expression of the mechanism with which these plants regulate the expenditure of the moisture necessary for metabolic processes. This mechanism and its expression differ in different life forms. A gradient of decreasing leaf persistence and increasing stem and branch photosynthesis is described by Whittaker & Niering (1965) for plants of the Sonoran Desert. They grade from species with persistent evergreen leaves, through semi-deciduous species with leaves (or leaf-bearing twigs of the semi-shrubs) persistent through less severe dry seasons but not more severe ones, to deciduous mesquite, palo verde, and semi-shrubs. These, in turn, grade into forms like ocotillo with quickly-produced and short-lived leaves that are soon lost after rain, to cacti which lack leaves. In the Middle East, Zohary (1953) distinguishes eight pheno-ecological types: evergreens which never become completely defoliated (*Acacia* type); evergreen, articulate-stem succulents which shed portions of older branches in summer (*Anabasis* type); evergreen spartoids which shed leaves in early winter (*Retama* type); winter green phanerophytes which shed all leaves in mid-summer (*Lycium* type); chamaephytes which considerably reduce their transpiring surface at the beginning of the dry period, including biseasonal annuals (*Reaumuria* type); annuals, cryptophytes and hemicryptophytes which finish their life cycle at the end of the rainy season (*Launaea* type); ephemerals which finish their life cycle long before the end of the rainy season (*Filago* type); and summer annuals which start to develop in spring, shed large leaves in early summer and retain green bract-like leaves up to the end of summer (*Salsola* type).

Detailed accounts of the relationships between climatic fluctuations and the phenological behaviour of plants in the Mohave Desert were given by Wallace & Romney (1972) and Ackerman & Bamberg (1972). Each species was found to have a characteristic response under a given climatic regime. Some species responded favourably and some unfavourably to the same conditions; each species had a favourable range of temperature for the initiation of phenophase. Within the same species, the response was variable: some plants grew and produced flowers and others not. A few plants responded to a summer rain of a sufficient amount by a short period of activity which, sometimes, included flowering and fruiting. It was also found that plants could survive long periods of unfavourable conditions (low moisture and high or low temperature) by dormancy or inactivity, and that their response was sudden and dramatic (i.e. within two or three days) when conditions became favourable.

References

Ackerman, T. L. & Bamberg, S. A. (1972). Phenological studies in the Mojave desert at Rock Valley, Nevada Test Site, Nevada. In: *Proceedings of a Symposium on Phenology and Seasonality*. 25th Annual Meeting of American Institute of Biological Sciences.

Ayyad, M. A. (1971). A study of solar radiation on sloping surfaces at Alexandria. *United Arab Republic Journal of Botany*, **14**, 65–73.

Ayyad, M. A. (1973). Vegetation and environment of the western Mediterranean coastal land of Egypt. I. The habitat of sand dunes. *Journal of Ecology*, **61**, 509–23.

Ayyad, M. A. & Ammar, M. Y. (1973). Relationship between local physiographic variations and the distribution of common Mediterranean desert species. *Vegetatio*, **27**, 163–76.

Ayyad, M. A. & Ammar, M. Y. (1974). Vegetation and environment of the western Mediterranean coastal land of Egypt. II. The habitat of inland ridges. *Journal of Ecology*, **62**, 439–56.

Ayyad, M. A. & El-Ghareeb, R. (1972). Microvariations in edaphic factors and species distribution in a Mediterranean salt desert. *Oikos*, **23**, 125–31.

Composite and interactive processes

Ayyad, M. A. & Hilmy, S. (1972). A study of morphological variations in *Asphodelus microcarpus*. In: *Proceedings of the First Egyptian Congress of Botany, Cairo, March 1972.*

Ayyad, M. A. & Hilmy, S. (1974). The distribution of *Asphodelus macrocarpus* and associated species on the western Mediterranean coastal land of Egypt. *Ecology*, **55**, 511–24.

Barbour, M. G. & Diaz, D. V. (1972). *Larrea* plant communities on bajada and moisture gradients in the United States and Argentina. *Bulletin of the Ecological Society of America*, **53** (2), 16.

Batanouny, K. H. & Batanouny, M. H. (1968). Formation of phytogenic hillocks. I. Plants forming phytogenic hillocks. *Acta Botanica Academiae Scientiarum Hungaricae*, **14**, 243–52.

Batanouny, K. H. & Batanouny, M. H. (1969). Formation of phytogenic hillocks. II. Rooting habit of plants forming phytogenic hillocks. *Acta Botanica Academiae Scientiarum Hungaricae*, **15**, 1–18.

Batanouny, K. H. & Abou El-Souod, S. (1972). Ecological and phytosociological study of a sector in the Libyan Desert. *Vegetatio*, **25**, 335–56.

Beadle, N. C. W. (1948). *The vegetation and pastures of New South Wales.* Government Printer, Sydney.

Beard, J. S. (1969). The natural regions of the deserts of Western Australia. *Journal of Ecology*, **57**, 677–712.

Beatley, I. C. (1967). Survival of winter annuals in the northern Mojave Desert. *Ecology*, **48**, 745–50.

Beatley, I. C. (1969). Biomass of desert winter annual plant populations in southern Nevada. *Oikos*, **20**, 261–73.

Bolshakov, A. F. & Rode, A. A. (1972). Soils of the solontzic complex in the northern part of the pre-Caspian lowland and their biological productivity. In: *Ecophysiological foundation of ecosystems productivity in arid zone* (ed. L. E. Rodin), pp. 122–4. Nauka, Leningrad.

Botschantzev, V., Kalenov, H., Mirochnitchenko, Yu., Pelt, N., Rodin, L. & Vinogradov, B. (1970). *Etudes geobotaniques des paturages du secteur ouest De Département de Medea de la République Algerienne Democratique et Populaire*, part 1. Nauka, Leningrad.

Box, T. W. (1961). Relationships between plants and soils of four range plant communities in south Texas. *Ecology*, **43**, 794–810.

Bronson, F. A., Miller, R. F. & McQueen, I. S. (1970). Plant communities and associated soil and water factors on shale-derived soils in northeastern Montana. *Ecology*, **51**, 391–407.

Brown, R. W. (1971). Distribution of plant communities in southeastern Montana badlands. *American Midland Naturalist*, **85**, 458–77.

Burygin, V. A. & Markova, L. E. (1972). Winter weather conditions and fodder production of ephemeral pastures in Uzbekistan. In: *Ecophysiological foundation of ecosystems productivity in arid zone* (ed. L. E. Rodin), pp. 114–16. Nauka, Leningrad.

Chapman, V. J. (1960). *Salt marshes and salt deserts of the world.* Interscience Publishers, New York.

Chapman, V. J. (1966). Vegetation and salinity. *Monographiae Biologicae*, **16**, 23–42.

Danin, A., Orshan, G. & Zohary, M. (1964). Vegetation of the Neogene sandy areas of the northern Negev of Israel. *Israel Journal of Botany*, **13**, 208–33.

Davis, P. H. (1953). The vegetation of the deserts near Cairo. *Journal of Ecology*, **41**, 157–73.

El-Ghonemy, A. A. & Tadros, T. M. (1970). Socio-ecological studies of the natural plant communities along a transect 250 km long between Alexandria and Cairo. *Bulletin of the Faculty of Science, Alexandria University*, **10**, 392–407.

Emberger, L. (1971). *Travaux de botanique et d'ecologie*. Masson, Paris.

Emberger, L. & Lemée, G. (1962). Plant ecology. *UNESCO Arid Zone Research*, **18**, 197–211.

Evenari, M. (1951). Ecological investigations in Palestine. II. On the vegetation of the Kurkar Hills. *Bulletin of the Research Council of Israel*, **1**, 48–58.

Evenari, M., Shanan, L., Tadmor, N. H. & Aharoni, Y. (1971). *The Negev: the challenge of a desert*. Harvard University Press, Cambridge, Massachusetts.

Fletcher, J. E. (1960). Some effects of plant growth on infiltration in the southwest. In: *Water yields in relation to environment in southwest United States* (ed. B. H. Warnock & J. L. Bardner), pp. 51–63. American Association for the Advancement of Science Committee on Desert and Arid Zone Research.

Flowers, S. & Evans, F. R. (1966). The flora and fauna of the Great Salt Lake region, Utah. In: *Salinity and Aridity, Monographiae Biologicae*, vol. 16, pp. 367–93.

Gates, D. H., Stoddart, A. L. & Cook, W. C. (1956). Soil as a factor influencing the plant distribution on salt-deserts of Utah. *Ecological Monographs*, **26**, 155–75.

Girgis, W. A. (1970). Phytosociological studies on the vegetation of the Maryut area project. *United Arab Republic Journal of Botany*, **13**, 235–54.

Goodall, D. W. (1972). Building and testing ecosystem models. In: *Mathematical models in ecology* (ed. J. N. R. Jeffers), pp. 73–94. Blackwell Scientific Publications, Oxford.

Goodall, D. W. (1973*a*). Ecosystem modelling in the Desert Biome. In: *Systems analysis and simulation in ecology*, vol. 3 (ed. B. C. Patten), pp. 73–94. Academic Press, New York.

Goodall, D. W. (1973*b*). Ecosystem simulation in the US/IBP Desert Biome. In: *Proceedings of the 1973 Summer Computer Conference*, Montreal, pp. 777–80a.

Greig-Smith, P. & Chadwick, M. J. (1965). Data on pattern within plant communities. III. *Acacia–Capparis* semi-desert scrub in the Sudan. *Journal of Ecology*, **53**, 465–74.

Halwagy, R. (1961). The vegetation of semi-desert north-east of Khartoum, Sudan. *Oikos*, **12**, 87–110.

Harniss, R. O. & West, N. E. (1973). Vegetation of the national reactor testing station, southeastern Idaho. *Northwest Science*, **47**, 30–43.

Hastings, J. R. & Turner, R. M. (1965). *The changing mile: an ecological study of vegetation change with time in the lower mile of an arid and semi-arid region*. University of Arizona Press, Tucson.

Heimann, H. (1966). Plant growth under saline conditions and the balance of the ionic environment. *Monographiae Biologicae*, **16**, 201–13.

Herbel, C. H., Ares, F. N. & Wright, R. A. (1972). Drought effects on a semidesert grassland range. *Ecology*, **53**, 1084–93.

Hillel, D. & Tadmor, N. (1962). Water regime and vegetation in the central Negev highlands of Israel. *Ecology*, **43**, 33–41.

Hills, E. S., Ollier, C. D. & Twidale, C. R. (1966). Geomorphology. In: *Arid lands: a geographical appraisal* (ed. E. S. Hills), pp. 53–76. Methuen, London.

Juhren, M., Went, F. W. & Philips, E. (1956). Ecology of desert plants. IV. Combined field and laboratory work on germination of annuals in the Joshua Tree National Monument, California. *Ecology*, **37**, 318–30.

Kassas, M. (1953). Landforms and plant cover in the Egyptian desert. *Bulletin de la Société de géographie d'Egypte*, **26**, 193–205.

Kassas, M. (1956*a*). The mist oasis of Erkwit, Sudan. *Journal of Ecology*, **44**, 180–94.
Kassas, M. (1956*b*). Landforms and plant cover in the Omdurman desert, Sudan. *Bulletin de la Société de géographie d'Egypte*, **29**, 45–58.
Kassas, M. (1957). On the ecology of the Red Sea coastal land. *Journal of Ecology*, **45**, 187–203.
Kassas, M. (1960). Certain aspects of landform effects on plant water resources. *Bulletin de la Société de géographie d'Egypte*, **33**, 45–52.
Kassas, M. (1966). Plant life. In: *Arid lands: a geographical appraisal* (ed. E. S. Hills), pp. 145–78. Methuen, London.
Kassas, M. (1970). Desertification versus potential for recovery in circum-Saharan territories. In: *Arid lands in transition* (ed. H. E. Dregne), pp. 123–42. American Association for the Advancement of Science, Washington DC.
Kassas, M. & Girgis, W. A. (1964). Habitat and plant communities in the Egyptian desert. V. The limestone plateau. *Journal of Ecology*, **52**, 107–19.
Kassas, M. & Imam, M. (1954). Habitat and plant communities in the Egyptian desert. III. The wadi bed ecosystem. *Journal of Ecology*, **42**, 424–41.
Kassas, M. & Imam, M. (1957). Climate and microclimate in the Cairo desert. *Bulletin de la Société de géographie d'Egypt*, **30**, 25–52.
Kassas, M., Imam, M. (1959). Habitat and plant communities in the Egyptian desert. IV. The gravel desert. *Journal of Ecology*, **47**, 289–310.
Kassas, M. & Zahran, M. A. (1962). Studies on the ecology of Red Sea coastal land. I. The district of Gebel Ataqa and El-Galala El-Bahariya. *Bulletin de la Société de géographie d'Egypt*, **35**, 129–75.
Kassas, M. & Zahran, M. A. (1965). Studies on the ecology of Red Sea coastal land. II. The district of El-Galala El-Qibliya to Hurgada. *Bulletin de la Société de géographie d'Egypt*, **38**, 155–93.
Kassas, M. & Zahran, M. A. (1967). On the ecology of the Red Sea littoral salt marsh, Egypt. *Ecological Monographs*, **37**, 297–316.
Kincaid, D. R., Gardner, J. L. & Schreiber, H. A. (1964). Soil and vegetation parameters affecting infiltration under semi-arid conditions. *International Association of Scientific Hydrology, Publication No. 65*, pp. 440–53.
Kirichenko, N. G. (1972). Primary productivity of the northern Aral Sea area plant communities in connection with weather and ecological conditions. In: *Ecophysiological foundation of ecosystems productivity in arid zone* (ed. L. E. Rodin), pp. 109–12. Nauka, Leningrad.
Lativ, M. (1967). Micro-environmental factors and species interrelationships in three batha associations in the foothill region of the Judean Hills. *Israel Journal of Botany*, **16**, 79–99.
Le Houérou, H. N. (1969). La vegetation de la Tunisie steppique. *Annales de l'Institut National de la Recherche Agronomique de Tunisie*, **42** (5), 1–622.
Le Houérou, H. N. (1972). An assessment of the primary and secondary production of the arid grazing lands ecosystems of North Africa. *Ecophysiological foundations of ecosystems productivity in arid zone* (ed. L. E. Rodin), pp. 168–72. Nauka, Leningrad.
Long, G. (1954). Contribution à l'étude de la végétation de la Tunisie centrale. *Annales de Service Botanique et Agronomique de Tunisie*, **27**, 1–388.
Long, G. (1955). The study of the natural vegetation as a basis for pasture improvement in the Western Desert of Egypt. *Bulletin de l'Institut du Désert d'Egypte*, **5**, 18–42.
Lyford, F. P. & Qashu, H. K. (1969). Infiltration rates as affected by desert vegetation. *Water Resources Research*, **5**, 1373–6.

Migahid, A. M. & Ayyad, M. A. (1959). An ecological study of Ras El-Hikma district. III. Plant habitats and communities. *Bulletin de l'Institut du Désert d'Egypte*, **31**, 74–98.

Migahid, A. M., Shafei, A., Abdel-Rahman, A. A. & Hammouda, M. A. (1959). Ecological observations in western and southern Sinai. *Bulletin de la Société de géographie d'Egypt*, **32**, 165–206.

Migahid, A. M., Batanouny, K. H. & Zaki, M. A. F. (1971). Phyto-sociological and ecological study of a sector in the Mediterranean coastal region in Egypt. *Vegetatio*, **23**, 113–34.

Noy-Meir, I. (1973). Desert ecosystems: environment and producers. *Annual Review of Ecology and Systematics*, **4**, 25–51.

Noy-Meir, I. (1974). Multivariate analysis of the semi-arid vegetation of southern Australia. II. Vegetation catenae and environmental gradients. *Australian Journal of Botany*, **22**, 115–40.

Noy-Meir, I., Tadmor, N. H. & Orshan, G. (1970). Multivariate analysis of desert vegetation. I. Association analysis at various quadrat sizes. *Israel Journal of Botany*, **19**, 561–91.

Obeid, M. & Mahmoud, A. (1971). Ecological studies in the vegetation of the Sudan. II. The ecological relationships of the vegetation of Khartoum Province. *Vegetatio*, **23**, 177–98.

Orshan, G. & Diskin, S. (1968). Seasonal changes in productivity under desert conditions. In: *Functioning of terrestrial ecosystems at the primary production level* (ed. F. E. Eckardt), pp. 191–201. Proceedings of the Copenhagen Symposium. UNESCO, Paris.

Orshan, G. & Zohary, M. (1963). Vegetation of the sandy deserts in the western Negev of Israel. *Vegetatio*, **11**, 112–20.

Patten, D. T. (1971). *Productivity and water stress in cacti.* US/IBP Desert Biome Research Memorandum RM 71-12. Utah State University, Logan.

Patten, D. T. (1972*a*). Growth and productivity of cacti in relation to environments in the Sonoran Desert, North America. *Ecophysiological foundation of ecosystems productivity in arid zone* (ed. L. E. Rodin), pp. 39–41. Nauka, Leningrad.

Patten, D. T. (1972*b*). *Productivity and water stress in cacti.* US/IBP Desert Biome Research Memorandum RM 72-17. Utah State University, Logan.

Patten, D. T. & Smith, E. M. (1973*a*). Early annual plant development in relation to environmental variabilities in the Sonoran Desert. *Bulletin of the Ecological Society of America*, **54** (1), 26.

Patten, D. T. & Smith, E. M. (1973*b*). *Phenology and function of Sonoran Desert annuals in relation to environmental changes.* US/IBP Desert Biome Research Memorandum RM 73-14. Utah State University, Logan.

Qashu, H. K., Evans, D. D., Wheeler, M. L. & Hanks, R. J. (1972). *Soil factors influencing water uptake by plants under desert conditions.* US/IBP Desert Biome Research Memorandum RM 72-37. Utah State University, Logan.

Quézel, P. (1965). *La végétation du Sahara du Tchad à la Mauritanie.* G. Fischer Verlag, Stuttgart.

Rogers, R. W. (1972). Soil surface lichens in arid and semi-arid southeastern Australia. III. The relationship between distribution and environment. *Australian Journal of Botany*, **20**, 301–16.

Rousyaeva, G. G. (1972). Dynamics of the above-ground plant biomass in *Artemisia vachanica* communities at the western Pamirs. In: *Ecophysiological foundation of ecosystems productivity in arid zone* (ed. L. E. Rodin), pp. 148–51. Nauka, Leningrad.

Saxena, S. K. (1972). The concept of ecosystem as exemplified by the vegetation of western Rajasthan. *Vegetatio*, **24**, 215–27.

Schoenenberger, A., Floret, C. & Soler, A. (1966). Carte phytoécologique de la Tunisie Septentrionale. III. Les unités forestières. *Annales de l'Institut National de la Researche Agronomique de Tunisie*, **39**, 90–125.

Schoenenberger, A. & Soler, A. (1967). Carte phyto-écologique de la Tunisie Septentrionale. III. Les unités forestierès. *Annales de l'Institut National de la Recherche Agronomique de Tunisie*, **40**, 151–93.

Shreve, F. (1942). The desert vegetation of North America. *Botanical Review*, **8**, 195–246.

Shreve, F. (1951). *Vegetation of the Sonoran Desert*. Carnegie Institute, Washington.

Slatyer, R. O. (1961). Methodology of a water balance study conducted on a desert woodland (*Acacia aneura*) community. *UNESCO Arid Zone Research*, **16**, 15–26.

Strahler, A. N. (1960). *Physical geography*. Wiley, New York.

Tadmor, N. H., Eyal, E. & Benjamin, R. (1972). Primary and secondary production of arid grassland. In: *Ecophysiological foundations of ecosystems productivity in arid zone* (ed. L. E. Rodin), pp. 173–7. Nauka, Leningrad.

Tadmor, N. H., Orshan, G. & Rawitz, E. (1962). Habitat analysis in the Negev of Israel. *Bulletin of the Research Council of Israel*, **11**, 148–73.

Tadros, T. M. (1953). A phytosociological study of halophilous communities from Mareotis (Egypt). *Vegetatio*, **4**, 102–24.

Tadros, T. M. & Atta, B. M. (1958*a*). Further contribution to the sociology and ecology of halophilous communities of Mareotis (Egypt). *Vegetatio*, **8**, 137–60.

Tadros, T. M. & Atta, B. M. (1958*b*). Plant communities of barley fields and uncultivated desert areas of Mareotis (Egypt). *Vegetatio*, **8**, 161–75.

Tadros, T. M. & El-Sharkawi, E. M. (1960). Phytosociological and ecological studies on the vegetation of Fuka-Ras El-Hekma area. *Bulletin de l'Institut du Désert d'Egypt*, **10**, 86–93.

Tevis, L. Jr. (1958*a*). Germination and growth of ephemerals induced by sprinkling a sandy desert. *Ecology*, **39**, 681–8.

Tevis, L. Jr. (1958*b*). A population of desert ephemerals germinated by less than one inch of rain. *Ecology*, **39**, 688–95.

Vesey-Fitzgerald, D. F. (1955). Vegetation of the Red Sea coast south of Jedda, Saudi Arabia. *Journal of Ecology*, **43**, 477–89.

Waisel, Y. (1971). Patterns of distribution of some xerophytic species in the Negev, Israel. *Israel Journal of Botany*, **20**, 101–10.

Wallace, A. & Romney, E. M. (1972). *Radioecology and ecophysiology of desert plants at the Nevada Test Site*. TID-25954. Laboratory of Nuclear Medicine & Radiation Biology, Los Angeles.

Walter, H. (1964). *Die Vegetation der Erde*, vol. 1. G. Fischer Verlag, Stuttgart.

Went, F. W. (1948). Ecology of desert plants. I. Observations on germination in the Joshua Tree National Monument, California. *Ecology*, **29**, 242–53.

Went, F. W. (1949). Ecology of desert plants. II. The effect of rain and temperature on germination and growth. *Ecology*, **30**, 1–13.

Went, F. W. (1953). The effects of rain and temperature on plant distribution in the desert. *Research Council of Israel, Special Publication No. 2*, pp. 230–7.

Went, F. W. (1955). The ecology of desert plants. *Scientific American*, **192**, 68–75.

Went, F. W. & Westergaard, M. (1949). Ecology of desert plants. III. Development of plants in the Death Valley National Monument, California. *Ecology*, **30**, 26–38.

West, N. E. & Caldwell, M. M. (1973). Snow as a factor in cool desert shrub community patterns and their dynamics. *Abstracts of the Annual Meetings of the Society for Range Management*, **26**, 22.

West, N. E. & Ibrahim, K. I. (1968). Soil–vegetation relationships in the shadscale zone of southeastern Utah. *Ecology*, **49**, 445–56.

Whittaker, R. H. & Niering, W. A. (1965). Vegetation of the Santa Catalina Mountains, Arizona: a gradient analysis of the south slope. *Ecology*, **46**, 429–52.

Wiedemann, A. M. (1971). Vegetation studies in the Simpson Desert, Northern Territory. *Australian Journal of Botany*, **19**, 99–124.

Williams, J. S. (1969). Gradients in species composition of desert vegetation near El-Paso, Texas. In: *Physiological systems in semi-arid environments* (ed. C. C. Hoff & M. L. Riedesel), pp. 273–83. University of New Mexico Press, Albuquerque.

Zahran, M. A. (1969). On the ecology of the east coast of the Gulf of Suez. I. Littoral salt marsh. *Bulletin de l'Institute du Désert d'Egypte*, **17**, 225–52.

Zohary, M. (1942). The vegetational aspects of Palestine soils. *Palestine Journal of Botany*, **2**, 200–46.

Zohary, M. (1945). Outline of the vegetation in Wadi Araba. *Journal of Ecology*, **32**, 204–13.

Zohary, M. (1947). A geobotanical soil map of western Palestine. *Palestine Journal of Botany*, **4**, 24–35.

Zohary, M. (1953). Hydroeconomical types in the vegetation of the Near East desert. In: *Proccedings of a Symposium on Hot and Cold Deserts* (ed. J. L. Cloudsley-Thompson), pp. 56–67. Institute of Biology, London.

Zohary, M. (1962). *Plant life of Palestine, Israel and Jordan*. Ronald Press, New York.

Zohary, M. (1972). *Geobotanical foundations of the Middle East*. G. Fischer Verlag, Stuttgart.

Zohary, M. & Feinburn, N. (1951). Outline of the vegetation of the northern Negev. *Palestine Journal of Botany*, **5**, 96–114.

Zohary, M. & Orshan, G. (1956). Ecological studies on the vegetation of the Near Eastern deserts. II. Wadi Araba. *Vegetatio*, **7**, 15–37.

Manuscript received by the editors June 1974.

3. Plant–plant interactions

M. G. BARBOUR

Introduction

This chapter title is very inappropriate if we take the conclusions of earlier biologists at face value. Darwin (1859) wrote that in 'absolute deserts, the struggle for life is almost exclusively with the elements'. More recently, Shreve (1942) concluded after years of work in deserts that there 'is the almost total lack of reaction by the plant on its habitat. The existence of a plant in a given spot for many years does nothing to make that spot a better [or worse?] habitat for some other plant or some other species'. The bracketed words are mine.

The general ecological literature today is not quite so dogmatic about excluding biotic factors in desert vegetation. We have come to accept one form of plant interaction at least: competition for moisture.* 'In sharp contrast to the forest, the struggle in the desert is not one of plant against plant for light and space, but one of all plants for moisture... because of competition for moisture, the plants are widely spaced' (Smith, 1966).

If competition for moisture were the only, or the major, form of plant interaction in deserts, then one could make three assumptions: plants tend to be distributed regularly, rather than randomly or clumped; the tendency for plants to be regularly distributed is accentuated as available moisture declines; and the entire root zone is occupied. Yet, recent reviews (Anderson, 1971; Anderson, Perry & Leiph, 1971; West & Tueller, 1971; Barbour, 1973*a*) show that these assumptions are not tenable. Some desert plants are positively associated (clumped); regular distributions are rare and do not seem to be more common in the most arid sites; and little is known about the root distribution of desert plants.

The objective of this chapter is to review briefly what is currently understood to be the degree of interaction between desert plants. Much of this information has been gained during the past 20 years. Examples are drawn from throughout the world, but this review remains biased towards North American deserts and *Larrea divaricata* Cav. (= *L. tridentata* (D.C.) Cor.) as a case-study organism.

* The terms competition, commensalism, amensalism, etc. used in this chapter are defined according to the Burkholder–Odum classification described by Malcolm (1966) and Odum (1971). Competition, for example, is defined as decreased rate of activity (−) for two species when interacting, and normal rate of activity (0) for both when not interacting.

The root systems of desert perennials

The vegetation cover of most deserts is less than 50%, and often less than 25% (see for example McGinnies, Goldman & Paylore, 1968). If competition, as a form of interaction, is going to occur, then it probably will not involve competition in the canopy level. Competition for moisture, for example, would be by root systems in the soil. How extensive are these root systems?

The root systems of some xerophytes or near-phreatophytes can be deep. Reviews by Cloudsley-Thompson & Chadwick (1964) and Oppenheimer (1960) summarize the familiar records of *Prosopis* and *Acacia* roots which penetrate to depths of 15 m or more. Roots of some non-xerophytes can be as deep, however: roots of alfalfa and sugar beets in Nebraska can penetrate 10 and 7 m deep respectively, and the record for alfalfa appears to be 40 m (Meinzer, 1927; Weaver & Clements, 1938).

However, root systems of many xerophytes are quite shallow. Zohary (1961) concluded, 'As a rule, plants of hamadas and rocky deserts have shallow roots adapted to the depth of rain penetration. Deep-rooting plants are characteristic only of sand dunes, aluvial soils, and run-on habitats'. Some succulents have very shallow, weak root systems. Roots of *Ferocactus wislizenii* of the Sonoran Desert seldom penetrate deeper than 2 cm (Cloudsley-Thompson & Chadwick, 1964). Cannon (1911) found that most *Larrea* roots in Arizona sites were located in the top 3–15 cm, unless the site was on a deep sand. R. H. Chew & A. E. Chew (1965), with considerably more data, reported that the entire root systems of most *Larrea* shrubs on their site was found in the top 10–25 cm of the soil.

The wide spacing of desert shrubs is often attributed to intense root competition for moisture, the implication being that the root systems are so extensive laterally that all topsoil is occupied by roots. Unfortunately, excavation data are scarce, but the few that exist indicate that this implication may be incorrect. Results of excavations by Cannon (1911), R. H. Chew & A. E. Chew (1965) and Migahid (1961) show that there may be considerable space between root systems as well as overlap of roots of close neighbours. Cannon's excavations showed *Larrea* roots could extend out radially 4 m, but not in all directions equally. A maximum extension of 4 m may be three times canopy extent, but this is equalled or exceeded by at least four Californian chaparral shrubs according to Hellmers, Horton, Jurhens & O'Keefe (1955).

The data of Migahid (1961) are especially revealing. From an aerial view, he shows the extent of roots in a 10 m² plot in Egypt. Although some 11 perennial species are present, total root cover appears to be only 8% of the plot area. In a 15 m-long bisect view, showing both canopy and root extent, canopy cover was 3%, root cover was 5%. He concluded that 'root competition is absent or greatly reduced' in such desert vegetation.

Friedman (1971), however, planted *Artemisia herba-alba* seedlings at different distances from mature *Zygophyllum dumosum* shrubs in the Negev desert and found evidence of root competition extending far beyond the canopy limits. The average canopy radius of the *Zygophyllum* shrubs was 50 cm, but growth of seedlings within 100 cm of the canopy center was suppressed and many were killed during the first year. During the second year, seedlings 200 cm from the canopy center showed growth suppression compared to controls. Friedman & Orshan (1974) have shown in the same region that perennials can modify seed production of winter annuals, even when the perennials are 1–4 m distant.

In the Mohave Desert, Fonteyn & Mahall (1978) selectively removed *Larrea* and *Ambrosia* shrubs and measured the effect on water potential of remaining *Larrea*. They concluded that *Larrea* and *Ambrosia* did compete for water.*

Harris (1967) has shown that the annual *Bromus tectorum*, in semi-arid grasslands about the Great Basin desert, has a faster growing root system than the native perennial *Agropyron spicatum*. *Bromus* exhausts soil moisture at greater and greater depths ahead of *Agropyron* roots. However, *Bromus* is only at a competitive advantage if both species must begin from seed. Without invoking disturbance of *Agropyron* by cattle or sheep grazing, it is difficult to see how Harris' data can explain the phenomenal rate of spread of *Bromus tectorum* in the United States.

Grazing pressure has also been invoked to partially explain why desert scrub has invaded many hectares of semi-arid grassland in Texas, New Mexico, and Arizona. Absence of fire and minor changes in climate have also been pointed to in this region as factors which have shifted the competitive balance between grasses and shrubs (see, for example, conflicting papers by Brown, 1950; Bogusch, 1952; Humphrey, 1958; Hastings, 1959; Buffington & Herbel, 1965). Certainly, the limited information we have about plant succession in the desert seems to show that the changes that do occur are due to physical factors like those above, rather than to plant–plant interactions (Shreve & Hinkley, 1937; Muller, 1940; Shantz & Turner, 1958; Hastings & Turner, 1965).

Additional information on the extent of root systems comes indirectly from reports of root:shoot ratios. Desert perennials are often assumed to have relatively high root:shoot ratios (i.e., much above 1), but in point of fact they do not. Table 3.1 summarizes ratios of various young and mature xerophytes and communities.

There are certainly not enough data in Table 3.1 to permit extrapolation to a root:shoot ratio for the 'average' xerophyte. However, from the species and communities listed, one may conclude that the root:shoot ratio varies over a wide range; that there is no 'typical' value, and that in a majority of

* Paragraph added in proof.

Table 3.1. *Root:shoot (R:S) weight ratios of various xerophytes and desert communities.* (Modified and expanded from Barbour, 1973*a*)

Species or community	Place	Age	R:S	Source
Atriplex lentiformis	California	6 months	0.21	Sankary, 1971
A. leucoclada	Syria	6 months	0.15	
Citrullus colocynthis	Syria	6 months	0.24	
Cucurbita palmata	California	6 months	0.40	
Haloxylon articulatum	Syria	6 months	0.90	
Salsola vermiculata	Syria	6 months	0.80	
Acantholimon venustum	Anatolia	5 months	0.86	Birand, 1961
Alhagi camelorum	Anatolia	4 months	1.17	
Astragalus micropterus	Anatolia	5 months	1.50	
Noea spinosissima	Anatolia	5 months	0.17	
Zygophyllum fabago	Anatolia	5 months	0.43	
Artemisia tridentata	Nevada	2 months	0.60	Wallace *et al.*, 1970
Grayia spinosa	Nevada	2 months	1.24	
Eurotia lanata	Nevada	1 month	0.18	
Lycium andersonii	Nevada	2 months	0.47	
Lycium pallidum	Nevada	2 months	1.01	
Coleogyne ramosissima	Nevada	2 months	0.12	
Yucca schidigera	Nevada	6 months	1.27	
Larrea divaricata (= *L. tridentata*)	Nevada	2 months	0.39	
Larrea divaricata (= *L. tridentata*)	S. west USA	40 days	0.60	Barbour, 1967
Larrea divaricata (= *L. tridentata*)	Arizona	1–65 yr	0.22 0.50	Chew & Chew, 1965
Acacia greggii	California	M[a]	1.20	Garcia-Moya & McKell, 1970
Brickellia incana	California	M	0.50	
Cassia armata	California	M	0.70	
Ephedra nevadensis	California	M	1.20	
Eriogonum fasciculatum	California	M	0.40	
Franseria dumosa	California	M	0.60	
Hymenoclea salsola	California	M	0.70	
Krameria grayi	California	M	0.40	
K. parvifolia	California	M	0.60	
Larrea divaricata (= *L. tridentata*)	California	M	0.30	
Salazaria mexicana	California	M	0.70	
Ambrosia dumosa	Nevada	M	1.01	Bamberg, Wallace Kleinkopf & Vollmer, 1973
Acamptopappus shockleyi	Nevada	M	0.50	
Atriplex confertifolia	Nevada	M	0.37	
A. canescens	Nevada	M	0.58	
Ephedra nevadensis	Nevada	M	0.72	
Grayia spinosa	Nevada	M	0.61	
Krameria parvifolia	Nevada	M	0.69	
Larrea divaricata (= *L. tridentata*)	Nevada	M	1.08	
Lycium andersonii	Nevada	M	0.73	
L. pallidum	Nevada	M	1.42	

Table 3.1 (*cont.*)

Species or community	Place	Age	R:S	Source
Chilopsis linearis	New Mexico	M	1.40	Ludwig, 1977
Ephedra trifurca	New Mexico	M	0.70	
Fallugia paradoxa	New Mexico	M	0.60	
Flourensia cernua	New Mexico	M	1.10	
Larrea divericata (= L. tridentata)	New Mexico	M	0.90	
Parthenium incanum	New Mexico	M	0.40	
Prosopis glandulosa var. *torreyana*	New Mexico	M	1.30	
Yucca elata	New Mexico	M	1.00	
Artemisia communities	Idaho	M	0.40 1.80	Pearson, 1966
Artemisia/*Poa* community	Syria	M	1.33	Rodin & Bazilevich, 1967
Hammada/*Poa*	Syria	M	3.29	
Artemisia/*Poa*/lichen	Syria	M	0.37 0.79	
Artemisia/*Anabasis*	USSR	M	6.70	
Artemisia community	USSR	M	3.55	
Artemisia/*Poa*	USSR	M	5.11	
Anabasis community	USSR	M	5.11	
Haloxylon/*Salsola*	USSR	M	7.33	
Eleagnus/*Tamarix*	USSR	M	0.96	
Populus/*Halimodendron*	USSR	M	0.82	

[a] M, mature.

cases it is less than 1. Compared to values for more mesic species and communities, the xerophyte ratios turn out to be relatively low.

The argument above that xerophytic root:shoot ratios may be modest has been properly developed from a biomass standpoint. If one were interested in water balance, it would be more appropriate to calculate root:leaf ratios as Anderson *et al.* (1971) point out. Desert shrubs, compared to mesophytes, may be exceptionally twiggy and retain few leaves (see, for example, recent studies of *Larrea* growth by Oechel, Strain & Odening, 1972; Burk & Dick-Peddie, 1973). A root:leaf ratio is still not quite appropriate, however, because only a fraction of the root mass is probably involved in water uptake.

While the examples cited above are not all-inclusive, I believe they illustrate that desert perennials do not necessarily possess extensive root systems, either in absolute extent or in terms of root:shoot length or weight ratios. Although the root:shoot length of young plants may be high, mature xerophytes appear to have rather shallow, laterally restricted root systems. It remains to be demonstrated that all of the topsoil beneath desert communities is occupied by roots, and that low plant density is primarily due to competition for moisture. This is not to say that competition for moisture cannot occur or, indeed, that it does not regularly occur. In a recent review, Noy-Meir (1973) concluded that competition may not be a factor in extreme deserts or in more mesic areas where frequent catastrophes (e.g. erosion) prevent the population density from building up to high enough densities, but that competition may indeed be a factor in mature arid shrub communities. It is in such shrub communities that observers have noted the releasing effect of shrub removal on normally suppressed herbaceous species (F. H. Wagner, personal communication). My conclusion is that so far we are relying on passive observation in concluding that competition for moisture is a factor, and there is need for experimental evidence like that of Fonteyn & Mahall (1978).

Allelopathy in desert plants

The ecological significance of allelopathic toxins in desert communities, or possibly in any other type of community, has not yet been made clear (see Sondheimer & Simeone, 1970; Barbour, 1973*b*).

The search for toxins which might be responsible for patterns of plant distribution in desert communities was stimulated by Bonner & Galston's work (1945) with *Parthenium incanum*. *Parthenium* shrubs appeared to inhibit growth of nearby shrubs of the same species through a water-soluble root secretion, transcinnamic acid. But, as Bonner (1950) and Borner (1960) later admitted, this effect was an artifact of sterile greenhouse procedures, and in nature the toxin was broken down too rapidly by soil microflora to be of ecological significance. Bennett & Bonner (1953) have shown that many other desert perennials possess water-soluble toxins which can cause wilting or death in such mesophytes as tomato, but the relationship this has to natural communities is difficult to perceive.

In response to Bonner's work, Muller & Muller (1956) searched for causes other than allelopathy to explain positive and negative associations of desert annuals with certain shrubs. They noted that leaf extracts of three shrubs, *Encelia farinosa*, *Franseria dumosa* and *Thamnosma montana*, all inhibited the growth of native annuals in greenhouse culture, yet in nature these same annuals occurred beneath the crowns of *Franseria* and *Thamnosma*. They suggested that the decisive factor in whether or not a shrub harbored annuals was whether or not it accumulated a mound of windblown soil and organic debris beneath it, such a mound being favorable for herb establishment. Among other factors, the branching pattern of *Encelia* prevented this sort of mound from accumulating, and hence prevented annuals from occurring beneath it. Allelopathy was not a factor.

Knipe & Herbel (1966) made aqueous extracts of *Larrea* twigs and leaves, and found the extracts had no effect on *Larrea* germination and early growth (as tested against a sucrose solution of the same water potential as the extract). Nutrient solution recycled for 1 month about the roots of *Larrea* seedlings had no effect on other *Larrea* plants. The authors did comment on the fact that extracts created a crust on the pot soil used, which impeded water infiltration rate. Adams, Strain & Adams (1970) have since shown that such crusts form in nature beneath *Larrea* canopies, as well as beneath those of *Prosopis juliflora* and *Cercidium floridum*. The absence of annuals around these perennials was directly caused by a presence of such hydrophobic layers, and the crusts were stable enough to remain for several years after the perennials had been removed by fire. Possibly such crusts prevent establishment of perennial seedlings as well. Dalton (1962) and Barbour (1967) have also made extracts of *Larrea* leaf and twig material, and reported no detrimental effects of *Larrea* germination or early growth.

Pattern analyses and available moisture

Certainly, when one looks about in the midst of a Mohave Desert community, one gets the impression of widely spaced shrubs at even intervals, somewhat like trees in an orchard. Both Shreve (1942) and Went (1955), two deservedly highly regarded desert researchers, have commented on the commonness of regular shrub arrangement, the latter ascribing it to water-soluble toxins exuded through roots, the former to competition for soil moisture. Actual sampling data to confirm or deny this impression of shrub distribution, however, are only now becoming available. Two sampling and data analysis techniques are most frequently used: contiguous quadrats and variance: block-size calculations (first developed by Thompson, 1958 and Greig-Smith, 1961); and random quadrats and Poisson distribution or variance: mean calculations (summarized by Greig-Smith, 1964). The former is the more sophisticated of the two, for it can reveal pattern at many different unit areas and is independent of quadrat size. However, like with other researchers, I have concluded that the statistical calculations for this method are not quite standardized. Samuel Zahl, Professor of Statistics at the University of Connecticut, is currently working on modifications of Greig-Smith's method (personal communication). The ecological interpretation of patterns revealed by the method is, in addition, open to debate.

Statistical evidence for the pattern of desert perennials is most abundant for *Larrea divaricata* (= *L. tridentata*) of North America. Data for other shrubs are still quite limited.

A study of four sites near Khartoum, Sudan (160 mm precipitation, mainly summer) by Greig-Smith & Chadwick (1965); also discussed by Cloudsley-Thompson & Chadwick (1964), showed random or clumped patterns for *Capparis decidua* and *Acacia ehrenbergiana*. The authors used a variance: block-size method. Anderson (1967) and Anderson, Jacobs & Malik (1969) showed by block analysis that distribution of several perennials in New South Wales, Australia, was either random or clumped: *Maireana pyramidata*, *Triodia irritans*, and *Atriplex vesicaria*. Anderson theorized that regularity is rare and will only result when the environment is perfectly homogeneous and if the plants are all genetically uniform.

Using a variance:mean ratio, McDonough (1965) concluded that *Opuntia bigelovii* at one site in the Mohave Desert was clumped at several scales of pattern. Beals (1968) used several statistical techniques to estimate pattern for *Cadaba rotundifolia* shrubs in the Ethiopian Desert, and most of them indicated a random distribution pattern. Quadrat maps for some Sonoran communities, drawn by Shantz & Piemeisel (1924), appear to show clumped shrub patterns.

Table 3.2 summarizes the results of five studies of *Larrea divaricata* (= *L. tridentata*), covering a total of 58 sites in all three warm desert regions of North America. Rainfall data are approximated from nearby US Weather Bureau stations.

Woodell, Mooney & Hill (1969) sampled 12 *Larrea*-dominated communities in the Mohave and Sonoran Desert regions of California. At each site, 25 randomly located points formed the centres of 93 m^2 circular quadrats, within which the number of shrubs for each species was counted. Pattern was determined by the variance:mean ratio. The authors also used a point-to-plant method, but those data are not included in Table 3.2.

Barbour (1969) sampled 16 *Larrea*-dominated communities throughout the southwest. At each site, *Larrea* shrubs were counted in 100 quadrats of 3 m^2 located in a restricted random fashion. Pattern was determined by a χ^2 test of fit to a Poisson series.

The third set of data in Table 3.2 comes from work I recently conducted in Arizona along a rainfall gradient from Yuma to Tucson (Barbour & Diaz, 1973). At each of 22 *Larrea*-dominated sites, a sample area of 1536 m^2 was subjectively located. This area was subdivided into 256 contiguous quadrats, each of 6 m^2. In each of these quadrats, the number of rooted *Larrea* shrubs was tallied, and the resulting data used to construct block-size:variance graphs. As block size increases it is possible for the pattern to change, and in some cases the first block sizes show a distribution pattern different from larger ones. Thus in ARIZ-13 (Table 3.2) *Larrea* shrubs were random in block size one through eight (6 m^2 through 48 m^2) but they were regular for all larger block sizes, and this is noted in the table as 'random/regular'.

One site, sampled by Turner (1962), is about 4 km southwest of Phoenix, Arizona. He sampled it with 48, randomly distributed quadrats of 5 m^2, and determined the pattern of *Larrea* and *Franseria dumosa* shrubs by the Poisson distribution. He concluded that both species were distributed at random.

A final group of eight sites in the Avra Valley, near Tucson, Arizona, was sampled by Wright (1970). He used strips of contiguous quadrats and measured canopy coverage, rather than density. His results are difficult to evaluate, but it appears that two of the sites showed regular *Larrea* distribution, four showed random pattern and two showed a clumped pattern.

Table 3.2. Larrea divaricata (= L. tridentata) *shrub patterns and available moisture in 58 sites throughout the warm desert in the United States*

Sources: (1), Woodell *et al.* (1969); (2), Barbour (1969); (3), Barbour & Diaz (1973); (4), Turner (1962); (5), Wright (1970). reg, regular; ran, random; cl. clumped; *r*, half average *Larrea* shrub height or diameter. Density is for *Larrea* shrubs only, and cover is % relative *Larrea* cover. Available moisture is mm precipitation m^{-3} *Larrea* canopy ha^{-1}.

Site name or code	Precipitation (mm)	Pattern	*r*	Shrubs ha^{-1}	Cover	Available moisture
Stovepipe Wells (1)	40	ran	0.35	60	?	4.23
Niland (1)	48	ran	0.50	85	?	1.13
C-70 (2)	68	reg	1.48	130	50	—
ARIZ-10 (3)	70	ran	0.36	215	6	—
ARIZ-11 (3)	85	ran	0.66	189	57	—
Inyokern (1)	91	reg	0.35	261	?	2.29
Yuma (1)	91	reg	1.20	107	100	0.12
ARIZ-12 (3)	95	cl	0.45	338	92	1.63
Cantil (1)	100	reg	0.65	206	?	0.44
ARIZ-13 (3)	103	ran/reg	0.58	266	63	—
Emigrant Springs (1)	111	reg	0.45	261	?	1.13
ARIZ-14 (3)	113	cl	0.45	403	47	—
N-80 (2)	114	cl	0.27	1790	100	0.80
Mohave (1)	128	ran	0.30	492	?	1.80
Haiwee (1)	143	ran	0.40	297	?	1.80
Randsburg (1)	150	cl	0.60	178	?	0.95
ARIZ-15 (3)	185	ran	0.48	670	59	—
Phoenix (4)	198	ran	?	?	?	—
ARIZ-16 (3)	198	cl	0.51	787	76	—
NM-73 (2)	203	cl	0.48	1760	8	—
Wildrose Canyon (1)	207	cl	0.35	535	?	2.45
Morongo Valley (1)	218	cl	0.80	485	?	0.21
ARIZ-17 (3)	218	cl	0.50	1131	86	—
ARIZ-18 (3)	218	cl	0.53	1118	96	0.31
ARIZ-19 (3)	218	cl	0.55	953	49	—
ARIZ-20 (3)	225	cl	0.78	1066	100	0.11
NM-76 (2)	229	ran/cl	0.38	2703	100	0.42
T-76 (2)	229	ran	0.30	3230	23	—
T-77 (2)	229	ran	0.30	3260	46	—
A-71 (2)	245	reg	0.56	1530	31	—
ARIZ-1 (3)	245	cl	0.70	1086	97	0.16
ARIZ-2 (3)	253[a]	cl	0.69	611	81	—
ARIZ-3 (3)	259[a]	cl	0.76	247	15	—
ARIZ-22 (3)	259[a]	cl	0.62	910	61	—
ARIZ-21 (3)	264[a]	ran	0.60	572	52	—
ARIZ-6 (3)	271[a]	ran	0.90	507	58	—
NM-72 (2)	263	cl	0.46	7460	47	—
ARIZ-7 (3)	275	ran/cl	0.64	1378	83	—
ARIZ-4 (3)	275	cl	0.52	3770	57	—
ARIZ-8 (3)	275	cl	0.69	462	19	—
T-70 (2)	275	ran	0.31	3630	53	—
ARIZ-9 (3)	275	cl	0.80	579	42	—
Avra Valley, 8 sites (5)	280	2 reg, 4 ran, 2 cl	?	?	?	—

[a] These sites are on a bajada above ARIZ-1. Their rainfall was estimated by adding 0.02 mm m^{-1} above ARIZ-1.

Table 3.2 (*cont.*)

Site name or code	Precipitation (mm)	Pattern	r	Shrubs ha^{-1}	Cover	Available moisture
T-71 (2)	280	ran	0.41	1730	23	—
T-72 (2)	280	cl	0.53	1820	22	—
T-73 (2)	280	cl	0.42	3030	46	—
T-74 (2)	280	ran	0.34	2390	92	0.61
T-75 (2)	280	cl	0.27	6190	73	—
Cabazon (1)	299	ran	0.70	237	?	0.89
NM-70 (2)	305	ran	0.22	1230	2	—
NM-71 (2)	305	ran	0.20	7460	47	—

All eight sites, lying relatively close together in one valley, experience the same rainfall.

One could conclude from Table 3.2, and what data are available for other perennials, that regularity is relatively uncommon and is not limited to the more arid sites. Of the 58 sites in Table 3.2, eight exhibited regular shrub distribution at low block sizes; five have rainfall of less than 114 mm yr^{-1}, and three receive 245–280 mm yr^{-1}. Clumping was observed at 25 sites, all within the range 95–280 mm yr^{-1}. The other 25 sites showed random shrub distribution, and they covered the entire spectrum of rainfall, 40–305 mm yr^{-1}. It is not possible to draw a strong correlation between rainfall and type of pattern.

Rainfall, of course, is not the same as 'available moisture'. The amount of moisture available per plant depends upon soil permeability, distribution of rainfall, plant density and size, and rate of evapotranspiration. When Woodell, Mooney & Hill (1969) arranged their 12 *Larrea* sites in order of increasing rainfall, they found a strong correlation between rainfall and type of pattern. As rainfall increased, pattern changed from regular to random to clumped. However, when their data appear with all other sites in Table 3.2, this kind of correlation disappears. In reviewing their work, Anderson (1971) noted that they had not taken into account differences in plant density. Anderson in effect divided their *Larrea* density data into rainfall to approximate the available moisture per plant per site. He assumed all plants at all sites were the same size. His division showed that regularity increased as moisture available per shrub increased, a conclusion opposite to that reached by the original authors.

T. J. King of Westminster School, London (personal communication, and see also King & Woodell, 1973) points out that shrub size was not equal at all 12 sites, and he has provided me with histograms of *Larrea* shrub diameters for each site. As a yet closer approximation of available moisture per plant, we could multiply average shrub volume by shrub density and divide this factor into rainfall. But how do we measure the volume of *Larrea*? Chew & Chew (1965) calculated volume of *Larrea* shrubs in southeastern Arizona by assuming a basic shape of an inverted cone with an elliptical cross section. Burk & Dick-Peddie (1973) in southern New Mexico used a cone with circular cross section. My own observations are that *Larrea* shrubs vary from conical to spherical to cylindrical, depending on region and substrate.

If a sphere is taken as the basic shape, then height measurements I have taken for many sites in Table 3.2 can be used to calculate volume. In each of those sites, roughly 100 random *Larrea* shrubs were measured for maximum height, and averages calculated. In Table 3.2 I have taken half the average height as the basis for r in the formula: volume $= 4/3\,\pi r^3$. For Woodell's 12 sites I used shrub diameter data provided by King, taking half the average diameter as the basis for r in the formula above. I realize that a cone may be a better estimate of canopy volume in some regions, but limited data prompted me to select the sphere. Nevertheless, since both sphere and cone are based on r^3, the available moisture data in Table 3.2 would show the same relationships from site to site whichever shape is used.

Looking only at Woodell's 12 sites in Table 3.2, I conclude that available moisture (mm precipitation m^{-3} *Larrea* canopy ha^{-1}) does not correlate well with type of pattern. Range of available moisture for four sites with regular distribution was 0.12–2.29; for five sites with random distribution, 0.89–4.23; and for three sites with clumped distribution, 0.21–2.45. This opposes the conclusions of the original authors.

But *Larrea* was not competing alone for moisture. Other species were present and their volumes and densities should somehow be entered into the equation. In the absence of sufficient data from Woodell *et al.* to permit this type of calculation, I selected from Table 3.2 all those sites which were essentially pure *Larrea* (90–100% relative cover) and determined available moisture as described above. There were only eight such sites, so conclusions can only be tentative, but again there seems to be no correlation: one regular site, 0.12; two random sites, 0.42 and 0.61; five clumped sites, 0.11–0.80.

I have recently sampled desert communities in northwestern Argentina dominated by *Larrea cuneifolia* (Barbour & Diaz, 1973). Table 3.3 summarizes data from 14 sites for which reliable precipitation data from nearby weather stations were available. Most of the 14 sites exhibited clumped distribution, so it is difficult to show a correlation between rainfall or available moisture and type of pattern.

Clearly, if available moisture is indeed the cause of pattern, we need to add more refinements to the calculations before a correlation can be demonstrated. Until then, we may do well to accept Anderson's (1971) conclusion: 'It seems that the prime hypothesis of regularity in desert shrub populations being the resultant of competition for a limiting water resource is not substantiated by...present data.'

Table 3.3. Larrea cuneifolia *shrub patterns and available moisture in 14 sites in northwestern Argentina*

Data are taken from Barbour & Diaz (1973). The abbreviations are as given in Table 3.2.

Site code	Precipitation (mm)	Pattern	*r*	Shrubs ha^{-1}	Cover	Available moisture
ARG-21	75	reg	0.35	1177	30	—
ARG-20	115	ran	0.37	1300	91	0.41
ARG-18	140	cl	0.40	1320	100	0.41
ARG-9	155	cl	0.58	953	100	0.17
ARG-4	165	cl	0.54	816	100	0.30
ARG-7	165[a]	cl	0.70	533	69	—
ARG-5	178[a]	cl	0.53	1027	58	—
ARG-6	193[a]	ran	0.47	2280	76	—
ARG-8	208[a]	cl	0.51	780	28	—
ARG-2	215	cl	0.79	469	31	—
ARG-25[b]	246	cl	0.87	2249	58	—
ARG-11	250	cl	0.48	1853	98	0.31
ARG-23[b]	256	cl	0.84	6522	50	—
ARG-14	320	cl	0.55	2405	98	0.18

[a] These sites are on a bajada above ARG-4. Their rainfall was estimated by adding 0.2 mm m^{-1} above ARG-4.

[b] Unpublished data.

Positive associations

Positive spatial associations might result from interactions such as commensalism, mutualism, and parasitism; they could also result from vegetative reproduction, or from a non-homogeneous habitat in which survival is favored in localized pockets rather than randomly over the entire surface.

Perhaps the most completely-documented case of commensalism between desert plants involves saguaro cactus (*Cereus giganteus*) and 'nurse plants'. According to long-term studies by Niering, Whittaker & Lowe (1963) and Steenbergh & Lowe (1969), virtually all successful saguaro seedlings are found in close proximity to a shade-producing object. In a small percentage of cases the object is inanimate, but in most cases it is a perennial plant. Observations by these authors, coupled with experiments by Turner, Alcorn, Olin & Booth (1966) show the positive effects of the nurse plant are shading (reduced temperature and rate of soil drying) and hiding of the young cactus from rodent herbivores. Some protection from frost may also result. In the Tucson area, Arizona, some 15 different species function as nurse plants. The frequency with which they appear as shade plants is an approximate indication of their relative abundance and their percentage cover. Turner *et al.* (1966) showed that soils from beneath different nurse plant species do differ in color, salinity, and nutrient status, but their data indicated that such soil differences had little effect on cactus seedling mortality.

As mentioned earlier in this paper, Muller (1953) and Muller & Muller (1956) showed that the nurse-plant effect extended to some annuals. Species of *Malacothrix* and *Chaenactis* are positively associated with the canopies of *Franseria dumosa* and *Thamnosma montana* shrubs. Apparently the reason for the association is that the growth form of these shrubs is a suitable trap for wind-blown organic debris. The debris collects beneath the canopies and this provides a better substrate for the annuals than soil in the open. Seeds of the annuals may also be trapped in abundance beneath the canopies.

In the Great Basin desert, *Purshia tridentata*, *Atriplex confertifolia*, and *Eurotia lanata* seedlings may also require nurse plants (reviewed by West & Tueller, 1971).

Epiphytism is another form of commensalism, and we have floristic accounts from several deserts to show that it is not uncommon (McGinnies *et al.*, 1968; Morello, 1958; Shreve & Wiggins, 1963). Coastal deserts that are subject to fog are especially high in numbers of epiphytes. However, there have been no biological or ecological studies of desert epiphytes and information on parasitic relationships is almost nil. The series of UNESCO arid zone volumes includes not one article on desert epiphytes or parasites; major reviews of desert biology by Cloudsley-Thompson & Chadwick (1964) and McGinnies *et al.* (1968) are silent on these subjects. Kuijt's (1969) world-ranging review of parasitic plants includes a bizarre picture of *Opuntia* possibly parasitizing an *Idria* in Baja California, but there is no text discussion of parasites in deserts.

Vegetative reproduction, producing clumps of individuals, is most obvious among species of *Opuntia*, but it may also be prevalent for shrubs as Wright (1970) points out in a recent review. Shrubs such as *Larrea* may have the lower branches covered by wind or water borne sediment, and these branches may take root and eventually produce new, independent individuals.

Sternberg (1976) showed that rings of *Larrea* shrubs in the Mohave Desert were enzymatically identical, hence clonal in origin. Some rings were nearly 8 m in radius and could have been over 5000 years old. Reproduction by seedlings may be rare.*

Additional floristic examples of positive associations could have been obtained from local floras and listed here, but that work would be outside the ecological objectives of this review. Clearly, the ecological investigation of desert epiphytes and parasites is an area that has received little attention up till now. An article which models the effect of parasites on two competing plant species (Chilvers & Brittain, 1972) perhaps points the way to more research in this area.

* Paragraph added in proof.

Summary

This chapter has reviewed the following types of plant–plant interactions in deserts: amensalism (allelopathy); competition (root competition for moisture); commensalism (nurse plant syndrome and epiphytes); and parasitism.

Strong ecological evidence for amensalism is lacking. Competition for moisture, seemingly an obvious form of desert plant interaction, remains to be demonstrated. It is not clear what fraction of soil volume is occupied by roots, because few excavations, tracer studies, or field competition experiments have been conducted. Root:shoot ratios of xerophytes, compared to mesophytes, are not high. Regular shrub distribution is rare, even in very arid sites. Using *Larrea divaricata* (= *L. tridentata*) as a case study organism, there is no strong correlation between rainfall – or available moisture per unit of canopy volume – and pattern of shrub distribution.

The requirement of some perennials and annuals for nurse plants appears to be non-specific and non-biotic; that is, the nurse plants provide shade, seclusion from herbivores, protection from frost, or they trap organic debris, and these services can be provided by inanimate objects. Very little information on the biology or ecology of desert epiphytes and parasites appears in the literature; apparently this is an area overlooked by most desert investigators.

References

Adams, S., Strain, B. R. & Adams, M. S. (1970). Water-repellent soils, fire, and annual plant cover in a desert scrub community of southeastern California. *Ecology*, **51**, 696–700.

Anderson, D. J. (1967). Studies on structure in plant communities. V. Pattern in *Atriplex vesicaria* communities in south-eastern Australia. *Australian Journal of Botany*, **15**, 451–8.

Anderson, D. J. (1971). Pattern in desert perennials. *Journal of Ecology*, **59**, 555–60.

Anderson, D. J., Jacobs, S. W. L. & Malik, A. R. (1969). Studies on structure in plant communities. VI. The significance of pattern evaluation in some Australian dry-land vegetation types. *Australian Journal of Botany*, **17**, 315–22.

Anderson, D. J., Perry, R. A. & Leigh, J. H. (1971). Some perspectives on shrub/environment interactions. In: *Wildland shrubs* (ed. C. M. McKell, J. P. Blaisdell & J. R. Goodwin), pp. 172–81. USDA Forest Service General Technical Report INT-1, Ogden, Utah.

Bamberg, S., Wallace, A., Kleinkopf, G. & Vollmer, A. (1973). *Plant productivity and nutrient interrelationships of perennials in the Mojave Desert.* US/IBP Desert Biome Research Memorandum, RM 73-10. Utah State University, Logan.

Barbour, M. G. (1967). Ecoclinal patterns in the physiological ecology of a desert shrub, *Larrea divaricata.* Ph.D. Dissertation, Duke University, Durham, North Carolina.

Barbour, M. G. (1969). Age and space distribution of the desert shrub *Larrea divaricata. Ecology*, **50**, 679–85.

Barbour, M. G. (1973*a*). Desert dogma reexamined: root/shoot productivity and plant spacing. *American Midland Naturalist*, **89**, 41–57.

Barbour, M. G. (1973*b*). Chemistry and community stability. In: *Air pollution damage to vegetation* (ed. J. A. Naegele), Advances in Chemistry Series No. 122, pp. 85–100.

Barbour, M. G. & Diaz, D. V. (1973). *Larrea* plant communities on bajada and moisture gradients in the United States and Argentina. *Vegetatio*, **28**, 335–52.

Beals, E. W. (1968). Spatial patterns of shrubs on a desert plain in Ethiopia. *Ecology*, **49**, 744–6.

Bennett, E. L. & Bonner, J. (1953). Isolation of plant growth inhibitors from *Thamnosma montana. American Journal of Botany*, **40**, 29–33.

Birand, H. (1961). Relations entre le développement des racines et des parties aériennes chez certains plantes xerophytes et leur résistance à la sécheresse. *UNESCO Arid Zone Research*, **16**, 175–82.

Bogusch, E. R. (1952). Brush invasion in the Rio Grande Plain of Texas. *Texas Journal of Science*, **4**, 85–91.

Bonner, J. (1950). The role of toxic substances in the interactions of higher plants. *Botanical Review*, **16**, 51–65.

Bonner, J. & Galston, A. W. (1945). Toxic substances from the culture media of guayule which may inhibit growth. *Botanical Gazette*, **106**, 185–98.

Borner, H. (1960). Liberation of organic substances from higher plants and their role in the soil sickness problem. *Botanical Review*, **26**, 393–424.

Brown, A. L. (1950). Shrub invasion of southern Arizona desert grassland. *Journal of Range Management*, **3**, 172–7.

Buffington, L. C. & Herbel, C. H. (1965). Vegetation changes on a semi-desert grassland range from 1858 to 1963. *Ecological Monographs*, **35**, 139–64.

Burk, J. & Dick-Peddie, W. A. (1973). Comparative production of *Larrea divaricata* Cav. on three geomorphic surfaces in southern New Mexico. *Ecology*, **54**, 1094–102.

Cannon, W. A. (1911). The root habits of desert plants. *Carnegie Institute of Washington Publication*, **131**, 1–96.

Chew, R. M. & Chew, A. E. (1965). The primary productivity of a desert shrub (*Larrea tridentata*) community. *Ecological Monographs*, **35**, 355–75.

Chilvers, G. A. & Brittain, E. G. (1972). Plant competition mediated by host-specific parasites – a simple model. *Australian Journal of Biological Sciences*, **25**, 749–56.

Cloudsley-Thompson, J. L. & Chadwick, M. J. (1964). *Life in deserts*. G. T. Foulis, London.

Dalton, P. D., Jr (1962). Ecology of the creosotebush *Larrea divaricata*. PhD Dissertation, University of Arizona, Tucson.

Darwin, C. R. (1859). *On the origin of species by means of natural selection*. J. Murray, London (1921).

Friedman, J. (1971). The effect of competition by adult *Zygophyllum dumosum* Bioss. on seedlings of *Artemisia herba-alba* Asso in the Negev desert of Israel. *Journal of Ecology*, **59**, 775–82.

Friedman, J. & Orshan, G. (1974). Allopatric distribution of two varieties of *Medicago laciniata* (L) Mill. in the Negev desert. *Journal of Ecology*, **62**, 107–14.

Garcia-Moya, E. & McKell, C. M. (1970). Contribution of shrub to the nitrogen economy of a desert-wash plant community. *Ecology*, **51**, 81–8.

Greig-Smith, P. (1961). Data on pattern within plant communities. I. The analysis of pattern. *Journal of Ecology*, **49**, 695–702.

Greig-Smith, P. (1964). *Quantitative and dynamic ecology*. Edward Arnold, London.

Greig-Smith, P. & Chadwick, M. J. (1965). Data on pattern within plant communities.

III. *Acacia–Capparis* semi-desert scrub in the Sudan. *Journal of Ecology*, **53**, 465–74.

Harris, G. A. (1967). Some competitive relationships between *Agropyron spicatum* and *Bromus tectorum*. *Ecological Monographs*, **37**, 89–111.

Hastings, J. R. (1959). Vegetation change and arroyo cutting in southeastern Arizona. *Journal of the Arizona Academy of Science*, **1**, 60–7.

Hastings, J. R. & Turner, R. M. (1965). *The changing mile*. University of Arizona Press, Tucson.

Hellmers, H., Horton, J. S., Juhrens, G. & O'Keefe, J. (1955). Root systems of some chaparral plants in southern California. *Ecology*, **36**, 667–78.

Humphrey, R. T. (1958). The desert grassland. *Botanical Review*, **24**, 193–252.

King, T. J. & Woodell, S. R. J. (1973). The causes of regular pattern in desert perennials. *Journal of Ecology*, **61**, 761–5.

Knipe, D. & Herbel, C. H. (1966). Germination and growth of some semi-desert grassland species treated with aqueous extract from creosotebush. *Ecology*, **47**, 775–81.

Kuijt, J. (1969). *The biology of parasitic flowering plants*. University of California Press, Berkeley.

Ludwig, J. A. (1977). Distributional adaptations of root systems in desert environments. In: *The belowground ecosystem: a synthesis of plant-associated processes* (ed. J. K. Marshall). Range Science Series No. 26. Colorado State University, Fort Collins.

Malcolm, W. M. (1966). Biological interactions. *Botanical Review*, **32**, 243–54.

McDonough, W. T. (1965). Pattern changes associated with the decline of a species in a desert habitat. *Vegetatio*, **13**, 97–101.

McGinnies, W. G., Goldman, B. J. & Paylore, P. (1968). *Deserts of the world*. University of Arizona Press, Tucson.

Meinzer, O. E. (1927). *Plants as indicators of ground water*. USGS Water Supply Paper 577. Washington, DC.

Migahid, A. M. (1961). The drought resistance of Egyptian desert plants. *UNESCO Arid Zone Research*, **16**, 213–33.

Morello, J. (1958). La provincia fitogeografica del monte. *Opera Lilloana* II, Tucuman, Argentina.

Muller, C. H. (1940). Plant succession in the *Larrea–Flourensia* climax. *Ecology*, **21**, 205–12.

Muller, C. H. (1953). The association of desert annuals with shrubs. *American Journal of Botany*, **40**, 53–60.

Muller, W. H. & Muller, C. H. (1956). Association patterns involving desert plants that contain toxic products. *American Journal of Botany*, **43**, 354–61.

Niering, W. A., Whittaker, R. H. & Lowe, C. H. (1963). The saguaro: a population in relation to environment. *Science*, **142**, 15–23.

Noy-Meir, I. (1973). Desert ecosystems: environment and producers. *Annual Review of Ecology and Systematics*, **4**, 25–51.

Odum, E. P. (1971). *Fundamentals of ecology*, 3rd edn. W. B. Saunders, Philadelphia.

Oechel, W. C., Strain, B. R. & Odening, W. R. (1972). Tissue water potential, photosynthesis, ^{14}C-labelled photosynthate utilization, and growth in the desert shrub *Larrea divaricata* Cav. *Ecological Monographs*, **42**, 127–41.

Oppenheimer, H. R. (1960). Adaptation to drought: zerophytism. *UNESCO Arid Zone Research*, **15**, 105–28.

Pearson, L. C. (1966). Primary productivity in a northern desert area. *Oikos*, **15**, 211–28.

Rodin, L. E. & Bazilevich, N. I. (1967). *Production and mineral cycling in terrestrial vegetation.* (English translation ed. G. E. Fogg). Oliver and Boyd, London.

Sankary, M. N. (1971). Comparative plant ecology of two Mediterranean-type arid areas, in Syria and California, with emphasis on the autecology of twenty dominant species. PhD Dissertation, University of California, Davis.

Shantz, H. L. & Piemeisel, R. L. (1924). Indicator significance of the natural vegetation of the south-western desert region. *Journal of Agricultural Research*, **28**, 721–802.

Shantz, H. L. & Turner, B. L. (1958). *Photographic documentation of vegetational changes in Africa over a third of a century.* University of Arizona Press, Tucson.

Shreve, F. (1942). The desert vegetation of North America. *Botanical Review*, **8**, 195–246.

Shreve, F. & Hinckley, A. L. (1937). Thirty years of change in desert vegetation. *Ecology*, **18**, 463–78.

Shreve, F. & Wiggins, I. L. (1963). *Vegetation and flora of the Sonoran Desert*, Vols. 1 & 2. Stanford University Press, Stanford, California.

Smith, R. L. (1966). *Ecology and field biology.* Harper & Row, New York.

Sondheimer, E. & Simeone, J. B. (eds.) (1970). *Chemical ecology.* Academic Press, New York.

Steenbergh, W. F. & Lowe, C. H. (1969). Critical factors during the first years of life of the saguaro (*Cereus giganteus*) at Saguaro National Monument. *Ecology*, **50**, 825–34.

Thompson, H. R. (1958). The statistical study of plant distribution patterns using a grid of quadrats. *Australian Journal of Botany*, **6**, 322–43.

Turner, F. B. (1962). Some sampling characteristics of plants and arthropods of the Arizona deserts. *Ecology*, **43**, 567–71.

Turner, R. M., Alcorn, S. M., Olin, G. & Booth, J. A. (1966). The influence of shade, soil, and water on saguaro seedling establishment. *Botanical Gazette*, **127**, 95–102.

Wallace, A., Romney, E. M. & Ashcroft, R. T. (1970). Soil temperature effects on growth of seedlings of some shrub species which grow in the transitional area between the Mojave and Great Basin deserts. *Biological Science*, **20**, 1158–9.

Weaver, J. E. & Clements, F. E. (1938). *Plant ecology*, 2nd edn. McGraw-Hill, New York.

Went, F. W. (1955). The ecology of desert plants. *Scientific American*, **192**, 68–75.

West, N. E. & Tueller, P. T. (1971). Special approaches to studies of competition and succession in shrub communities. In: *Wildland shrubs* (ed. C. M. McKell, J. P. Blaisdell & J. R. Goodin), pp. 165–71. USDA Forest Service General Technical Report INT-1, Ogden, Utah.

Woodell, S. R. J., Mooney, H. A. & Hill, A. J. (1969). The behaviour of *Larrea divaricata* (creosote bush) in response to rainfall in California. *Journal of Ecology*, **57**, 37–44.

Wright, R. A. (1970). The distribution of *Larrea tridentata* (DC) Coville in the Avra Valley, Arizona. *Journal of the Arizona Academy of Science*, **6**, 58–63.

Zohary, M. (1961). On hydro-ecological relations of the near east desert vegetation. *UNESCO Arid Zone Research*, **16**, 199–212.

Added in proof:

Fonteyn, P. J. & Mahall, B. E. (1978). Competition among desert perennials. *Nature*, **275**, 544–5.

Sternberg, L. (1976). Growth forms of *Larrea tridentata. Madroño*, **23**, 408–17.

Manuscript received by the editors October 1973

4. Animal–animal interactions

F. H. WAGNER & R. D. GRAETZ

Introduction

The paired relationships in which each species interacts with the others in an ecosystem have been variously termed 'consortisms' and 'coactions' in the ecological literature. These interactions may be to the benefit of one or both species involved, they may be to the detriment of one or both, or they may be without effect on one or both.

At the population level of ecological integration, 'benefit' and 'detriment' imply increase or reduction of population levels from what they would be without the action of the consortisms involved. Because of compensating population mechanisms, individual animals may be killed by predators or parasites, or spared by symbiotic relationships, without there being any population response. Two species may utilize the same resource, but if in so doing they do not reduce the resource sufficiently to affect the populations of one or both, they cannot be considered in competition.

Intraspecific interactions between individuals of a species may be viewed in the same perspective. Their use of a resource, including the territorial partitioning of space, is not appropriately considered competitive unless they reduce that resource to the point of reducing birth rates, and/or increasing mortality and emigration rates, and ultimately reducing population levels.

The questions of scientific interest about desert species relate both to particular populations and to abstract generalities about classes or groups of desert species (such as trophic groups, taxonomic groups, etc.). We wish to know if consortisms are significant constraints on the populations of desert species. If so, what is their relative importance vis-à-vis physical factors and intraspecific competition for resources, the latter to include agonistic behavior which may constitute competition for space? And more generally, are consortisms and intraspecific competition more or less influential as population constraints on desert species than on species of more mesic systems: grasslands, temperate forests, tropical systems? What role do these constraints play in shaping the structure of desert communities, and is that structure similar to, or different from, the communities of more mesic systems as a result? Evolution serves as a time scale for all of these questions as we wish to know the degree to which adaptive change has mitigated the relationships.

An ultimate scientific goal is to quantify these relationships as precisely as possible. In the case of particular populations, their demographic parameters become dependent variables in functional relationships, the interacting species constituting the independent variables. Such functions are included in mathematical models which enable one to predict the behavior of the population under specified conditions; and make it possible to devise management strategies with which it is possible to increase, preserve, or reduce the population according to the needs of the situation.

The parameters for such 'tactical' models (May, 1973) are provided by empirical research. Because they require long-term, intensive measurement and considerable resources, they have been developed for only a few species around the world. To our knowledge, only two exist for desert species: the black-tailed jackrabbit (*Lepus californicus*) by Stoddart (1977) and the coyote (*Canis latrans*) by Connolly & Longhurst (1975) in western USA. Sufficient data may exist to permit their development on an additional six or seven: two or three species of locusts in Africa and Asia (*Schistocerca gregaria, Locusta migratoria, Locustana pardalina*), and perhaps one or two in Australia (*Austroicetes cruciata, Chortoicetes terminifera*), and two species of Australian kangaroos (*Megaleia rufus* and *Macropus fuliginosus*). But Varley, Gradwell & Hassell (1973) decry the almost universal lack of demographic data needed to model population behaviour.

General or theoretical statements about classes of populations may also be expressed as mathematical models. Such models tend to be more abstract and less realistic representations of particular species and situations. Accordingly, their predictions are likely to be less precise statements about the behavior of individual populations to the degree that they depart from the realism of those particular cases. Their value lies in at least qualitative statements (or, more properly, deductions) about the behavior that can be expected of populations and species not yet studied. If they are in a very early stage of development, their statements may be taken as hypotheses to be tested by subsequent research. But if they have matured by surviving a lengthy period of test and experimentation, their predictions may be the basis for human action just as are the predictions of highly realistic models of individual populations.

The parameters for theoretical models come from several sources. They may come from empirical research, though perhaps abstracted or generalized from several particular populations. They may be derived by formal deduction from other models; or they may simply be stated as intuitively probable patterns on the basis of varied, former experience.

Despite the considerable effort over a period of some decades on the physiological, morphological, and behavioral adaptations of desert animals, much less research effort has been directed to population and community phenomena, and that mostly in the last decade. Few species have been studied sufficiently to permit development of 'tactical' models. And to our knowledge, no theoretical or 'strategic' models (May, 1973) have been developed specifically for desert species, although some characteristics of desert species and communities have been deduced with more general population and community models.

Hence, this chapter can review the existing data base, but is constrained to draw largely qualitative generalizations at this stage of our understanding about desert systems. And it can, with that base, test some of the hypotheses proposed from more general models.

Interspecific competition

Hypotheses from theoretical predictions

Birch (1957) defined interspecific competition as the situation in which two species require the same resource. If in attaining it, they reduce it to the point where the population level of one or both is reduced below what it would be without the presence of the other, competition occurs. The essential criterion of competition is, therefore, some reduction in the population of at least one of the consorts.

Contemporary general models of community structure and function provide hypotheses of the importance of competition in desert systems. Brown (1975, pp. 334–5) has reviewed at some length the predictions by MacArthur (1972, pp. 170–2) and May & MacArthur (1972) that species diversity is positively correlated with the productivity of resources, and inversely correlated with variability in productivity. Since deserts are characterized both by low productivity and extreme variability, they should, on the basis of this reasoning, have lower diversity than more mesic systems. And among deserts, the more arid a system is, the lower its expected diversity.

The implication for an understanding of competition is that high diversity is facilitated by a high degree of specialization in resource use, narrow niche breadth, and rather thorough exploitation of resources to the point of competitive pressure. But in an unproductive system with scant, undeveloped resources, such specialization and prodigal resource use is both risky and maladaptive. Survival is more probable for a small number of broad-niche species which can resort to a variety of resources and use those resources conservatively, and which remain at densities that are low relative to them. The result would be relatively low levels of competition.

These predictions could be tested by measuring the levels of competition among desert animals, and by comparing them with the competition levels in more mesic systems. However, we do not have effective, empirical measures of competitive pressures on populations which can be compared between taxonomically very different species, or between very different communities. But we can at this stage attempt to evaluate the prevalence of competition between desert species and compare it subjectively with the prevalence of competition in more mesic systems.

Empirical evidence of competition

There are several criteria for inferring the existence of interspecific competition, some of which are discussed by Diamond (1978). We will now list these and review evidence for each in desert systems.

(1) The existence of niche segregation and resource partitioning between species in a community implies that their evolutionary predecessors may have competed and diverged to permit coexistence. Such patterns have been found in virtually every group of desert species examined, with separation occurring along a variety of resource scales: food size and nutritional quality; various elements of habitat (vegetation structure, topography, soil texture); and time, both diel and seasonal.

Thus rodent species in North American deserts segregate by dietary preference, often facilitated by morphological differences (Kenagy, 1973*a*; Brown & Lieberman, 1973); by such habitat features as soil texture (Hardy, 1945; Hoover, Whitford & Favill, 1977) and vegetation structure (Rosenzweig & Winakur, 1969; Rosenzweig, 1973; Beatley, 1976; Holbrook, 1978), and by diel and seasonal activity cycles (Kenagy, 1973*b*; O'Farrell, 1974). Mares (1977) reported topographic segregation by three species of phyllotine rodents in the Monte of northwestern Argentina.

On a broader geographic scale, Genarro (1968) postulated that the northern range limit of the kangaroo rat *Dipodomys merriami* in New Mexico, USA, is set by competition with *D. ordii*, as is the northern limit of the grasshopper mouse *Onychomys torridus* set by *O. leucogaster*.

Newsome (1975) considered the feeding and habitat segregation of the red and grey kangaroos in arid central Australia to be so profound as to be 'prime example of competitive exclusion'. Small (1971) observed habitat segregation and some degree of partitioning in prey preferences by coyotes (*Canis latrans*), grey fox (*Urocyon cinereoargenteus*), and bobcat (*Lynx rufus*) in west-central Arizona, USA.

Several authors have observed variations in habitat preference between bird species in North American deserts (Dixon, 1959; Austin, 1972; Tomoff, 1974). Keast (1959) reported the same for Australian desert birds.

Pianka (1975) reported niche separation in lizards of the Kalahari, Australian, and North American deserts along several resource gradients: food, habitat, and time. Milstead (1957) concluded that habitat segregation avoided competition for a common food resource among several species of whiptail lizards (*Cnemidophorus*) in the Chihuahuan Desert of North America.

Several authors have reported similar separations in a variety of arthropod species: North American ants (Cole, 1934; Whitford & Ettershank, 1975; Whitford, Johnson & Ramirez, 1976; Schumacher & Whitford, 1976; Davidson, 1977*a*, 1977*b*; Chew, 1977; Whitford, 1978); African locusts (Uvarov, 1954); tenebrionid beetles in the Namib Desert (Holm & Edny, 1973), and virtually the entire invertebrate fauna in the Pre-Saharan zone of southern Tunisia (Heatwole, Muir & Zell, 1975).

(2) A more compelling basis for inferring the existence of competition is variations in niche boundaries in a species when it is first brought into sympatry by the arrival of a second species, or as it occurs in different communities with different numbers and kinds of potentially competing species. Brown (1975) noted the latter phenomenon in different communities of North American rodents, as did Davidson (1977*a*) in harvester ants.

(3) Character displacement is closely related to the last criterion. Brown (1975) observed displacement in body size of North American rodents while Creusere & Whitford (1976) observed displacement in song among Anurans. Huey, Pianka, Egan & Coons (1974) observed evidence of behavioral and morphological character displacement in Kalahari fossorial lizards. Davidson (1977*a*) observed variations in worker-size polymorphism as competitive pressures varied.

(4) Density compensation may be a manifestation of niche shift expressed in terms of the density of a competitor. Brown (1975) observed higher population densities of some rodent species in areas which did not have competitive species. Newsome & Corbett (1975) observed an inverse correlation between the density of exotic *Mus musculus* and the number of native rodent species in Australian deserts. The high lizard diversity of Australian deserts may be the result of a meager mammalian and snake diversity, while lizard diversity may be somewhat suppressed by competition with birds in the Kalahari and North American deserts (Pianka, 1975).

(5) Although the above evidence in total is persuasive, it is nevertheless circumstantial in that it represents situations where the interactions may have occurred in the past. Once resolved, it is conceivable that competition no longer exists as a pressure on the contemporary populations. Furthermore, they do not involve any measurements which show depletion of the resource presumably competed for.

An effective demonstration that competition currently exists is in the manipulation of one competitor, either artificially or through natural, environmental change, followed by a response in the population of one or more other species. Probably the most elegant experiments of this sort to date in desert systems are those of Brown & Davidson (1977) and of Whitford as cited in Brown & Davidson's paper. Brown & Davidson constructed replicated exclosures from which (*a*) rodents were removed and ant populations were measured, (*b*) ant populations were removed and rodent numbers were measured, (*c*) both rodents and ants were removed and seed stores in the soil were measured, (*d*) both rodent and ant populations were maintained as controls and seed stores were measured.

Table 4.1. *Response of ant populations to rodent removal and rodent populations to ant removal in the Sonoran Desert of North America.* (From Brown & Davidson, 1977)

Experimental variable	Control	Rodents removed	Ants removed	Increase over control (%)
No. ant colonies	318	543	—	71
Rodents				
Number	122	—	144	18
Biomass (g)	4.13	—	5.12	24

These studies, carried out in the Sonoran Desert of western USA, showed ant population increases where rodents were excluded and rodent population increases where ants were excluded (Table 4.1). Where both ants and rodents were excluded, seed stores increased by a factor of 5.5 over those of the controls.

Brown and Davidson further showed that both ant and rodent species densities in the deserts of western USA are a function of productivity as represented by their parameter of dependable precipitation. Both in a series of areas along a south-to-north gradient of increasing moisture, the number of ant species did not increase while the rodent species increased much faster than was the case with an array of areas all at similar latitude. They inferred that the declining temperatures of the more northerly areas were inimical to development of complex ant communities, and the competitive void then permitted greater development of a complex rodent fauna.

Walter Whitford (unpublished work cited in Brown & Davidson, 1977) conducted similar studies in the Chihuahuan Desert to the east of the Sonoran. While he did not measure the rodent response, he too recorded an increase in ant populations where rodents were excluded, and an increase in seed stores where both groups were excluded.

Analogous results have been reported for carnivores in western USA by Wagner (1972). The poison sodium monofluoroacetate (1080) came into general use in western USA in the late 1940s to control coyote populations and reduce predation on domestic sheep. In the more northerly states of the West, large amounts of 1080 were used and coyotes were materially reduced almost immediately and maintained at reduced densities through the 1950s and 1960s. In the more southerly states, less poison was used and the evidence does not suggest generalized reduction in coyote numbers.

Wagner developed indices to the population densities of bobcat, grey and red foxes (*Vulpes fulva*), and badger (*Taxidea taxus*). In the northerly states, these species increased almost immediately following coyote reduction (Fig. 4.1). In the southerly states, they did not increase. The coyote is suspected

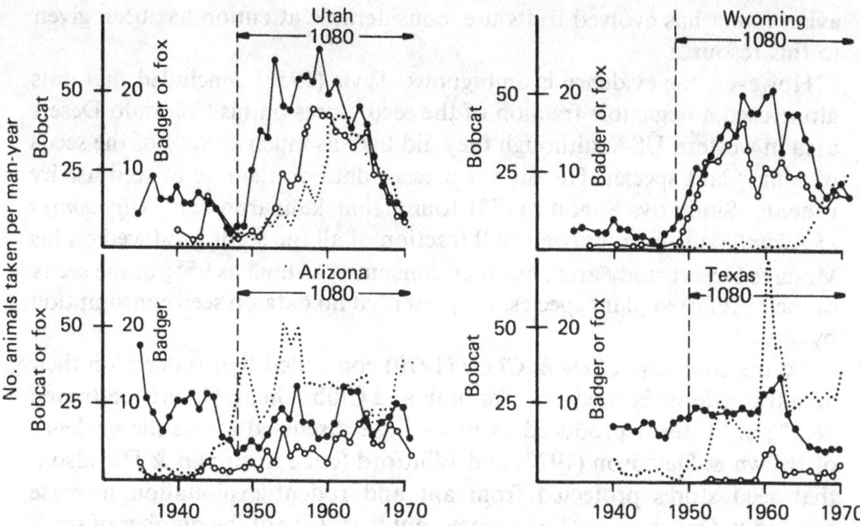

Fig. 4.1. Annual population trends of bobcats, badgers, and foxes in two northern and two southern states as shown by indices from the records of the US Fish and Wildlife Service, Division of Wildlife Service. After Wagner (1972). ○—○, Badger; ●—●, bobcat; . . ., fox.

not to have exhausted the food base of the other species, but to have applied competitive pressure through aggressive behavior.

Milstead (1965) observed populations of three species of *Cnemidophorus* lizards in the Chihuahuan Desert in 1951–52, and again in 1962. The earlier period had been preceded by a prolonged drought, and the three species were present on his study area in comparable densities. By the second period, precipitation had returned to normal and *C. tigris* had increased markedly while the other species had declined to an equal degree. Milstead speculated that *C. tigris* may have been sufficiently favored by the better rainfall pattern that it gained a demographic advantage over the other species. By increasing, it exerted heavier competitive pressure on them and reduced their number.

A similar pattern was observed by Muir & Heatwole (1976) in the soil microarthropods of Pre-Saharan southern Tunisia. During a 3-yr period with rising precipitation, niches occupied by soil mites were usurped by small bugs of the family Lygaeidae. The change was inferred by them to result from superior competition from the lygaeids which themselves were enhanced by the increased moisture.

(6) Since competition implies use of a resource to the point of limiting one or more of the competitors, it could be illuminating to analyze degrees of resource use. Since seeds constitute a large, high-quality fraction of the primary production of arid areas, and a sizeable rodent, ant, and winter

avian fauna has evolved to its use, considerable attention has been given to this resource.

However, the evidence is ambiguous. Tevis (1958) concluded that ants alone used a negligible fraction of the seed stores on his Colorado Desert area in western USA, although they did take as much as 90% of the seeds of some plant species. He did not present data on the rate of seed use by rodents. Similarly, Soholt (1973) found that kangaroo rats (*Dipodomys merriami*) consumed only a small fraction of all the seeds produced on his Mohave Desert study area, but they consumed as high as 95% of the seeds of their preferred plant species. He presented no data on seed consumption by ants.

To the contrary, Chew & Chew (1970) concluded that rodents on their area transitional between the Sonoran and Chihuahuan Deserts destroyed 86.5% of all seeds produced. And we have discussed above the evidence of Brown & Davidson (1977) and Whitford (cited in Brown & Davidson) that seed stores protected from ant and rodent exploitation increase markedly. Davidson (1977*a*) pointed out that it is not the number of seeds present which may be inportant, but whether that number constitutes a sufficiently high resource density to permit foraging granivores a positive energy balance.

Evidence on the consumption of other components of desert primary production is somewhat less than one would desire, but in general suggests that a minor fraction is taken as live plant material (cf. Chew & Chew, 1970). Yet Key (1959) suggests that Australian grasshoppers and locusts place sufficiently heavy pressure on the vegetation to be significant competitors with sheep.

In several deserts, plant material appears to be largely used as detritus after it has died. Termites are important in this regard. Nutting, Haverty & La Fage (1975) and Johnson & Whitford (1975) have measured rates of cellulose consumption on two North American areas and found these to be 92 and 50% respectively of the rate of cellulose production. In the latter case, only the consumption of subterranean termites was measured. According to Nutting (personal communication), Sonoran Desert termites tend simply to consume their resource entirely, then produce large numbers of alates which disperse to new sources of wood.

In general, we will see numerous cases in Chapter 5 on animal population dynamics of species fluctuating annually with variations in precipitation. Very commonly these seem to be linked to variations in the food base: variations in primary production in the case of herbivores, and variations in herbivore numbers in the case of predators. It is difficult to avoid the impression that food availability is one, common source of restraint on animal numbers in the desert and, by implication, that food resources are often used fully.

Comments on diversity in deserts

Eric Pianka has discussed diversity of desert ecosystems at some length in Volume 1. But the close association between interspecific competition and diversity makes it perhaps appropriate to touch briefly on a few aspects of the topic here which he did not treat in his discussion.

The basic question to be explored here has been touched on previously: have evolutionary patterns in deserts produced low-diversity communities, as predicted by current competition and diversity theory? The mechanism is the assumed survival value of maintaining small numbers of low-density species which do not press the limits of their resources, and which do not experience high levels of interspecific competition. The latter point is important to the general question under discussion: the prevalence and influence of interspecific competition in desert systems. Alternatively, what other forces may have played a part in shaping the diversity patterns that do exist?

The answers depend in part on what one calls a desert. Different investigators use a variety of criteria for setting the upper moisture limits: potential evapotranspiration:precipitation ratios; 'effective' moisture indices; temperature–rainfall equations; or simply mean annual precipitation levels set variously at 300, 250, 200, or 150 mm. The deserts of the world exist in a continuum between whatever point is selected for this upper bound, and precipitation levels that approach zero.

At the arid extreme – that is, at 50 mm of precipitation or less – there is no question that diversity is quite low. One may see dunal vistas of the Sahara's Grand Erg Oriental with but a single plant (*Calligonum comosum*) within view. Louw (1972) describes dúnal areas of the Namib where no plant may grow for years, and the meagre animal pyramid is supported by wind-carried detritus. Brinck (1956) describes similar areas in the Namib and Kalahari. George (1977) describes rocky areas (*hamada*) in the Moroccan Sahara where no flowering plants grow, and a vestigial fauna is supported by blue-green algal films on the rocks. Grenot (1974) comments on the general impoverishment of the Saharan flora.

The explanation for this low diversity must remain somewhat conjectural, and it is not clear that it is a function of a lengthy evolutionary process in which diverse biotas, intense resource exploitation, and high levels of competition have been weeded out. It seems more likely that at these extremes of the physical environment, the probability that any species can evolve the specialized adaptations needed for survival is low just as with other extreme environments like thermal springs, permanent ice sheets, and extreme salinity. This statement is made with some reserve, however, in light of Hamilton's (1971) observations on tenebrionid beetles in the Namib. The six species of *Cardiosis* occupy broader niches in the northern desert, where they are not sympatric, than in the south where they are sympatric. The difference, he suggests, may be due to competition.

At somewhat higher rainfall (100–200 mm), but still well within the range accepted by most observers as being desert or arid, there is the persuasive evidence reviewed above for the role of competition in shaping community diversity. Diversity in some groups does appear to be a positive function of resource productivity. Thus Brown & Davidson (1977) found both rodent and ant species densities to be positively correlated with their measure of dependable precipitation. And within comparable deserts, there is a surprising degree of functional similarity among rodent communities (MacMahon, 1976).

Pianka (1975) found that lizard species density correlated positively with long-term mean, annual precipitation, and with the standard deviation of that mean. Since the standard deviation is an *absolute* measure of variation, it is correlated with mean precipitation, as Pianka found, unless *relative* variation is itself exceptionally variable. Hence the three values were intercorrelated among the sites which Pianka analyzed.

In a number of cases, diversity of some desert groups appears to be a function of resource diversity. Thus vegetative diversity has been pointed out to be function of soil texture diversity by numerous authors (Shreve, 1942; Yang & Lowe, 1956; Cross, Goldstein, Lowe & Morello, 1977), although more recently, Phillips & MacMahon (1978) have reported the relationship to depend importantly on the percentage of sand particles in the soil.

Once the vegetation pattern is established, avian diversity is then a function of the diversity of vegetative structure (Dixon, 1959; Keast, 1959; Austin, 1972; Tomoff, 1974). In a particularly interesting set of measurements, Otte (1977) showed that the number of grasshopper species in both the Sonoran Desert and the Argentine Monte increased linearly as the number of plant species increased. His regressions indicated an average of one grasshopper species per two plant species in the Sonoran, and one per four in the Monte.

These observations imply a measure of species packing per unit of resource, the degree of packing doubtless limited by competition. Hence species diversity in deserts does appear to be shaped in part by competition, and *within* desert systems, apparently is inversely correlated with the degree of aridity.

But in the broader context of all terrestrial systems, many groups of organisms have as high a diversity in deserts as, or higher than, more mesic systems. Whittaker & Niering (1965) found the vegetation of lower mountain slopes to be more diverse than that of higher (and more mesic) elevations in the Sonoran Desert. They also found the Sonoran vegetation more diverse than that of the eastern forests of North America and comparable elevations in the maritime climate of southern California.

The number of plant species on the three 'hot' desert sites studied by the US/IBP Desert Biome project is 37 (Jornada bajada in the Chihuahuan Desert), 73 (Silver Bell site in the Sonoran), and 51 (Rock Valley site in the

Mohave Desert). In the Dar ez Zaoui site of the Tunisian Pre-Saharan Project, where mean, annual rainfall is *c.* 100 mm, the total number of species is 53 (Wagner, 1977*a*). These values compare favourably with the number of species per vegetation stand analyzed by Curtis (1959) and coworkers in Wisconsin (mesic north-central USA). Within the 1420 stands analyzed throughout the state in a wide range of environments, the number of species varied mostly between 30 and 60 per stand.

Similarly, the diversity of many desert animal groups compares favorably. That lizards and rodents attain some of their greatest diversity in deserts has been widely recognized (e.g., Chew & Chew, 1970; Pianka, 1975).

Cloudsley-Thompson (1976, p. 47) commented on the high species density of insects in deserts, remarking that a high percentage of the known orders of terrestrial insects are found in this type. The numbers of above-ground invertebrate species (mostly arthropods) on the US/IBP sites have ranged between approximately 600 (Rock Valley, Nevada, at 150 mm mean, annual precipitation) and 1000 (Curlew Valley, Utah, at 250 mm annual precipation.) On the Tunisian Pre-Saharan site the total number of arthropod and mollusc species is approximately 1200. By comparison, the number of arthropod species on the US/IBP Grassland site is 1242 (George M. Van Dyne, personal communication). For three douglas fir (*Pseudotsuga mensiezi*) stands in Oregon (northwestern USA), Edmunds (1974) reported a total of 696 invertebrate species.

The pattern has been shown for some groups which have been given intense scrutiny. Moldenke (1976) found bee diversity in the Mohave (*sic*) and Colorado Deserts to exceed that in all but three of the other 14 California biotic regions he investigated. It is our impression that bombyliid and asilid flies, tenebrionid beetles, ants, and possibly some groups of wasps and orthopterans are more diverse in deserts than in many, more mesic systems.

There are, of course, exceptions and birds are a notable example. Peterson (1975) showed a declining northeast–southwest gradient in the number of avian species in the USA, with lowest numbers in the arid and semi-arid southwestern region. But overall, diversity of the biota as a whole in all but the severest deserts may be exceeded only by that of the tropics.

One can only conjecture upon the reasons for this. But most deserts lie at relatively low latitudes, between 35° and 20°. As a result, their high diversities may be one reflection of the general pole-to-equator gradient in biotic complexity. Whether the causal mechanism is great geologic age, or high temperatures which permit long seasons of activity and relatively rapid generation turnover, or some other effect of heavy radiation load; those mechanisms play a part alongside competition in shaping desert community structure.

One last aspect of diversity bears brief mention. There is considerable temporal variation in the species that can be observed on desert areas because of interspecific differences in diel and seasonal activity cycles; the high mobility and nomadic tendencies of numerous desert species and the proclivity of many species to remain dormant in seed, egg, and diapausal stage during drought periods. Thus, the Tunisian Pre-Saharan studies have shown profound seasonal turnover in invertebrate communities (Muir & Heatwole, 1976).

Species densities and the relative numbers of constituent species are widely known to vary annually with year-to-year variations in precipitation. Such variations have been observed in mammals (Whitford, 1976), in birds (Miller & Stebbins, 1964, p. 15; George, 1977), and in lizards (Milstead, 1965; Whitford & Creusere, 1977). M'Closkey (1972), who has observed similar annual variations in rodent communities of California coastal sage scrub, points out the hazards of drawing inferences about diversity on the basis of one-time measurements.

One implication would seem to be that the species density and diversity of any given desert area could be expected to vary around some equilibrium value geared to the precipitation norm for that area. But annual values would fluctuate around the equilibrium depending on annual rainfall patterns. The patterns of variation around the equilibrium in different areas and taxonomic groups would seem to be interesting questions for study.

Predation

Theoretical considerations

Like the other consortisms, predation is a two-way relationship. The predator of course acts as a negative influence on the prey, and prey a positive influence on the predator. Either species may affect the other's population in one or more of several ways: (*a*) cause it to fluctuate, (*b*) regulate it around some long-term equilibrium density (*sensu* Wagner, 1969), (*c*) act as a destabilizing force on population density, and/or (*d*) serve as one determinant of the mean density around which equilibrium is maintained.

The questions about predation in desert ecosystems for which we seek answers are analogous to those we ask about interspecific competition. In the case of particular populations, we wish to develop tactical models which incorporate mathematical statements about the effects of predators on a prey population (if the latter is the dependent variable whose behavior the model is to simulate) or conversely of prey on predators (if the latter is the species of interest). With such models, we wish to predict the behavior of the populations in question under specified conditions.

In the case of general theory, we wish to know the relative importance of predation, vis-à-vis the other environmental factors, as an influence on desert species generally, and how this influence compares with that in other ecosystems. Or again, we wish to know generally the relative importance of prey as a food resource in affecting predator numbers. Such generalizations could come inductively from explicit understanding of numerous, particular populations; or they could be deduced from strategic models, the variables and relationships of which were from strategic models, the variables and relationships of which were from empirical observation of particular populations. Thus, both approaches require considerable empirical understanding.

We also seek an understanding of the role of predation in desert community structure. What role might predation play in reducing prey competition for resources, thereby increasing prey diversity and perhaps also affecting the plant trophic level, or whichever level supports the prey? Alternatively, what role do the prey play in determining the size and diversity of the predator community?

In the case of the prey, the most important parameter to measure in addressing these questions is the predation mortality rate: the proportion of the prey population killed by each predatory species, and by all of them collectively. Such mortality can be measured directly through the use of radiotelemetry (Stoddart, 1970); or estimated by knowing the number of prey killed per predator per unit of time, the number of predators, and the number of prey (Wagner & Stoddart, 1972); or with such life-table techniques as *k*-factor analysis (Varley *et al.*, 1973; Stoddart, 1977).

These mortality rates are then measured each year for a number of years, while the numbers of predators and alternate prey are measures simultaneously. After a period of such measurement, prey mortality rate due to predation can be expressed as functions of prey members, predator numbers, and perhaps the numbers of alternate prey. Such functions are used as the building blocks of predictive models.

This approach has probably been used most effectively with the winter moth (*Operophtera brumata*) in England (Varley *et al.*, 1973) where the functional relationships of several predatory and parasitic species were incorporated in a model for the species. To our knowledge, it has been accomplished for only one desert prey species: the black-tailed jackrabbit in northern Utah (western USA) by Stoddart (1977).

In the same manner, predator reproductive and/or mortality rates may be expressed as functions of prey and predator numbers. This too was accomplished by Varley *et al.* (1973) for the predators of the winter moth. We are aware of but a single desert predatory species for which it has been accomplished: the coyote in western USA (Connolly & Longhurst, 1975; F. H. Wagner *et al.*, unpublished).

Because we have so few thorough studies of predator–prey systems, we are nowhere near being able to propose the kinds of generalizations we seek about predation in desert systems and must instead rely on more equivocal evidence. In general, predation research has not developed and used the array of criteria for inferring that predation is a significant population population constraint as we have seen in the case of interspecific competition. Several occur to us which could perhaps be used in future research, and which are somewhat analogous to the competition criteria:

(1) differences in morphological features and behavioral characteristics, including habitat use patterns, between populations of the same prey species in areas with and without certain predatory species. The criterion is analogous to character displacement and niche segregations of populations with and without competitors. But like character displacement and niche segregation, these differences would imply evolutionary adjustments in the past to mitigate the effects, and do not necessarily imply strong, current effects;

(2) differences in density between prey populations of the same species with and without predators. The criterion is analogous to density compensation, and like it is circumstantial. But as Diamond (1978) points out in the case of density compensation, given enough replicates the probability of cause and effect is greatly enhanced;

(3) fluctuations in prey numbers following fluctuations in predator numbers. The latter could occur naturally, or be man-induced either inadvertently or through advertent experimentation. The reverse can also be used for inferring the role of prey in the population dynamics of predators: fluctuations in predator numbers following fluctuations in prey.

While there is occasional allusion to these effects in desert species, they have not had the systematic treatment given to the analogous ideas in competition. Consequently, we are not in any position to infer that predation is or is not a generally influential constraint on the populations of desert species. To be sure, carnivory occurs everywhere, but it does not necessarily follow that the effects are of any great significance. The fraction of the prey population removed could be so small that it is inconsequential. Or predation may act upon animals in excess of habitat or behavioral thresholds which would be lost whether or not predators took them.

As Diamond (1978) points out, Andrewartha & Birch (1954) wrote 25 years ago that interspecific competition was not a constraint of widespread importance on animal populations. But he is now able to write that it *is* an influence of general significance. The change in view results from the burst of research activity and conceptualization in competition and community structure in the intervening quarter century. While a great deal of predation research has been carried out during this same period, and the empirical base greatly expanded, there has not been as much flowering of concepts and coalescence of theory as in the competition area. This has certainly been true of desert systems.

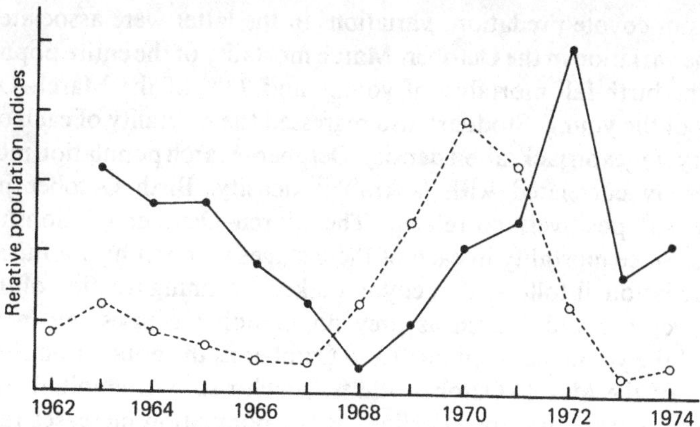

Fig. 4.2. Population trends of jackrabbits and coyotes in a Great Basin Desert area. After Wagner (1977*b*). ○—○, Jackrabbit; ●—●, coyote.

Empirical evidence of effects on prey

Studies on predator–prey systems

The black-tailed jackrabbit is a widely distributed species in western North America. In some areas of the Great Basin Desert, as in the Curlew Valley US/IBP Desert Biome research site, it far exceeds all other wild, vertebrate herbivores in biomass (Wagner, 1971). While it is preyed upon by a number of raptorial species and carnivorous mammals, by far the dominant species in Curlew Valley, in terms of biomass and number of jackrabbits taken, is the coyote (Wagner, 1971). Both species fluctuate (Fig. 4.2) over a range of about 15–60-fold (highest densities divided by the lowest), depending on the season of measurement. The coyote lags behind the jackrabbit by one or two years.

The investigators who have studied this system for 17 years have measured jackrabbit mortality at four life-history stages: intra-uterine, young (between birth in spring or summer and October), adults (between March and October), and the entire population (between October and March) (Stoddart, 1970; Wagner & Stoddart, 1972; Gross, Stoddart & Wagner, 1974). Mortality rates have been calculated for each of these four time intervals. Mortality due to coyote predation has been measured with radiotelemetry (Stoddart, 1970), and estimated with a knowledge of coyote and jackrabbit numbers, and estimates of the number of jackrabbits killed per coyote per unit of time (Wagner & Stoddart, 1972). This predation loss was shown to be a major fraction of the entire mortality.

After a number of years of data had accrued, Stoddart (1977) subjected the measurements to *k*-factor analysis (Varley *et al.*, 1973) and showed that most of the year-to-year variation in jackrabbit morality was associated with variations in coyote predation: variations in the latter were associated with 87% of the variation in the October–March mortality of the entire population, 76% of the birth-fall mortality of young, and 72% of the March–October mortality of the young. Stoddart also regressed the mortality of each of these life-history stages on jackrabbit density. October–March population mortality was inversely correlated with jackrabbit density. Birth–October juvenile mortality was positively correlated. The March–October relationship was parabolic. Since mortality in each of these stages is so strongly a function of coyote predation, it follows that coyotes take a declining fraction of the prey between October and March as prey population increases; an increasing fraction of the young between birth and October as the population rises; and a fraction of the March–October adults, which rises in the initial stages of population increase and then declines as the population increases further.

In total, Stoddart concludes that the two species behave significantly as a Lotka–Volterra system, with the rises and falls in prey numbers largely a function of variations in coyote predation. The latter exhibits what Holling (1959) has termed a functional response in its feeding activity, and a lagging numerical response. Therefore, the fluctuation pattern, and the stability of the population in the sense of its propensity always to return toward a mean or equilibium density even though overshooting it, appear largely attributable to coyote predation.

Furthermore, the mean density of the jackrabbit population can be said to be largely determined by that predation, at least for the present. If coyote predation were reduced or removed, the mean density over a period of years would rise. This statement must be tempered with the proviso that, with prolonged higher jackrabbit density, it might reduce its food supply or incur a disease outbreak which would then reduce it. Hence, one could not predict a permanent population increase if coyote predation were removed. But an initial increase seems likely, and the statement seems reasonable that coyotes at present are the major constraint on jackrabbit mean density over a period of years.

The converse would also be true. If coyote predation were increased, one could reasonably expect reduced mean jackrabbit densities over a period of years. There appears to be some evidence to support this view. Jackrabbits declined during the early 1960s, existed at low densities in the mid-1960s, and increased from 1968–1970 (Fig. 4.2). They again decreased in the early 1970s, and remained at low levels during the mid-1970s. Coyotes declined during the 1960s, reached a low in 1968, increased to a peak in 1972, then declined again during the 1970s. But they did not decline to as low levels in the 1970s as those of the 1960s. Possibly as a result, jackrabbits descended to lower levels in the 1970s than those of the 1960s.

In addition, Stoddart has measured a jackrabbit and coyote population in the nearby state of Idaho where coyote control is very light and densities are high. In this area, jackrabbit densities are substantially lower than in his Curlew Valley research area.

Newsome & Corbett (1975) described a similar pattern of rodent outbreaks in arid and semi-arid Australia, although it was different in several fundamental respects. Rodent populations decline during drought periods, contracting into isolated pockets of more favorable habitat. With the rodent decline, predator populations – dingo (*Canis familiaris dingo*), barn owl (*Tyto alba*), snakes, and the introduced European fox (*Vulpes vulpes*) and domestic cat – decline as well. Given a series of years with above average rainfall and widespread habitat improvement, rodents increase to 'plague' proportions. As they do, the predators increase but because of their high reproductive rates, the rodents increase faster and escape predatory restraint. As the weather returns to normal, the rodents once again decline. Predator numbers have now risen to high levels, and with reduction in rodents, the predator:prey ratio is high and heavy predatory pressure is applied. The predators now hasten the decline of rodents, and could effectively block any rodent recovery if by chance there were a favorable rainfall year.

This system differs from the jackrabbit–coyote system in that it is the weather-induced waxing and waning of habitat which is the underlying cause of fluctuations in prey numbers. The predators follow along, but their role is largely one of modifying the form of the fluctuation by hastening and perhaps deepening the decline of the prey. According to Stoddart's interpretation, however, coyote predation is the predominant force effecting, and determining the pattern of, fluctuation in the jackrabbit.

Myers & Parker (1975) describe a somewhat similar pattern in the introduced European rabbit (*Oryctolagus cuniculus*) in Australia. Here again, rabbit populations contract to a few, favorable sites for their warrens during drought periods. During periods of increased rainfall, they increase in numbers and extend their distribution. When dry conditions return, populations decline gradually through reduced birth rates, or more abruptly through dispersal. As in the case of the rodents, predator populations rise and fall with these prey fluctuations.

A common thread in all of these cases may be the variability of prey populations in ecosystems of relative biotic simplicity. While there are several species of rodents in Curlew Valley, the jackrabbit equals or exceeds their biomass even during periods of its population low. For the coyote, it is a relatively simple prey system, and while there are alternate predatory species, the coyote exceeds their importance by far. For the jackrabbit it is a relatively simple predatory system. Wagner (1971) has suggested that if there were an abundance of suitable, alternate prey, coyote numbers and their pressures on jackrabbits could be stabilized. By the same token, if there were numerous, alternate and effective jackrabbit predators, the constraints on their numbers could be diversified, complementary, and perhaps less variable.

Similarly, Newsome & Corbett (1975) comment on the low diversity of the rodent fauna of arid Australia. In their view, if dingos could survive on alternate food when rats (*Rattus villosissimus*) become scarce (and presumably not decline in numbers), they could prevent surviving rat populations from increasing when rainy periods improve habitat.

In general, deserts with rainfall of 100 mm or more appear complex, as discussed above. Hence these three simple predator–prey systems may not be the rule. The variability of the precipitation does produce great variation in plant production, and subsequently in herbivore populations, as Wagner will discuss in Chapter 7. Whether or not that variation is somewhat muted by the substantial predatory diversity more typical of deserts remains to be seen.

It may be no coincidence that most of the historic accounts of extreme black-tailed jackrabbit plagues relate to the northerly, more simple Great Basin Desert (Palmer, 1897; Nelson, 1909). While there are some reports of modest fluctuation in the more southerly US 'hot' deserts (Chihuahuan, Sonoran, Mohave), apparently the fluctuations have not been as extreme as those that have occurred in the north. Miller & Stebbins (1964) did not observe any 'extreme' change in jackrabbit numbers in their Mohave Desert study area during a 15-yr period.

Turner, Hoddenbach, Medica & Lannom (1970) measured survival rates of the lizard *Uta stansburiana* in enclosures with or without 2.5 leopard lizards ha^{-1} (*Crotaphytus wislizenii*), a predator on *Uta*. In the predator-free enclosures, survival of *Uta* hatchlings from summer to the following spring was 0.43; that of *Uta* adults was 0.40 and 0.55. In enclosures with leopard lizards, the survival of both *Uta* hatchlings and adults was 0.15. These authors concluded that *Crotaphytus* predation may be a significant influence on *Uta* demography over the latter's range in the Mohave Desert.

Yom-Tov (1970) reported much lower densities of the snail *Trochoidea seetzeni* on the south slopes of wadis in the Negev Desert than on the north slope. He attributed this to heavy predation by dormice (*Eliomys melanurus*) and gerbils (*Gerbillus dasyurus*) which prefer the south slopes.

Hsiao & Kirkland (1973) reported increases in the sagebrush defoliator (*Aroga websteri*) in the US Great Basin Desert in the second year of their study. They attributed this to a lower level of hymenopteran parasitism in this year.

Predator–prey ratios in deserts

Cloudsley-Thompson (1976, p. 53) and others have commented on the high proportion of predatory species in desert faunas. This impression seems to come in part from the high density of arachnids and reptiles in deserts, and perhaps from the fact that a preponderance of desert birds are carnivorous (Gullion, 1960). One also gets the impression that the predatory dipteran

Table 4.2. *Percentage of predatory and parasitic species among invertebrates of mesic and more arid regions*

Locale	Biotic type	Faunal group	Predators and parasites %	Source
Mesic regions				
Arctic North America	West coastal region	Insects	24	Weiss (1926)
Northeastern USA	Statewide	Insects	25–28	Weiss (1926)
Eastern USA	Deciduous forest	Foliage insects	33	Whittaker (1952)
Southeastern USA	Abandoned fields in deciduous forest	Arthropods	30	Hurd & Wolf (1974)
Northwestern USA	Conifer forest	Invertebrates	35	Edmunds (1974)
Arid and semiarid regions				
Great Basin, USA	Desert, grassland, scrub, woodlands	Insects	53	Horning & Barr (undated)
Great Basin, USA	Desert	Invertebrates	34–44	Osborne (1975)
Southern Nevada, USA	Mohave Desert	Arthropods on *Larrea* foliage	37–40	Mispagel (1974)

families Bombyliidae and Asilidae are especially numerous, partially verified by Paramonov's (1959) statement that these are the dominant families in the Australian dipterofauna.

A more systematic test of this impression may be made by comparing the actual percentages of predatory species in desert faunas with those of more mesic areas (Table 4.2). These suggest that in more mesic aras – whether coniferous or deciduous forest, early successional stage or late, or the Arctic – the percentage of predators in the invertebrate fauna is typically a fourth to a third. In more arid types, the percentage of predatory and parasitic forms is more typically in the range 30–60. Orians & Solbrig (1977) have also commented on the high percentage of predatory forms in the faunas of the North American Sonoran Desert and the Argentine Monte.

It is not clear from these data whether the higher percentages of predatory forms is a characteristic of arid areas *per se*, or whether it is once again the indirect result of deserts occurring largely at fairly low latitudes. A considerable body of evidence indicates a pole-to-equator increase in the proportion of carnivorous species in the fauna (cf. Pianka, 1966). Naumov (1922, p. 274) described southward increases in predatory pressures in the Soviet Union while Turner (1977) has mentioned the possibility of increased predator pressures on lizards toward the equator.

Adaptations for mitigating predatory influence

A number of authors have pointed out the prevalence of certain adaptations in desert species which they surmise to be the result of strong predatory pressures. Thus, while Buxton (1923) and Sumner (1925) have questioned the view, numerous authors have emphasized the cryptic colorations of desert species and concluded that this is adaptive. Miller & Stebbins (1964) have pointed out that the pressures for concealing coloration in desert species must be acute because of the prevalence of predators and the openness of the habitat.

Lea (1964) has commented at some length on the phenomenon of phase variation in the brown locust (*Locustana pardalina*) of the southern African karoo. When the young locusts develop during periods of high density, they become active and brightly colored. When confronted by small predators they bunch together and use their aposematic coloration. When the young grow during periods of low population density, they develop into very inactive animals with dull, cryptic coloration. Lea surmises that the pattern evolved under heavy predation by small predators.

Tomoff (1977) reported on the high percentage of birds in North American Sonoran and Argentine Monte which nest in cavities, spinescent trees and shrubs, and cacti. He suggests that this may be the result of strong predatory pressures on nesting birds.

An interesting set of observations has developed concerning predatory pressures around desert water holes. Cade (1965) observed regular raptor hunting around a water hole in the Namib Desert, as did Fisher, Lindgren & Dawson (1972) in Australian deserts and Beck, Engen & Gelfand (1973) in the Sonoran. Elder (1956) observed frequent use by carnivorous mammals in the Sonoran. All of these authors, like Miller & Stebbins (1964) suspected that prey species were vulnerable to attack while concentrated around these areas. Beck *et al.* (1973) found the raptors to be diurnal in their visits, while Gambel's quail (*Lophortyx gamebelii*) were crepuscular, and he postulated that the quails' pattern has developed to avoid the raptors. Cade & Fisher *et al.* (1972) similarly attributed prey behavior patterns around water holes to predatory pressures. The latter went so far as to surmise that predation may have worked as a selective force in developing some desert species which do not require free water. In their view, human development of water areas in Australia may have reduced predatory pressures by enabling the prey to disperse more widely.

Empirical evidence of effects on predators

The reciprocal of the predation question we have been addressing – the significance of predation as a constraint on the populations of desert prey species – is the significance of prey numbers as a determinant of predator populations. If predators maintain their populations close to the limits of available food supplies, the latter may be the major determinant of predator densities and as they fluctuate, so too will the predators fluctuate. Alternatively, if the major constraints on predator populations are the physical environment, suitability of habitat, their own social behavior, or disease and their own predators, their densities could be maintained well below the limits of their food supply. In this case, their prey could fluctuate without their fluctuating in response.

Demonstrating the influence of prey on predator populations depends minimally on measurements of prey and predator numbers over a period of years, which show parallel fluctuations. Demonstrating cause and effect, however, depends on measurements of predator reproductivity, mortality, and dispersal rates, and elucidating which of these are associated with variations in the food base. To our knowledge, these measurements have been made more completely in coyote populations of the North American Great Basin Desert than in any other desert mammalian predator. Clark (1972) initially showed fluctuations in black-tailed jackrabbit populations (Fig. 4.2).

Following the development of techniques for measuring mortality rates at several life-history stages (Knudsen, 1976; Wagner *et al.* unpublished) have shown a correlation between October–October annual mortality rates of the entire population and jackrabbit density. Since most of the mortality is man-induced, the relationship seems to involve increased feeding activity as food becomes scarce. With increased activity, coyotes become more vulnerable to guns, traps, and poisons and their mortality rate rises.

Wagner (1977*b*, 1978) pointed out that where a predator fluctuates causally from year-to-year with its prey, this implies that the long-term mean prey density is a determinant of long-term mean predator density. Hence, Curlew Valley jackrabbits are not only the main determinants of coyote fluctuations, but also of their mean density.

Fluctuations in response to changes in mammalian prey populations have been reported in other desert predatory species. Joseph Platt (personal communication) observed fluctuations in raptor populations in response to jackrabbit fluctuations in Curlew Valley, the same area in which Clark, Knudsen, and Wagner *et al.* measured coyote responses to jackrabbit change. Egoscue (1975) reported a decline in kit fox (*Vulpes macrotis*) abundance as jackrabbits declined in another Great Basin area 160 km south of Curlew Valley. Newsome & Corbett (1975) describe fluctuations in mammalian and avian predators, and even snakes, as Australian desert rodents fluctuate.

Fluctuations have also been reported in desert insectivorous species which correlated with annual variations in rainfall. It is usually assumed that the insect species, which are food for the predators, are fluctuating with the rainfall variations, and that the predators are fluctuating in response to vagaries of their food supply. Thus Miller & Stebbins (1964) describe fluctuations in several species of passerine birds in the North American

Mohave Desert , while George (1976) describes the same for Sahara species. Turner *et al.* (1970) report population fluctuations in the lizard *Uta stansburiana* in the Mohave Desert which are associated with annual variations in winter rainfall. These fluctuations result from annual variations in the reproductive rate associated with moisture variations (Turner, Medica & Smith, 1974), and are assumed to be related to differences in the abundance of insects associated with annual plants. Mayhew (1966) has shown similar, annual variations in reproductive activities of *Uma notata* in the Colorado Desert of southern California, and has also attributed these to variations in the abundance of insects associated with annuals.

Agonistic behavior

The evidence regarding the role of intraspecific aggressive behavior in the population mechanisms of desert animals is ambiguous. There has been a tendency among investigators to postulate that desert species should evolve highly aggressive social structures which space individuals widely and maintain low densities in order to assure a food supply from sparse and highly variable resources.

One can compile a sizeable body of evidence to support this hypothesis. Rodents of the Heteromyidae are often cited as examples. Kangaroo rats, kangaroo mice, and pocket mice do tend to be solitary and intolerant of conspecifics. Maza, French & Aschwanden (1973) described the home-range pattern of the pocket mouse *Perognathus formosus* in the Mohave Desert. They concluded that home range was sufficiently large to provide adequate food for the individual and served to limit population density. Mares (1977) concluded that *Phyllotis griseoflavus* was the most aggressive of the three species in this genus which he observed in northwestern Argentina, and this species also occupied the most arid habitats. Fleming (1974) similarly concluded that in Costa Rica, *Heteromys desmarestianus*, an arid-land species, was more intolerant of conspecifics than *Liomys salvini*, a denizen of tropical forests.

Schroder & Rosenzweig (1975) removed individual kangaroo rats from a Sonoran Desert community. There was a strong tendency, however, for the spaces opened thereby to be filled promptly by individuals of the same species suggesting that intraspecific competition may have been a stronger force than interspecific competition.

Similarly, there is abundant evidence of aggressive behavior and territorial defense in many species of desert lizards (Mayhew, 1968; Stamps, 1977). Milstead (1957, 1970) surmised that the function of territory is to space out the animals in relation to food, and that the process limits population growth. And of course, a variety of territorial patterns exist in desert birds.

But this seeming support for the hypothesis may be a function of perspective.

The question is not whether aggressive behavior and spacing mechanisms occur widely in desert species, but whether they are more pronounced and influential. Eisenberg (1967) has systematically studied several species of five genera of Heteromyidae, two of Dipodidae, five of Gerbillinae, and several of Cricetinae from three continents, some from desert inhabitants, and some from non-desert forms. He failed to find any consistent pattern among the desert forms.

Thus social organizations of the arid-adapted Gerbillinae span a range from relatively solitary species to more tolerant, communal forms. Two dipodids – *Jaculus orientalis* and *Allactaga elator* – are semitolerant while the heteromyids are consistently solitary, whether forest or desert types. Within arid-land forms of *Peromyscus* (*Cricetinae*), *P. crinitis* is intolerant while *P. eremicus* is more tolerant.

Eisenberg surmises that the intolerance of the Heteromyidae may have evolved as a defense of its food caches rather than an adaptation to arid environments. Consequently the species of this family, most of which cache in all habitats, are also aggressive and solitary in all habitats.

It is also possible to rationalize *a priori* to the contrary: why a rigidly territorial and site-oriented social organization might be maladaptive in a desert species. Since resources are spatially variable, it could be more adaptive for species not to be site-oriented, and to have considerable freedom to move about.

There is a considerable body of evidence to support this notion. Naumov (1972, pp. 347–75) describes the territorial pattern of shaggy-legged jerboas in the Kyzyl Kum Desert of Soviet Asia:

> When feeding is finished – and it occupies only the first part of the night – the jerboa wanders widely, running across plots belonging to other jerboas. During this period it does not feed but merely 'tests' food, apparently seeking new pastures...In deserts, with their quick shifts in vegetation and invertebrate animal populations, this type of territory exploitation is biologically desirable, but on superficial observation it creates an impression of anarchy in the use of the land.

In her lengthy review of lizard social behavior, Stamps (1977) points out that *Uta stansburiana* is very aggressive in semiarid areas of Texas, but is less aggressive with considerable home-range overlap in desert areas of Nevada and California. In some species, territorial patterns may dissolve during periods of food scarcity. Stamps (1977, p. 313) concludes that 'Unpredictable fluctuations in food and other resources appear to rule out specific site defense even when other factors...are favorable.'

Among desert forms, the horned lizard (*Phrynosoma cornutum*) feeds on ants which, in deserts, have highly clumped distributions. This species does not defend a home range, and some individuals are highly nomadic (Stamps, 1977). The large, carnivorous forms also appear less agressive and territorial. The leopard lizard (*Crotaphytus wislizenii*) is completely non-territorial, often with shifting home ranges and nomadic behavior. In related reasoning, Wiens (1976) discusses at some length the view that territoriality and aggressiveness should be less marked in systems where resources are too scarce to meet energy demands of this type of social organization.

Most authors agree that agonistic behavior tends to be species specific. This point is most commonly emphasized by the herpetologists (e.g. Mayhew, 1968; Milstead, 1970). But there do appear to be exceptions of this generalization, primarily in birds and mammals. Thus Anderson & Anderson (1973) state that cactus wrens (*Campylorhynchus brunneicapillus*) in the Sonoran Desert go about their daily activites without interacting with the other birds in their locale, with one conspicuous exception. They do engage in brisk altercations with curve-billed thrashers (*Toxostoma curvirostre*).

MacMillen (1964) discusses the strong, interspecific role of the woodrat (*Neotoma lepida*) in a semidesert scrub area of southern California. Individuals of this species were found to be strongly dominant over those of the six rodent species in the area, and surprisingly tolerant of their own conspecifics, in staged captive encounters. In the field, woodrats establish territories in summer which orient around cactus (*Opuntia occidentalis*) plants, the only source of water during the dry season. These areas are defended against the intrusion of rodents from other species. Populations of those species which need free water decline at this season because, in MacMillen's view, they are denied access to the cacti by *Neotoma*. Thus *Neotoma* aggression appears to be an important determinant of the population density of other rodent species in this area.

We have earlier discussed the population increases of mammalian carnivores in western USA as coyote populations were reduced by government predator control. Wagner (1977*a*) reviewed several sources of evidence suggesting that these and other interactions of North American Carnivora are behavioral rather than competition for a common food resource. Food preferences of the responding species are in most cases rather different, while at the same time there are observations of interspecific aggressive encounters. The larger canids generally appear dominant over the smaller canids, felids, and mustelids.

General discussion

There is both the need and desire to generalize the observations we have reviewed. Such generalizations could be developed either by induction from the available data, or deduced from strategic models. The validity of the conclusions derived by the latter route would still need to be tested or 'validated' with the existing data base. Thus the adequacy of that empirical base is an inevitable determinant of the validity of our generalizations, whether induced or deduced.

The adequacy of the base leaves much to be desired. Only a small fraction of desert species have been studied, and of these a much smaller fraction have been studied in such a way as to estimate the critical parameters, or address the key questions. In particular, the data are inadequate for the vast number of invertebrate species, and generalizations based on the more numerous vertebrate studies may not be applicable to the great variety of invertebrate forms.

Having posed these reservations there does appear to be considerable evidence of interspecific competition among desert animals. Niche segregation may be seen in virtually every group that has been examined. Except for Pianka's work on lizards, we do not yet have measurements of the degree of niche overlap in desert species which could be compared with animals in other biomes. But diversity appears high in all but the deserts with precipitation below about 100 mm. And even at the most arid extremes, there is evidence of resource partitioning.

There is also evidence for rather full utilization of some resources: seeds by ants and rodents, cellulose by termites, jackrabbits by predators, and habitat or space by rodents, lizards, and carnivorous mammals. The rainfall-induced fluctuation of insectivorous species, especially birds and lizards, suggests a shortage of insect foods.

The combination of character displacement in lizards and rodents; density compensation in rodents; and what might appropriately be called 'species compensation' between rodents and ants, and lizards and birds, offers yet another array of evidence. To this may be added the 'experimental' evidence from the exclosure studies by Brown & Davidson (1977), predator-control in western USA, and the rainfall variations which favor one competitor and disfavor others (Milstead, 1965).

There is one important study which reaches conclusions rather different from the general picture being developed here. Thomas (1975) investigated the ecology of 30 species of tenebrionid beetles on the Rock Valley research site of the US/IBP Desert Biome program, an area approaching 1 km^2. His data showed 23 of the 30 species maintaining breeding populations on the site. Thomas found some degree of temporal and habitat segregation among these species, but he was more impressed with the degree of overlap and coexistence. In his opinion, the degree of segregation was not sufficient to permit coexistence if there were a real shortage of resources. The animals are detritivores, and in Thomas's view were not exhausting the resource. Rather, their numbers were being kept at low to moderate densities by periodic dry years. Above-average rainfall permitted population increases, primarily by enhancing survival of larvae. In essence, Thomas views these desert beetles as *r*-selected species, the populations of which are continuously depressed by a capricious environment. Their numbers are thus kept at sufficiently low densities that they do not exhaust their resources, do not compete as a result, and therefore coexist with considerable niche overlap.

With the exception of Thomas's study, there appears in total to be considerable evidence that interspecific competition is widespread among desert animals. And the evidence of competition between such disparate groups as lizards and birds, and rodents and ants, suggests that we may have only begun to understand wide-ranging competitive pressures which MacArthur (1972) called 'diffuse competition.'

Evidence for predation is even less conclusive than that for competition. Fewer generalizations, even of the most tentative sort, are warranted. Carnivory obviously exists everywhere in deserts, and there are numerous studies detailing the food habits of carnivores, or predator and parasite lists operating on a given species. But there are very few data which give any clue to its relative importance as a population constraint on desert species, data which could be compared with results from other biomes. Protective adaptations are widespread, although no one can suggest at this stage that they are any more or less prevalent in deserts than in other systems. One generalization does appear valid: the percentage of predatory and parasitic species in desert faunas is higher than in more mesic, and/or more northerly, biomes and may be equalled or exceeded only by the tropics.

Agonistic behavior, too, is widespread in desert birds, mammals, and lizards. But it is not clear whether it is any more or less influential as a population constraint in deserts than in other systems. Nor is it clear that desert species are any more aggressive than non-desert species. The few examples cited of interspecific aggression are in fact cases of interspecific competition, and add to the evidence reviewed above on that constraint.

In sum, one gets the impression from the data available to date that biotic influences may collectively be important constraints on the populations of desert species. Except for the arid extremes, deserts appear similar to the tropics in their high diversity and percentage of predatory species, and perhaps because of the same influences which produce the latitudinal clines in these two characteristics. That similarity exists despite the fact that deserts and tropics are at opposite extremes of the range of environmental variability.

There is, therefore, reason to question two rather widespread impressions of desert ecosystems: (*a*) that desert animals are limited primarily by the physical environment and relatively little by biotic constraints; and (*b*) that desert organisms may be limited primarily by predation and behavioral self-limitation in order to minimize resource exploitation and competition for scarce and variable resources.

The impression that biotic interaction may be minimal in deserts probably stems from firstly, the low densities and wide dispersion of organisms; secondly, the prevalence of nocturnal and fossorial life styles; and thirdly, the long periods of dormancy in the form of torpor, aestivation, and diapause.

References

Anderson, A. H. & Anderson, A. (1973). *The cactus wren*. University of Arizona Press, Tucson.

Andrewartha, H. G. & Birch, L. C. (1954). *The distribution and abundance of animals*. University of Chicago Press, Chicago.

Austin, G. T. (1972). Breeding birds of desert riparian habitat in southern Nevada. *Condor*, **72**, 431–6.

Beatley, J. C. (1976). Environments of kangaroo rats (*Dipodomys*) and effects of environmental change on populations in southern Nevada. *Journal of Mammalogy*, **57**, 67–93.

Beck, B. B., Engen, C. W. & Gelfand, P. W. (1973). Behavior and activity cycles of Gambel's quail and raptorial birds at a Sonoran Desert waterhole. *Condor*, **75**, 466–70.

Birch, L. C. (1957). The meanings of competition. *American Naturalist*, **91**, 5–18.

Brinck, P. (1956). The food factor in animal desert life. In: *Bertil Hanstrom: zoological papers in honor of his sixty-fifth birthday, November 20th, 1956* (ed. K. G. Wingstrand), pp. 120–7. Zoological Institute, Lund, Sweden.

Brown, J. H. (1975). Geographical ecology of desert rodents. In: *Ecology and evolution of communities* (ed. M. L. Cody & J. M. Diamond), pp. 315–41. Belknap Press of Harvard University Press, Cambridge, Massachusetts.

Brown, J. H. & Davidson, D. W. (1977). Competition between seed-bearing rodents and ants in desert ecosystems. *Science*, **196**, 880–2.

Brown, J. H. & Lieberman, G. A. (1973). Resource utilization and coexistence of seed-eating desert rodents in sand dune habitats. *Ecology*, **54**, 788–97.

Buxton, P. A. (1923). *Animal life in deserts: a study of the fauna in relation to the environment*. Edward Arnold, London.

Cade, T. D. (1965). Relations between raptors and columbiform birds at a desert water hole. *Wilson Bulletin*, **77**, 340–5.

Chew, R. M. (1977). Some ecological characteristics of ants on a desert-shrub community in southeastern Arizona. *American Midland Naturalist*, **98**, 33–49.

Chew, R. M. & Chew, A. E. (1970). Energy relationships of the mammals of a desert shrub (*Larrea tridentata*) community. *Ecological Monographs*, **40**, 1–21.

Clark, F. W. (1972). Influence of jackrabbit density on coyote population change. *Journal of Wildlife Management*, **36**, 343–56.

Cloudsley-Thompson, J. (1976). *Deserts and grasslands*. Part I, *Desert life*. Doubleday, Garden City, New York.

Cole, A. C. Jr (1934). An ecological study of the ants of the southern desert shrub region of the United States. *Annals of the Entomological Society of America*, **27**, 388–405.

Connolly, G. E. & Longhurst, W. M. (1975). *The effects of control on coyote populations: a simulation model*. University of California Division of Agricultural Sciences Bulletin No. 1872. Davis.

Creusere, F. M. & Whitford, W. G. (1976). Ecological relationships in a desert Anuran community. *Herpetologica*, **32**, 7–18.

Cross, J., Goldstein, G., Lowe, C. H. & Morello, J. (1977). Gradient analysis of bajada communities. In: *Convergent evolution in warm deserts: an examination of strategies and patterns in deserts of Argentina and the United States* (ed. G. H. Orians & O. T. Solbrig), pp. 92–100. Dowden, Hutchinson & Ross, Stroudsburg, Pennsylvania.

Curtis, J. T. (1959). *The vegetation of Wisconsin: an ordination of plant communities*. University of Wisconsin Press, Madison.
Davidson, D. W. (1977*a*). Species diversity and community organization in desert seed-eating ants. *Ecology*, **58**, 711–24.
Davidson, D. W. (1977*b*). Foraging ecology and community organization in seed-eating ants. *Ecology*, **58**, 725–37.
Diamond, J. M. (1978). Niche shifts and the rediscovery of interspecific competition. *American Scientist*, **66**, 322–31.
Dixon, K. L. (1959). Ecological and distributional relations of desert scrub birds of western Texas. *Condor*, **61**, 397–409.
Edmunds, R. L. (ed.) (1974). *An initial synthesis of results in the coniferous forest biome 1970–1973*. US/IBP, Ecosystem Analytical Studies, Coniferous, Forest Biome Bulletin No. 7. University of Washington, Seattle.
Egoscue, H. J. (1975). Population dynamics of the kit fox in western Utah. *Bulletin of the Southern California Academy of Sciences*, **74**, 122–7.
Eisenberg, J. F. (1967). A comparative study in rodent ethology with emphasis on evolution of social behavior, I. *Proceedings of the United States National Museum*, **122** (3597), 1–51.
Elder, J. B. (1956). Watering patterns of some desert game animals. *Journal of Wildlife Management*, **20**, 368–78.
Fisher, C. D., Lindgren, E. & Dawson, W. R. (1972). Drinking patterns and behavior of Australian desert birds in relation to their ecology and abundance. *Condor*, **74**, 111–36.
Fleming, T. H. (1974). Social organization of two species of Costa Rican heteromyid rodents. *Journal of Mammalogy*, **55**, 543–61.
Genarro, A. L. (1968). Northern geographic limits of four desert rodents of the genera *Peromyscus*, *Dipodomys*, and *Onychomys* in the Rio Grande valley. *American Midland Naturalist*, **80**, 477–93.
George, U. (1977). *In the deserts of this Earth*. Harcourt Brace Jovanovich, New York & London.
Grenot, C. J. (1974). Physical and vegetational aspects of the Sahara Desert. In: *Desert biology*, **2** (ed. G. W. Brown), pp. 103–64. Academic Press, New York & London.
Gross, J. E., Stoddart, L. C. & Wagner, F. H. (1974). *Demographic analysis of a northern Utah jackrabbit population*. Wildlife Monographs No. 40. Wildlife Society, Washington DC.
Gullion, G. W. (1960). The migratory status of some western desert birds. *Auk*, **77**, 94–5.
Hamilton, W. J., III (1971). Competition and thermoregulatory behavior of the Namib Desert tenebriomid beetle genus *Cardiosis*. *Ecology*, **52**, 810–22.
Hardy, R. (1945). The influence of types of soil upon the local distribution of some mammals in southwestern Utah. *Ecological Monographs*, **15**, 71–108.
Heatwole, H., Muir, R. J. & Zell, E. D. (1975). *Invertebrate populations*. US/IBP Desert Biome, Tunisian PreSaharan Project Programme Report No. 3, pp. 35–54. Utah State University, Logan.
Holbrook, S. J. (1978). Habitat relationships and coexistence of four sympatric species of *Peromyscus* in northwestern New Mexico. *Journal of Mammalogy*, **59**, 18–26.
Holling, C. S. (1959). The components of predation as revealed by a study of small-mammal predation of the European pine sawfly. *Canadian Entomologist*, **91**, 293–320.
Holm, E. & Edny, E. B. (1973). Daily activity of Namib Desert arthropods in relation to climate. *Ecology*, **54**, 45–56.

Hoover, K. D., Whitford, W. G. & Flavill, P. (1977). Factors influencing the distribution of two species of *Perognathus*. *Ecology*, **58**, 877–84.
Horning, D. S. & Barr, W. F. (undated). *Insects of Craters of the Moon National Monument Idao.* University of Idaho, College of Agriculture, Miscellaneous Series No. 8.
Hsiao, T. H. & Kirkland, R. L. (1973). *Demographic studies of sagebush insects as functions of various environmental factors.* US/IBP Desert Biome Research Memorandum RM 73–34, 2.3.3.7-1 to 2.3.3.7-28. Utah State University, Logan.
Huey, R. B., Pianka, E. R., Egan, M. E. & Coons, L. W. (1974). Ecological shifts in sympatry: Kalahari fossorial lizards (*Typhlosaurus*). *Ecology*, **55**, 304–16.
Hurd, L. E. & Wolfe, L. L. (1974). Stability in relation to nutrient enrichment in arthropod consumers of old-field successional ecosystems. *Ecological Monographs*, **44**, 465–82.
Johnson, K. A. & Whitford, W. G. (1975). Foraging ecology and relative importance of subterranean termites in Chihuahuan Desert ecosystems. *Environmental Entomology*, **4**, 66–70.
Keast, A. (1959). Australian birds: their zoogeography and adaptations to an arid country. In: *Biogeography and ecology in Australia* (ed. A. Keast, R. L. Crocker & C. S. Christian), pp. 89–114. Dr W. Junk, The Hague.
Kenagy, G. J. (1973*a*). Adaptations for leaf eating in the Great Basin kangaroo rat, *Dipodomys microps*. *Oecologia*, **12**, 383–412.
Kenagy, G. J. (1973*b*). Daily and seasonal patterns of activity and energetics in a heteromyid rodent community. *Ecology*, **54**, 1201–19.
Key, K. H. L. (1959). The ecology and biogeography of Australian grasshoppers and locusts. In: *Biogeography and ecology in Australia* (ed. A. Keast, R. L. Crocker & C. S. Christian), pp. 192–210. Dr W. Junk, The Hague.
Knudsen, J. J. (1976). Demographic analysis of a Utah–Idaho coyote population. MS Thesis, Utah State University.
Lea, A. (1964). Some major factors in the population dynamics of the brown locust *Locustana pardalina* (Walker). In: *Ecological studies in southern Africa* (ed. O. H. S. Davis), pp. 269–83. Dr W. Junk, The Hague.
Louw, G. N. (1972). The role of advective fog in the water economy of certain Namib Desert animals. *Symposia of the Zoological Society of London*, **31**, 297–314.
MacArthur, R. H. (1972). *Geographical ecological patterns in the distribution of species.* Harper & Row, New York, Evanston, San Francisco & London.
MacMahon, J. A. (1976). Species and guild similarity of North American desert mammal faunas: a function analysis of communities. In: *Evolution of desert biota* (ed. D. W. Goodall), pp. 133–48. University of Texas Press, Austin, Texas & London.
MacMillen, R. E. (1964). *Population ecology, water relations, and social behavior of a southern California semi-desert rodent fauna.* University of California Publications in Zoology No. 71.
Mares, M. A. (1977). Water balance and other ecological observations on three species of *Phyllotis* in northwestern Argentina. *Journal of Mammalogy*, **58**, 514–20.
May, R. M. (1973). *Stability and complexity of model ecosystems. Monographs in Population Biology No. 6.* Princeton University Press.
May, R. N. & MacArthur, R. H. (1972). Niche overlap as a function of environmental variability. *Proceedings of the National Academy of Sciences USA*, **69**, 1109–13.
Mayhew, W. W. (1966). Reproduction in the arenicolous lizard *Uma notata*. *Ecology*, **47**, 9–18.

Mayhew, W. W. (1968). Biology of desert amphibians and reptiles. In: *Desert Biology*, vol. 1 (ed. G. W. Brown), pp. 195–356. Academic Press, New York & London.

Maza, B. G., French, N. R. & Aschwanden, A. P. (1973). Home range dynamics in a population of heteromyid rodents. *Journal of Mammalogy*, **54**, 405–25.

M'Closkey, R. T. (1972). Temporal changes in populations and species diversity in a California desert community. *Journal of Mammalogy*, **53**, 657–76.

Miller, A. H. & Stebbins, R. C. (1964). *The lives of desert animals in Joshua Tree National Monuments*. University of California Press, Berkeley & Los Angeles.

Milstead, W. W. (1957). Some aspects of competition in natural populations of whiptail lizards (Genus *Cnemidophorus*). *Texas Journal of Science*, **9**, 410–47.

Milstead, W. W. (1965). Changes in competing populations of whiptail lizards (*Cnemidophorus*) in southwestern Texas. *American Midland Naturalist*, **73**, 75–80.

Milstead, W. W. (1970). Late summer behavior of the lizards *Sceloporus merriami* and *Urosaurus ornatus* in the field. *Herpetologica*, **26**, 343–54.

Mispagel, M. E. (1974). An ecological analysis of insect populations on *Larrea tridentata* in the Mohave Desert. MS Thesis, Californian State University, Long Beach.

Moldenke, A. R. (1976). California pollination ecology and vegetation types. *Phytologia*, **34**, 305–61.

Muir, R. J. & Heatwole, H. (1976). *Temporal activity patterns of animals*. US/IBP Desert Biome, Tunisian PreSaharan Project Programme Report No. 5, pp. 93–114. Utah State University, Logan.

Myers, K. & Parker, B. S. (1975). Effect of severe drought on rabbit numbers and distribution in a refuge area in semiarid north-western New South Wales. *Australian Wildlife Research*, **2**, 103–20.

Naumov, N. P. (1972). *The ecology of animals* (ed. N. D. Levine, English translation by F. K. Plous, Jr). University of Illinois Press, Urbana & Chicago.

Nelson, E. W. (1909). *The rabbits of north America*. United States Department of Agriculture, Bureau of Biological Survey, American Fauna, No. 29. Washington, DC.

Newsome, A. E. (1975). An ecological comparison of the two arid-zone kangaroos of Australia, and their anomalous prosperity since the introduction of ruminant stock to their environments. *Quarterly Review of Biology*, **50**, 389–424.

Newsome, A. E. & Corbett, L. K. (1975). Outbreaks of rodents in semi-arid and arid Australia: causes, preventions, and evolutionary considerations. In: *Rodents in desert environments* (ed. I. Prakash & P. K. Gosh), pp. 17–53. Dr W. Junk, The Hague.

Nutting, W. L., Haverty, M. I. & LaFage, J. P. (1975). *Demography of termite colonies as related to various environmental factors: population dynamics and role in detritus cycle*. US/IBP Desert Biome Research Memorandum, RM 75–31, pp. 53–78. Utah State University, Logan.

O'Farrell, M. J. (1974). Seasonal activity patterns of rodents in a sagebrush community. *Journal of Mammalogy*, **55**, 809–23.

Orians, G. H. & Solbrig, O. T. (1977). Degree of convergence of ecosystem characteristics. In: *Convergent evolution in warm deserts: an examination of strategies and patterns in deserts of Argentina and the United States* (ed G. H. Orians & O. T. Solbrig), pp. 225–55. Dowden, Hutchinson & Ross, Stroudsburg, Pennsylvania.

Osborne, W. (1975). *Invertebrates*. US/IBP Desert Biome Research Memorandum RM75-1, 23–47. Utah State University, Logan.

Otte, D. (1977). Grasshoppers. In: *Convergent evolution in warm deserts: an exami-*

nation of strategies and patterns in deserts of Argentina and the United States (ed. G. H. Orians & O. T. Solbrig), pp. 142–7. Dowden, Hutchinson & Ross, Stroudsburg, Pennsylvania.

Palmer, T. S. (1897). *The jackrabbits of the United States.* United States Department of Agriculture, Division of Biological Survey. Washington, DC.

Paramonov, S. J. (1959). Zoogeographical aspects of the Australian dipterofauna. In: *Biogeography and ecology in Australia* (ed. A. Keast, R. L. Crocker & C. S. Christian), pp. 164–91. Dr W. Junk, The Hague.

Peterson, S. R. (1975). *Ecological distribution of breeding birds.* Proceedings of a Symposium on Management of Forest and Range Habitats for Non-game Birds, United States Department of Agriculture and Forest Services Technical Report WO-1, 59–80.

Phillips, D. L. & MacMahon, J. A. (1978). Gradient analysis of a Sonoran Desert bajada. *Southwestern Naturalist*, **23**, 669–80.

Pianka, E. R. (1966). Latitudinal gradients in species diversity: a review of concepts. *American Naturalist*, **100**, 33–46.

Pianka, E. R. (1975). Niche relations of desert lizards. In: *Ecology and evolution of communities* (ed. M. C. Cody & J. M. Diamond), pp. 292–314. Belknap Press of Harvard University Press, Cambridge, Massachusetts.

Rosenzweig, M. L. (1973). Habitat selection experiments with a pair of coexisting heteromyid rodent species. *Ecology*, **54**, 111–7.

Rosenzweig, M. L. & Winakur, J. (1969). Population ecology of desert roden communities: habitats and environmental complexity. *Ecology*, **50**, 558–72.

Schoener, T. W. (1977). Competition and the niche. In: *Biology of the Repitilia*, vol. 7, *Ecology and behavior A* (ed. C. Gans & D. W. Tinkle) pp. 35–136. Academic Press, London, New York & San Francisco.

Schroder, G. D. & Rosenzweig, L. (1975). Perturbation analysis of competition and overlap in habitat utilization between *Dipodomys ordii* and *Dipodomys merriami. Oecologia*, **19**, 9–28.

Schumacher, A. & Whitford, W. G. (1976). Spatial and temporal variation in Chihuahuan Desert and faunas. *Southwestern Naturalist*, **21**, 1–8.

Shreve, F. (1942). The desert vegetation of North America. *Botanical Review*, **8**, 195–246.

Small, R. L. (1971). Interspecific competition among three species of Carnivora on the Spider Ranch, Yavapai Co., Arizona. MS Thesis, University of Arizona, Tucsan.

Soholt, L. F. (1973). Consumption of primary production by a population of kangaroo rats (*Dipodomys merriami*) in the Mojave Desert. *Ecological Monographs*, **43**, 357–76.

Stamps, J. A. (1977). Social behavior and spacing patterns in lizards. In: *Biology of the Reptilia*, vol. 7, *Ecology and behavior A*. (ed. C. Gans & D. W. Tinkle), pp. 265–334. Academic Press, London, New York & San Francisco.

Stoddart, L. C. (1970). A telemetric method for detecting jackrabbit mortality. *Journal of Wildlife Management*, **34**, 501–7.

Stoddart, L. C. (1977). *Population dynamics, movement and home range of black-tailed jackrabbits* (Lepus californicus) *in Curlew Valley, Northern Utah.* United States Energy Research and Development Administration, Contract No. E(11-1)-1329, Progress Report. Utah State University, Logan.

Sumner, F. B. (1925). Some biological problems of our southwestern deserts. *Ecology*, **6**, 352–71.

Tevis, L., Jr (1958). Interrelationships between the harvester ant *Veromessor pergandei* (Mayr) and some desert ephemerals. *Ecology*, **39**, 695–704.

Thomas, D. B., Jr (1975). Dynamics of a species assemblage of desert Tenebrionid beetles. MS Thesis, California State University, Long Beach.

Tomoff, C. S. (1974). Avian species diversity in desert scrub. *Ecology*, **55**, 396–403.

Tomoff, C. S. (1977). Birds. In: *Convergent evolution in warm deserts: an examination of strategies and patterns in deserts of Argentina and the United States* (ed. G. H. Orians & O. T. Solbrig), pp. 140–2. Dowden, Hutchinson & Ross, Stroudsburg, Pennsylvania.

Turner, F. B. (1977). The dynamics of populations of squamates, crocodilians, and rhynchocephalians. In: *Biology of the Reptilia*, vol. 7. *Ecology and behavior A.* (ed. C. Gans & D. W. Tinkle), pp. 157–264. Academic Press, London, New York & San Francisco.

Turner, F. B., Hoddenbach, G. A., Medica, P. A. & Lannom, J. R. (1970). The demography of the lizard, *Uta stansburiana* Baird and Gerard, in southern Nevada. *Journal of Animal Ecology*, **39**, 505–19.

Turner, F. B., Medica, P. A. & Smith, D. D. (1974). *Reproduction and survivorship of the lizard,* Uta stansburiana, *and the effects of winter rainfall, density and predation on these processes.* US/IBP Desert Biome Research Memorandum RM 74-21, 117–28. Utah State University, Logan.

Uvarov, B. P. (1954). The desert locust and its environment. In: *The biology of deserts* (ed. J. L. Cloudsley-Thompson), pp. 85–9. Institute of Biology, London.

Varley, G. C., Gradwell, G. R. & Hassell, M. P. (1973). *Insect population ecology: an analytical approach.* Blackwell Scientific Publications, Oxford.

Wagner, F. H. (1969). Ecosystem concepts in fish and game management. In: *The ecosystem concept in natural resource management* (ed. G. M. Van Dyne), pp. 254–307. Academic Press, New York & London.

Wagner, F. H. (1971). Predator–prey instability and diversity of the Curlew Valley ecosystem. In: *Proceedings of the 1971 Meeting of the Southwestern and Rocky Mountain Division, American Association for the Advancement of Science, Tempe, Arizona.* Abst. Utah State University, Logan.

Wagner, F. H. (1972). *Coyotes and sheep: some thoughts on ecology, economics and ethics.* Utah State University, 45th Honor Lecture.

Wagner, F. H. (1977*a*). *Integrative summary.* US/IBP Desert Biome Tunisian Pre-Saharan Project Programme Report No. 6, pp. 301–19. Utah State University, Logan.

Wagner, F. H. (1977*b*). *Coyotes and non-game wildlife. Proceedings of a symposium on coyotes, wildlife and meat production.* 143rd Annual Meeting of the American Association for the Advancement of Science, Denver, Colorado. Utah State University, Logan.

Wagner, F. H. (1978). Some concepts in the management and control of small mammal populations. In: *Populations of small mammals under natural conditions: A symposium held at the Pymatuning Laboratory of Ecology, May 14–16, 1976* (ed. D. P. Synder), pp. 192–202. University of Pittsburg Pymatuning Laboratory of Ecology Special Publication Series, No. 5.

Wagner, F. H. & Stoddart, L. C. (1972). Influence of coyote predation on black-tailed jackrabbit populations in Utah. *Journal of Wildlife Management*, **36**, 329–42.

Weiss, H. B. (1926). The similarity of food habit types on the Atlantic and western Arctic coasts of America. *American Naturalist*, **60**, 102–4.

Whitford, W. G. (1976). Temporal fluctuations in density and diversity of desert rodent populations. *Journal of Mammalogy*, **57**, 351–69.

Whitford, W. G. (1978). Foraging in seed-harvester ants *Pogonomyrmex* spp. *Ecology*, **59**, 185–9.

Whitford, W. G. & Creusere, F. M. (1977). Seasonal and yearly fluctuations in Chihuahuan Desert lizard communities. *Herpetologica*, **33**, 54–65.
Whitford, W. G. & Ettershank, G. (1975). Factors affecting foraging activity in Chihuahuan Desert harvester ants. *Environmental Entomology*, **4**, 689–96.
Whitford, W. G., Johnson, P. & Ramirez, J. (1976). Comparative ecology of the harvester ants *Pogonomyrmex barbatus* (F. Smith) and *Pogonomyrmex rugosus* (Energy). *Insectes Sociaux*, **23**, 117–32.
Whittaker, R. H. (1952). A study of summer foliage insect communities in the Great Smoky Mountains. *Ecological Monographs*, **22**, 1–44.
Whittaker, R. H. & Niering, W. A. (1965). Vegetation of the Santa Catalina Mountains, Arizona: a gradient analysis of the south slope. *Ecology*, **46**, 429–52.
Wiens, J. A. (1976). Population responses to patchy environments. *Annual Review of Ecology and Systematics*, **7**, 81–120.
Yang, T. W. & Lowe, C. H. Jr (1956). Correlation of major vegetation climaxes with soil characteristics in the Sonoran Desert. *Science*, **123**, 542.
Yom-Tov, Z. (1970). The effect of predation on population densities of some desert snails. *Ecology*, **51**, 907–11.

Manuscript received by the editors September 1978

5. Plant–animal interactions

R. D. GRAETZ

Introduction

The arid and semi-arid areas of the world have in the past supported a diverse herbivorous fauna (Cloudsley-Thompson & Chadwick, 1964) but now in many instances flora and fauna have been depleted by the actions of man (e.g. Newsome & Corbett, 1977). The fact that the land used for centuries by man's domesticated livestock still consist of diverse communities of palatable plants indicates the existence of compensatory interactions (Ellison, 1960). These interactions have been found to be intriguingly dynamic, reciprocal and intricate (e.g. Williams, 1968).

I shall describe the characteristics of animal–plant interactions using studies largely done in the arid lands of Australia. For simplicity I shall limit myself to herbivory only, in which I shall include zoochory, and exclude the non-consumptive interactions such as shelter, animal-assisted pollination and so on. It is not a thorough-going review because there are few studies adequate on both depth and scope and I am one of those who – 'admit that as a consequence of specialized botanical or zoological training they are not fitted to deal with the plant/herbivore interface' (Harper, 1969).

My approach has been to adopt the concept that a community be defined as a set of species rather than a set of individuals (Goodall, 1966; Whittaker & Woodwell, 1972), for although plant and animal populations will be discussed, it is the interdependence of species both spatially and temporally through herbivory that is vertebral. Thus I will not artificially separate out and discuss individual processes involved (defoliation, trampling, nutrient redistribution, etc.) but rather try to describe the functioning and behaviour of the whole system. Though this approach is restricted, in that the bulk of the literature deals directly with domesticated animals, perturbations due to the introduction of man's exotic herbivores sometimes provide insights into the balances existing in the ancestral system.

Desert ecosystems

Water, above all else, influences all biological activity in arid lands (Noy-Meir, 1973), the extreme and episodic nature of climatic events being reflected both in the numbers of any population and in the patterns of activity of its individuals. Thus the spatial and temporal variability of rainfall has exaggerated importance in arid ecosystems however low the overall turnover of biomass.

Composite and interactive processes

To dwell on this stochastic nature of desert environments tends to obscure the success of the individual species that inhabit them. The desert ephemeral plant represented for 90% of the time as a dormant seed is as successful as the hardy, xerophytic tree with a life-span of 300 years or more. The only criteria of evolutionary success are occupancy and reproduction.

An environment characterized by extremes of both abiotic and biotic influences may well mould perennial plant communities into relicts representing frozen history to the observer today. This can be observed where the duration of man's cultural impost has been less than the life span of the perennial components of arid plant communities. Williams (1975) has estimates of the life-spans for some perennial species on the Australian arid rangelands. Grasses may live from 20–50 years, shrubs 100 years and, small trees, about 300 years. Field observations, e.g. Hall, Specht & Eardley (1964), suggest that many Australian landscapes are populated by relict, museum-type plant communities in which post-European settlement practices have limited or prevented the recruitment of seedlings (Correll & Lange, 1966).

Finally, successful separation of the effects of the biotic perturbations of herbivory and competition from other species is a difficult task in the desert environment (West & Tueller, 1972) where climatic influences may partially or completely over-ride all others.

Herbivore–plant interactions in desert ecosystems

Ants

The herbivores to be discussed in this chapter are: ants, rodents, termites, grasshoppers and large mammals. It better serves the purpose of this essay to group the herbivores according to the plant parts eaten. Ants therefore, which are usually conspicuous in arid lands, have been studied because of the readily observed zoochory (Carroll & Janzen, 1973). Genera such as *Messor*, *Monomorium*, *Chelaner*, *Pheidole* and *Meranoplus* live mainly on harvested plant seeds which, in some cases, are fed directly to their brood (Cloudsley-Thompson & Chadwick, 1964). Other genera, such as *Pogonomyrex* harvest leaf material directly, denuding areas up to 10 m diameter around nests.

The ecological significance of the transportation, scarification, caching, and consumption of seeds for the structure and functioning of the plant community, is, however, very difficult to assess fully (Janzen, 1971) for it is necessary to compare these losses with all the others that comprise the reproductive wastage. Few, if any, experimental studies have done this.

One study, rather more complete than most, is that of Briese (1974). This Australian study examined the structure and function of any populations within an experimenal locality wherein a semi-arid perennial shrub community of *Atriplex vesicaria* had been converted and maintained through the use of grazing animals to a perennial grassland. The site, representative of a larger, well-defined area (Leigh & Noble, 1972) showed a high diversity of ant species; 35 species were present in 500 m^2. These results generally reflect the insect diversity of semi-arid Australia (Taylor, 1972). Harvester species, of which there were six representing the genera *Meranoplus*, *Pheidole* and *Chelaner*, occupied some 48% of the total colonies which were randomly distributed across the site. These species as a group showed the greatest seasonal fluctuations in harvesting activity and, compared with the carnivorous and omnivorous groupings, the shortest periods of total foraging activity. Comparative diets for these harvester species within the two vegetation types were constructed by including a weighting factor for the number of colonies active during the sampling period. Measurements of seed production were made also. The results suggest that it is unlikely for the harvesters to influence the absolute abundance of vegetation even though they harvested 20–30% of the total seed production. Previously reported figures for total seed taken in a desert habitat were of the order of only 1% (Tevis, 1958). Whilst differential seed selection was regularly observed in the field and laboratory, foraging was concentrated more on the commoner species which reverses the findings of Tevis (1958). Stronger diet selection and activity patterns were shown when there was an abundance of freshly fallen seed. As seed abundance declined, the ant species took a wider range of seeds, the dietary pattern loosely following the seasons.

Briese (1974) infers that the influence on community structure of these opportunistic, generalist feeders would be to reduce the relative competitive pressure of the commoner plant species on the less common. This influence would be slight however, since the combined effect of the diet selection by all the seed harvester species over extended periods tended to even the impact on the seed populations.

An impact of ants as secondary herbivores was also noted by Briese (1974). There was a strong positive association between the ant *Iridomyrmex* sp. and various saprophagic coccid species. Approximately 30% of the live *A. vesicaria* shrubs were infested with coccids and a further 20% showed signs of previous attack. The coccids were actively tended by the *Iridomyrmex* ants so preventing normal predation. The coccids appeared to reduce the vigour of the perennial shrub, or attack only debilitated individuals, and there are historical records suggesting wide scale destruction of this plant species occurred in the past (Froggatt, 1910). Despite the detail and completeness of the observation programme of Briese (1974), no experimental manipulation (cf. Harper, 1969) was possible over an appropriate time interval to test the predictions outlined above.

Rodents

A characteristic of deserts across the world is the presence of small-mammal populations (jerboas, kangaroo rats, and pocket mice). They are almost exclusively nocturnal in habit and live in small localized populations. Twelve or more genera in at least four different families have evolved in the various deserts of the world (Cloudsley-Thompson & Chadwick, 1964). In the arid regions of Australia two marsupial genera *Antechinomys* and *Sminthopsis* have evolved a similar life form.

The diet of the rodents varies with habitat, and may be omnivorous (e.g. Kenagy, 1972, 1973). Plant seeds, however, form the major single part of the diet for most of the time. Their foraging and food-storage behaviour is usually related to the plant community wherein they live. Physiological adaptions to seed eating appear to be widespread. R. M. Chew & A. E. Chew (1970) found that for the rodents studied, 49% of caloric intake was from seeds and the efficiency of plant-food utilization was highest for seeds.

Seeds, of course, represent an attractive food supply because energy, essential nutrients, and vitamins are more concentrated in them than in any other plant part. Seeds are slow to decompose, and, importantly for arid regions, their high fat content provides a significant source of metabolic water.

In undisturbed communities, the diet consists largely of the seeds of annual and ephemeral species with less emphasis on those from perennials. Given the reproductive strategies of the two types of plants, these animals may not strongly influence the composition or abundance of desert plant communities. The reciprocal relationship is apparent on over-grazed and degraded rangelands in North America. Here the perennial grass and shrub species have been suppressed or eliminated and have been replaced by annual species. The rodent populations increase but fluctuate wildly with the seasons often further degrading the rangelands (Packard, 1977).

Recent summaries of the role of small mammals in North American and Australian rangelands (Packard, 1977; Newsome & Corbett, 1975) list the studies over the last half century but in all instances the evidence is circumstantial. No parallel long-term studies on the plant populations have been carried out. Generally, the evidence presented here and in other studies (Smigel & Rosensweig, 1974; French *et al.*, 1974) suggests the animals are flexibly selective in diet, having a more specialized diet when seed is abundant, opportunistically shifting with changing abundance (and size) of seed and other plant material. This selectivity is also influenced by the presence of competing species.

Table 5.1. *Feeding preferences of termites from mulga lands near Alice Springs, Australia.* (From Watson *et al.*, 1973)

Species	Wood	Main diet grass	Debris
Mastotermitidae			
Mastotermes darwiniensis Froggatt	*	—	—
Rhinotermitidae			
Schedorhinotermes actuosus (Hill)	*	*	*
Termitidae			
Amitermes perarmatus (Silvestri)	*	—	—
Amitermes abruptus Gay, *A capito Hill*	*	—	*
Amitermes obtusidens Mjöberg	—	*	*
Amitermes agrilus Gay, *A. dentosus* Hill, *A. vitiosus* Hill	*	*	*
Drepanotermes (five species)	—	*	*
Microcerotermes distinctus Silvestri, *M. serratus* (Froggatt)	*	—	*
Nasutitermes sp. near *N. magnus* (Froggatt)	—	*	—
Tumulitermes (three species)	—	*	*

Birds

Seeds are a major dietary item for some birds in deserts. For some plant species birds are the major dispersal vector. However, the numbers of seed-eating birds that live entirely in deserts are few and compared with small mammals, are unimportant. Many species, however, temporarily migrate into desert regions to feed upon a flush of seed or fruit. The role of these birds in dissemination and the predisposition of seeds to germinate has been recognized, but data are lacking to assess their role in subsequently influencing plant populations.

Termites

Termites occur extensively in semi-arid and arid lands (Lee & Wood, 1971*a*) and the diet of termites consists largely of cellulose derived from living or dead plants. A study of termites by Watson, Lendon & Low (1973) within mulga (*Acacia aneura*) grassland near Alice Springs, Australia, demonstrated that whilst different species do utilize different sources of plant material, debris, intact plants, etc., their diets overlap (Table 5.1). The preferences shown by any one species were site-specific, apparently in response to the availability of the plant material. This study estimated that the Alice Springs site, stocked with cattle at a biomass of 10–15 kg liveweight ha^{-1} also supported an equal biomass density of harvester termites, and of red kangaroos (*Megaleia rufa*) at 0.16 kg ha^{-1}. The termites were estimated to take in excess of 100 kg ha^{-1}

yr^{-1} dry weight of plant material – much more than that taken by the grazing animals which was estimated at 7–10 kg ha^{-1}. The dietary overlap between termites, domestic stock, and other native herbivores has yet to be elucidated.

The influence of termites in arid plant communities also has an indirect component as well as the more immediate one of herbivory. Termites, by their activities, influence soils and pedogenic processes through changes in the physical structure of soils and in redistribution of nutrients within soils (Lee & Wood, 1971*a*, *b*). Termite galleries, which often occupy up to 20% of the ground surface area, are extremely hard, resisting the penetration of water or lodging of seeds, remaining intact for 50–100 years after being abandoned. They also represent considerable stores of plant nutrients derived from digested forage that has been used for cementing the galleries and mounds. This forage has usually been gathered from over a wide area, and since the mound is usually repaired as rapidly as it is eroded, it represents a significant redistribution of soil nutrients.

Lee & Wood (1971*b*) have calculated the amount of nutrients withheld in mounds present in two sub-tropical grassland communities. This redistribution would probably be more marked in the infertile, more arid soils (Charley & Cowling, 1968).

Termites exert an influence that is strongly localized within arid plant communities which themselves are usually characterized by spatial pattern (e.g. Anderson, 1971). Termites can thus occupy a pivotal position with respect to the stability of an ecosystem.

Watson & Gay (1970) have described an exemplar situation wherein the interaction of a termite population, domestic stock, and climatic variability combined to produce a largely irreversible degradation of a mulga-grassland community. A grass-eating termite (*Drepanotermes perniger*), widespread in arid Australia, constructs large termitaria just below the soil surface of open grassland areas and only rarely in the *A. aneura* groves. Unwise felling of these trees by pastoralists resulted in an increase in grass growth in the termite populations. This new relationship is apparently unstable for when drought limits grass growth, both plants and termite populations crash. Soil erosion, aggravated by the presence of domestic stock, exposes the cement-hard termitaria which then are rarely recolonized by plants.

Grasshoppers and Locusts

Grasshoppers and locusts are well represented in arid lands often serving as major food sources for primary carnivores. The diversity of form and behaviour and the economic impact of some has generated a comparatively extensive literature. Records of locust plagues in desert lands have been part of man's earliest recorded history.

The nature and scope of the interaction of these insects with arid plant communities is reflected in the following interpretation of the adaptations of locusts and grasshoppers to arid conditions. Locusts have generated an impressive anecdotal record; the sudden population explosions, the migration patterns of nymphs and winged adults and the subsequent denudation of large areas characterize this record. Grasshoppers, or non-migratory acridid species, on the other hand, appear to be more tightly and stably coupled within arid-land food chains. Their diets are often highly specific to one or two plant species throughout life. Their numbers are determined primarily by available food and predation. Thus grasshopper species exhibit successional trends closely related to the vegetation succession. Such grasshopper species usually attain higher, more stable densities in a given community than other species with less specific dietary requirements and broader distributions (Blocker, 1977).

Accurate estimates of the rate of intake of plant material are difficult to make. Bernays & Chapman (1973) calculated that the intake of green grass for an Australian locust (*Chortoicetes terminifera*) was between 25–30% of its body weight a day so as to obtain a necessary water intake of *c.* 13% of body weight per day. Blocker (1977) has calculated that, under grass-range conditions, for a density of 6–7 individuals m^{-2}, consumption equalled that of a grazing steer. However, total impact as a primary herbivore involves more than just the amount ingested since many graminivorous species attack leaves close to the ground, felling the leaf and leaving most of it to decompose.

Locusts are significantly different in population dynamics and adaptation to arid conditions to warrant further discussion. Most locusts show little observable morphological specialization to arid habitats. An example is quoted by Cloudsley-Thompson & Chadwick (1964). The 'desert locust' (*Schistocerca gregaria*) has a distribution that roughly coincides with the hot Palaearctic deserts and so it could be classed as a true desert dweller. When populations are very low, the locust is restricted to wadis or other moist islands within the desert landscape and only during swarming does it temporarily extend its territory. It appears to be poorly adapted physiologically to arid conditions and is only found in them because of sporadic migratory behaviour.

This pattern of seemingly hazardous behaviour common to most locusts has been largely unravelled for the species *C. terminifera* (Clark, 1969, 1972) and it illustrates the interplay of a variable climate on arid plant communities. The study area was primarily the 'channel country', the Cooper basin, of south-west Queensland; an area characterized by a very variable rainfall and floods with consequent flushes of growth of annual ephemeral plant species. The 192000 km^2 area has been stocked with cattle for about 100 years (Clark, 1969).

The area supports a complex of about 44 acridid species. None of the species are as abundant or as widespread as *C. terminifera*. Those species

Table 5.2. *The essential requirements of different life cycle stages of* C. terminifera *for survival and development.* (From Clark, 1978)

Life cycle stage	Essential resource
Egg	Water from soil. Amount required equivalent to 2.25 times the weight of the egg
Young nymph (after hatching) for approximately the first two instars	Soft growing annual grasses or legumes e.g. *Dactyloctenium radulans, Medicago* spp. etc.
Adult female	Access to green plants in order to mature sexually under high temperatures

which have highly specific diet (and habitat) occasionally reach high densities in localized areas only: for example, *Urnisa guttulosa* which is strongly restricted to the red sandhill habitat. The spectrum is completed by others such as *Ailopus* sp., *Austracis* sp., *Pycnostyctus* sp., which are not habitat-restricted and so occur in comparatively low numbers only.

C. terminifera population sizes are controlled primarily by quantity of rainfall in space and time. This dependence is catalogued in Table 5.2. Under optimum conditions, eggs of *C. terminifera* must absorb two and a half times their original weight to hatch, being able to absorb water from the soil only over the range 0–10 bars (Wardhaugh, 1973; Clark, 1978). Soil drought may limit the rate of development or may kill the embryo completely.

Once hatched the nymphs display high diet specificity towards one or two green, vigourously growing annual grass species. In this area the species are mainly *Dactyloctenium radulans, Eragrostis cilianensis* when a large range of other forage is available. More exact figures illustrating the selectivity of this species are available for another less arid but comparable site (Table 5.3).

The quality and quantity of the plant species available coupled with selective dietary behaviour exert a strong control over the population at the nymphal stage. Even when starved the nymphs take only sub-optimal meals from other species and, if these much less favoured species are in a dry state, die from desiccation (Bernays & Chapman, 1973).

The life cycle of *C. terminifera*, from hatching to the development of the young winged adult, takes 30–40 days. Once this adult is 5–7 days old it is capable of flying efficiently (Lambert, 1972). However, on a diet of dry plants young adult females fail to mature sexually and cannot reproduce until the next rainfall (Clark, 1972). Reproductive patterns become spatially diversified by rainfall. Analogous observations have been made of the breeding behaviour of desert-dwelling birds (Keast & Marshall, 1954; Serventy & Marshall, 1957).

Young locust adults launch themselves just past sunset. *C. terminifera* has a comparatively low airspeed and appears adapted for sustained flight rather than speed. Flight time is temperature-dependent (Clark, 1969, 1971) and is repeated each night until females become mature and ready to lay. The net

Table 5.3. *Percentage frequency of occurrence of green foods in the forecut of* C. Terminifera *in relation to their relative abundance in the habitat.* (From Barnays & Chapman, 1973)

	Cynodon area		*Chloris* area	
Food genus	Abundance in habitat[a]	Abundance in foreguts	Abundance in habitat[a]	Abundance in foreguts
Chloris	6	5	75	61
Cynodon	83	59	0	2
Lepidium	33	0	42	2
Solanum	28	3	27	7
Bassia	0	0	23	2
Dichondra	44	19	0	10
Medicago	6	5	12	18
Oxalis	11	3	8	2
Boerhavia	17	0	29	0
Eclipta	39	0	12	0

[a] Abundance of each food in the habitat expressed as the percentage of quadrats in which it was recorded.

result of these flights is determined by the prevailing and potential weather patterns. The height is determined by conditions within the local boundary-layer: the presence and height of nocturnal inversion layers, and the direction and velocity of the wind. Wind velocities greater than the insects' flying speed (*c.* 2 m s^{-1}) cause the insect to fly downwind.

An hypothesis relating these patterns of behaviour to the reproductive strategy of *C. terminifera* has been proposed (Clark, 1971). The random orientation of flying locusts at low windspeeds, that is, under dry anticyclonic conditions with very small pressure gradients, results in widespread dispersal. A downwind orientation at higher windspeeds results in a strong displacement of the insects and is usually associated with the approach and passage of low-pressure systems, convergent winds and the probable development of storms. The consequent increase in turbulence would concentrate the insects in flight with a further concentration within the zone of convergence of the weather systems. Thus during these periods the locusts have an enhanced chance of locating rain and so of finding suitable breeding conditions to perpetuate these dispersed populations.

Large mammals

For larger mammalian herbivores it seems that the undisturbed desert lands once carried diverse small mobile populations, most of which have subsequently been displaced or exterminated by domesticated or feral competitors introduced by man. The Old World deserts, for example, carried widely distributed populations of oryx (*Oryx leucoryx*), the Addax antelope (*Addax nasomaculatus*), various gazelles (*Gazella* spp.), the dromedary (*Camelus dromedarius*) and Bactrian camel (*C. bactrianus*). In the New World the arid and semi-arid regions carried the bison (*Bison bison*), the pronghorn (*Antilocapra americana*), deer species (*Odocoileus* spp.) and the big-horn (*Ovis camadensis*). And in Australia there existed an analogously diverse marsupial fauna of kangaroos (*Megaleia*, *Osphranter*, *Macropus*), rat-kangaroo (*Bettongia*, spp.) and hare wallabies (*Lagorchestes* spp.). The populations in the undisturbed state of the latter were small as judged from the observations of early explorers (Newsome, 1971*b*; Newsome & Corbett, 1977). Consider now the displacement of the native marsupial populations in arid Australia through the introduction of sheep and cattle. Though the disturbed state only is documented, it is possible to infer some aspects of coupling of the marsupial populations to the arid-zone vegetation.

The introduction of domestic stock into the arid lands of Australia approximately a century ago precipitated competitive interactions which resulted in the extinction of several small or medium-sized species (Newsome, 1971*b*) and the increase in numbers and distribution of the large herbivorous kangaroos (Newsome, 1962; Kirsch & Poole, 1972). The almost complete extinction of the small marsupial populations (*Bettongia*, *Largorchestes* etc.) was probably due to a loss of shelter and a consequent increase in predation rather than a competition for forage. The hypotheses to explain the prospering of the kangaroos, the group most likely to suffer the severest competition for food from the alien ruminants, allow extrapolation back to the undisturbed state (Newsome, 1975). The two herbivores considered here are the 'red' kangaroo (*Megaleia rufa*) and the 'euro' (*Osphranter robustus*). They are the only two endemic kangaroos of the arid zone and are sympatric over most of this geographical area (Frith & Calaby, 1969). Studies have shown that for both species natural predation is insignificant in determining population sizes today though man can seriously reduce populations locally, or for a time.

The euro has been extensively studied in the Pilbara pastoral district of north-western Western Australia where its numbers have risen so as to be declared a pest (e.g. Ealey, 1967*a*). These studies have been compiled by Newsome (1975). In aboriginal times, euro populations were small and localized around refugia in granite outcrops to escape heat and drought. These sedentary populations so-formed were not cohesive (Ealey, 1967*b*) and during the hot rainless periods there was apparently a high mortality. In refugia, the animals could largely avoid desiccation by staying in cool caves during the day and by feeding at night on the perennial 'soft' spinifex plant (*Triodia pungens*). Free water requirements were probably low, for the euro has the ability to withstand considerable dehydration and recycle urea (Brown, 1964), though thirst was apparently limiting in prolonged droughts. Even during the driest periods the forage (*T. pungens*) contains 30% w/w of water rising to

60% after rainfall (Ealey, Bentley & Main, 1965). The advent of the irregular monsoonal rains lifts these physiological constraints. The euros disperse and breed, because ephemeral forage and free water would now be available on the plains. As the dry winter progresses the animals retreat again to the rocky drought refugia. This pattern of existence is still observable particularly in the unsettled inland desert areas adjacent to the pastoral areas.

The advent of sheep *c.* 1870 began to change this pattern (Ealey, 1967*a*). The euro, originally uncommon and confined to the rocky hills, became abundant and moved out onto the plains, sheltering under trees (Ealey, 1962). By the 1930s euro populations were many times that of the sheep and the over-grazed pastoral properties, principally damaged by domestic stock, were being abandoned. European man had provided permanent water supplies at regular intervals across the land and his domestic stock had eaten out the native grasses such that the drought-resistant spinifex (*T. pungens*) came to dominate both the hills and the plains (Suijendorp, 1955). *T. pungens* is a sub-maintenance diet for sheep because of its low protein content but not so for the euro (Ealey & Main, 1967). Thus European man has greatly extended the range and abundance of the euro and allowed it to invade new habitats.

The red kangaroo (*M. rufa*) is also present in the Pilbara area and still coexists with the euro on the coastal areas where the climate is more favourable and spinifex domination is less severe. But the red kangaroo is absent from the more arid inland areas where the euro numbers had greatly increased as a result of sheep grazing.

The red kangaroos and the euros are similar in size and are sexually dimorphic, with similar digestive, water and reproductive physiologies. However, they diverge markedly in other aspects. For the euro, *T. pungens* forms a major dietary component at all times, rising in dry times (> 60%) and falling after rain (*c.* 20%) (Ealey & Main, 1967). In the same area the diet of the red kangaroo consists largely of green perennial or annual grasses which are now restricted in occurrence to creek frontages or run-on areas (Storr, 1968). Similar dietary patterns have been observed in other arid areas (Chippendale, 1962, 1968; Griffith & Barker, 1966) and explain the sedentary versus nomadic behaviour of the two species.

The red kangaroo was well represented till *c.* 1930 in the Pilbara but has been rare since the severe droughts of 1944 and 1945 (Ealey, 1967*a*). Sheep numbers also continue to fall indicating the largely irreversible damage due to over-grazing (Ealey, 1967*a*).

Newsome (1975) considers that the population shifts in both plant and animals may be explained by considering the dietary behaviour of the three herbivores involved. The data of Storr (1968) obtained for the coastal areas allows the following explanation for the more arid interiors. The dietary impact of the euro and sheep apparently overlapped very little after rain when a variety of ephemeral and annual species were available (Table 5.4).

Table 5.4. *Diet of euros, red kangaroos and sheep*. (From Newsome, 1975)

	Number of species	Abundance (% ground cover)		Mean nitrogen content (% dry wt)		Mean epidermal areas in faeces (%)					
						Euro		Red kangaroo		Sheep	
		Drought	Good seasons	Drought	Good seasons	Drought	After rains	Drought	Good seasons	Drought	Good seasons
Triodia pungens	1	42.3	53.6	0.65 } 0.8	1.1 } 1.5	} 19	70	12	12	15	12
Other spinifex	2	2.1	16.2	0.9	1.8						
Aristida spp.	2	0	0	0.7	0.7	33	1	29	2	1	4
Cenchrus spp.	2	45.2	14.3	0.8	1.4	34	7	34	30	33	37
Other grasses	3+	1.7	2.7	0.8	1.8	5	20	10	53	14	19
Non-grasses	9+	8.7	1.5	1.5	1.7	9	2	15	3	37	28

However, the diets of the red kangaroo and the sheep were similar. During protracted rainless periods the diets of all three converged, with the sheep including in its diet the less palatable browse species.

The nomadic red kangaroo with a high dietary dependence on green grass was thus gradually excluded by the two more-adaptable species. However, as the over-grazing continued, the sheep has finally almost excluded itself by altering the plant community to one on which only the euro can survive and breed.

Discussion

The task now is to draw together these examples and distil whatever understanding we can from them. From the literature examined here we can qualify in part the niches of the desert-dwelling herbivores for the particular communities where they were studied (Whittaker & Woodwell, 1972). We can only speculate about much of the quantitative aspects of the herbivore niche – total seed or foliage flows, niche overlap with other competing herbivores or decomposers, etc. – for the experiments have not yet been conducted. The qualitative evidence alone can do much to increase our understanding of desert ecosystems since any generalities may be interpreted as constraints to the optimization of the functioning of such ecosystems (Cody, 1974) and may thus be global in application.

The most useful intra-community niche parameter is the herbivore niche breadth: the spatial and temporal use of the food resource. By comparing the information presented here with generalized, predicted herbivore/food-resource characteristics we may describe more clearly the position of three herbivores within their respective ecosystems. These generalized responses have been derived from models and variously tested (e.g. Levins, 1968, 1969*a*, *b*; Krebs, 1972; Emlen, 1973; Cody, 1974). For example, a food resource present in great abundance or with a high renewal rate would tend to favour herbivores with specialist dietary requirements against those with more generalist requirements. Similarly a mixture of very similar food resources would tend to favour a generalist versus a specialist feeder. Spatial and temporal variability in any given food resource are correlated in influence but the herbivore response is dependent on size (Emlen, 1973). Small herbivores that are limited by mobility in an environment where the food resource is 'patchy' or 'coarse-grained', usually have specialist, preferential diets. Larger herbivores on the other hand can regard the same variability as being 'fine-grained' only, and so have generalist diets utilizing the food resource more in proportion with its abundance.

A species whose population is limited by predation can be expected to show greater dietary preference than one limited by food resources. Coexistence of herbivorous species in any one habitat dictates certain specialist/generalist combinations but these are strongly influenced by population turnover rates (Cody, 1974).

Examining now the seed-eating ants we find that they appear to live well with the food resource available, harvesting only 20% of the total seed fall (Briese, 1974). They are not generalist in diet at all times; rather, they appear to be flexibly selective, have a wide niche breadth, and concentrate on the freshly fallen seed. The seedfalls in this study were abundant (averaging 0.25 g m^{-2}) and were localized underneath the parent plant so that this dietary behaviour was very efficient in energy expenditure allowing the most rapid acquisition and storage of seeds. Thus the ants may regard the food resource environment as 'coarse' or 'fine-grained' and adjust as a colony accordingly.

Further, both Briese (1974) and Culver (1972) have shown that there is almost no niche overlap between harvester species. Niche differentiation appears to be due to qualitative and temporal differences in food harvesting. This lack of competitive pressure between species would further buffer the ant populations from environmental fluctuations, the prime stabilizing factor being the very loose coupling of these herbivores to the food source.

Ant population/biomass studies are lacking to test this hypothesis. However, ant colonies may be regarded in foraging and other behaviour as 'individuals' (Carroll & Janzen, 1973). They appear to be long-lived, and predation at founding stage appears to limit colony numbers (Briese, 1974), much like the behaviour of perennial plants (Williams, 1977). As social insects, however, they possess the adaptively significant characteristic of recovery from the large population declines that occur, however infrequently, when all forage and stored food is depleted.

The niche characteristics of small mammalian herbivores differ from those of the harvester ants. Although they show flexible dietary behaviour depending upon seed abundance interspecific competition appears to place constraints on the width of the food- and space-resource niche. The partitioning of these two resources appears to be determined by seed and animal size along with animal mobility and territorial defence (Smigel & Rozensweig, 1974; Brown & Lieberman, 1973).

Populations of small mammals vary widely with the seasons and include the occasional irruption or plague (Newsome & Corbett, 1975; Krebs *et al.*, 1973). They appear to be more tightly coupled to the food supply than ants showing marked seasonal population fluctuations (French *et al.*, 1974; Packard, 1977). Populations persist through, and recover from, such unfavourable periods because of the longevity and reproductive capacity of individuals.

Termites appear to have a similar forage-niche characteristic to ants but, because of divergent physiologies, may obtain the basic cellulose from a wider variety of sources. Similarly colonies and populations may be temporarily buffered from an unfavourable environment through food storage and social organization. Intra-space and inter-specific interactions are poorly understood particularly as they may affect the siting of, and foraging from, termitaria. The influence of the physical presence of termitaria on vegetation may last for centuries (Clos-Arceduc, 1956).

The highly specific dietary behaviour of the plague locust (*C. terminifera*) and the water-dependent developmental stages enables it to exploit a highly variable arid environment. These two characteristics give information about the present and imminent environments and are thus good predictors of a successful population dispersal. As such, they are analogous to information-collating germination controls found in the seeds of arid-zone plants.

Plague locusts have little overall significance as primary herbivores in the functioning of undegraded arid ecosystems. The sporadic nature of the populations and the migrating habit reduce their significance in undisturbed ecosystems to less than that of termites, for example. However, their significance can be greatly increased in the ecosystems that have been strongly perturbed by man's cultural impact where they may perpetuate instability over long periods of time.

The interaction between euro, red kangaroo and sheep can be explained quite well by consideration of niche width and overlap. The pre-man euro populations appear to have been controlled by the availability of food, water and shelter. Unfortunately, there is little evidence available on interspecific competition. The food-resource niche breadth was apparently wide. They coexisted with the more dietary-specialized red kangaroo. There was little niche overlap with respect to food or space, for the red kangaroo was nomadic and migratory, following rainstorms, etc. (Newsome, 1975). Their sympatric and yet discrete distribution across the whole of arid Australia can be predicted (Cody, 1974) by also considering their highly adapted reproductive strategies for the avoidance of high-risk breeding (Newsome, 1965*a*, *b*; Sadlier, 1965). The forced introduction of sheep with a strong food niche overlap with the red kangaroo resulted in the latter's displacement. Finally, through changes in the spectrum of food available, sheep and euro competed for food, but never space, until the sheep itself had been largely excluded.

In conclusion, the temporal and spatial variability of forage availability in desert areas may be expected to encourage the evolution of herbivores with wide generalist dietary habits (Emlen, 1973). Although temporally flexible, however, it appears that desert-dwelling herbivores all have recognizable specialization in diet.

Herbivores like the ants and termites, for whom the food-resource environment is 'coarse-grained' and that are largely without predation or competitive influence live well within the resource further buffering their populations by food storage (May, 1973). In many ways they reflect the adaptations of perennial desert plants (Noy-Meir, 1973; Woolhouse, 1974; Charnoff & Schaffer, 1973).

The locusts, with the most inflexible diets, suffer as a consequence of great population fluctuations. However, the aggressive expansion followed by withdrawal to refugia is as successful as any other herbivore/plant coupling. Small mammals have affinities with the larger mammals but predation and competition enforce differing constraints. It is feasible that a yet unmeasured interplay between migration and extinction within the 'patchy' food and predator environments, may stabilize and buffer the survival of a species within desert ecosystems (e.g. Levins, 1969*a*, *b*, 1970; Byant, 1973; May, 1973).

References

Anderson, D. J. (1971). Pattern in desert perennials. *Journal of Ecology*, **59**, 555–60.

Bernays, E. A. & Chapman, R. F. (1973). The role of food plants in the survival and developments of *Chortoicetes terminifera* (Walker) under drought conditions. *Australian Journal of Zoology*, **21**, 575–92.

Blocker, H. D. (1977). The impact of invertebrates as herbivores on arid and semi-arid rangeland in the United States. In: *The impact of herbivores on arid & semi-arid rangelands. Proceedings of the 2nd US/Australia Rangeland Panel*, pp. 357–76. Australian Rangeland Society, Perth.

Briese, D. T. (1974). Ecological studies of an ant community in a semi-arid habitat. Unpublished PhD thesis, Australian National University, Canberra.

Brown, G. D. (1964). The nitrogen requirements of macropod marsupials. Unpublished PhD Thesis, University of Western Australia.

Brown, J. H. & Lieberman, G. A. (1973). Resource utilization and coexistence of seed-eating desert rodents in sand dune habitats. *Ecology*, **54**, 788–97.

Bryant, E. H. (1973). Habitat selection in a variable environment. *Journal of Theoretical Biology*, **41**, 421–9.

Carroll, C. R. & Janzen, D. H. (1973). Ecology of foraging ants. *Annual Review of Ecology and Systematics*, **4**, 231–58.

Charley, J. L. & Cowling, S. W. (1968). Changes in soil nutrient status resulting from overgrazing and their consequences. *Proceedings of the Ecological Society of Australia*, **3**, 28–38.

Charnov, E. L. & Schaffer, W. M. (1973). Life history consequences of natural selection: Coles result revisited. *American Naturalist*, **107**, 791–3.

Chew, R. M. & Chew, A. E. (1970). Energy relationships of the mammals of a desert shrub (*Larrea tridentata*) community. *Ecological Monographs*, **40**, 1–21.

Chippendale, G. M. (1962). Botanical examination of kangaroo stomach contents of cattle rumen contents. *Australian Journal of Science*, **25**, 21–2.

Chippendale, G. M. (1968). The plants grazed by red kangaroos, *Megaleia rufa* (Desmarest), in Central Australia. *Proceedings of the Linnean Society of New South Wales*, **93**, 98–110.

Clark, D. P. (1969). Night flights of the Australian plague locust (*Chortoicetes terminifera*) (Walk.)) in relation to storms. *Australian Journal of Zoology*, **17**, 329–52.

Clark, D. P. (1971). Flights after sunset by the Australian plague locust, *Chortoicetes terminifera* (Walk.) and their significance in dispersal and migration. *Australian Journal of Zoology*, **19**, 159–76.

Clark, D. P. (1972). The plague dynamics of the Australian plague locust *Chortoicetes terminifera* (Walk.). *Proceedings of the International Study Conference. Current and future problems of Acridology, London 1970*, pp. 275–87.

Clark, D. P. (1978). The significance of the availability of water in limiting invertebrate numbers. In: *Studies of the Australian arid zone*, 3, *Water in Rangelands* (ed. K. M. W. Howes), pp. 198–207. CSIRO, Melbourne.

Clos-Arceduc, M. (1956). Etude sur photographies aériennes d'une végétale saheliénne: la brousse tigrée. *Bulletin de l'Institut francaise d'Afrique Noire, Serie A*, **18**, 677–84.

Cloudsley-Thompson, J. L. & Chadwick, M. J. (1964). *Life in deserts.* Foulis, London.

Cody, M. L. (1974). Optimization in ecology. *Science*, **183**, 1156–64.

Correll, R. L. & Lange, R. T. (1966). Some aspects of the dynamics of vegetation in the Port Augusta–Iron Knob area, South Australia. *Transactions of the Royal Society of South Australia*, **90**, 41–3.

Culver, D. C. (1972). A niche analysis of Colorado ants. *Ecology*, **53**, 126–31.

Ealey, E. H. M. (1962). Biology of the euro. Unpublished PhD Thesis, University of Western Australia.

Ealey, E. H. M. (1967*a*). Ecology of the euro, *Macropus robustus* (Gould) in north-western Australia. I. The environment and change in euro and sheep population. *CSIRO Wildlife Research*, **12**, 9–25.

Ealey, E. H. M. (1967*b*). Ecology of the euro, *Macropus robustus* (Gould) in north-western Australia. II. Behaviour, movement, and drinking patterns. *CSIRO Wildlife Research*, **12**, 27–51.

Ealey, E. H. M., Bentley, P. J. & Main, A. R. (1965). Studies on the water metabolism of the hill kangaroo, *Macropus robustus* (Gould) in north-western Australia. *Ecology*, **46**, 473–9.

Ealey, E. H. M. & Main, A. R. (1967). Ecology of the euro, *Macropus robustus* (Gould), in north-western Australia. III. Seasonal changes in nutrition. *CSIRO Wildlife Research*, **12**, 53–65.

Emlen, J. M. (1973). *Ecology: an evolutionary approach.* Addison-Wesley, Massachusetts.

Ellison, L. (1960). Influence of grazing on plant succession of rangelands. *Botanical Review*, **26**, 1–78.

French, N. R., Maza, B. G., Hill, H. O., Aschwanden, A. P. & Kaaz, H. W. (1974). A population study of irradiated desert rodents. *Ecological Monographs*, **44**, 45–72.

Frith, H. J. & Calaby, J. H. (1969). *Kangaroos.* Cheshire, Melbourne.

Froggatt, W. W. (1910). Insects which damage saltbush. *Agricultural Gazette of New South Wales*, **21**, 465–71.

Goodall, D. W. (1966). The nature of the mixed community. *Proceedings of the Ecological Society of Australia*, **1**, 84–96.

Griffiths, M. & Barker, R. (1966). The plants eaten by sheep and by kangaroos grazing together in a paddock in south-western Queensland. *CSIRO Wildlife Research*, **11**, 145–67.

Hall, E. A. A., Specht, R. L. & Eardley, C. M. (1964). Regeneration of the vegetation on Koonamore Vegetation Reserve, 1926–1962. *Australian Journal of Botany*, **12**, 205–64.

Harper, J. L. (1969). The role of predation in vegetation diversity. *Brookhaven Symposia in Biology*, **22**, 48–62.

Janzen, D. H. (1971). Seed predation by animals. *Annual Review of Ecology and Systematics*, **2**, 465–92.

Keast, J. A. & Marshall, A. J. (1954). The influence of drought and rainfall on reproduction in Australian desert birds. *Proceedings of the Zoological Society of London*, **124**, 493–9.

Kenagy, G. J. (1972). Saltbush leaves: excision of hypersaline tissue by kangaroo rat. *Science*, **178**, 1094–6.

Kenagy, G. J. (1973). Adaptations for leaf eating in the Great Basin kangaroo rat, *Dipodmys microps*. *Oecologia*, **12**, 383–412.

Kirsch, J. A. W. & Poole, W. E. (1972). Taxonomy and distribution of grey kangaroos, *Macropus giganteus* Shaw 1790 and *M. fuliginosus* Desmarest 1817, and their subspecies. *Australian Journal of Zoology*, **20**, 315–39.

Krebs, C. J. (1972). *Ecology: the experimental analysis of distribution and abundance*. Harper & Row, New York.

Krebs, C. J., Gaines, M. S., Keller, B. L., Myers, J. H. & Tamarin, R. H. (1973). Population cycles in small rodents. *Science*, **179**, 35–41.

Lambert, M. R. K. (1972). Some factors affecting flight in field populations of the Australian plague locust, *Chortoicetes terminifera* (Walk.) in New South Wales. *Animal Behaviour*, **20**, 205–17.

Lee, K. E. & Wood, T. G. (1971*a*). *Termites and soils*. Academic Press, London.

Lee, K. E. & Wood, T. G. (1971*b*). Physical and chemical effects on soils of some Australian termites, and their pedological significance. *Pedobiologia*, **11**, 376–409.

Leigh, J. H. & Noble, J. C. (1972). *Riverine Plain of New South Wales: its pastoral and irrigation development*. CSIRO Division of Plant Industry, Canberra.

Levins, R. (1968). *Evolution in changing environments*. Monographs in population biology, 2. Princeton University Press.

Levins, R. (1969*a*). Some demographic and genetic consequences of environmental heterogeneity for biological control. *Bulletin of the Entomological Society of America*, **15**, 237–40.

Levins, R. (1969*b*). The effect of random variations of different types on population growth. *Proceedings of the National Academy of Sciences, USA*, **62**, 1061–5.

Levins, R. (1970). Extinction. In: *Some mathematical problems in biology*, vol. 2 (ed. M. Gerstenhaber), pp. 77–107. American Mathematical Society, Washington, DC.

May, R. M. (1973). Stability in randomly fluctuating versus deterministic environments. *Americna Naturalist*, **107**, 621–49.

Newsome, A. E. (1962). The biology of the red kangaroo, *Macropus rufus* (Desmarest) in central Australia. Unpublished MSc thesis, Adelaide University.

Newsome, A. E. (1965*a*). The abundance of red kangaroos, *Megaleia rufa* (Desmarest) in central Australia. *Australian Journal of Zoology*, **13**, 269–87.

Newsome, A. E. (1965*b*). The distribution of red kangaroos, *Megaleia rufa* (Desmarest), about sources of persistent food and water in central Australia. *Australian Journal of Zoology*, **13**, 289–99.

Newsome, A. E. (1971*a*). The ecology of red kangaroos. *Australian Zoologist*, **16**, 32–50.

Newsome, A. E. (1971*b*). Competition between wildlife and domestic stock. *Australian Veterinary Journal*, **47**, 577–86.

Newsome, A. E. (1975). An ecological comparison of the two arid zone kangaroos of Australia, and their anomalous prosperity since the introduction of ruminant stock to their environment. *Quarterly Review of Biology*, **50**, 389–424.

Newsome, A. E. & Corbett, L. K. (1975). Outbreaks of rodents in semi-arid and arid Australia: causes, preventions and evolutionary considerations. In: *Rodents in desert environments* (ed. I. Prakash & P. Ghosh). Dr W. Junk, The Hague.

Newsome, A. E. & Corbett, L. K. (1977). The effects of native, feral and domestic animals on the productivity of the Australian rangelands. In: *The impact of herbivores on arid and semi-arid rangelands. Proceedings of 2nd US/Australian Rangeland Panel*, pp. 331–51. Australian Rangeland Society, Perth.

Noy-Meir, I. (1973). Desert ecosystems: environment and producers. *Annual Review of Ecology and Systematics*, **4**, 25–41.

Packard, R. L. (1977). Effects of herbivores on seed usage, dispersal, and reproduction with particular reference to mammals. In: *The impact of herbivores on arid and semi-arid rangelands. Proceedings of the 2nd US/Australian Rangeland Panel*, pp. 211–26. Australian Rangeland Society, Perth.

Sadleir, R. M. E. S. (1965). Reproduction in two species of kangaroo (*Macropus robustus and Megaleia rufa*) in the arid Pilbara region of Western Australia. *Proceedings of the Zoological Society of London*, **145**, 239–61.

Serventy, D. L. & Marshall, A. J. (1957). Breeding periodicity in Western Australian birds with an account of unseasonal nesting in 1953 and 1955. *Emu*, **57**, 99–126.

Smigel, B. W. & Rosensweig, M. L. (1974). Seed selection in *Dipodomys merriami* and *Perognanthus penicillatus*. *Ecology*, **55**, 329–39.

Storr, G. M. (1968). Diet of kangaroos (*Megaleia rufa* and *Macropus robustus*) and Merino sheep near Port Hedland, Western Australia. *Journal of the Royal Society of Western Australia*, **51**, 25–32.

Suijdendorp, H. (1955). Changes in the pastoral vegetation can provide a guide to management. *Journal of Agriculture, Western Australia*, **4**, 683–7.

Taylor, R. W. (1972). Biogeography of insects of New Guinea and Cape Yorke Peninsula. In: *Torres Strait Symposium* (ed. D. Walker), pp. 213–30. Australian National University Press, Canberra.

Tevis, L. (1958). Interrelations between the harvester ant *Veromessor pergandei* and some desert ephemerals. *Ecology*, **39**, 695–704.

Wardhaugh, K. G. (1973). A study of some factors affecting egg development in *Chortoicetes terminifera* (Walk.). Unpublished PhD thesis, Australian National University, Canberra.

Watson, J. A. L., Lendon, C. & Low, B. S. (1973). Termites in mulga lands. *Tropical Grasslands*, **7**, 121–6.

Watson, J. A. L. & Gay, F. J. (1970). The role of grass eating termites in the degradation of a mulga ecosystem. *Search*, **1**, 43–6.

West, N. E. & Tueller, P. T. (1972). Special approaches to studies of competition and succession in shrub communities. In: *Wildland shrubs – their biology and utilization* (ed. C. M. McKell, J. P. Blaisdell & J. R. Goodin), pp. 165–71. USDA Forest Service General Technical Report. INT.1.

Whittaker, R. H. & Woodwell, G. M. (1972). Evolution of natural communities. In: *Ecosystem structure and functions*, pp. 137–56. Oregon State University Press.

Woolhouse, H. W. (1974). Longevity and senescence in plants. *Science Progress, Oxford*, **61**, 123–47.

Williams, O. B. (1968). That uneasy state between animal and plant in the manipulated situation. *Proceedings of the Ecological Society of Australia*, **3**, 167–74.

Williams, O. B. (1977). Reproductive wastage in rangeland plants, with particular reference to the role of herbivores. In: *The impact of herbivores on arid and semi-arid rangelands. Proceedings of the 2nd US/Australia Rangeland Panel*, pp. 227–48. Australian Rangeland Society, Perth.

Manuscript received by the editors April 1975

6. Effects of biotic components on abiotic components

K. E. LEE

The principal effects of the biotic components on the abiotic components of arid ecosystems relate to the soil, which is the zone of contact between biotic and abiotic systems. These effects may conveniently be considered under two main headings: effects of plants on soils; and effects of animals on soils. Effects on soils are taken to include not only direct physical effects, but also processes that influence the chemical constitution, water relationships, and hydrology of soils.

Effects of plants on soils

Chemical effects

In arid areas, a characteristic of perennial plants is that they are discontinuously distributed, with much bare ground in between. Chemical effects on soils are thus concentrated around individual plants or groups of plants, contrasting with unaffected soils between plants.

Many of the plants accumulate soluble salts. Wood (1925) compared NaCl and Cl^- concentrations in leaves of nine species of *Atriplex* (Chenopodiaceae) in Australia with their concentrations in the soils on which they grew. The ranges of concentrations in the leaves were 8.80–32.90% NaCl, 5.35–20.00% Cl^-, and for soil 0.02–0.87% NaCl, 0.01–0.53% Cl^-. In one species (*A. nummularia*), he found that the concentration of all soluble salts was 25%, including 7.78% Cl^-, 1.57% Ca^{2+}, 6.20% K^+, and 6.00% Na^+. Osmotic pressure due to NaCl alone was 65 bar for *A. paludosa* and 40 bar for *A. vesicaria*, and Wood compared these values with results of Harris, Gortner, Hoffman & Valentine (1921), who found osmotic pressure of 67.5 bar in *A. canescens* and 153 bar in *A. confertifolia* from Arizona deserts.

Jessup (1969) extended Wood's (1925) observations, measuring the concentrations of soluble salts in soils under *A. vesicaria* and *Maireana astrotricha* canopies, under bare ground between the bushes, and also under grass, under *Bassia*, and bare 'scald' patches, in north-eastern South Australia. He found that the concentration of Cl^- and of total soluble salts was significantly higher under *A. vesicaria* and *M. astrotricha* at 0–2.5 cm and 15 cm depth than between bushes, under grass, or *Bassia*. Under 'scald' patches, where about 10 cm of soil had been removed by wind erosion, corresponding samples to those at 15 cm depth under bushes had the same salt concentrations as samples between bushes. He also sampled at 30 cm, 45 cm, and 60 cm depth and found in all cases that Cl^- concentration reached a maximum at about 60 cm and did not change significantly below that depth.

The extra Cl^- in the topsoil under bushes is apparently derived from leaching of the dead leaves shed by the bushes, and the maximum at 60 cm was taken to indicate the maximum depth of water penetration in the soil profile. During dry periods the bushes shed leaves, apparently to reduce transpiration, leading to an increase in Cl^- concentration in the topsoil and a corresponding decrease in the surrounding root zone of the bushes. This accumulation would have a further advantage to the bushes during periods of moisture stress, since it would inhibit the growth of less salt-tolerant plants such as grasses. Rain results in leaching of the soluble salts accumulated under the bushes and more even distribution throughout the surrounding soil.

Roberts (1950) similarly investigated soluble salt distribution under *A. confertifolia* and *Sarcobatus vermiculatus* and between bushes in soils under 280 mm annual rainfall in south-west Utah and 180 mm annual rainfall in California. He demonstrated increases in concentration of soluble salts relative to soil between the bushes, of three to six times under *A. confertifolia* at 0–5 cm depth at both sites, and an increase of about 14 times under *S. vermiculatus* at the same depth at the Utah site.

Fireman & Hayward (1952) worked in the Escalante Desert, Utah, an area with saline and saline–alkali soils (i.e., soils with sufficient Na^+ or other soluble salts to interfere with the growth of most plants). They studied soils under *A. tridentata*, which has a 'generalized' system of lateral roots in the topsoil and a deep taproot; *A. confertifolia*, which has widespread shallow lateral roots and a taproot penetrating to 120–240 cm; and *S. vermiculatus*, which has a moderate development of coarse shallow roots to 50–75 cm and a strong taproot to 450 cm or more. They found a significant correlation between crown size and increase in pH, total soluble salts and exchangeable Na^+ in the soil. The effect of increasing Na^+ under *S. vermiculatus* was to produce marked effects on soil tilth as indicated by permeability measurements, moisture retention, settling volume and wet sieving. Excess gypsum leached through the soil in laboratory tests resulted in replacement of most of the Na^+ and consequent improvement in tilth and permeability. Roots, stems, branches and leaves of all three plant species were analysed for Na^+, K^+, Mg^{2+}, Ca^{2+}, Cl^-, and SO_4^{2-} and for malic, citric, oxalic and total organic acids. The total ionic content of each plant species was positively correlated with the soluble salt content of the underlying soils. Ion concentrations differed for various plant parts, and the species exhibited large differences in salt accumulation when grown side by side in saline and non-saline soils. *S. vermiculatus* had the greatest effect on soils and also had by far the greatest proportion of anions in the form of organic acids, especially in the leaves.

Table 6.1. *Nitrogen and phosphorus content of soils from some arid and humid zones.* (Data from Charley & Cowling, 1968)

	Nitrogen		Phosphorus	
Soils	No. of samples	Mean N content (%)	No. of samples	Mean P content (ppm)
Australia				
Arid	77	0.06	70	240
Humid	138	0.22	208	620
Other countries				
Arid	38	0.11	643	701
Humid	—	—	1270	730

Fireman & Hayward (1952) postulated that when the leaves fall they decompose slowly, liberating cations and anions that form more or less soluble salts. These would initially be Na, Mg, and Ca salts of Cl^-, SO_4^{2-}, malic, citric, and oxalic anions. The organic anions would then decompose, forming CO_3^{2-} and HCO_3^-. These ions precipitate as calcium and magnesium salts, leaving Na^+ and K^+ salts which react with the exchange complex to displace Ca^{2+} and Mg^{2+} forming soils high in exchangeable Na^+ and K^+, with high pH and deterioration of soil structure.

The presence of high concentrations of soluble salts in leaves of desert plants, with consequent high osmotic pressures, provides a mechanism for uptake of water through the leaves from unsaturated air. The significance of this capability and its relationship to the nature of the leaf epidermis are discussed below.

Charley & Cowling (1968) measured the content and vertical distribution of nitrogen and phosphorus in some Australian arid-zone soils, and compared their findings with data from arid- and humid-zone soils of several other countries (Table 6.1). They found that arid zone soils generally have lower nitrogen content than humid zone soils and that Australian arid zone soils have especially low nitrogen contents. Unlike nitrogen, phosphorus content appears to be only moderately affected by climatic factors and is most affected by the amount of phosphorus in the parent materials. Charley & Cowling pointed out that mean phosphorus content of Australian humid-zone soils is rather low compared with soils elsewhere, while Australian arid-zone soils have exceptionally low phosphorus levels. The latter may be attributed either to massive leaching in previous, more intense weathering cycles, or to low levels of phosphorus in the unweathered rock, or to the rock constituents having themselves been through a succession of weathering and depletion cycles. Low phosphorus levels which limit plant growth may be at least partly responsible for the low nitrogen content of the soils. Organic carbon content

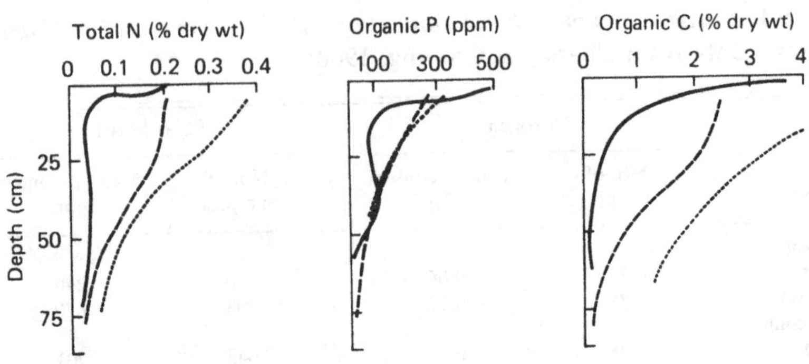

Fig. 6.1. Nitrogen, organic phosphorus, and organic carbon depth functions in soils of humid, sub-humid, and arid areas. (From Charley & Cowling, 1968). ——, Arid; – –, sub-humid; - - -, humid.

of arid-zone soils is also low, and again lower in Australian soils than in equivalent soils in other countries.

Concentrations of nitrogen, phosphorus, and organic carbon as a function of depth in arid-zone soils show extreme accumulation near the surface and depletion at depth as compared with concentrations in soils of more humid zones (Fig. 6.1). This correlates with the intermittent wetting and generally slight depth of water penetration in arid-zone soils, with consequent concentration of biological activity close to the surface.

Charley & Cowling discussed organic matter production, its effects on nutrient turnover, and the results of overgrazing on saltbush (*Atriplex*) country in Australia. They concluded that in a stable *Atriplex* community on solonetzic soils in western New South Wales there is a delicate balance between nutrient supply and plant growth. Annual leaf and fruit fall exceed the weight of leaves held in the community and litter breakdown and nutrient cycling are rapid during the short periods when adequate moisture is available. Microbial production of nitrate falls sharply with increasing depth and is very low below 15 cm, where substrate concentrations and populations of nitrifying organisms are both low. Sudden flushes of plant growth associated with unusually heavy rain, a well known phenomenon in arid areas, have led to a belief that soils in arid areas are potentially fertile, requiring only adequate water to show their productivity. However, Charley & Cowling (1968) quote two Australian examples where two unusually wet years in succession resulted in a flush of plant production in the first year, but slight production in the second year, and they conclude that this is a consequence of the relatively small pool of available plant nutrients and their concentration close to the soil surface.

Damage to vegetation resulting from overgrazing in western New South Wales has resulted in erosion of surface soil, exposing the saline B horizon, and failure of saltbush to re-establish even with sowing and adequate watering. Loss of the nutrient pool associated with as little as 5 cm of soil from the surface results in a serious decline in fertility. Some evidence suggests that plants draw a large part of their phosphorus needs from organic phosphorus compounds, which in these soils must be concentrated near the surface and would be readily lost by erosion. Loss of available phosphate from the functional pool may then partly limit nitrogen accumulation by restricting the capability of native legumes that grow on degraded soils to fix atmospheric nitrogen. The end result of overgrazing then would be the initiation of a series of interacting processes that result in long-term depletion of available nutrients from the soil.

Fixation of nitrogen by algae, which form thin crusts on the surface in arid areas, may be a significant influence in the absence of nitrogen-fixing organisms commonly found in more humid areas. Rogers & Lange (1971) note that, in arid areas of Australia grazed by sheep, a crust of lichens extends uninterrupted over vast areas. Rogers, Lange & Nicholas (1966) investigated algal fixation of nitrogen in Australian arid land where a crust of lichens, some blue-green algae, and other cryptogams covered up to 30% of the surface. They cultured 12 of the most common species in the laboratory in an atmosphere enriched with ^{15}N. Nine of the species showed no significant nitrogen fixation, two (*Lecidea crystalifera* and *Parmelia conspersa*) fixed very small amounts of nitrogen, while *Collema coccophorus* fixed sufficient nitrogen that it may contribute significantly to the soil nitrogen balance.

Physical effects

Water uptake

The characteristic shrubs of Australian arid regions contrast with the succulents (Crassulaceae, Euphorbiaceae, Cactaceae) of African and American deserts. The dominant desert shrubs in Australia are Chenopodiaceae, especially *Atriplex* spp. and *Maireana* spp., and xerophily depends on reduction of transpiration to a minimum, peculiar modifications in structure, and physiological modifications, rather than water storage within the plant. Wood (1925) tested the ability of some Australian plants to take up water from the air. He placed shoots of *Atriplex nummularia* in a sealed jar at 85% relative humidity and then outside the jar in dry air, in a dark room to prevent photosynthesis, at a constant temperature of 29.5 °C. He measured the weight of water absorbed by the shoots in the jar and loss due to transpiration outside the jar over a 3-day period, and also salt concentration in the shoots. The water lost by transpiration outside the jar about equalled that gained when in the jar, and salt concentration varied inversely with water content.

The experiment was repeated with four other species of *Atriplex*, *Maireana sedifolia*, *M. aphylla*, *Rhagodia spinescens*, and *Bassia obliquicuspis* (all Chenopodiaceae with a non-cutinized epidermis); and also, for comparison, with *Eucalyptus corynocalyx*, *Sterculia diversifolia*, and *Acacia decussata* (with a cutinized epidermis). All species of *Atriplex* tested absorbed water through the leaves. The remaining Chenopodiaceae absorbed some water but less than *Atriplex* spp., and those with a cutinized epidermis not at all. Wood included that in non-cutinized plants, water absorption into epidermal cells was unimpeded and was due to the high osmotic pressure of the cells, resulting from their high soluble salt content, while water could not pass through the surfaces of cutinized epidermal cells. In *Atriplex* spp. water absorption through the leaves was probably more important than absorption through the roots.

The ability of desert plants to take up water from air at 85% relative humidity must greatly influence their survival in dry periods. Even in extreme conditions when daytime temperatures may reach 45 °C and relative humidity falls to 5%, a drop in temperature at night to about 25 °C would increase relative humidity to about 85%, and such temperature changes from day to night are common in desert environments. Although the direct effect of water uptake is on plant growth, the plant tissue is eventually deposited on the soil and increases the amount of energy cycled in the plant–soil system.

Wind erosion

Marshall (1971) investigated the protective effect of shrubs (*Maireana aphylla*) against soil erosion by wind in a semi-arid, shrub-dominated rangeland in southern Australia. He recognized two ways in which perennial shrubs protect the soil surface: first, their presence as immovable roughness of the surface reduces the soil surface area available for erosion by an amount equal to the basal area of the shrubs, and second, and more importantly, the shrubs retard wind flow. A wind-tunnel investigation of cylindrical roughness elements with diameter: height ratios from 0.5 to 5 showed that for practical purposes the overall shearing stress of regular and random arrays of similar element density was very similar.

Assuming this generalization applies to field conditions, Marshall examined the *M. aphylla* community to determine density of shrubs and lateral cover (which relates the frontal area of a shrub exposed to the wind to the mean area covered by the shrub). A series of wind profiles was measured on a nearby site, and the overall drag coefficient was determined. From these data he calculated the critical lateral cover for the shrub community, that is, the minimum below which there would be significant wind erosion between bushes. For the shrub association tested, Marshall concluded that the soil was adequately protected from wind erosion. But a plot of projected cover of elements with varying diameter: height ratios showed that much more soil surface would be lost under large diameter:height ratio roughness elements than in the case of small ratios. Roughness elements of diameter:height ratios between 1 and 2 appeared to offer the best compromise between efficiency of protection and moderate roughness density.

Water erosion

Craddock & Pearse (1938), working on granitic mountain soils in Idaho, measured surface run-off and soil removal by erosion. They used artificial rain storms of 47–50 mm h^{-1}, applied to 96 plots, each of approximately 20 m^2, with slopes of 17–22°, and which had been badly depleted by fire or overgrazing. These are not true arid lands, but are subject to torrential summer rains. Four range types were included: (*a*) wheatgrass, which represents the grassland climax; (*b*) downy chess; (*c*) lupin-needlegrass, and (*d*) annual weeds, representing progressive stages in range deterioration.

Rainfall intensities of 47–50 mm h^{-1} can be expected at least once each summer in the area. Surface run-off from the wheatgrass plots was 0.5% of the water applied and soil loss by erosion was 7.53 kg ha^{-1}. Downy chess provided moderately effective cover, lupin-needlegrass rather less, and under annual weeds run-off was 60.8% of water supplied and loss by erosion 18900 kg ha^{-1}. Analyses of variance showed that range type was the most important factor influencing run-off and erosion, having more than twice the effect of the next most important factor (rainfall intensity). Samples taken from 0–1.25 cm depth before and after the artificial rainstorms showed differential removal of fine particle sizes as compared with sand and gravel, and also more organic matter and organic nitrogen in the eroded material than in the original soil. The surface soil lost 6% of its silt, 7% of its clay, 20% of its organic matter, and 20% of its organic nitrogen, indicating a greater loss of fertility than is apparent from the amount of soil lost.

Algal crusts, commonly observed on the soil surface in arid areas, appear to reduce erosion losses, but no quantitative data are available. Fletcher & Martin (1948) studied algal crusts on desert soils in Arizona and concluded that they promote infiltration, decrease erosion, and aid in the establishment of plant seedlings. They quoted similar observations by Booth (1941) over 'hundreds of acres of badly eroded land in the south-central United States'. Rogers & Lange (1971) also concluded that deterioration of algal crusts in arid Australia, due to heavy stock trampling, resulted in increased soil erosion.

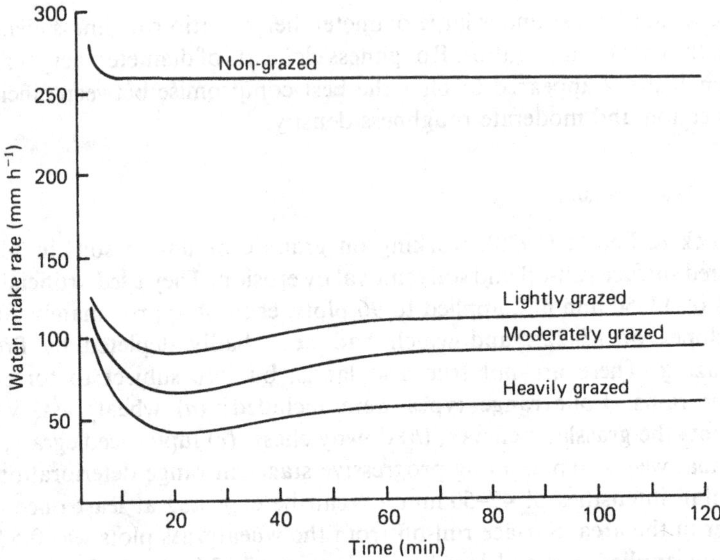

Fig. 6.2. Average water intake rates obtained with sprinkling infiltrometer. (From Rhoades *et al.*, 1964).

Effects of animals on soils

Compaction and cementing of surface layers

Compaction of surface soil due to stock treading and consequent restriction of water infiltration and aeration are well known in the higher rainfall areas of the world (Wind & Schothorst, 1964; Gradwell, 1968). Less is known of such phenomena in arid and semi-arid regions.

Rhoades, Locke, Taylor & McIlvain (1964) summarized results of 20 years' experiments in a semi-arid region (580 mm mean annual rainfall, potential evapotranspiration 1830 mm yr^{-1}) of stabilized sand dunes in Oklahoma. Treatments comprised stocking with cattle at rates of 0, 8.1, 6.9, and 4.8 ha per animal unit, on native vegetation dominated by *Artemisia filifolia* (38% cover) with a basal cover (8%) of grasses and forbs, producing about 1120 kg ha^{-1} yr^{-1} dry weight. Infiltration rates varied inversely with stocking rates. Using a sprinkler method infiltration totalled 59, 92.5, 112, and 259 mm h^{-1} for grassed areas in heavily-, moderately-, lightly-, and non-grazed plots (Fig. 6.2). Results for ring infiltrometers correlated closely with those for the sprinkler method but were 2.33–3.16 times greater (Fig. 6.3).

Ellison (1960) discussed measurements of volume–weight of soils in areas trampled by cattle compared with the same soils in undisturbed areas. Hedrick (1948) found variations in volume–weight of the surface 2.5 cm associated with different levels of grazing of annual vegetation in California,

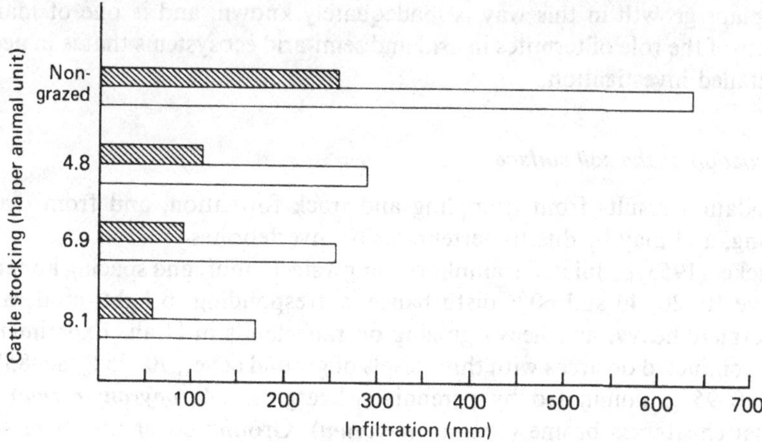

Fig. 6.3. Comparison of total water intake during a 1-h period using sprinkling infiltrometers (open bars) and double-ring type infiltrators (hatched bars). (From Rhoades *et al.*, 1964).

but no consistent correlation was apparent. Kelting (1954) showed an increase of about one third in bulk density of the surface 7.5 cm in grazed pasture compared with adjacent ungrazed pasture. Lodge (1954) recorded significant differences in bulk density between soils of two protected and heavily grazed areas on the Canadian prairies, but not on two other soils.

No generalizations can be made about the compacting of soil by stock trampling in arid areas. There seems no doubt that it can be an important factor reducing infiltration of water, but its importance apparently varies greatly with soil type and plant cover. The low density of animals possible in arid regions may make compaction by trampling important only around watering points and along tracks. The effects of trampling on surface run-off and erosion may be important; these are discussed later.

Termites frequently dominate (in terms of biomass) the soil fauna of arid and semi-arid regions. In constructing nests, feeding galleries, and associated structures, they select, transport, rearrange, and cement soil particles, often producing massive structures that interfere with water infiltration. In extreme cases soil development under such structures is inhibited and previously formed horizons may disappear, as was recorded by Boyer (1958) under and around mounds of *Macrotermes subhyalinus* in west African grasslands. In a semi-arid grassland area of south-eastern Australia Watson & Gay (1970) recorded up to 350 nests ha^{-1}, covering up to 20% of the ground surface, built by *Drepanotermes perniger* and *D. rubriceps*. The nests have a cemented cap below the ground surface, which is exposed by erosion after the termites have died out in unfavourable (dry) years. The cemented caps resist water penetration and seed lodgement for at least 35 years and probably longer after they are exposed. The significance of termites in reducing water infiltration and plant growth in this way is inadequately known, and is one of many aspects of the role of termites in arid and semi-arid ecosystems that is in need of detailed investigation.

Denudation of the soil surface

Denudation results from trampling and track formation, and from over-grazing, and may be due to vertebrates or invertebrates.

Packer (1953) simulated trampling, using a steel 'hoof' and spacing impacts to give 10, 20, 40 and 60% disturbance, corresponding to light, moderate, moderately heavy, and heavy grazing on rangelands in Utah. Experiments were conducted on areas with three levels of ground cover, 70–75%, 80–85%, and 90–95%, dominated by perennial wheatgrass (*Agropyron inerme*) or annual cheatgrass brome (*Bromus tectorum*). Ground cover and bare soil openings were measured before and after treatment. All trampling treatments increased the amount of bare ground and increased the size of bare soil openings. There was a positive correlation between these increases and intensity of trampling up to 40% disturbance, but no appreciable difference between 40% and 60% disturbance. On plots with 90–95% ground cover none of the trampling treatments reduced ground cover below the limit (70%) previously determined as necessary to minimize erosion and run-off on undisturbed range. In most arid and semi-arid areas plant cover is less than 70%, so trampling must always increase denuded areas and so influence soil loss by erosion.

Insects are probably the most important agents in denudation of arid and semi-arid rangelands. In South Africa, harvester termites, mainly Hodotermitinae, but also Nasutitermitinae and Macrotermitinae, may almost entirely denude grasslands (Naudé, 1934; Coaton, 1951, 1954; Hartwig, 1955, 1966; Anonymous, 1960). Denudation by termites in South Africa has been found to follow reduction in plant cover as a result of over-grazing by stock, with a subsequent increase in termite populations to 'saturation level'. Before harvester termite control was instituted in some areas of Zululand, there was an annual decrease of 25% in carrying capacity (Anonymous, 1960). Saunders (1969) reported that in inland areas of Cape York Peninsula and north Queensland, termites could reduce the vegetation on land of low cattle-carrying capacity to a point where the land was almost valueless. Wallace (1970) noted denudation of grazing lands in south-west Queensland by *Drepanotermes* spp. Locusts and grasshoppers can also denude large areas, but their effects are sporadic, and though they may be disastrous in the short-term, the long-term and more continuous effects of termites may be more important and should be more carefully investigated.

Wight & Nichols (1966) studies vegetation around the denuded ant hills of western harvester ants (*Pogonomyrmex occidentalis*) and compared them with adjacent areas unaffected by ants in the Big Horn Basin, Wyoming,

Table 6.2. *Production and percentage foliage cover of* Atriplex nuttalli *saltbush in concentric sampling zones around denuded areas of mounds of western harvester ants* (Pogonomyrmex occidentalis). *All results are means of five samples. (Adapted from Wight & Nicholls, 1966)*

	Production (g m²)			Foliage cover (%)		
Ant mound no.	Zone 1 450–600 cm	Zone 2 600–750 cm	Zone 3 750–900 cm	Zone 1 450–600 cm	Zone 2 600–750 cm	Zone 3 750–900 cm
1	278	188	89	58	34	14
2	203	73	127	38	14	26
3	274	136	64	49	28	16
4	236	83	88	35	17	16
5	192	68	76	31	15	14
Means	237	110	89	42	21	16

where the mean annual rainfall is 197 mm and the dominant forage plant is *Atriplex nuttallii.* They measured plant production in three concentric zones defined by circles of approximately 450, 600, 750 and 900 cm radius centred on ant hills and compared it with production on areas of the same size, unaffected by ants. Total production was very little different (Table 6.2). They speculated that denudation of the ant hill serves to increase soil moisture in the area around the hills, due to lack of transpiring plants.

It seems more likely that in any circumstances where water is limiting for plant growth, any local feature that tends to concentrate the supply by shedding or collecting water, such as a bare ant hill or an ephemeral stream course, will increase production by increasing water supply to the plants where water is concentrated. Most harvesting termites do not build mounds, but live in subterranean nests, while other herbivorous insects that denude soils, such as grasshoppers and locusts, are migratory and live on the soil surface.

Burrowing

A wide variety of vertebrates and invertebrates make burrows or nests in soils, and these activities are responsible for the most important effects of animals on soils, namely changes in porosity, water-holding capacity, infiltration, and physical overturning of soil horizons.

Vertebrates

Many qualitative observations but few quantitative investigations have been reported. Vertebrates, except perhaps in small areas with great concentrations of burrows, probably have little effect on porosity or infiltration. Their main effects are in the physical disturbance and overturning of soil profiles.

Koford (1960) estimated that blacktail prairie dogs, at a maximum (spring and summer) density of 12.5 adults and 10 young ha^{-1} (2.5–3.5 g m^{-2}) bring to the surface about 1 kg of soil m^{-2} yr^{-1}. Borst (1968) calculated that Californian ground squirrels (*Citellus beecheyi*), in natural grasslands of southern California, apparently overturn and mix the surface and subsurface soil horizons to 75 cm depth in about 360 years, and he considered that their activities account for the presence of juvenile soils on old landscapes, as the time required to develop textural B horizons is more than 360 years. Thorp (1949) concluded that the surface soil over one third of an area of experimental plots at Akron, Colorado, had been converted from silt loam to loam as a result of the borrowing activities of rodents and badgers.

Invertebrates

Satchell (1958) calculated that earthworm burrows may account for up to 67% of the total air space in soil and concluded that they enhance aeration, drainage, and water absorption, and sometimes limit soil erosion by reducing surface run-off. However, earthworms are usually rare or absent in arid and semi-arid environments. Madge (1969) studied the effects of two species of earthworms that cast on the soil surface, *Hyperiodrilus africanus* and *Eudrilus eugeniae*, in seasonally dry grasslands in southern Nigeria. *Hyperiodrilus africanus* is most common in shaded areas and *E. eugeniae* in open grasslands. The weight of casts is about 17.6 g m^{-2} yr^{-1}, and assuming a bulk density of about 1.5 this represents annual accumulation of a layer 1.2 mm deep. Physical analyses of casts and soils led Madge to conclude that the cast soil originated from deep in the profile. His analyses of soil clearly illustrate horizon differentation, but at such a high rate of turnover and surface casting there should be a uniform profile to the limit of the earthworms' activities.

Insects, especially ants and termites, are the dominant animals of soils in arid and semi-arid regions. Bucher & Zuccardi (1967) studied the effects on soils of the mound-building ant *Atta vollenweideri* in a semi-arid region of Tucuman Province, Argentina. The mounds averaged 5.5 m in diameter, 0.4 m high, and their mean density was 2.9 ha^{-1}. Mean volume of soil in the mounds was 23 m^3 ha^{-1}, and mean weight 3×10^4 kg ha^{-1}. Chemical and physical analyses of soil from mounds indicated that the soil was derived from a $CaCO_3$-rich horizon at a depth > 150 cm. Bucher & Zuccardi calculated that *A. vollenweideri* brought to the surface about 0.85 m^3 ha^{-1} yr^{-1} or 1100 kg ha^{-1} yr^{-1} of soil. This corresponds to a layer 0.085 mm deep over the whole soil surface, and since the transported soil is derived from deep in the profile it implies inversion of soil profiles at a rate at least equivalent to that recorded for termites (see below).

Termites affect soils and pedogenetic processes in two principal ways.

(1) Physical effects. In constructing mounds, nests, or gallery systems, soil particles are selected, transported, cemented together, and mixed with organic matter.
(2) Chemical and biochemical effects. Organic debris or living plant tissue is collected, often over extensive foraging areas, transported to mounds, nests, or subterranean depots, and subjected to intense degradation when digested by termites.

Physical effects

Particle-size analyses of termite mounds and other structures compared with the soil horizons from which materials are derived show a general trend of selection in favour of finer size fractions, but considerable variation between species and within most species (Lee & Wood, 1971*a*, *b*). Some examples for Australian termites are given in Table 6.2. A few species (e.g. African *Apicotermes* spp.) do display precise and consistent selection of particle size fractions (Stumper, 1923). The soil selected and brought to the surface is usually from deep soil horizons (Lee & Wood, 1971*a*, *b*), resulting in a slow but continuous physical overturning of soil profiles.

Nye (1955) in West Africa, Williams (1968) and Lee & Wood (1971*b*) in northern Australia, have estimated rates of accumulation at the surface of soil derived from deep horizons by termites in seasonally arid tropical grasslands. Their estimates are: West Africa (Nye), 2.5 mm 100 yr^{-1}; Northern Australia (Williams), 1.25 mm 100 yr^{-1}; Northern Australia (Lee & Wood), 10.0 mm 100 yr^{-1}; Northern Australia (Lee & Wood), 2.0 mm 100 yr^{-1}. This amount of soil is less, probably by a least one order of magnitude, than that moved to the surface by large populations of lumbricid earthworms in temperate regions. However, the common 'casting worms' have their burrows mainly in the A horizon and do not move large amounts of soil from deep horizons to the surface.

'Humivorous' termites, which ingest soil from upper horizons and digest organic matter that it contains may fairly closely resemble earthworms in their effects on soils. Hébrant (1970) has calculated that humivorous *Cubitermes* spp. in Congo savannas may ingest and excrete more than 10 kg of soil m^{-2} yr^{-1}, which is comparable with Madge's figures for earthworms in Nigeria (see above).

The soil incorporated into mounds and other termite constructions is resorted, packed, and often stuck together with organic materials (usually excreta). Figs. 6.4 and 6.5 illustrate two types of packing commonly found in termite-modified soil. The soil particles in Fig. 6.4 are very tightly packed, forming a massive, hard structure, with very few voids, and containing little organic matter (1.6% organic C). The structure in Fig. 6.5 is also tightly

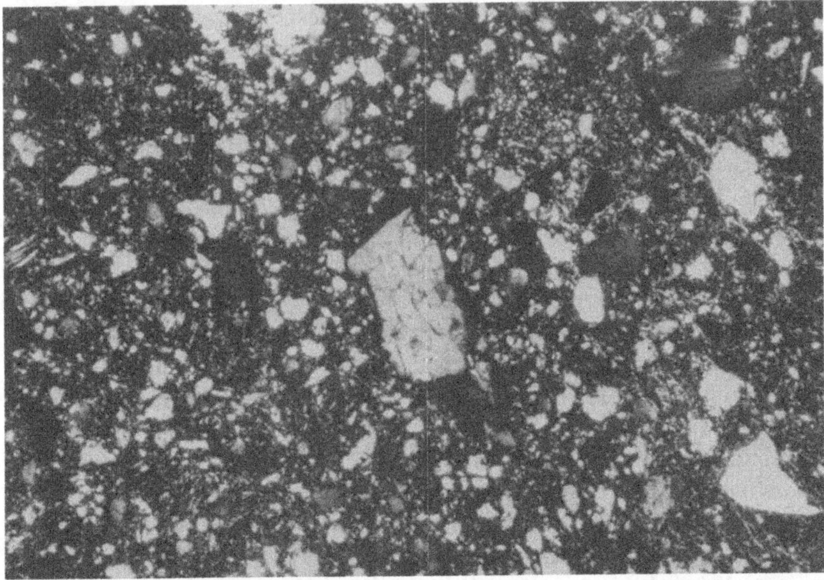

Fig. 6.4. Thin section of gallery wall in a mound of *Drepanotermes rubriceps*, showing closely packed mainly mineral particles with virtually no voids. Viewed under crossed Nicol prisms; width of frame approximately 2.15 mm. (From Lee & Wood, 1971*a*).

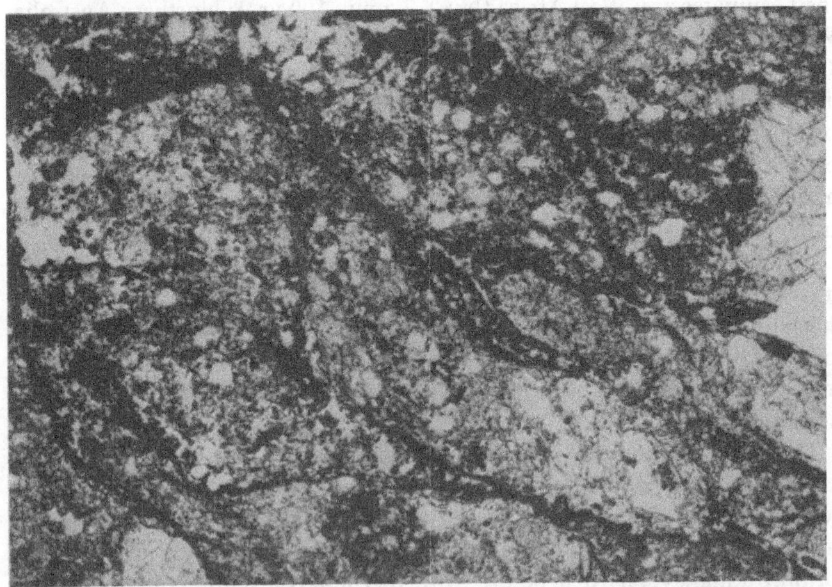

Fig. 6.5. Thin section of outer wall of mound of *Amitermes laurensis*, showing irregular lens-like pellets of soil, interlayered and cemented with layers of excreted organic matter. Viewed under crossed Nicol prisms; width of frame approximately 2.15 mm. (From Lee & Wood, 1971*a*).

packed, rather less massive and hard, and consists of irregular lens-like pellets of soil, interlayered and cemented with layers of excreted organic matter (organic C content of sample 3.0%)

Chemical and biochemical effects

Changes in chemical and biochemical characteristics of soil due to termites are discussed in detail by Lee & Wood (1971*a*, *b*). Termite-modified soils generally display increases in organic carbon, nitrogen, exchangeable calcium, magnesium, and potassium, and total exchangeable cations relative to unmodified soils (Table 6.3). The increases are apparently due to addition by the termites of excreted organic matter derived from digestion of plant tissue to soil used for construction, and there is a fairly close correlation between organic matter content and exchangeable cation concentrations.

The incorporation of plant nutrients in termite mounds may significantly influence the cycling of nutrients in grassland ecosystems. As long as mounds are occupied they are repaired as soil is eroded from them. Eventually mounds are abandoned and erosion returns the soil they contain to the A horizon of the soil. Assuming that there is an equilibrium between new mound construction and old mound erosion, the quantity of plant nutrients in mounds will be reasonably constant. The mounds of Australian termites do not support plant growth, so nutrients in mounds are withheld from circulation in the plant/soil system. In two seasonally arid northern Australian grassland ecosystems Lee & Wood (1971*b*) calculated the amount of organic carbon, nitrogen, phosphorus and exchangeable plant nutrients withheld in mounds as a percentage of the total in the mounds/A_1 horizon system. The results (Table 6.4) indicate that a large proportion of the total nutrients in an ecosystem may be withheld in mounds, but more work is necessary to find out whether this is a general phenomenon and to assess its significance.

The digestive processes of termites, which involve symbiotic intestinal protozoa and/or bacteria, result in utilization of > 90% of total carbohydrates in the plant material ingested. There is evidence that at least some termites can digest lignin, and > 80% utilization has been reported. In Macrotermitinae, which have very complex symbiotic relationships involving intestinal symbionts and basidiomycetes that grow on structures made from faeces, followed by re-ingestion of faeces and fungi, almost all the available energy in food appears to be utilized. The excreted end products of termite digestion, principally lignin, carbohydrates, and alkali-soluble polyphenolic materials similar to the humic and fulvic acids that can be extracted from soils, are used for construction, either alone to form 'carton' or incorporated with soil. The plant nutrients referred to above are also included in the excreted end products.

Table 6.3. *Particle size and some chemical analyses of soil samples and termite-modified soil materials from the same sites in northern Australian grasslands*

In each case analyses of unmodified soils are followed by those of termite-modified soil. All results on basis of oven-dry weight. Soil horizons identified as the origin of the termite-modified material on basis of physical and chemical analyses, and clay mineralogy. (Data from Lee & Wood, 1971*b*)

	Particle–size analyses				Chemical Analysis						
									Exchangeable cations (milliequivalents %)		
Soil horizon (depth in cm) and termite species	Coarse sand (%)	Fine sand (%)	Silt (%)	Clay (%)	Organic matter (%)	Organic C (%)	N (%)	Ratio C:N	Ca	Mg	K
B (40–50)	20	45	14	18	2.7	0.2	0.020	10.0	0.1	0.8	0.06
Nasutitermes triodiae											
(1) Outer part of mound	18	41	12	25	7.3	2.7	0.100	27.0	0.2	3.2	1.0
(2) Nursery of mound	9	37	19	29	21.0	10.0	0.440	22.7	7.5	6.9	1.6
B (20–30)	17	63	2	16	3.2	0.2	0.022	9.1	2.1	0.8	0.30
Drepanotermes rubriceps											
Mound	14	59	3	20	5.2	1.4	0.085	16.5	5.7	1.8	0.50
A (0–8)	12	56	23	7	2.7	0.8	0.055	14.5	0.1	0.7	0.05
Tumulitermes pastinator											
Outer part of mound	4	45	27	19	8.2	3.7	0.17	21.8	4.2	5.3	0.29

Table 6.4. *Soil, organic carbon, nitrogen, and exchangeable cations in termite mounds as a percentage of total quantities in the mound/A_1 horizon system in two northern Australian grassland ecosystems*

Mound-building termites sampled	Depth of A_1 horizon (cm)	Soil in mounds (%)	Quantities in mounds as % of total mound/A_1 soil						
			Organic C	N	P	Exch. Ca	Exch. Mg	Exch. K	Exch. Na
Nasutitermes triodiae	8	2.0	9.6	5.3	5.0	9.1	22.0	13.1	2.7
Amitermes vitiosus } *Tumulitermes hastilis* }	10	3.1	6.3	4.6	3.0	5.2	5.2	6.6	3.5

Conclusions

The natural biota of arid ecosystems exist in a fragile and constantly varying relationship with harsh and unpredictable abiotic environmental pressures that for much of the time allow little latitude for survival. Despite the fragility of the relationship the biota have a critical role in modifying the effects of abiotic processes that most influence their own existence. Shading by plants, imbibition of water from the air, protection of the soil surface from wind and water erosion, accumulation and cycling of nutrients and soluble salts, organic matter decomposition, and physical overturning of soil profiles by burrowing animals all contribute to the development and maintenance of biota and soil fertility. Over-grazing by vertebrates and invertebrates, tramping, cementing of surface layers, and consequent erosion and prevention of seed lodgement, contribute to fertility losses.

There is a need for much more research to elucidate more clearly the role of the biota in arid-land ecosystems. Present knowledge is confined to a limited range of plants and animals, and it is clear that management practices are critical in limiting damage by introduced grazing mammals to ecosystems that are susceptible to far-reaching changes when subjected to extraneous pressure.

References

Anonymous (1960). Destruction of grazing by harvester termites. *Farming in South Africa*, **35**, 6–9.

Booth, W. E. (1941). Algae as pioneers in plant succession and their importance in erosion control. *Ecology*, **22**, 38–46.

Borst, G. (1968). The occurrence of crotovinas in some southern Californian soils. *Transactions of the 9th international congress of soil science, Adelaide*, vol. 2, pp. 19–27. Angus & Robertson, Sydney.

Boyer, P. (1958). Influence des remaniements par le termite et de l'érosion sur l'évolution pédogenétique de la termitière épigée de *Bellicositermes rex*. *Comptes Rendus Hebdomadaires des Séances de l'Académie des Sciences*, **247**,749–51.

Bucher, E. H. & Zuccardi, R. B. (1967). Signification de los hormigueros de *Atta vollenweideri* Forel como alteradores del suelo en la Provinces de Tucuman. *Acta Zoologica Lilloana*, **23**, 83–96.

Charley, J. L. & Cowling, S. W. (1968). Changes in soil nutrient status resulting from overgrazing and their consequences in plant communities of semi-arid areas. *Proceedings of Ecological Society of Australia*, **3**, 28–38.

Coaton, W. G. H. (1951). The snouted harvester termite; natural mortality as an aid to chemical control. *Farming in South Africa*, **26**, 263–7.

Coaton, W. G. H. (1954). Veld reclamation and harvester termite control. *Farming in South Africa*, **29**, 243–8.

Craddock, G. W. & Pearse, C. K. (1938). Surface runoff and erosion on granite mountain soils of Idaho as influenced by range cover, soil disturbance, slope, and precipitation intensity. *United States Department of Agriculture Circular, No. 482*. US Government printing office, Washington, DC.

Ellison, L. (1960). Influence of grazing on plant succession of rangelands. *Botanical Review*, **26**, 1–78.

Fireman, M. & Hayward, H. E. (1952). Indicator significance of some shrubs in the Escalante Desert, Utah. *Botanical Gazette*, **114**, 143–55.

Fletcher, J. E. & Martin, W. P. (1948). Some effects of algae and molds in the rain-crust of desert soils. *Ecology*, **29**, 95–100.

Gradwell, M. W. (1968). Compaction of pasture topsoils under winter grazing. *Transactions of the 9th international congress of soil science, Adelaide*, vol. 3, pp. 429–35. Angus & Robertson, Sydney.

Harris, J. A., Gortner, R. A., Hoffman, W. F. & Valentine, A. T. (1921). Maximum values of osmotic concentration in plant tissue fluids. *Proceedings of the Society for Experimental Biology and Medicine*, **18**, 106–9.

Hartwig, E. K. (1955). Control of snouted harvester termites. *Farming in South Africa*, **30**, 361–6.

Hartwig, E. K. (1966). The nest and control of *Odontotermes latericius* (Haviland) (Termitidae: Isoptera). *South African Journal of Agricultural Science*, **9**, 407–18.

Hébrant, F. (1970). Etude de flux energetique chez deux éspèces du genre *Cubitermes* Wasmann (Isoptera, Termitinae), termites humivores des savanes tropicales de la région éthiopienne. Doctorate of Science Memoir, Université Catholique de Louvain, Laboratoire d'écologie animale.

Hedrick, D. W. (1948). The mulch layer of California annual ranges. *Journal of Range Management*, **1**, 22–5.

Jessup, R. W. (1969). Soil salinity in saltbush country of north-eastern South Australia. *Transactions of Royal Society of South Australia*, **93**, 69–78.

Kelting, R. W. (1954). Effects of moderate grazing on the composition of a native tall-grass prairie in central Oklahoma. *Ecology*, **35**, 200–7.

Koford, C. B. (1960). The prairie dog of the North American plains, and its relations with plants, soil, and land use. In: *Ecology and management of wild grazing in temperature zones*, pp. 327–41. International Union for the Conservation of Nature and Natural Resources, Morges, Switzerland.

Lee, K. E. & Wood, T. G. (1971*a*). *Termites and soils*. Academic Press, London & New York.

Lee, K. E. & Wood, T. G. (1971*b*). Physical and chemical effects on soils of some Australian termites, and their pedological significance. *Pedobiologia*, **11**, 376–409.

Lodge, R. W. (1954). Effects of grazing on the soils and forage of mixed prairie in southwestern Saskatchewan. *Journal of Range Management*, **7**, 166–70.
Madge, D. S. (1969). Field and laboratory studies on the activities of two species of tropical earthworms. *Pedobiologia*, **9**, 188–214.
Marshall, J. K. (1971). Drag measurements in roughness arrays of varying density and distribution. *Agricultural Meteorology*, **8**, 269–92.
Naudé, T. J. (1934). Termites in relation to veld destruction and erosion. *South African Department of Agriculture Plant Industry Series 2*. Bull. 134, pp. 1–20. Government Printer, Pretoria.
Nye, P. H. (1955). Some soil-forming processes in the humid tropics. IV. The action of the soil fauna. *Journal of Soil Science*, **6**, 73–83.
Packer, P. E. (1953). Effects of trampling disturbance on watershed condition runoff and erosion. *Journal of Forestry*, **51**, 28–31.
Rhoades, E. D., Locke, L. F., Taylor, H. M. & McIlvain, E. H. (1964). Water intake on sandy range as affected by 20 years of differential stocking rates. *Journal of Range Management*, **17**, 185–90.
Roberts, R. C. (1950). Chemical effects of salt-tolerant shrubs on soils. *Proceedings of 4th international congress of soil science, Amsterdam*, vol. 1, pp. 404–6. Hoitsema, Groningen.
Rogers, R. W. & Lange, R. T. (1971). Lichen populations on arid soil crusts around sheep watering places in South Australia. *Oikos*, **22**, 93–100.
Rogers, R. W., Lange, R. T. & Nicholas, D. J. D. (1966). Nitrogen fixation by lichens of arid soil crusts. *Nature, London*, **209**, 96–7.
Satchell, J. E. (1958). Earthworm biology and soil fertility. *Soils and Fertilizers*, **21**, 209–19.
Saunders, G. W. (1969). Termites on northern beef properties. *Queensland Agricultural Journal*, **95**, 31–6.
Stumper, R. (1923). Sur la composition chimique des nids de l'*Apicotermes occultus* Silv. *Comptes Rendus Hebdomadaires des Séances de l'Académie des Sciences*, **177**, 409–11.
Thorp, J. (1949). Effects of certain animals that live in soils. *Scientific Monthly*, **68**, 180–91.
Wallace, M. M. H. (1970). Insects of grasslands. In: *Australian grasslands* (ed. R. M. Moore), pp. 361–70. Australian National University Press, Canberra.
Watson, J. A. L & Gay, F. J. (1970). The role of grass-eating termites in the degradation of a mulga ecosystem. *Search*, **1**, 43.
Wight, J. R. & Nichols, J. T. (1966). Effects of harvester ants on production of a saltbush community. *Journal of Range Management*, **19**, 68–71.
Williams, M. A. J. (1968). Termites and soil development near Brock's Creek, Northern Territory. *Australian Journal of Science*, **31**, 153–4.
Wind, G. P. & Schothorst, C. J. (1964). The influence of soil properties on suitability for grazing and of grazing on soil properties. *Transactions of the 8th international congress on soil science, Bucharest*, vol. 3, pp. 571–80. Academy Romanian Socialist republic, Bucharest.
Wood, J. G. (1925). The selective absorption of chlorine ions; and the absorption of water by the leaves in the genus *Atriplex*. *Australian Journal of Experimental Biology and Medical Science*, **2**, 45–56.

Manuscript received by the editors October 1973

7. Population dynamics

F. H. WAGNER

Introduction

As Solomon (1971) and I (Wagner, 1969) have suggested, the major, general questions of interest in population ecology are concerned with the numbers of organisms, and are: (1) the demographic mechanisms and environmental interactions producing variation in numbers; (2) the demography and environmental interactions producing population equilibrium or the absence of long-term, mean trend in numbers (i.e. regulation); and (3) those same two aspects in determining the long-term, mean numbers at which equilibrium is maintained. The subject expands into the field commonly termed population biology when these questions are also viewed in an evolutionary context, with emphasis on the genetic changes and adaptive patterns wrought by the environment over somewhat longer time scales.

For this review, I wish to address these questions about desert species. What are the patterns of fluctuation, mechanisms of equilibrium, constraints limiting density, and adaptive demographic traits in desert species in general? And how are these characteristics similar to, or different from, their counterparts in other ecosystems?

Certain aspects of the topic, particularly the demography, have been reviewed recently in some detail (Mayhew, 1968; French, Stoddart & Bobeck, 1975; Prakash & Ghosh, 1975; Turner, 1977). This review singles out certain aspects of the overall topic which seem to be of particular interest and have not been treated extensively in recent reviews, and for which there are sufficient data to permit at least preliminary hypotheses.

Short-term fluctuation

Magnitudes of variation

It is by now well-recognized that deserts are among the most variable ecosystems owing to the high temporal and spatial variation in rainfall, and to the close coupling of biotic variation to these vagaries of the weather. A number of authors have separately explored the magnitude of variation in primary production that is associated with rainfall variation (Hutchings & Stewart, 1953; Blaisdell, 1958, Paulsen & Ares, 1962; Cable & Martin, 1975; Le Houérou & Hoste, 1977; Novikoff, 1977; Wagner, 1978*a*), and that of a variety of consumers. But no one has, to my knowledge, compared the variability between trophic levels in desert systems, and within trophic levels

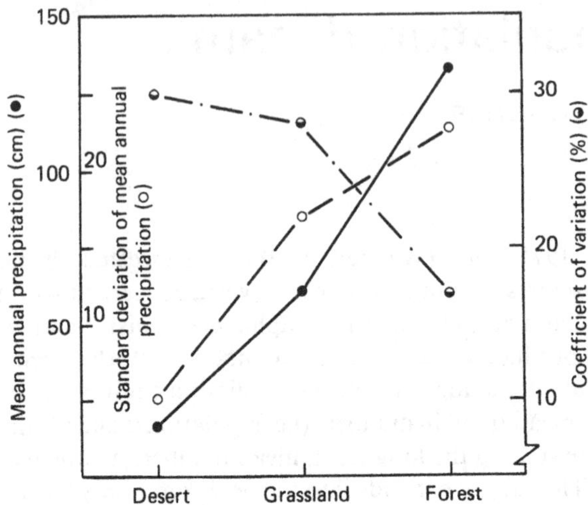

Fig. 7.1. Relationships between mean, annual precipitation, and measures of absolute (standard deviations of the means) and relative (coefficients of variation) variation in the means at the US Forest Services Desert Experimental Range, Utah (Hutchings & Stewart, 1953), a shortgrass prairie near Hays, Kansas (Hulett & Tomanek, 1969), and a tulip poplar (*Liriodendron*) forest near Oak Ridge, Tennesssee (Robert L. Burgess, personal communication). (After Wagner, 1978*a*.)

for several systems. It would seem of interest to view the ecosystem as a whole, determining whether the magnitude of variation in rainfall is transmitted uniformly throughout the system, or whether it is either damped or amplified as it ascends the trophic levels. And it seems desirable to compare the variability that does exist at each level with that of the other ecosystems.

Magnitude of variation in precipitation

A number of authors have explored the relationships between the magnitude of annual precipitation and its variation. Absolute variation, as measured by the standard deviation of the mean annual rainfall, rises as the mean annual rises. But *relative* variation, as measured by the coefficient of variation about the mean annual, varies inversely with the mean. This has been discussed by Conrad (1941) for the world as a whole, by Wagner (1978*a*) for ecosystems with varying degrees of moisture in North America (Fig. 7.1), and for the province of British Columbia, Canada (Longley, 1952). Within arid regions, it has been reported for the state of Arizona, USA (Hastings & Turner, 1965) and for Tunisia (Le Houérou, 1976).

Hence, precipitation varies through a wider absolute range of values in mesic regions, but those values are relatively more clustered around their means than in arid regions. Precipitation is less predictable in the latter, and becomes less so the greater the aridity. In Tunisia, the coefficient of variation ranges from 30% in the most mesic regions to 80% in the arid extremes of the Sahara (Le Houérou, 1976).

Magnitude of variation in primary production

A number of time series are available for net, annual primary production in arid and more mesic regions. The longest series are those of government experiment stations, but shorter series are available from other research projects (Table 7.1).

Several general statements seem warranted from these statistics. Variability of perennial primary production increases as moisture levels decline and the variability of those levels increases. The values for annuals are too spotty to determine whether the same trend is true for them. But the variability of annuals is clearly greater than that of perennials, a fact long recognized by desert ecologists. Since annuals become an increasingly high fraction of the total primary production with progressively more extreme aridity, their variability will increasingly weight the variability of the total production. Hence, both because the perennials become more variable, and because the highly variable annuals become an increasing fraction of the total production, variability in primary production appears to increase with progressively greater aridity. Comparison between the variability of precipitation and its respective primary production (Table 7.1) is somewhat equivocal. Although perennial production in Rock Valley is markedly more variable than the rainfall, the two coefficients are sufficiently close in the other three areas to temper any generalization that perennial production in deserts might be more variable than the precipitation. The two values for annuals *are* considerably higher than the ones for their respective precipitation, and it seems quite possible that this is a general pattern for this group of plants. Consequently, in those deserts where annuals provide a material fraction of the total primary production, variability of the latter may be greater than that of the precipitation.

Magnitudes of variation in animal populations

Assessing the variability of desert animal populations is beset with the problem of high mobility in some desert species. The extreme case is the desert locust (*Schistocerca gregaria*) a population of which may occupy an area for a single generation, then disperse as far as another continent, and its succeeding generations never return. The species has, as Naumov (1972, p. 391) puts it, 'no permanent nesting place'. There are, of course, similar patterns in other species of locusts in arid areas.

Among birds, George (1976) describes an area in the Moroccan Sahara

Table 7.1. *Coefficients of variation for total annual precipitation and annual above-ground, net primary production in several desert and more mesic sites*

Locale	Mean annual (mm)	Coefficient of Variation (%) (years)			Source
		Precipn	Perenn.	Annuals	
Rock Valley, Mohave Desert, USA	138	45(14)	60(6)	175(6)	F. B. Turner (personal communication)
Southern Tunisia, Pre-Saharan Zone	131[a]	39(28)	39(5)[b]	100(5)	Novikoff (1977)
Pine Valley, Great Basin Desert, USA	171	30(13)	36(13)	—[c]	Hutchings & Stewart (1953)
Dubois, Idaho, Great Basin, USA	281	20(23)	21(13)	—[c]	Blaisdell (1958)
Range for deserts	104–281	20–39	21–74	100–170	—
Short-grass plains, Kansas, USA	606	28(24)	41(24)	130(24)	Hulett & Tomanek (1969)
Deciduous forest, Tennessee, USA	1386	17(10)	12(10)	—	Robert Burgess (personal communication)

[a] For village of Medenine.
[b] For dominant shrub, *Rhanterium suaveolens*, only.
[c] Annuals are not a significant part of the mature, Great Basin vegetation.

which had a nesting population of 400 spotted sandgrouse (*Pterocles senegallus*) in one year, and only 80 the next. He also observed 300–400 desert warblers (*Sylvia nana*) in the wetter year, and only 8 in the following dry year. Miller & Stebbins (1964) report the presence of LeConte thrashers (*Toxostoma lecontei*) in Joshua Tree National Monument (American Mohave Desert) in average or above-average precipitation years when insect food is adequate, but its absence from the Monument in dry years. Daumas (1971) recounts the extreme nomadism of the ostrich (*Struthio camelus*) along the northern fringe of the Algerian Sahara. Australian investigators in particular have emphasized the degree of nomadism among their avifauna (Keast & Marshall, 1954; Rowley, Braithwaite & Chapman, 1973). Keast (1959) reported that 26–30% of the Australian avifauna can be classed as true nomads.

A number of authors have reported nomadism among the larger mammalian species: Daumas (1971) for the 'el rinne' gazelle (probably *Gazella leptoceros*) in the Sahara, Newsome (1965*a*) for the red kangaroo (*Megaleia rufa*) in central Australia. On a smaller scale, Whitford (1976) described an influx of more mesic rodent species into a bajada in the North American Chihuahuan Desert during a year of high rainfall.

These cases help to give the impression of high variability among desert animal species. But they represent a small fraction of the thousands of species in the world's deserts, and there is some reason to believe that the emphasis on them may have developed a stereotype of greater faunal variability than exists. Serventy (1971) has already suggested that the image of nomadism among desert birds has been strongly colored by its prevalence in Australia, and that in fact it is much more restricted in the proportion of species displaying it and in the extent of movement in other parts of the world.

In more sedentary species, we can examine the variation in the same way we examined it in precipitation and vegetation, and compare it with them. These results (Table 7.2) suggest variability, at least in a number of vertebrate species, which is comparable with that of perennials and less extreme than that of annuals and the vegetation as a whole. Much of the desert fauna may, therefore, be less variable than the common stereotype would lead one to believe. There is additional evidence from shorter periods of studies and other sources.

Among the mammals, French *et al.* (1975) divided the world's rodent species into three groups according to their demographic characteristics: reproductive and survival rates, densities typically attained, and variability. The desert-inhabiting Heteromyidae and Dipodidae, plus the genus *Neotoma* in the Cricetinae, are demographically the most conservative on the basis of their classification. Intermediate in their classification are the desert-dwelling Cricetinae and Gerbillinae and all are considerably more conservative than the temperate-latitude microtines and murines.

For arid, northern Sudan, where *Jaculus jaculus* and *Gerbillus pyramidum*

Table 7.2. *Coefficients of variation for annual size of various animal populations*

Species	Locale	Coefficient of variation (%)	Source
Mammals			
Pocket mouse (*Perognathus parvus*)	Great Basin Desert, USA	29	O'Farrell *et al.* (1975)
Pocket mouse (*P. formosus*)	Mohave Desert, USA	54	French *et al.* (1974)
Pocket mouse (*P. longimembris*)	Mohave Desert, USA	53	French *et al.* (1974)
Kangaroo rat (*Dipodomys microps*)	Mohave Desert, USA	61	French *et al.* (1974)
Kangaroo rat (*D. merriami*)	Mohave Desert, USA	124	French *et al.* (1974)
Jackrabbit (*Lepus californicus*)	Great Basin Desert, USA	92	Stoddart (1977)
Coyote (*Canis latrans*)	Great Basin Desert, USA	48	Stoddart (1977)
Birds			
Cactus wren (*Campylorhincus brunneicappilus*)	Sonoran Desert, USA	33	Anderson & Anderson (1973)
Gambel quail (*Lophortyx gambelii*)	Sonoran Desert, USA	80	Swank & Gallizioli (1954)
Scaled quail (*Calipepla squamata*)	Desert grassland, USA	35	Brown *et al.* (1978)
Scaled quail (*Calipepla squamata*)	Desert grassland, USA	73	Campbell *et al.* (1973)
Lizards			
Whiptail lizard (*Cnemidophorus tigris*)	Mohave Desert, USA	24	Turner (1977)
Leopard lizard (*Crotaphytus wislizenii*)	Mohave Desert, USA	17	Turner (1977)
Side-blotched lizard (*Uta stansburiana*)	Mohave Desert, USA	42	Turner (1977)
Side-blotched lizard (*Uta stansburiana*)	Short-grass plains, USA	29	Turner (1977)
Insects			
Tenebrionid beetle (*Philolithus densicollis*)	Great Basin Desert, USA	129	Hinds & Rickard (1973)
Plague locust (*Chortoicetes terminifera*)	W. New South Wales, Australia	124	Clark (1974)

are dominant species, Happold (1975) writes 'There is no evidence for cyclical periods of great abundance, even though a very wet year, like 1964, could have resulted in high population numbers.' Maza, French & Aschwanden (1973) consider heteromyids to be 'among those rodents that have relatively low replacement rates and greater population stability'.

The intent here is in no way to deny the many reports to be discussed below that desert rodent populations fluctuate, generally with variations in rainfall; but that those variations are less extreme than the fluctuations of temperate murines and microtines. This seems surely to be true of the heteromyids and dipodids, and to a lesser degree the desert-inhabiting cricetines and gerbillines. However, Naumov (1975) reports drastic fluctuations in *Meriones unguiculatus* in Mongolia, and marked variations in *M. libycus* in central Asia and the Middle East. As exceptions that may prove the rule, the most persistent reports of 'plagues' in desert rodent populations come from Australia. The species of the low-diversity, Australian rodent fauna which are involved in these plagues are in the genera *Rattus*, *Notomys*, and *Pseudomys*, along with the exotic *Mus musculus* (Newsome & Corbett, 1975). All are murines.

Wagner & Stoddart (1972) and Cross, Stoddart & Wagner (1974), while reporting the black-tailed jackrabbit (*Lepus californicus*) in parts of the North American Great Basin Desert to be markedly cyclic, still found the species less volatile than the snowshoe hare (*L. americanus*) has been reported to be in a number of northern US and Canadian studies. In the North American 'hot' deserts, the species appears considerably more stable. Robert M. Chew (Personal Communication) found little variation in their numbers on the US/IBP Desert Biome site in Rock Valley, Nevada, a Mohave Desert area. In that same desert, Miller & Stebbins (1964) similarly observed little change in jackrabbit numbers in Joshua Tree National Monument over a 15-year period of censuses.

Clark (1972) and Stoddart (1977) showed the coyote (*Canis latrans*) also to be cyclic in their Great Basin study area. But the range of variation is no more marked than that reported for the species in the southern Canadian provinces by Keith (1963).

Birds are difficult to generalize, largely because of the shortage of long-term studies. The Gambel quail (*Lophortyx gambelii*) traditionally has been considered a highly variable, 'annual' species. But even its numbers do not appear as variable as the annual vegetation on which it depends (Table 7.2), or markedly greater than desert rodents. The cactus wren (*Campylorhynchus brunneicapillus*), the one passerine species for which I was able to find a reasonably long series of censuses (Table 7.2), does not appear to vary markedly.

Data from two shorter-term studies are available. In three years of censuses on a study area in Big Bend National Park in Texas (Chihuahuan Desert), Dixon (1959) counted 51, 30, and 40 territorial males ha^{-1} divided among

11 species. In three successive years, Raitt & Pimm (1976) estimated 42.6, 130.2, and 60.4 breeding birds per 100 km^{-2} during May on a Chihuahuan Desert bajada study area in New Mexico.

Much has been written about rainfall-related variations in reproductive rates (Mayhew, 1966; Turner, Medica & Smith, 1974; Ballinger, 1977) and population size (Milstead, 1965; Turner, 1977; Whitford & Creusere, 1977) in desert lizards. Yet these conspicuous components of the desert fauna are among the least variable (Table 7.2).

A general sense of the variability of desert invertebrate populations is the most difficult to develop because the species are so numerous and varied. The Acrididae are one of the most studied groups of desert arthropods, and the highly mercurial character of locust populations has probably colored the general impression of desert arthropod variability. The locusts themselves are difficult to generalize in any quantitative way because of the extreme spacial and temporal variability. A great deal has been written about locust ecology, but a brief résumé here will clarify the problems of characterizing any discrete populations.

Since Uvarov's (1921) pioneering hypothesis on phase transformation in locusts, they have come to be distinguished from the grasshoppers by their two-phase polymorphism: a sedentary, solitary phase, and a migratory, gregarious phase. The grasshoppers generally only have a solitary phase although there appears to be some degree of continuous variation between the two groups rather than there being two totally discrete entities.

The solitary phase tends to persist during periods of unfavorable (i.e. dry) conditions, in some cases surviving in refugia of favorable habitat (Uvarov, 1954; Kennedy, 1956). During such dry periods, egg maturation rates are low because soil moisture is low, and nymph survival rates are low because plant food is inadequate.

Given above average rainfall and increased plant production, populations increase. There is some disagreement over the time required for plague numbers to form, Kennedy (1956) claiming that several seasons are required while Gupta & Prakash (1975, p. 386) maintain that they form in a single generation.

However long it takes, swarming is initiated by the return of dry conditions (Kennedy, 1956). As vegetation dies and habitat constricts, the now-abundant solitary forms are forced to crowd into the remaining vestiges of habitat. In doing so, contacts between individuals are transmitted through the nervous system, and the physiology is shifted to form the high social, very active and mobile, gregarious phase. The excitement grows until the animals take wing in a swarm and emigrate with the aid of prevailing winds.

The problem of quantification lies in determining where and when a population should be measured. Should it be during the solitary phase before concentration; during the build-up phase, but before habitat shrinkage and, if during, at what point? Or should it be measured within the swarm at the down-wind point which has not previously had locusts? This problem is reflected in the bewildering range of densities cited for the species in the literature (Table 7.3). Clearly, the traditional, mesic-zone practice of periodically censusing a population in a discrete area, and describing the pattern and magnitude of fluctuation is not applicable to these species.

Clark (1974) did provide eight years of census data for *Chortoicetes terminifera* in an area of eastern Australia where, in his opinion, emigration and immigration were not a confounding variable. This population fluctuated through about two orders of magnitude, and I calculated its coefficient of variation at 124%, not exceptionally variable by comparison with the other populations shown (Table 7.2).

The Lepidoptera may be another highly variable group in deserts. Miller & Stebbins (1964) described high densities of sphinx moth (Sphingidae) caterpillars in a year of abundant rain. Yet this order of insects appears characteristically to be highly variable in other parts of the world, a number of species having been cited in the ecological literature as classical examples of the variability of insect populations. Thus, Williamson (1972, p. 12) demonstrated that the four species of moths in German coniferous forests reported on by Schwerdtfeger (1941), and the marshland-inhabiting scarlet tiger moth (*Panaxia dominula*) Oxford in England, fluctuate through 2–4 orders of magnitude. Similarly, Baltensweiler (1968) has shown the grey larch tortrix (*Zeiraphera griseana*) to oscillate through four orders of magnitude in Swiss coniferous woodlands. The extreme fluctuations of the spruce budworm (*Choristoneura fumiferana*) in Canadian coniferous forests, although over a longer time scale, are well known (Morris, 1963).

Although beetle populations appear to fluctuate less markedly than Lepidoptera, Thomas (1975) placed considerable emphasis on the variability of tenebrionid populations in his Rock Valley study area (Mohave Desert), and concluded that they are *r*-selected forms (*sensu* MacArthur & Wilson, 1967) which have evolved to respond rapidly to a capricious environment. But the one long-term (9-year) study of desert tenebrionids shows *Philolithus densicollis* populations to be only slightly more variable than some desert mammal and bird populations (Table 7.2), the highest annual density exceeding the lowest by a factor of 31, or 1.3 orders of magnitude. Thomas's own census data on 14 species over three and four years show high : low ratios in only three species as high as 10–31 : 1, and in eight species they are less than 5 : 1. Here again, these species appear less variable than the mountain pine beetle (*Dendroctomus ponderosae*) and spruce beetle (*D. refipennis*) of Montana coniferous forests in western North America (Roe & Amman, 1970; Schmid & Frye, 1977).

Mispagel (1974) censused three species of Membracidae (Homoptera) on the Desert Biome's Rock Valley research site, the same as Thomas's, over

Table 7.3. *Locust densities reported from different areas, life-history stages, and phase of population build-up*

Species	Locale	Life-history stage	Outbreak phase	No. per 1000 m²	Source
Schistocerca gregaria	W. India	Adults	Solitary	0.3–0.7	Roonwal (1953)
Chortoicetes terminifera	E. Australia	Adults	Solitary	5	Clark (1974)
S. gregaria	N. Niger	—	Early outbreak	20[a]	Roffey & Popov (1968)
C. terminifera	E. Australia	Adults	High density	230	Clark (1974)
S. gregaria	N. Niger	—	Early outbreak	400[b]	Roffey & Popov (1968)
S. gregaria	W. India	Adults	Swarming	10	Roonwal (1953)
S. gregaria	N. Niger	—	Swarming	10000–20000	Roffey & Popov (1968)
S. gregaria	W. India	5th nymphal stage	Swarming	1000000	Roonwal (1953)
S. gregaria	W. India	1st nymphal stage	Swarming	25000000	Roonwal (1953)

[a] Average for entire region. [b] Density in congregation areas.

a 3-year period during which there was considerable variation in rainfall. The populations hardly varied between years, the highest, annual, combined-adult populations exceeding the lowest by a factor of only 1.1. Mispagel concluded that rainfall variations had no effect on the populations of these species.

In southern Tunisia, on the research site of the Desert Biome's Pre-Saharan Project, rainfall in 1974–75 was about 80% above normal, that in 1975–76 was about five times normal. Total arthropod biomass during each season of 1976 exceeded the 1975 biomass in the respective season. The disparity was greatest in the fall, the 1976 value exceeding that for 1975 by a factor of about two. In spring, summer, and winter, 1976 values exceeded those of 1975 by 80% or less (Muir, 1977).

In sum, precipitation and primary production in desert regions are clearly more variable than in more mesic systems. A large fraction of desert animal species – perhaps a majority – undergo annual population fluctuations correlated with the annual variations in rainfall and vegetative production. The variability among many of the vertebrates, as measured by the coefficient of variation about the mean, annual population size, appears less extreme than that of the primary production. In many cases it is less marked than that of ecologically or taxonomically comparable species in more mesic ecosystems. While it is too early to generalize for invertebrates, it does appear that the latter statement can be made for a number of insect species.

Mechanisms of fluctuation

Rainfall correlations

In those desert species which undergo short-term fluctuation, the vast majority of them fluctuate with variations in precipitation. The point is made repeatedly by so many authors that it does not seem useful to amass a systematic list of references to the fact here. Some of the more detailed analyses which show the pattern of interaction may be of interest, however.

O'Farrell, Olson, Gilbert & Hedlund (1975) obtained a correlation coefficient (r) of 0.99 between summer densities of their pocket mouse (*Perognathus parvus*) populations and the preceding October–April precipitation. The study was conducted in the northern Great Basin where most of the precipitation occurs between these months.

The Sonoran Desert has both winter and summer rainfall seasons. In 13 years of Gambel quail censuses in Arizona, Swank & Gallizioli (1954) found a close correlation ($r = 0.80$) between the late summer or fall quail population level and the total precipitation of the preceding December–April.

The Australian plague locust (*Chortoicetes terminifera*) goes through 2–3 generations per year in eastern Australia (Clark, 1974). The number of adults produced in one generation is directly correlated with the total rainfall during a 3-week period at the time of hatching of that generation.

Hinds & Rickard (1973) were able to associate 90% of the annual population variation in *Philolithus densicollis*, a tenebrionid beetle, with three meteorological variables: (*a*) November minimum soil temperatures; (*b*) October precipitation both during the fall in which the eggs were laid; and (*c*) October maximum temperature two years later at the time the mature beetles emerged. These relationships were obtained in a northern Great Basin study area. Although based on only four years of data, Thomas (1975) demonstrated an r^2 of 0.99 between the population size of another tenebrionid, *Centrioptera muricata*, and the sum of spring and late-summer precipitation. *C. muricata* has a 1-year life cycle. Thomas's Mohave Desert study area has a predominantly winter rainfall season.

There are some species which appear to fluctuate largely independent of rainfall. This appears true of the black-tailed jackrabbit in the Great Basin (Stoddart, 1977), and those predators which fluctuate largely in response to changes in jackrabbit numbers (Egoscue, 1975; Wagner, 1978*b*).

Demographic mechanisms

Among the vertebrates, the population fluctuations appear to be importantly a function of variations in reproductive rates. This is widely reported for rodents (Beatley, 1969; Maza *et al.*, 1973; Naumov, 1975; Cloudsley-Thompson, 1976). The same is true of the red kangaroo (*Megaleia rufa*) in central Australia (Newsome, 1965*b*).

Swank & Gallizioli's (1954) 13 years of Gambel quail data from the Sonoran Desert are still one of the best data sets for desert birds. Application of a second-order equation to relate the observed number of young per adult in fall to the previous December–April rainfall gives a correlation coefficient of 0.83 (Fig. 7.2). Sowls (1960) observed a similar pattern for five years of Arizona Gambel quail data. Gullion's (1960) seven years of Mohave Desert data on the same species provide an r of 0.89 with a simple linear test of young per adult in fall on October–March precipitation.

Variations in fall age ratios associated with some measure of precipitation have been observed in other quail species of semi-arid areas of southwestern USA (McMillan, 1964; Campbell, Martin, Ferkovich & Harris, 1973). But none is as variable as in Gambel quail where the percentage of young in some years is 0 (Gullion, 1960). In exceptionally dry years, the species apparently does not breed at all, as Miller & Stebbins (1964) observed.

Perhaps the greatest precipitation-linked variation in avian reproduction has been reported for Australian species (Keast & Marshall, 1954; Braithwaite & Frith, 1969; Rowley *et al.*, 1973). According to Keast & Marshall (1954), some species may inhibit reproduction for 'a succession of seasons' during

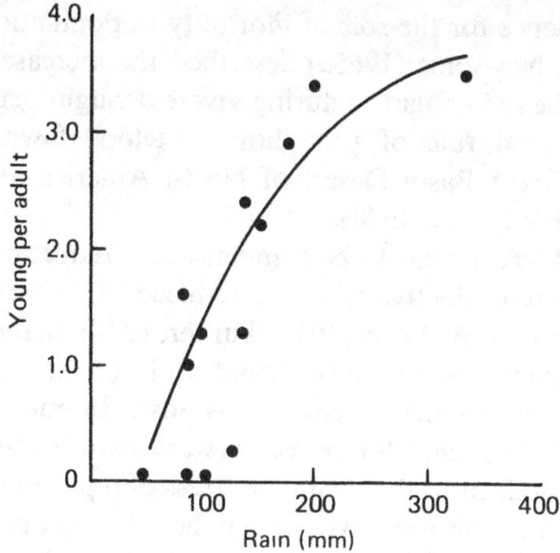

Fig. 7.2. Relationship between percentage young in the fall population of Gambel quail (*Lophortyx gambelii*) and previous December–April precipitation. Data from Swank & Gallizioli (1954). $Y = -1.11 + 0.0269x - 0.00004x^2$, $r = 0.830$.

a prolonged drought period. Similarly, Nelson (1973) reports that blackwinged stilts (*Himantopus himantopus*) and avocets (*Recurvirostra avosetta*) do not breed in a Jordanian desert *qa* (pond) in years when the winter has been dry and the water level quite low in spring.

Numerous authors have pointed to year-to-year variations in lizard reproduction, which are associated with rainfall variations (Mayhew, 1966; Turner, Hoddenbach, Medica & Lannom, 1970; Turner *et al.*, 1974; Ballinger, 1977; Turner, 1977; Whitford & Creusere, 1977). Turner *et al.* (1974) obtained correlation coefficients of 0.87–0.89 for the number of clutches per season for adult female *Uta stansburiana* regressed on various parameters of winter rainfall and supplemental irrigation, and 0.61–0.88 for young females. The studies were conducted in the Rock Valley site of the Mohave Desert. The desert tortoise (*Gopherus agassigi*) is another species which may not breed in drought years (Miller & Stebbins, 1964).

Among the invertebrates there is evidence that population fluctuation is partly a function of variations in fecundity. This appears to be the case for the locusts (Edney, 1974) and of the tenebrionid *Philolithus densicollis* (Hinds & Rickard, 1973). But there are also a number of species whose fluctuations are influenced by mortality at various life-history stages. Thus locust outbreaks are strongly a function of egg and nymphal survivorship (Uvarov, 1954; Dempster, 1963; Clark, 1974; White, 1976). Similarly, survivorship at key life-history stages is important to desert tenebrionids (Hinds & Rickard, 1973; Thomas, 1975). Thomas identified the importance of survival in the early and late larval stages, and accounted for 99% of the variations in population trend with these two variables in mortality over a 4-year period.

There is also evidence for the role of mortality in population fluctuations of larger mammals. Newsome (1965*a*) described the increased mortality of pouched young in the red kangaroo during severe drought periods in central Australia. The survival rate of pronghorn antelope fawns (*Antilocapra americana*) in the Great Basin Desert of North America also varies with precipitation (Beale & Smith, 1970).

In some species, there appear to be compensatory reductions in mortality rates in years when reproductive rates are reduced. This seems to be fairly well established in lizards (Ballinger, 1977; Turner, 1977). In the Gambel quail in Arizona, Sowls (1960) observed more equal adult sex ratios in the autumns of dry years when reproductive effort was low. In moist years, when reproductive output was high, adult sex ratios were distorted in favor of males suggesting higher female mortality from the stress of reproduction. However, there was evidence that cohorts of young hatched during dry years survived less well during their tenure in the population than cohorts produced in favorable years.

Environmental causation

A large fraction of the desert population fluctuations associated with rainfall appears to be linked through the food supply. However, it is not clear for many herbivorous species, particularly rodents and birds, whether the relationship is one of variations in the quantity or quality of food. Thus, as Nichols, Conley, Batt & Tipton (1976) pointed out: 'Plant estrogens (Pinter & Negus, 1965), plant gonadotropins (Bodenheimer & Sulman, 1946), vitamins (Beatley, 1969), and dietary water (Bradley & Mauer, 1971; Beatley, 1969) have been suggested as possible initiating stimuli for rodent production.'

There has been a tendency for some years to implicate vitamin A shortages in the rainfall-associated fluctuations of North American gallinaceous birds since Lehman (1953) first suggested this mechanism. Hungerford (1964) actually measured vitamin A stores in Arizona Gambel quail livers, and showed variation with rainfall and vegetative growth. In an interesting invertebrate analogy, White (1976) postulated that locust outbreaks are abetted by improvement in quality of vegetation associated with above average rainfall, specifically in improving the nitrogen content.

The idea of a qualitative food constraint rather than a quantitative one would seem to be in accord with certain aspects of the situation. Desert rodents are predominantly granivorous as are Gambel quail. In the Mohave and Great Basin Deserts, the seeds are significantly supplied by winter annuals which bear fruit in spring. Hence the seed resources are renewed in spring, and those species which are active throughout the year must overwinter largely on seed production of one or more previous years. Since reproductive activity begins in many species in late winter or early spring, there is hardly time for the reproductive process to respond to what will be a favorable or unfavorable seed crop. On the other hand several authors have reported that green vegetation becomes a portion of the diet in late winter and spring (Bradley & Mauer, 1971; Reichman & van de Graaf, 1975). Conceivably it is this qualitative addition to the diet rather than any quantitative shortage of food which is important (van de Graaf & Balda, 1973; O'Farrell *et al.*, 1975), but the question is still open.

The situation in lizards is somewhat different. In these species clutch (or litter) size and clutch frequency vary with precipitation as discussed above. As also mentioned, reproductive output is a function of the female's fat reserves. Ballinger (1977) actually measured differences in body fat stores in wet and dry years when reproductive efforts by *Urosaurus ornatus* were strong and weak. Furthermore, he and other investigators (Mayhew, 1966; Whitford & Creusere, 1977) also measured variations in arthropod abundance which were correlated with the fatstore and reproductive variations. The implication here seems to be that the lizards are responding to the abundance of the food resource, and therefore are substantially food-limited. Furthermore Mayhew (1966) and Hoddenbach & Turner (1968) imply that the insect abundance is importantly determined by the abundance of annual plants, on which they feed almost entirely. This might also imply that the insects are food-limited.

In other cases where predators fluctuate with the abundance of their food supply, there is implied a quantitative food shortage. This is the case with the coyote (Wagner, 1978) and the kit fox (Egoscue, 1975).

Habitat variations also appear to be a causal link between rainfall and population fluctuations. This is clearly the case with Australian waterfowl which only breed where and when water areas are adequate (Braithwaite & Frith, 1969) as with Nelson's (1973) Jordanian shorebirds. Whitford (1976) reported that populations of the woodrat (*Neotoma albigula*) increased following increased precipitation because the density of Apache plume (*Fallugia paradoxa*) shrubs was increased. The rats use the shrubs for nesting sites.

Temporal patterns of short-term fluctuation

Annual species and the random-numbers model

In analyzing the variability of desert animal populations, we have examined what might be termed the 'vertical' dimension of fluctuation, that is, the numerical similarity or disparity between successive values for population size. That variability provides insights into (*a*) the inherent demographic characteristics of a species, (*b*) the variability of the environment in which it exists, and (*c*) the homeostatic mechanisms both intrinsic to the species and in the environment which tend to mute the effects of stochastic elements in the system.

Composite and interactive processes

It is also of interest to engage in various forms of time–series analysis to examine the 'horizontal' or temporal dimension of fluctuation in desert populations. Such analyses depend on long-term censuses, and in the interests of generalization, censuses of numerous populations and species. There have been few such long-term censuses of desert populations and no intensive or comprehensive analysis can be made here. But it is of interest to single out a few populations for brief consideration which might elicit interest in similar, more extensive analyses in the future when more data are available.

One can conceive of the possible temporal patterns of population fluctuation falling along a continuum bounded by two hypothetical, extreme models. The first is a random-numbers series where, in effect, each successive population size would occur at random, independent of the preceding and following population size. This null hypothesis of population fluctuation has been posed by Cole (1951) who then analyzed a number of population time series for their similarity to a random-number series.

One characteristic of random-numbers series is the absence of any serial correlation, and any animal population which approached the random-numbers model would have little or no serial correlation. The populations which Cole analyzed departed slightly from perfect congruence with a random-numbers series, for one reason, precisely because they contained a measure of serial correlation.

That a series of values for some phenomenon occurs at random does not imply that those values are not causally related to, and correlate with, some other phenomenon. The values are random only with respect to each other, and can be independently correlated with some other series of random values, e.g. weather.

These lengthy, prefatory remarks are made to provide perspective for considering the population patterns of certain desert populations. As mentioned above, Swank & Gallizioli (1954) observed a close correlation ($r = 0.80$) of fall population size in Gambel quail on the previous winter's precipitation in the Sonoran Desert. O'Farrell *et al.* (1975) similarly found a close relationship ($r = 0.99$) between pocket mouse (*Perognathus parvus*) abundance and winter precipitation in the northern Great Basin. Hinds & Rickard (1973) related the number of tenebrionid beetles (*Philolithus densicollis*) to three weather parameters ($r = 0.95$) in the same area as that of O'Farrell *et al.* Thomas (1975) related population size of another tenebrionid (*Centrioptera muricata*) to the sum of spring and late-summer rainfall ($r = 0.99$) during four years in the Mohave Desert.

The important point here is that the weather variables are correlated with population *size* in these examples. Environmental influences act upon reproductive and mortality rates, and therefore on population *rates* of change. In the absence of measurements for reproductive and mortality rates, it is usually appropriate to correlate the values for environmental variables with rates of change calculated from successive measures of population size. In a population with relatively limited variation in reproductive and mortality rates due to other factors, a single influential factor will be correlated with rates of change, and not population size *per se*. And in such situations, there is usually a significant amount of serial correlation in successive annual censuses, at least of vertebrate populations. Insect populations have such high potential rates of change that they commonly fail to show serial correlation in arithmetic census values, although Williamson (1972, p. 12) has shown that when converted to logarithmic form, what previously appeared to be highly formless insect fluctuations prove to have marked serial correlation.

When vertebrate populations like the above show correlations between a meteorological variable and population size, there are several implications which disclose aspects about their ecology. The first is that there is no serial correlation, and indeed I tested the Gambel quail data and found none ($r = 0.04$). The implication is that regardless of how high or low a population is in any year, that level has no influence on the next year's level.

This implication carries several subsequent ones. The population's fall–spring mortality rate must be heavily density dependent. High or low fall populations must be cut back to about the same level each spring so that the breeders are free to produce a fall population that is largely a function of rainfall, and not of spring numbers. The further implication is that there is something approaching a spring threshold, possibly of habitat, or even of seasonal refugia.

A number of authors have spoken of such refugia. Gullion (1960) remarked that Gambel quail populations are very disjunct in their distribution, there being populations in valley or arroyo situations, and others on hills and ridges. The valley habitat appears to be more secure, and its populations fluctuate less violently than the hill populations. In arid areas of Australia, Myers & Parker (1975) describe habitat refugia into which populations of the introduced rabbits (*Oryctolagus cuniculus*) shrink during dry periods, and from which they re-populate the terrain during moist times. Newsome & Corbett (1975) describe small scattered pockets of surviving rodent populations in much the same pattern as that of the rabbits. Naumov (1975) describes refugia for *Rhombomys opimus* in Turkemanian SSR which occupy 5–10% of the terrain. He calls these 'centers of survival' from which the remaining landscape is re-populated during more favorable moisture periods. Uvarov (1954) speaks of similar 'ecological islands' for the desert locust, remarking that such islands have been expanded with the extension of agriculture and the locust problem thereby exacerbated.

The dependency on refugia would seem to imply that species with these patterns are not well adapted to deserts. Implicitly, they cannot survive in most of the desert environment during periods of dry, or perhaps even average moisture conditions. Uvarov makes this point for the locust. Despite it prominence, abundance, and economically deleterious role in desert regions, he considers *Schistocerca gregaria* poorly adapted to desert conditions. And it may be significant that the Australian species described by Myers & Parker (1975), and by Newsome & Corbett (1975), are either introduced species or members of the less well desert-adapted Murinae.

The Gambel quail populations fluctuate like annual plants, and to a substantial degree it can be considered an annual species. The same is true of O'Farrell's pocket mice, and the two tenebrionid populations. This is not unexpected in insects with 1-year life cycles which are in fact annual species. It is not so characteristic of vertebrate species with some measure, even though small, of generation carry-over. In short, these particular desert species' populations approach the random-numbers model.

Serially correlated, moderately stochastic species

By no means do all desert species behave as annuals. The species in Table 7.2 with coefficients of variation below 50% generally have populations which vary moderately, tending to fluctuate back and forth around an arithmetic mean. The same would be true of the relatively constant avian populations reported by Dixon (1959) and Raitt & Pimm (1976), and of Mispagel's (1974) three species of Membracidae. Desert lizard populations, despite all that has been written about the effects of precipitation on their reproductive rates, prove to fluctuate most moderately.

The time series are short on these species, and hence I did not calculate serial correlations. But it is doubtless appreciable, the species behaving more like perennials than annuals. There is nevertheless some stochastic variation associated with weather. But either these species' sensitivity to those variations is moderate or the full force of the variations is muted by generation overlap and/or the species' own homeostatic mechanisms. Since they vary only moderately with variations in precipitation, they would appear moderately well adapted to the desert environment.

Cyclic species and the sine curve model

The extreme on the fluctuation continuum which is opposite to the random-numbers extreme is the cycle, with the sine curve as its abstract model. The points on a sine curve are almost perfectly correlated serially, there being no stochastic variation in them. It is therefore completely deterministic, and the pattern in a cyclic population must be determined either by a smoothly oscillating succession of forces within the biotic system itself, or by perfectly cyclic forces extrinsic to the system. Examples of both have been reported for desert species.

Gross *et al.* (1974) and Stoddard (1977) have shown the black-tailed jackrabbit in portions of the Great Basin to be cyclic, following approximately a 10–12 year oscillation. Stoddart attributes the jackrabbit fluctuations to varying predatory pressures by the coyote. At different stages of the jackrabbit life cycle within the year, coyote predation is positively density dependent, negatively so, and parabolic (the form of the functions expressing predation mortality over jackrabbit density). All of them combined, however, produce a Lotka–Volterra oscillation. The jackrabbit cycle is therefore produced by cyclic predation pressure. Furthermore, the coyote is cyclic (Wagner, 1978*b*) by virtue of its dependency on the jackrabbit. The latter, in effect provides a cycle of food availability.

What appears to be the almost complete insensitivity of these species to variations in weather might imply that they are extremely well adapted to the desert physical environment. Their fortunes appear to be more determined by aspects of biotic environment which are independent of the physical, in particular each other.

Gupta & Prakash (1975, p. 364) contend that desert locust populations in the Thar Desert undergo an 11-year cycle, with variations from 7–14 years, that goes back to 1860. The build-up, peak, and early decline of the swarms occurs over seven of the 11 years. The low densities persist for four years. In their opinion, the cycle is produced by a cycle in the weather, possibly linked to the sun-spot cycle.

In total, there appear to be desert species arrayed along nearly the entire continuum of fluctuation patterns. There appear to be more examples in the center or toward the stochastic end of the spectrum. But we need studies of many more species before we can perceive the nature of the curve for desert ecosystems. That perception is a desirable goal for research on desert populations.

Life-history patterns of desert animals

r and K selection

The much-discussed subject of *r* and *K* selection seems to have had one of its earliest roots in Dobzhansky's (1950) observation that temperate-latitude species tend to have higher reproductive rates and lower survival rates than those of tropical zones. He rationalized this on the grounds that the unvarying climate of the tropics places little constraint on the animal species which then grow to year-round high densities and proliferate into a diversity of competing forms. Competition is high and specialization becomes an important contributor to fitness.

In temperate or higher latitudes, the environment is alternately favorable and unfavorable, and populations alternate periods of reproduction during favorable seasons with periods of mortality during unfavorable seasons. In this situation, fitness is conferred by high reproductive rates which enable species to recover their numbers quickly and strongly during that period when resources are abundant.

In a remarkable case of convergent thinking, MacArthur & Wilson (1967) deduced the characteristics required for species to colonize islands, and once colonized, avoid extinction. Effective colonizers are highly fecund and vagile species which can populate an area quickly. Those species, which are then able to endure on an island once it is colonized, trade off demographic volatility for competitive specialization. MacArthur and Wilson termed these two classes of species *r*-selected and *K*-selected, and noted the analogy with Dobzhansky's temperate and tropical dichotomy. In a sense, animals in temperate environments must recolonize the terrain each year. Those in equable tropical climes remain at high densities, relative to resources, and must compete with a plethora of other species. If Dobzhansky's generalization is a correct representation of the character of temperate and tropical organisms, then they can be said in general to be *r*-selected and *K*-selected, respectively. This entire theme has been greatly elaborated by Gadgil & Bossert (1970), Pianka (1970), King & Anderson (1971), Roughgarden (1971), and others.

A number of authors have pointed out that desert ecosystems should have more *r*-selected species. Deserts are probably the most variable systems on a year-to-year time scale; and as [illegible]inck (1956) has pointed out, their seasonal variability compares with that of higher-latitude regions. Hence, deserts are an excellent arena in which to test the theoretical prediction of *r*-selection in harsh and variable environments.

Several investigators have reasoned that this is the case. Although Doyen & Tschinkel (1974) concluded that tenebrionid beetles are *K*-selected forms, Thomas (1975) considered the desert species which he studied in the Desert Biome Rock Valley site to be *r*-selected. This conclusion was not reached by comparing the demography of his species with that of related or similar species in less-variable environments. Rather it was reached on the grounds that his populations fluctuated in response to rainfall variations, that the moisture constraint appeared to maintain the populations at densities below which they pressed the limits of their food supply, and that as a result a large number of species (24) were able to coexist in a small area (*c.* 1 km^2) with minimal niche segregation. Fleming (1974) studied two species of Costa Rican heteromyid rodents: *Liomys salvini*, a denizen of seasonal, deciduous tropical forests, and *Heteromys desmarestianus* which inhabits moist, relatively non-seasonal tropical forests. Because *L. salvini* had larger mean litter size, earlier age of sexual maturity, and more litters per month of breeding, he concluded that it was more *r*-selected than *H. desmarestianus*. Kikkawa (1974) compared the behavior and ecology of avian communities in moist, subtropical forests and semiarid habitats of eastern Australia. The communities in the dryer formations, though more diverse, contained species with larger mean clutch sizes than those of the forested areas.

Reports like these could also be arrayed which reach the opposite conclusion. But before doing so, it seems desirable to approach the question from a more general perspective. The hypothesis under test is whether desert environments tend to select for *r* strategies, and the evidence we seek in order to accept or reject the hypothesis is whether or not contemporary desert species are predominantly *r*-selected forms. Several points would seem to need explication before weighing the evidence. First, *r*- and *K*-selectedness are, as Gadgil & Solbrig (1972) contended, relative criteria. A species is *r*-selected only in comparison with another. But the question arises as to what should be compared.

Pianka (1970) has grouped organisms into *r* and *K* categories, and these categories turn out to be the contemporary invertebrates and vertebrates. In effect he compares these two groups of organisms, and in listing 'variable and/or unpredictable' climates *vs.* 'fairly constant and/or predictable' ones as correlates, he seems to imply that the basic demographic patterns of invertebrates and vertebrates have been selected by these two climates. Carrying this to its logical conclusion, we might ask the question whether invertebrates are more successful, and constitute a larger fraction of the fauna, in deserts than in the tropics or temperate forests. We might ask the converse for vertebrates.

Although I have not attempted any such comparison, it seems improbable that invertebrates are any less successful in the 'fairly constant and/or predictable' environment of the tropics than in the 'variable and/or unpredictable' climates of the desert. And while the tropics may be more favorable for vertebrates, particularly aquatic and amphibious forms, it is not clear that they constitute a markedly lower fraction of desert faunas than of tropical faunas. More fundamentally, it seems to me that such a rationale is not a meaningful one. The basic demographic patterns of invertebrates and vertebrates undoubtedly evolved eons ago, probably before they even emerged from their ancestral, aquatic environment, and under forces which today remain at best conjectural. They have, of course, continued to evolve since emergence. But the broad differences which separate Pianka's two groups surely developed before terrestrial existence, and certainly before development of modern-day environments. Hence it has not seemed to me appropriate to contrast the demography of contemporary insects and mammals, to label them as *r*- and *K*-selected, and somehow infer that the gross differences have been selected by contemporary or recent environments.

For this reason, comparisons would seem to be most appropriate between orders and families within a class; and better still, between genera and species.

This is the case with the above vertebrate comparisons, but Thomas's (1975) rationale, thought-provoking and imaginative as it is, seems to derive from the same frame of reference as Pianka's dichotomy.

The most effective comparison can be made among rodents because of the abundance of evidence. In North America, desert rodent communities are dominated by species of the family Heteromyidae (Bradley & Mauer, 1973; French *et al.*, 1974; Whitford, 1976; Olding & Cockrum, 1977). Numerous authors have commented on the conservative demographic characteristics of the heteromyids (Sumner, 1925; Maza *et al.*, 1973; Whitford, 1976). French *et al.* (1975) reviewed the demographic characteristics of the world's rodents and erected three categories in descending order of demographic volatility. The heteromyids were the most conservative: small litter size (in comparison with other rodents), restricted breeding seasons, relatively late sexual maturity, relatively high survival rates. That these characteristics produced a slower population response (lower *r*) than that of the next high category, the cricetines, was shown by Whitford (1976). During a period of above average rainfall on his Chihuahuan Desert research site, the cricetines increased more than the heteromyids.

The heteromyids are also common in the tropics. While they have demographic characteristics similar to those of the desert species, and clearly more conservative than those of higher-latitude cricetines and microtines (Fleming, 1974), it is not clear whether there is a difference between desert and tropical heteromyids. Fleming's results, based on small samples for two species, are hardly a basis for generalizing. But it is clear that the dominant North American desert rodent species are demographically more conservative (or could be termed more *K*-selected) that temperate-zone cricetines and microtines, and therefore fail to support the *r*-selected hypothesis.

In North Africa and Asia, desert rodent communities are dominated by the Dipodidae and the Gerbillinae (Happold, 1975; Misonne, 1975; Naumov, 1975; Prakesh, 1975). While French *et al.* (1975) place the dipodids in the same, conservative category as the heteromyids, the prominent gerbillines are placed in the intermediate category with the cricetines. Thus, desert rodent populations in this part of the world may not be as conservative demographically as those of the New World, but they are nevertheless more conservative than the microtine- and murine-dominated communities of temperate climes.

There appears to be some evidence of comparable patterns in lagomorphs, although there are complicating features. Litter sizes decrease from north to south in the black-tailed jackrabbit (Gross *et al.*, 1974), reaching their smallest size in the US in the southern deserts. Litter size also decreases from north to south in the cotton-tail rabbit (*Sylvilagus*), reaching its smallest reported size in the Audubon cottontail (*S. audubonii cedrophilus*) and desert cottontail (*S. audubonii arizonae*) of the Sonoran Desert (Sowls, 1957; Stout,

1970). The number of litters per year increases southward in both the jackrabbit and cottontail. But the increase is not sufficient to offset the decline in litter size (Gross *et al.*, 1974; Sowls, 1957); and consequently if the *r* value were calculated on litter size and number of litters per female, productivity would be lowest in the southern deserts. The major unknown is the contribution of early maturity of young in the south. Sowls (1957) found about a third of the juvenile females collected near the end (late summer) of the restricted, seasonal breeding period to be pregnant. Something similar may occur in jackrabbits.

The evidence for birds is not as abundant as that for small mammals, but tends to point in the same direction. With a few exceptions, the Columbiformes have a highly uniform, determinate clutch size of two and their productivity is a function of the number of broods per female per season. Since the breeding season increases in length toward the equator, they would tend to have higher productivity in low-latitude desert regions than in temperate latitudes. This presumes that breeding intensity per unit of time remains the same at all latitudes. The latitudinal effect is just that, however, and not a desert effect: presumably desert populations of the same or related species would have no higher productivity than populations in more mesic regions at the same latitude.

The related sandgrouse (*Pterocles* spp.) are characteristic inhabitants of African–Asian deserts. They, too, have two-egg clutches, and apparently only one per year. Whether one or several, species with two-egg clutches would hardly qualify as '*r*-selected' species in comparison to the class Aves in general.

The bustards (*Choriotis* spp) are large gamebirds of the African, Asian, and Australian deserts. They typically have clutch sizes of one or two, and in view of their great size might not reproduce at one year of age. Prakash & Ghosh (1975, p. 427) place the clutch size of the Great Indian bustard at a 'single egg – rarely 2'.

The Gambel quail, with a mean clutch size in the Sonoran Desert of 12 (Gorsuch, 1934), and sexual maturity at one year of age, is a prolific species. But it is no more so than related species of quail in more mesic regions: 13 in scaled quail (*Callipepla squamata*) in the Oklahoma plains (Schemnitz, 1961); 14 in the bobwhite (*Colinus virginianus*) in humid, southeastern USA (Murray & Frye, 1957); and 14 in the California quail (*Lophortyx californicus*) in the Mediterranean climate of western California valleys (Lewin, 1963).

There are a number of interesting cases among the passerines. Bodenheimer (1954) has observed that clutch sizes in desert birds are commonly smaller than those of related species in more mesic areas. As an example, he contrasts the 5-egg clutch of the mourning wheatear (*Oenanthe lugens*) in the Negev with the 42-egg performance of the related robin (*Erythropygia* sp.) in the Mediterranean region. A more appropriate comparison might have been with other species of *Oenanthe* in southern Europe where clutch sizes generally run five to six. In either case, the Israeli desert species has no greater and perhaps smaller, clutch size than related, non-desert forms.

Walsberg (1978) describes an interesting situation in the case of the phainopepla (*Phainopepla nitens*). He studied populations which nest twice each year: once in coastal woodlands of southern California in summer (May–July) and once in spring (March–April) on their winter range in the Sonoran Desert. Clutches in the coastal area are two to three, those in the desert 'virtually always' two. The species is obviously a conservative one, demographically. Anderson & Anderson (1973) determined the mean clutch size of the cactus wren in the Sonoran Desert to be 3.3 and 3.4 in two study areas, with two clutches per female common. In the house wren (*Troglodytes aedon*) of mesic, eastern North America, clutches generally range from five to eight, with two commonly produced in a season.

In a case which may be contrary to the above, Ohmart (1973) found the mean clutch size of the roadrummer (*Geococcyx californianus*) in the Sonoran Desert to be 4.6. Clutches of the other North American cuckoos which, unlike the Old World species, are not parasitic generally range from two to five (Peterson, 1941).

In total, this evidence suggests that desert mammals and birds have comparatively low *r* values. It is not clear that they are lower than their counterparts in the tropics at comparable latitudes – there is evidence pointing in both directions as well as suggesting no differences. But it is clear that their demography tends to be more conservative than that of more mesic, temperate-zone species.

I suspect that the mechanism involving the lower *r* values of desert species is part of the broader latitudinal trends in demography which embrace birds and mammals generally. They appear to be independent of differences in seasonality and variability of the environment, and perhaps of the biotic influences which have been posed to explain the latitudinal cline in avian clutch size. Because deserts generally exist at low latitudes, the demographies of their species tend to be conservative, much as Dobzhansky (1950) generalized for tropical species. It is perhaps not appropriate to speculate on causal mechanism here, but it seems clear that desert warm bloods are not significantly '*r*-selected.'

The situation with lizards is rather different. The subject has been treated as extensively with this group as it has with rodents. Tinkle (1969), Tinkle, Wilbur & Tilley (1970), and Turner (1977) have summarized the demographic characteristics of tropical and temperate lizard species. It is instructive to compare these general characteristics with selected characteristics of Pianka's (1970) *r*-selected forms (Table 7.4). The description of tropical forms by Tinkle *et al.* is strikingly close to Pianka's characteristics of *r*-selected types. In essence, Tinkle *et al.* have generalized tropical lizards to be *r*-selected in

Table 7.4. *Comparison of the general demographic characteristics of tropical lizards (Tinkle* et al. *1970) with selected traits of Pianka's (1970) r-selected species*

Trait	Pattern in tropical lizards[a]	Pattern in *r*-selected species[a]
Age at maturity	'Early maturity'	'Early reproduction'
Reproductive pattern	'Small clutches' 'Usually multibrooded' 'Higher per season fecundity'	'High r_{max}'
Longevity	'Short life expectancy'	'Short, usually less than 1 year'
Body size	'Smaller body-size at first breeding'	'Small body size'

[a] The quoted descriptions of Tinkle *et al.* compare these traits of tropical lizards with those of temperate species. The descriptions of Pianka are related to animals in general, and these characteristics are generalized to invertebrates in comparison with vertebrates.

comparison with temperate species, and contrary to Dobzhansky's generalizations and conventional wisdom in *r*- and *K*-selection.

There is some evidence that desert species follow patterns similar to those of tropical forms. Thus Tinkle (1967) found reproductive rates of *Uta stansburiana* higher in semiarid west Texas than at a higher latitude and altitude in western Colorado. Turner (1977) reported that clutch sizes of *Cnemidophorus tigris* are larger in his southern Nevada, Mohave Desert research site than farther north in the USA.

With somewhat different results, Derickson (1976) compared populations of *Sceloporus undulatus* and *S. graciosus* in southern Kansas and southwestern Utah, respectively. The two areas are at comparable latitude, but the Kansas area, in a grassland region, receives about 7–8 times as much precipitation as does the Utah Mohave Desert region. The Kansas species had mean clutch sizes of six to seven, produced three clutches per season, and matured earlier than the Utah species which had clutch size means of 3.1–4.4, and produced two clutches per season.

The mechanisms underlying these patterns can be suggested with fairly good probability of being correct, and once again are related to latitude but in a way different from that of birds and mammals. Low-latitude species mature earlier than those of temperate zones, possibility because of the stimulus of higher temperatures for these ectotherms. Growth rate apparently does not keep pace with the maturation rate, and consequently low-latitude species mature at smaller body sizes (Tinkle, 1969; Tinkle *et al.* 1970) although there appear to be some exceptions to this (Turner, 1977).

Clutch size is importantly determined by the physiological reserves of the female, lager clutches being produced both by larger females and by females with larger fat stores (Mayhew, 1966; Tinkle, 1967; Derickson, 1976;

Ballinger, 1977; Turner, 1977). In the equable climate of the tropics, lizards tend to reproduce year-round. In doing so, females will tend not to recover their fat stores (Tinkle, 1969). Thus temperate species produce larger clutches, both because of the delayed maturity and larger female body size; and because the breeding season is limited, and there is a period during the year when the female can recoup her fat stores. But the tropical species more than compensate for the smaller clutch sizes with larger numbers of clutches, and in total have the higher-reproductive rate. Combined with the earlier maturity, the tropical species have considerably higher *r* values.

There is some suggestion that desert species, while in general following the latitudinal pattern, may have lower *r* values than their counterparts in more mesic areas at the same latitude. Because of the strong dependence of reproductive output on the reserves of the female, the reproductive rate is highly sensitive to food availability, particularly in deserts. Thus Mayhew (1966), Turner *et al.* (1974), Ballinger (1977), and Whitford & Creusere (1977) all observed marked variations between years in reproductive output, apparently the result of variations in rainfall which produced variations in insect abundance as described above. The perpetually lower insect abundance in deserts may explain the differences observed by Derickson (1976) between Kansas and Utah *Sceloporus* populations. Turner (1977) observed that lizard population densities in deserts are characteristically lower than in more productive ecosystems.

In sum, lizard populations in the tropics tend to have higher *r* values, contrary to Dobzhansky's generalizations and contemporary views on *r*- and *K*-selection. The causation appears latitudinal and may be physical rather than biotic. Desert species, existing at relatively low latitudes, appear to follow somewhat the same pattern, but may have lower *r* values than their counterparts in more mesic areas because of food scarcity. Hence the concept of *r*-selection seems of doubtful application in their case.

While somewhat anecdotal, it seems pertinent to mention one final reptilian species. The desert tortoise (*Gopherus agassizi*) is a denizen of North American hot deserts, apparently surviving well in relatively undisturbed, climax-type situations (Berry, 1978). The species has a demographic pattern virtually as conservative as that of man. Although specimens have reached sexual maturity at 12–13 years of age in captivity, it may be atained in the wild only at 15–20 years (Berry, 1978). Females lay a clutch of 4–6 eggs once mature, and may have a maximum longevity of 50–100 years. Berry (1978) concludes that this is a *K*-selected species.

There is no point in attempting to generalize the *r* and *K* situation in the invertebrates at this time. The species are too numerous, the comparative demographic data too few. It is tempting to speculate that the stepped-up metabolism of lizards at low latitudes might also occur in invertebrates. The species might then shorten generation time and increase *r* values as do the lizards. But Edney (1974) argues that many desert arthropods are slow developers, partly because of food scarcity and their ability to survive while eating infrequently.

If the demographic characteristics of desert animals do not appear to be uniquely adapted to their environments, as *r*- and *K*-selection theory predicts, we are then left with the burden of proposing an alternative hypothesis. The question to be answered is: what evolutionary process or processes have produced or determined the demographic patterns we observe in contemporary desert species? I would suggest that there might be at least four, not all of equal importance, and not all necessarily operative in each species. Ideally, we would like to ascertain what fraction of the demographic patterns we observe is contributed by each. No such quantification is yet possible, but we do have a basis for postulating the importance of the processes relative to each other – in essence, a list of priorities.

The first candidate I would propose, with latitude held constant, is what I choose to term 'phylogenetic momentum.' I purposely use this term distinct from Wilson's (1975) 'phylogenetic inertia' because the implicit meaning, while close, is not precisely the same. The term inertia implies resistance to a force, and in Wilson's use the implied force is selection. I do not wish to take as an *a priori* given that selection is, or is not, operating. Furthermore, I shall argue that the traits under consideration evolved early in the evolution of the species, perhaps at the generic, family, or order stage, and long predate the contemporary environments in which they occur. Those traits, I suggest, have persisted through the evolutionary precursors of contemporary species to the present, and thus cannot logically be said to have been predominantly shaped by contemporary environments. At best, they can only have been modified slightly by recent selection, but the basic patterns were forged long before. In this perspective, the traits have carried through the phylogenetic precursors over time, and I prefer the more dynamic term 'momentum' to the more static one 'inertia'.

The literature on clutch and litter size, and other aspects of reproductive rates, emphasizes the differences between species and struggles to explain (or more accurately, rationalize) those differences. I am more impressed with the similarities between taxonomically related species than the differences. Heteromyid rodents have relatively short breeding seasons, and have mean litter sizes of 2.5–4 irrespective of their environments, whether desert or tropical. The cricetines have comparable litter sizes, but breed through longer periods of the year, again largely irrespective of their environments.

The differences between these two groups are not environmental but taxonomic. The point is implied in Fleming's (1974) statement: 'Annual productivity of temperature (*sic*) and tropical cricetid or murid rodents appears to be higher than that of their heteromyid counterparts'. And it was implied most comprehensively by French *et al.* (1975) when they grouped the world rodents into three demographic categories which proved to be taxonomic rather than environmental groups. Their conclusion was: 'By and large, the groups set up on a taxonomic basis also form natural groups on a demographic scheme'.

The same is true of the several species of quail distributed from east to west across southern USA. Mean clutch sizes are about 12–14, whether of the bobwhite in the subtropical southeast, the scaled quail in the semi-arid plains, the Gambel quail in the desert, or the California quail in the Mediterranean climates of West Coast valleys and foothills.

The same could be said of almost any group of related species. Since there appears to be uniformity within genera or families, and in some cases within orders, it seems reasonable to postulate that those patterns evolved in the past with the evolution of those genera, families, or orders. And since the similarities exist across broad variations in environments, it seems unlikely that selection by contemporary environments has been a major force in molding the contemporary demographic patterns.

Now the emphasis so far has been on similarity, and of course there are variations. And perhaps the second most important process is what, out of ignorance, will be called 'latitude'. Within the endotherms, there are obviously important latitudinal variations in demography, some discussed above for desert species. Most of the attempts to explain these phenomena have been made by ornithologists who have attempted to rationalize the well-known clines in clutch size. There is no point in reviewing these here, except to say that they tend to focus somewhat provincially on altricial birds, without projecting broader-perspective theories that also embrace precocial birds and mammals. There also tends to be an automatic assumption that differences within species, or between closely related species, are genotypic and little likelihood accorded the possibility that they might be phenotypic. It seems to me that an alternate hypothesis is needed. Certain facts would appear to be relevant to the question, and might ultimately contribute to an explanation. The latitudinal variations in both photoperiod and temperature are well known to have profound effects on the physiology of endotherms. It has long been recognized by animal husbandry specialists that when temperate-latitude domestic animals are transported to tropical latitudes, they grow more slowly, mature later, and experience reduced breeding intensity (Hafez, 1968). The transport of temperate-latitude forages to these animals in the tropics has demonstrated that the change is not a nutritional one. Although less marked in tropical breeds which have been selected for the prevailing climates, some degree of these patterns is still evident. Mills (1942) has suggested that warm bloods have optimum environmental temperatures for discharging their heat loads. At temperatures higher than the optimum, heat tends to accrue and numerous adjustments are made in response.

Moreau, Wilk & Rowan (1947) and Wagner (1957) have observed that birds at higher latitudes experience more marked prebreeding gonad recrudescence and build-up of fat stores than do individuals at lower latitudes. The difference is assumed to be the result of different photoperiodic regimes.

In total, there appears to be a gearing down of many physiological functions in endotherms toward the equator, apparently due either to the thermal or photoperiodic environments, or both. When we contemplate the latitudinal clines in clutch size in a wide variety of birds, both altricial and precocial; the latitudinal clines in mammalian litter size in rodents, lagomorphs, and carnivores; and the evidence of delayed sexual maturity, it is difficult to avoid the suspicion that we are observing particular cases of what seems to be a broad class of phenomena.

Desert heteromyids have smaller litters and shorter breeding seasons than temperate microtines, perhaps for these reasons and not because of selection by the desert environment. The heteromyids, after all, have the same characteristics in the tropics. Rabbits of the genus *Sylvilagus* have smaller litter sizes in the Sonoran Desert than do those of northeastern USA. The same can be said of black-tailed jackrabbits in the Sonoran Desert (Vorhies & Taylor, 1933) *vs.* those in the northern Great Basin (Gross *et al.*, 1974); of Gambel quail clutches in the Sonoran Desert *vs.* those of bobwhites in northern USA; and of cactus wrens in Arizona *vs.* house wrens in northern USA.

Within lizards, and perhaps more broadly in ectotherms, there seems to be an analogous but reverse pattern. Toward the equator, sexual maturity occurs earlier and breeding takes place through an increasing fraction of the year until it is apparently year-round in the tropics. Because the non-breeding time is minimal, and there is a continual reproductive demand on the female's energy stores, clutches or litters remain smaller than in temperate climes, but the total reproductive output at low latitudes is higher.

The latter statements appear true of warm-desert species. If those species were compared with temperate ones, one would be tempted to take this as evidence for *r* selection. But these characteristics hold for tropical as well as desert species. The patterns again appear to be associated with correlates of latitude, perhaps the thermal environment, and transcend diverse environments at any one latitude.

We have by-passed the question of whether these differences are genotypic. One would assume that they are at least in part. But there is no way of determining at this stage how the genetic differences were brought about. Although I do not regard it as an *a priori* certainty, perhaps the differences were produced by selection. If they were, they must have involved selective forces related to latitudinal clines in the physical environment which transcend broad differences in both physical and biotic environments at a given latitude.

A third process contributing to the demographic patterns of desert animals is more speculative than the first two. We have seen several examples where, in closely related species at similar latitudes, desert species are more conservative demographically than their more mesic counterparts. This seemed to be true of some lizards, Gambel quail, and the wheatear. This is a small number of species, and can hardly be considered a general pattern on the basis of the data so far available. If it is real in lizards, it may reflect shortage of food in deserts, as suggested above, and may be phenotypic or genotypic. Possibly something similar operates in the warm bloods.

A fourth process is that of *r*- and *K*-selection. We examined three cases, again between related species at similar latitudes, where desert species had more volatile demography than their tropical or more mesic counterparts. This is also a small number of examples, and hardly basis for concluding that *r*- and *K*-selection are generally operative.

In total, it appears that the major forces determining the demographic characteristics of desert animals are their phylogenetic momentum and whatever influences are associated with latitude. The space remaining for *r*- and *K*-selection, and the evidence for its existence, cast serious doubt on its significance, if not its reality.

Precipitation as a stimulus for reproductive timing and intensity

Evidence on the role of precipitation as a timing mechanism for the breeding season is confusing; it may have been somewhat over-stereotyped, and perhaps defies simple generalization. The phenomena may best be viewed as falling along a continuum between total response to precipitation as a *zeitgeber* irrespective of the season of occurrence, and total response to the photoperiod with no evident influence from rainfall.

Evidence for the first extreme seems to have come more from the Australian experience than anywhere else, and may have promoted a general impression that the phenomenon is more common in deserts than it is. Thus Braithwaite & Frith (1969) and Rowley *et al.* (1973) report numerous species of waterfowl and corvids breeding at any season in which rainfall occurs. According to Keast (1959) there is a spring seasonality to bird nesting along the mesic east and southwest coastal strips, and a summer season in the subtropical north with its summer rainfall. But over virtually the entire remainder of the continent, nesting is irregular and aseasonal. In rare years with both winter and summer rain, there may be two seasons. Marshall (1961) has stressed the point that vertebrates in deserts have abandoned response to photoperiod and have adopted more appropriate stimuli, namely precipitation, to ensure breeding success.

However, Serventy (1971) concludes that this emancipation from the photoperiod is not so evident in other parts of the world. And other authors, while acknowledging that there may be some response of breeding intensity to the amount and timing of precipitation, still maintain that there is a basic photoperiodic seasonality. Thus, according to Cloudsley-Thompson (1976, p. 109) 'Although the timing of the reproductive cycle in birds depends primarily upon daylength, breeding among desert species is frequently coordinated with either rainfall or the visual stimulus provided by green vegetation.'

In Iran, where much of the desert area lies between 30–35 °N, and the rainfall season is winter, the major breeding season for the rodents is spring (Misonne, 1975). To the east in India, where the Thar Desert falls between 25–30°, the rainfall has shifted to a monsoonal one. Yet according to Prakash (1975) most rodents breed from March–September, although some reproduce year-round. There is some evidence of a response to the rains and food availability in the seasonal species, reproductive activity being most intense in late summer. But Gupta & Prakash (1975, p. 460) argue that the intensive breeding effort in spring, plus continued activity, albiet at a reduced rate in early summer before the rains, indicates a persisting importance of the photoperiod.

In northern Sudan, between 16–22° N, the rainfall season is also a monsoonal one. But at this latitude, rodent reproduction has now shifted around largely to the latter part of the year (Happold, 1975). *Jaculus jaculus* breeds largely in October–November, although some activity may persist to February. *Gerbillus pyramidum* begins earlier, in June, but also continues to February. In this locale, April–June is the hot, dry period. Seemingly, at this latitude the photoperiodic influence has largely waned and breeding is now timed primarily to coincide with, or follow, the rains.

The US desert areas are largely above 30° N, some as high as about 46°. Reproduction in these areas is still basically seasonal but some adjustments to the rainfall seasons can be seen in mammalian populations. The Great Basin and Mohave Deserts are winter rainfall areas with summer the dry period. In the Great Basin, black-tailed jackrabbits at 41–42° breed from late winter or early spring to about July (Gross *et al.*, 1974). Bradley & Mauer (1971) observed the reproductive activities in *Dipodomys merriami* in the Mohave Desert at about 36° from February–August but most pregnancies in March–May. This general pattern holds for other species (Bradley & Mauer, 1973), although some cricetines breed year-round.

In the Sonoran Desert at about 32°, jackrabbits reproduce in virtually every month of the year (Vorhies & Taylor, 1933). But the Sonoran has a bimodal rainfall pattern, with both winter and summer seasons. Vorhies and Taylor observed the greatest intensity of reproduction in spring and late summer, with reduced activity in winter and early summer. To the contrary, both Sowls (1957) and Stout (1970) reported the breeding season of the Audubon and desert cottontail rabbits in the Sonoran Desert to be January–August, both concluding that there was no adjustment to the rainfall seasons. Ohmart (1973) reported two clutches of eggs in the roadrunner (*Geococcyx californi-*

anus) in the Sonoran Desert. The second follows the summer rains and is larger than the first, possibly as an adaptation to the more abundant food supply. He observed a similar breeding-season bimodality in the brown towhee (*Pipilo fuscus*), rufous-winged sparrow (*Aimophila carpalis*), and the curve-billed thrasher (*Toxostoma curvirostre*).

Some degree of photoperiodic seasonality appears to persist in avian breeding as far south as about 27° N. Short (1974) observed 85 species of birds as summer or permanent residents in the thorn forest ecotone at the southern edge of the Sonoran Desert in Mexico. He obtained evidence of breeding in 59 species, of which 38 definitely nested in summer, 21 before July. The rainfall pattern here is bimodal as elsewhere in the Sonoran.

Some species reproduce year-round, apparently oblivious to precipitation, aridity, and photoperiod, although generally at low latitudes. This is true of many of the columbiforms, the cricetine rodents, and according to Cloudsley-Thompson (1976) the double-banded courser (*Hemerodromus africanus*) in South Africa.

Among lizard species that have been studied in North America, the breeding season lengthens from north to south, and widens to year-round effort in the tropics (Tinkle, 1969). In some species, drought may delay reproductive activity (Mayhew, 1966).

In summary, the emancipation of desert vertebrate reproductive seasons from photoperiodic timing, and its complete aseasonal linkage to precipitation appears to be a significant reality only in Australia, and perhaps below about 20° latitude. Above this point, reproductive activities still occur largely between winter and fall. But there may be some shifts in timing or intensity within the limits of this period which correspond to rainfall timing. Some species appear to be insensitive to the timing of precipitation, including year-round breeders.

Population regulation and limitation

Concepts of regulation and limitation

Despite the half-century debate on this topic, there is still widespread disagreement and misunderstanding of it. The problem centers on semantic difficulties with the terms, different investigators implying different phenomena in the terms regulate, limit, and control. The semantic difficulties, in turn, arise from failures to pose explicit definitions in demographic terminology or notation for these terms.

Although his meaning has often been misconstrued, Nicholson (1933, 1954*a*, *b*) recognized early that the maintenance of equilibrium in a population, and determination of the mean density at which that equilibrium is maintained, are two distinct, if related processes. The term equilibrium merely implies that there is no net change over time in mean density, although there can be short-term fluctuations around a mean. That relative constancy of numbers can be experienced at a wide range of densities: very low, intermediate or high.

Others besides Nicholson (Wagner, Besadny & Kabat, 1965; Klomp, 1966; Solomon, 1971) have perceived the distinction between these two processes, but somehow that distinction seems not to have gained widespread recognition and acceptance. One of the reasons is that these verbal descriptions are not sufficiently explicit to clarify the distinction. I attempted such a distinction previously (Wagner, 1969), and will reiterate it briefly here.

In an equilibrium population, with no net change over a period of years, the mean instantaneous rate of change between successive annual censuses is zero ($\bar{r} = 0$). The maintenance of that condition is properly termed 'regulation' (Klomp, 1966), and influences which accomplish this regulate the population. The school of thought to which I subscribe contends that such regulation is attained by influences which intensify their effect when a population is above its mean, and ease their effect when it is below its mean, i.e. density-dependent influences.

There is, it should be recognized, the alternate school of thought which contends that populations can avoid excessively high or low numbers when under the influence of chance (i.e. density-independent) factors. They do so by having numerous subpopulations scattered over a region with heterogeneous environments. Each subpopulation has a probability of extinction, and some do disappear. But each subpopulation has a higher probability of survival, and most do. When one disappears, its areas can be repopulated by immigration from one or more of the other fortunate subpopulations which have not succumbed to the low probability of extinction. In short, a single population has a certain probability of surviving and of failing. If that population is subdivided into numerous subpopulations, each with independent probabilities of surviving or failing which are equal to those of the original single population, the probabilities of extinction for the species become vanishingly small by thus 'spreading the risk'. In order to evaluate the regulatory mechanisms of desert species, we must evaluate either their density-dependent patterns or the means by which they 'spread the risk'. There is fragmentary evidence of each of these which we will evaluate shortly.

Demographic definition of the determination of mean density, which I shall henceforth call 'limitation', is somewhat more elusive. When a population goes to equilibrium at some mean density, its mean rate of change has been reduced by one or more (usually a complex of) environmental influences from the genetic maximum for that species in the locale to zero. Hence, the determination of mean density can in a sense be said to be the reduction of a population's rate of change from its genetic maximum to zero, and those influences which effect that reduction limit it.

It is here where much of the confusion arises. Most authors who accept the regulation paradigm agree that it is achieved by density-dependent influences, and the importance of these becomes fixed in one's thought. But when regulation and limitation are considered a single phenomenon, then the determination of density somehow is assumed to be possible only under the influence of density dependence. If we hold that a population is limited by any influence which reduces r to zero, this can be achieved by factors which vary their effect independent of density, but on the average over time have a mean depressive effect on r by either reducing reproductive or survival rates. Thus, a population is *regulated* by density-dependent factors, and more precisely by factors which depress r more when density is high than when it is low. A population is *limited* by the sum-total of all factors which depress r: density dependent, density independent, even those with a positive-feedback effect, and those with non-monotonic density functions (Wagner, 1969).

One interesting consequence of this pattern is that a population can be regulated by one set of influences, but largely limited by another. A prime example of this is in the winter moth (*Operophtera brumata*) as shown by Varley, Gradwell & Hassell (1973). In analyzing the array of factors operating on this species in British oak woods by k-factor analysis (a k-factor is actually a form of mortality rate), these investigators found only one factor out of a complex of six which operated in a density-dependent pattern. This was predation on the pupae, and its relative importance as measured by its mean k value over the period of study, was only second highest in the complex of six.

The most important influence (that with the highest mean k) was overwinter mortality termed 'overwinter disappearance'. It operated in a distinctly density independent mode, and while the specific cause was not identified it was thought to be related somehow to the weather. Hence the winter moth can be said to be predator regulated, but primarily weather limited (if in fact overwinter disappearance is produced by a weather influence).

Varley *et al.* (1973) did not extend k-factor analysis to this approach for analyzing limitation. Their main purpose was to isolate the factor responsible for most of the variation in the moth population (winter disappearance) and to develop functional equations for each of the factors affecting the moth which could then be used in predictive models. But in my opinion, their k-factor analysis has a great, untapped potential for elucidating the patterns of limitation on animal numbers.

Ideally, we could proceed to elucidate the regulatory and limiting influences on desert species if we had k-factor analyses on a large number of them. In fact, we have such analyses on only two: the black-tailed jackrabbit (Stoddart, 1977) and the coyote (Wagner *et al.*, unpublished). Hence we are obliged to proceed by subjective evaluation of the data sets cited repeatedly above. But it seems desirable here to hold out k-factor analysis, or some analogue, as the ultimate procedural goal for analyzing desert animal populations. And it seems desirable to explicate the distinction between regulation and limitation in order that we may unambiguously proceed with a consideration of these two processes in desert species.

Regulation of desert animal populations

A number of authors have sought, and reported, evidence of density-dependent patterns in desert species. Chew (1975) studied density dependence in *Perognathus formosus* experimentally by confining populations artificially at low, medium, and high densities in enclosures, and measuring their performance. Although the experiment was confounded somewhat by annual variations in food availability, there was some evidence of reduced reproduction at high densities. Turner (1977) summarized a number of density dependent patterns in lizard demography. Tinkle (1967) found natality to be inversely related to density in *Uta stansburiana* populations.

A classic case of density-dependence is the phase transformation and emigration of desert locusts (Dempster, 1963). Their numbers increase during periods of rain and vegetation expansion. As aridity returns and vegetation dies, the solitary forms are progressively compacted in smaller and smaller areas, and at higher and higher densities. The increasing inter-individual contacts now stimulate transformation to the gregarious phase which ultimately reduce density in the area of origin by emigrating.

I have discussed above the deduction that an annual population pattern implies stringent density dependence in fall–spring mortality. This conclusion would seem to apply to all of those species which appear to depend on a habitat threshold or population refugium.

There is also some evidence of spreading the risk. Thomas (1975) concluded that tenebrionid beetles in his Mohave Desert study area were maintained at levels below any shortage of food by periodic unfavorable (dry) weather patterns. In effect, the only influences of significance which operated on them were density independent. During the breeding season, they wandered widely over the landscape depositing small numbers of eggs in numerous places. The strategy seemed to be one of distributing eggs widely so that some would be situated in places receiving the spotty, low probability events which characterize desert precipitation.

Somewhat the same appears true of the desert locust. While Uvarov (1954) has commented on the importance of permanent refugia, it seems likely many locust populations colonize by change. At the time of outbreak, the swarms appear to take wing into the south winds at the front of a low-pressure cell (Rainey, 1951). Since the winds are convergent in a low, the swarms get caught up into the cell and are carried down in it with the prevailing wind flow. The survival advantage is that lows produce rain, and so the swarm eventually descends in an area which is rained upon, where accordingly there is soil moisture to enhance egg development, and where subsequently there is plant growth to feed the nymphs.

There is always a risk that the rain will not be sufficient to allow completion of the life cycle. Some swarms are carried out to sea. Others reach unfavorable mountainous areas or areas which are too humid. Hence there is a considerable element of risk attached to the future of any one swarm. But where enough swarms form across both the northern and southern fringes of the Sahara, plus most of the Middle East and western India, the probability is sufficiently high that enough will succeed to ensure survival of the species.

Nomadism, in those species which display it, is another means for improving the risk. And in sum, there appear to be several regulatory patterns in desert species which prevent them from fluctuating to excessively high or dangerously low levels.

Limitation of desert animals

As I have discussed above, it appears that *k*-factor analysis, or some analogue, has great potential for quantifying the relative importance of limiting influences. There are frequent allusions in the ecological literature to 'resource limitation', 'predator limitation', 'self limitation', etc. But in my experience, populations often are limited by complexes of factors and the research goal should be to elucidate these complexes, to quantify the relative importance of each, and then generalize across species lines. Only in this way can we identify the primary importance of resources or predators or intrinsic mechanisms, and place them in perspective with the other factors operating. Since research on desert species has not progressed very far in this direction, we are left with a more qualitative approach to the problem. We can at best identify some of the factors which are operating, and in some cases made a subjective assessment of their importance.

One could argue that any environmental influence, which by varying over a range of values from year-to-year produces correlated fluctuations in a population, also, on the average, limits the density of that population to some degree (Wagner, 1978*b*). If the extreme values for the influence allow population increase, then it could be argued that the mean or modal values are less favorable. Since they may collectively distribute around their mean, the majority of values for the influence and its long-term average are unfavorable and limit the population.

This does not imply that the varying factor is the only limiting influence. Nor does it imply in any way that a population, eased of the pressure of this factor, would continue increasing indefinitely. With some increase, density-dependent pressures would doubtless increase and ultimately establish a new equilibrium. The contention here is only that a factor, which by varying produces fluctuations in a population can be considered as one influence which limits a population to some degree. Hence by identifying environmental factors which cause fluctuation, we can identify some limiting influences.

We have stressed the correlations between rainfall variation and population fluctuation in desert species in previous discussions. In some cases, the effects of rainfall appear direct. Desert locusts need soil moisture for egg development (Uvarov, 1954). Tenebrionids need proper soil temperature and moisture for larval survival (Hinds & Rickard, 1973; Thomas, 1975). In these cases, the mean weather pattern in a local can be considered to impose some degree of limitation.

But much of the rainfall-induced fluctuation appears due to the mediating availability of food. This was surely true for lizards which must be, on the basis of the evidence examined, widely limited at least in part by food supply. We were uncertain whether the similar patterns in herbivorous species were due to variations in the quantity or quality of food. But whichever it is, these species evidently are food-limited to a considerable degree.

In Chapter 4 on animal–animal interactions, additional evidence was examined of food limitation: shortage of seeds for rodents and ants, shortage of cellulose for termites. The jackrabbit–coyote oscillations we have discussed imply significant food limitation on coyote numbers.

In total, food shortage appears to be a significant and pervasive limiting influence on desert animal populations, a point already stressed by Brinck (1956). It is possible, in the light of Brincks's discussion, that the significance of food shortage may increase with progressively greater aridity and hence be greatest in the driest deserts.

There are also obvious cases of habitat limitation. This would appear to be true of those species which depend on refugia in times of drought stress. With more favorable microhabitats the lower threshold density would presumably be higher, and perhaps the outbreak numbers as well during favorable periods. This is suggested by Gorsuch's (1934) early historical accounts of bountiful Gambel quail populations.

Interesting cases of habitat limitation have been reported several times for woodrats (*Neotoma* spp.) which utilize certain succulents and shrub types for both nesting and water sources (MacMillen, 1964; Stones & Hayward, 1968; Brown, Lieberman & Dengler, 1972; Whitford, 1976). In analogous situations, Tomoff (1974) commented on the importance of thorny succulents and shrubs as nesting habitat for Sonoran Desert birds. Raitt & Maze (1968) measured the number of breeding birds and species in a Chihuahuan Desert area and compared their results with those of other studies. The cross-desert comparison showed that, as the number of species increased total avian density increased. Since species density is importantly a function of habitat diversity (Tomoff, 1974), one could infer that as new habitat elements were added new species and additional numbers of birds are added.

In Chapter 4, considerable evidence was reviewed which pointed to interspecific competition, predation, and agonistic behavior as limiting influences on a variety of species. In total, it seems clear that desert species are limited by complexes of influences, some factors being more influential on some species than others. A balanced assessment of the relative importance of different factors, and combinations thereof, remains to be made before we can generalize effectively. But the problems are tractable and that assessment will be possible at some future date.

References

Anderson, A. H. & Anderson, A. (1973). *The cactus wren*. University of Arizona Press, Tucson.

Ballinger, R. E. (1977). Reproductive strategies: food availability as a source of proximal variation in a lizard. *Ecology*, **58**, 628–35.

Baltensweiler, W. (1968). The cyclic population dynamics of the grey larch tortrix, *Zeiraphera griseana* Hübner (= *Semasia diniana* Guenée) (Lepidoptera: Tortricidae). In: *Insect abundance. Symposia of the Royal Entomological Society of London, 4* (ed. T. R. E. Southwood), pp. 89–97. Blackwell, Oxford.

Beale, D. M. & Smith, A. D. (1970). Forage use, water consumption, and productivity of pronghorn antelope in western Utah. *Journal of Wildlife Management*, **34**, 570–82.

Beatley, J. C. (1969). Dependence of desert rodents on winter annuals and precipitation. *Ecology*, **50**, 721–4.

Berry, K. H. (1978). Livestock grazing and the desert tortoise. *Transactions of the North American Wildlife and Natural Resources Conference*, **43**, in press.

Blaisdell, J. P. (1958). *Seasonal development and yield of native plants on the upper Snake River plains and their relation to certain climatic factors.* United States Department of Agriculture Technical Bulletin No. 1190. Washington, DC.

Bodenheimer, F. S. (1954). Problems of physiology and ecology of desert animals. In: *Biology of deserts* (ed. J. L. Cloudsley-Thompson), pp. 162–7. Institute of Biology, London.

Bodenheimer, F. S. & Sulman, F. (1946). The estrous cycle of *Microtus guentheri* D. and A. and its ecological implications. *Ecology*, **27**, 255–6.

Bradley, W. G. & Mauer, R. A. (1971). Reproduction and food habits of Merriam's kangaroo rat, *Dipodomys merriami. Journal of Mammalogy*, **52**, 497–507.

Bradley, W. G. & Mauer, R. A. (1973). Rodents of a creosote bush community in southern Nevada. *Southwestern Naturalist*, **17**, 333–44.

Braithwaite, L. W. & Frith, H. J. (1969). Waterfowl in an inland swamp in New South Wales III. Breeding. *CSIRO Wildlife Research*, **14**, 65–109.

Brinck, P. (1956). The food factor in animal desert life. In: *Bertil Hanstrom: zoological papers in honor of his sixty-fifth birthday November 20th, 1956* (ed. K. G. Wingstrend), pp. 120–7. Zoological Institute Lund, Sweden.

Brown, O. E., Cochran, C. L. & Waddell, T. E. (1978). Using call counts to predict hunting success for scaled quail. *Journal of Wildlife Management*, **42**, 281–7.

Brown, J. H., Lieberman, G. A. & Dengler, W. F. (1972). Woodrats and cholla: dependence of a small mammal population on the density of cacti. *Ecology*, **53**, 310–3.

Cable, D. R. & Martin, S. C. (1975). *Vegetation responses to grazing, rainfall, site condition, and mesquite control on semidesert range*. Department of Agriculture and Forest Services Research Paper RM-194. Washington, DC.

Campbell, H., Martin, D. K., Ferkovich, P. E. & Harris, B. K. (1973). *Effects of hunting and some other environmental factors on scaled quail in New Mexico.* Wildlife Monographs No. 34. Wildlife Society, Washington, DC.

Chew, R. M. (1975). *Effect of density on the population dynamics of* Perognathus formosus *and its relationships within a desert ecosystem.* US/IBP Desert Biome Research Memorandum RM 75-18, pp. 19–25. Utah State University, Logan.

Clark, F. W. (1972). Influence of jackrabbit density on coyote population changes. *Journal of Wildlife Management*, **36**, 343–56.

Clark, D. P. (1974). The influence of rainfall on the densities of adult *Chortoicetes terminifera* (Walker) in central western New South Wales, 1965–73. *Australian Journal of Zoology*, **22**, 365–86.

Cloudsley-Thompson, J. (1976). *Deserts and grasslands*, part 1. *Desert life*. Doubleday, New York.

Cole, L. C. (1951). Population cycles and random oscillations. *Journal of Wildlife Management*, **15**, 233–52.

Conrad, V. (1941). The variability of precipitation. *Monthly Weather Reviews*, **69**, 5–11.

Daumas, E. (1971). *The ways of the desert*. (Translated from the 1850 original French version of Sheila M. Ohlendorf.) University of Texas Press, Austin, Texas, & London.

Dempster, J. P. (1963). The population dynamics of grasshoppers and locusts. *Biological Reviews*, **38**, 490–529.

Derickson, W. K. (1976). Ecological and physiological aspects of reproductive strategies in two lizards. *Ecology*, **57**, 445–58.

Dixon, K. L. (1959). Ecological and distributional relations of desert scrub birds of western Texas. *Condor*, **61**, 397–409.

Dobzhansky, T. (1950). Evolution in the tropics. *American Scientist*, **38**, 209–21.

Doyen, J. T. & Tschinkel, W. F. (1974). Population size, microgeographic distribution and habitat separation in some tenebrionid beetles (Coleoptera). *Annals of the Entomological Society of America*, **67**, 617–26.

Edney, E. B. (1974). Desert arthropods. In: *Desert biology*, vol. 2 (ed. G. W. Brown), pp. 311–84. Academic Press, New York & London.

Egoscue, H. H. (1975). Population dynamics of the kit fox in western Utah. *Bulletin of the Southern California Academy of Science*, **74**, 122–7.

Fleming, T. H. (1974). The population ecology of two species of Costa Rican heteromyid rodents. *Ecology*, **55**, 493–510.

French, N. R., Maza, B. G., Hill, H. O., Aschwanden, A. P. & Kaaz, H. W. (1974). A population study of irradiated desert rodents. *Ecological Monographs*, **44**, 45–72.

French, N. R., Stoddart, D. M. & Bobeck, B. (1975). Patterns of demography in small mammal populations. In: *Small mammals: their productivity and population dynamics* (ed. F. B. Golley, K. Petrusewicz & L. Ryszkowski), pp. 73–102. Cambridge University Press, Cambridge, London, New York & Melbourne.

Gadgil, M. & Bossert, W. H. (1970). Life historical consequences of natural selection. *American Naturalist*, **104**, 1–24.

Gadgil, M. & Solbrig, O. T. (1972). The concept of *r*- and *K*-selection: evidence from wildflowers and some theoretical considerations. *American Naturalist*, **106**, 14–31.

George, U. (1976). *The deserts of this earth*. Harcourt Brace Jovanovich, New York & London.

Gorsuch, D. M. (1934). *Life history of the Gambel quail in Arizona*. University of Arizona Bulletin V(4), Biological Sciences Bulletin No. 2. Tucson.

Gross, J. E., Stoddart, C. S. & Wagner, F. H. (1974). *Demographic analysis of a northern Utah jackrabbit populations.* Wildlife Monographs No. 40. Wildlife Society, Washington, DC.

Gullion, G. W. (1960). The ecology of Gambel's quail in Nevada and the arid Southwest. *Ecology*, **41**, 518–36.

Gupta, R. & Prakash, I. (1975). *Environmental analysis of the Thar Desert.* English Book Depot, Dehra Dun, India.

Hafez, E. S. E. (1968). *Adaptation of domestic animals.* Lea & Febiger, Philadelphia.

Happold, D. C. D. (1975). The ecology of rodents in the northern Sudan. In: *Rodents in desert environments* (ed. I. Prakash & P. K. Ghosh), pp. 15–45. Dr W. Junk, The Hague.

Hastings, J. R. & Turner, R. M. (1965). *The changing mile. An ecological study of vegetation change with time in the lower mile of an arid and semiarid region.* University of Arizona Press, Tucson.

Hinds, W. T. & Rickard, W. H. (1973). Correlations between climatological fluctuations and a population of *Philolithus densicollis* (Horn) (Coleoptera: Tenebrionidae). *Journal of Animal Ecology*, **59**, 215–19.

Hoddenbach, G. A. & Turner, F. B. (1968). Clutch size of the lizard *Uta stansburiana* in southern Nevada. *American Midland Naturalist*, **80**, 262–5.

Hulett, G. K. & Tomanek, G. W. (1969). Forage production on a clay upland range site in western Kansas. *Journal of Range Management*, **22**, 270–6.

Hungerford, C. R. (1964). Vitamin A and productivity in Gambel's quail. *Journal of Wildlife Management*, **28**, 141–7.

Hutchings, S. S. & Stewart, G. (1953). *Increasing forage yields and sheep production on intermountain winter ranges.* United States Department of Agriculture Circular 925. Washington, DC.

Keast, A. (1959). Australian birds: Their zoogeography and adaptations to an arid country. In: *Biogeography and ecology in Australia* (ed. A. Keast, R. L. Crocker & C. S. Christian), pp. 89–114. Dr W. Junk, The Hague.

Keast, J. A. & Marshall, A. J. (1954). The influence of drought and rainfall on reproduction in Australian desert birds. *Proceedings of the Zoological Society of London*, **124**, 493–9.

Keith, L. B. (1963). *Wildlife's ten-year cycle.* University of Wisconsin Press, Madison.

Kenedy, J. S. (1956). Phase transformation in locust biology. *Biological Reviews*, **31**, 349–70.

Kikkawa, J. (1974). Comparison of avian communities between wet and semiarid habitats of eastern Australia. *Australian Wildlife Research*, **1**, 107–16.

King, C. E. & Anderson, W. W. (1971). Age specific selection. II. The interaction between *r* and *K* during population growth. *American Naturalist*, **105**, 137–56.

Klomp, H. (1966). The dynamics of a field population of the pine looper. *Advances in Ecological Research*, **3**, 207–305.

Lehman, V. W. (1953). Bobwhite population fluctuations and vitamin A. *Transactions of the North American Wildlife Conference*, **18**, 199–246.

Le Houérou, H. N. (1976). Tunisia. *Ecological Bulletin (Stockholm)*, **24**, 127–35.

Le Houérou, H. N. & Hoste, C. H. (1977). Rangeland production and annual rainfall relations in the Mediterranean Basin and in the African Sahelo-Sudanian Zone. *Journal of Range Management*, **30**, 181–9.

Lewin, V. (1963). Reproduction and development of young in a population of California quail. *Condor*, **65**, 249–78.

Longley, R. W. (1952). Measures of the variability of precipitation. *Monthly Weather Review*, **80**, 111–17.

MacArthur, R. H. & Wilson, E. O. (1967). *The theory of island biogeography*. Princeton University Monograph in Population Biology No. 1.

MacMillen, R. E. (1964). *Population ecology, water relations, and social behavior of a southern California semi-desert rodent fauna*. University of California Publications in Zoology, 71. Berkeley.

Marshall, A. J. (1961). *Biology and comparative physiology of birds*. Academic Press, New York & London.

Mayhew, W. W. (1966). Reproduction in the Arenicolous lizard *Uma notata*. *Ecology*, **47**, 9–18.

Mayhew, W. W. (1968). Biology of desert amphibians and reptiles. In: *Desert biology* (ed. G. W. Brown), pp. 195–356. Academic Press, New York & London.

Maza, B. G., French, N. R. & Aschwanden, A. P. (1973). Home range dynamics in a population of heteromyid rodents. *Journal of Mammalogy*, **54**, 405–25.

McMillan, I. I. (1964). Annual population changes in California quail. *Journal of Wildlife Management*, **28**, 702–11.

Miller, A. H. & Stebbins, R. C. (1964). *The lives of desert animals in Joshua Tree National Monument*. University of California Press, Berkeley & Los Angeles.

Mills, C. A. (1942). *Climate makes the man*. Harper & Bros, New York & London.

Milstead, W. W. (1965). Changes in competing populations of whiptail lizards (*Cnemidophorus*) in southwestern Texas. *American Midland Naturalist*, **73**, 75–80.

Misonne, X. (1975). The rodents of the Iranian deserts. In: *Rodents in desert environments* (ed. I. Prakash & P. K. Ghosh), pp. 47–58. Dr W. Junk, The Hague.

Mispagel, M. E. (1974). An ecological analysis of insect populations on *Larrea tridentata* in the Mohave Desert. MS Thesis, California State University, Long Beach.

Moreau, R. E., Wilk, A. L. & Rowan, W. (1947). The moult and gonad cycles of three species of birds at five degrees south of the equator. *Proceedings of the Zoological Society of London*, **117**, 345–64.

Morris, R. F. (ed.) (1963). The dynamics of epidemic spruce budworm populations. *Memoirs of the Entomological Society of Canada*, **31**, 1–332.

Muir, R. J. (1971). *Estimates of total macroarthropod biomass*. US/IBP Desert Biome, Tunisian PreSaharan Project Progress Report No. 7, pp. 200–7. Utah State University, Logan.

Murray, R. W. & Frye, O. E. (1957). *The bobwhite quail and its management in Florida*. Florida Game and Fresh Water Fish Commission Game Publication No. 2. Tallahassee, Florida.

Myers, K. & Parker, B. S. (1975). Effect of severe drought on rabbit numbers and distribution in a refuge area in semiarid north-western New South Wales. *Australian Wildlife Research*, **2**, 103–20.

Naumov, N. P. (1972). *The ecology of animals* (ed. N. D. Levine, translated by F. K. Plons, Jr). University of Illinois Press, Urbana & Chicago.

Naumov, N. P. (1975). The role of rodents in ecosystems of the northern deserts of Eurasia. In: *Small mammals: their productivity and population dynamics* (ed. F. B. Golley, K. Petrusewicz & L. Ryszkowski), pp. 299–309. Cambridge University Press, Cambridge, London, New York & Melbourne.

Nelson, B. (1973). *Azraq: desert oasis*. Ohio University Press, Columbus.

Newsome, A. E. (1965*a*). The abundance of red kangaroos, *Megaleia rufa* (Desmarest) in central Australia. *Australian Journal of Zoology*, **13**, 269–87.

Newsome, A. E. (1965*b*). Reproductions in natural populations of the red kangaroo, *Megaleia rufa* (Desmarest), in central Australia. *Australian Journal of Zoology*, **13**, 735–59.

Newsome, A. E. & Corbett, L. K. (1975). Outbreaks of rodents in semiarid and arid Australia: causes, preventions, and evolutionary considerations. In: *Rodents in desert environments* (ed. I. Prakash & P. K. Ghosh), pp. 117–53. Dr W. Junk, The Hague.

Nichols, J. D., Conley, W., Batt, B. & Tipton, A. R. (1976). Temporally dynamic reproductive strategies and the concept of *r*- and *K*-selection. *American Naturalist*, **110**, 995–1005.

Nicholson, A. J. (1933). The balance of animal populations. *Journal of Animal Ecology*, **2**, 132–78.

Nicholson, A. J. (1954*a*). Compensatory reactions of populations to stresses and their evolutionary significance. *Australian Journal of Zoology*, **2**, 1–8.

Nicholson, A. J. (1954*b*). An outline of the dynamics of animal populations. *Australian Journal of Zoology*, **2**, 9–65.

Novikoff, G. (1977). *Vegetation measurements*. US/IBP Desert Biome, Tunisian PreSaharan Project Progress Report No. 7, 19–34. Utah State University, Logan.

O'Farrell, T. P., Olson, R. J., Gilbert, R. O. & Hedlund, J. D. (1975). A population of Great Basin pocket mice, *Perognathus parvus*, in the shrub steppe of south-central Washington. *Ecological Monographs*, **45**, 1–28.

Ohmart, R. D. (1973). Observations on the breeding adaptations of the roadrunner. *Condor*, **75**, 140–9.

Olding, R. J. & Cockrum, E. L. (1977). Estimation of desert rodent populations by intensive removal. *Journal of Arizona Academy of Science*, **12**, 94–108.

Paulsen, H. A. & Ares, F. N. (1962). *Grazing values and management of black grama and tobosa grasslands and associated shrub ranges of the southwest*. United States Department of Agriculture Forest Service Technical Bulletin No. 1270. Washington, DC.

Peterson, R. T. (1941). *A field guide to western birds*. Houghton Mifflin, Boston.

Pianka, E. R. (1970). On *r*- and *K*-selection. *American Naturalist*, **104**, 592–7.

Pinter, A. J. & Negus, N. C. (1965). Effects of nutrition and photoperiod on reproductive physiology of *Microtus montanus*. *American Journal of Physiology*, **208**, 633–8.

Prakash, I. (1975). The population ecology of the rodents of the Rajasthan Desert, India. In: *Rodents in desert environments* (ed. I. Prakash & F. K. Ghosh), pp. 75–116. Dr W. Junk, The Hague.

Prakash, I. & Ghosh, P. K. (eds) (1975). *Rodents in desert environments*. Dr W. Junk, The Hague.

Rainey, R. C. (1951). Weather and the movements of locust swarms: a new hypothesis. *Nature, London*, **168**, 1057–60.

Raitt, R. J. & Maze, R. L. (1968). Densities and species composition of breeding birds of a creosotebush community in southern New Mexico. *Condor*, **70**, 193–205.

Raitt, R. J. & Pimm, S. L. (1976). Dynamics of bird communities on the Chihuahuan Desert, New Mexico. *Condor*, **78**, 427–42.

Reichman, O. J. & van de Graaf, K. M. (1975). Association between ingestion of green vegetation and desert rodent reproduction. *Journal of Mammalogy*, **56**, 503–6.

Roe, A. L. & Amman, G. D. (1970). *The mountain pine beetle in lodgepole pine forests*. United States Department of Agriculture Forest Service Research Paper INT-71. Washington, DC.

Roffey, J. & Popov, G. (1968). Environmental and behavioral processes in a desert locust outbreak. *Nature, London*, **219**, 446–50.

Roonwal, M. L. (1953). Food preference experiments on the desert locust, *Schistocerca gregaria* (Forskal) in its permanent breeding grounds in Mekran. *Journal of the Zoological Society of India*, **5**, 44–58.

Roughgarden, J. (1971). Density-dependent natural selection. *Ecology*, **52**, 453–68.
Rowley, I., Braithwaite, L. W. & Chapman, G. S. (1973). The comparative ecology of Australian corvids. III. Breeding seasons. *CSIRO Wildlife Research*, **18**, 67–90.
Schemnitz, S. D. (1961). Ecology of the scaled quail in the Oklahoma panhandle. *Wildlife Monographs* No. 8.
Schmid, J. M. & Frye, R. H. (1977). *Spruce beetle in the Rockies*. United States Department of Agriculture Forest Service General Technical Report RM-49. Washington, DC.
Schwerdtfeger, F. (1941). Über die Uraschen des Massenwechsels der Insekten. *Zeitschrift für Angewandte Entomologie*, **28**, 254–303.
Serventy, D. L. (1971). Biology of desert birds. In: *Avian biology*, vol. 1. (ed. D. S. Farner & J. R. King), pp. 287–339. Academic Press, New York & London.
Short, L. L. (1974). Nesting of southern Sonoran birds during the summer rainy season. *Condor*, **76**, 21–32.
Solomon, M. E. (1971). Elements in the development of population dynamics. In: *Dynamics of populations. Proceedings of the Advanced Study Institute on 'Dynamics of Numbers in Populations, Oosterbeek, The Netherlands, 7–18 September 1970* (ed. P. J. den Boer & G. R. Gradwell), pp. 29–40. Centre for Agricultural Publishing and Documentation, Wageningen.
Sowls, L. K. (1957). Reproduction in the Audubon cottontail in Arizona. *Journal of Mammalogy*, **38**, 234–42.
Sowls, L. K. (1960). Results of a banding study of Gambel's quail in southern Arizona. *Journal of Wildlife Management*, **24**, 185–90.
Stoddart, L. C. (1977). *Population dynamics, movement and home range of black-tailed jackrabbits* (Lepus californicus) *in Curlew Valley, Northern Utah*. US Energy Research and Development Administration, Contract No. E(11-1)-1329, Progress Report. Utah State University, Logan.
Stones, R. C. & Hayward, C. L. (1968). Natural history of the desert woodrat, *Neotoma lepida. American Midland Naturalist*, **80**, 458–76.
Stout, G. G. (1970). The breeding biology of the desert cottontail in the Phoenix region, Arizona. *Journal of Wildlife Management*, **34**, 47–51.
Sumner, F. B. (1925). Some biological problems of our southwestern deserts. *Ecology*, **6**, 352–71.
Swank, W. G. & Gallizioli, S. (1954). The influence of hunting and rainfall upon Gambel's quail populations. *Transactions of the North American Wildlife Conference*, **19**, 283–96.
Thomas, D. B., Jr (1975). Dynamics of a species assemblage of desert tenebrionid beetles. MS thesis, Californian State University, Long Beach.
Tinkle, D. W. (1967). *The life and demography of the side-blotched lizard*, Uta stansburiana. University of Michigan Museum of Zoology Miscellaneous Publication No. 132.
Tinkle, D. W. (1969). The concept of reproductive effort and its relation to the evolution of life histories of lizards. *American Naturalist*, **103**, 501–16.
Tinkle, D. W., Wilbur, H. M. & Tilley, S. G. (1970). Evolutionary strategies in lizard reproduction. *Evolution*, **24**, 55–74.
Tomoff, C. S. (1974). Avian species diversity in desert scrub. *Ecology*, **55**, 396–403.
Turner, F. B. (1977). The dynamics of populations of squamates, crocodilians and rhyncocephalians. In: *Biology of the Reptilia, Ecology and Behavior A* vol. 7 (ed. C. Gans & D. W. Tinkle), pp. 157–264. Academic Press, London, New York & San Francisco.
Turner, F. B., Hoddenbach, G. A., Medica, P. A. & Lannom, J. R. (1970). The

demography of the lizard, *Uta stansburiana* Baird and Girard, in southern Nevada. *Journal of Animal Ecology*, **39**, 505–19.

Turner, F. B., Medica, P. A. & Smith, D. O. (1974). *Reproduction and survivorship of the lizard,* Uta stansburiana, *and the effects of winter rainfall, density and predation on these processes.* US/IBP Desert Biome Research Memorandum RM 74-26, pp. 117–28. Utah State University, Logan.

Uvarov, B. P. (1921). A revision of the genus *Locusta* L. (*Pachytylus* Fieb.) with a new theory as to the periodicity and migrations of locusts. *Bulletin of Entomological Research*, **12**, 135–63.

Uvarov, B. P. (1954). The desert locust and its environment. In: *Biology of deserts* (ed. J. L. Cloudsley-Thompson), pp. 85–9. Institute of Biology, London.

van de Graaf, K. M. & Balda, R. P. (1973). Importance of green vegetation for reproduction in the kangaroo rat, *Dipodomys merriami merriami. Journal of Mammalogy*, **54**, 509–12.

Varley, G. C., Gradwell, G. R. & Hassell, M. P. (1973). *Insect population ecology. An analytical approach.* Blackwell Scientific Publications, Oxford, London, Edinburgh & Melbourne.

Vorhies, C. T. & Taylor, W. P. (1933). *The life histories and ecology of jackrabbits,* Lepus alleni *and* Lepus californicus *ssp. in relation to grazing in Arizona.* University of Arizona Agricultural Experimental State Technical Bulletin 49, Tucson.

Wagner, F. H. (1969). Ecosystem concepts in fish and game management. In: *The ecosystem concept in natural resource management* (ed. G .M. Van Dyne), pp. 259–307. Academic Press, New York & London.

Wagner, F. H. (1978). Some concepts in the management and control of small mammal populations. In: *Populations in small mammals under natural conditions. A Symposium held at the Pymatuning Laboratory of Ecology, May 14–16, 1976* (ed. D. P. Snyder), pp. 192–203. University of Pittsburg Pymatuning Laboratory of Ecology Special Publication Series, 5.

Wagner, F. H. (1979). Integrating and control mechanisms in arid and semiarid ecosystems – considerations for impact assessment. *Proceedings of a Symposium on Biological Evaluation of Environmental Impact. 27th Annual AIBS Meeting, New Orleans.* Council on Environmental Quality, Washington, D.C. In press.

Wagner, F. H., Besadny, C. D. & Kabat, C. (1965). *Population ecology and management of Wisconsin pheasants.* Wisconsin Department of Conservation Technical Bulletin No. 34. Madison, Wisconsin.

Wagner, F. H. & Stoddart, L. C. (1972). Influence of coyote predation on black-tailed jackrabbit populations in Utah. *Journal of Wildlife Management*, **36**, 329–42.

Wagner, H. O. (1957). Variations in clutch sizes at different latitudes. *Auk*, **74**, 243–57.

Walsberg, G. E. (1978). Brood size and the use of time and energy by the phainopepla. *Ecology*, **59**, 147–53.

White, T. C. R. (1976). Weather, food and plagues of locusts. *Oecologia*, **22**, 119–34.

Whitford, W. G. (1976). Temporal fluctuations in density and diversity of desert rodent populations. *Journal of Mammalogy*, **57**, 351–69.

Whitford, W. G. & Creusere, F. M. (1977). Seasonal and yearly fluctuations in Chihuahuan Desert lizard communities. *Herpetologica*, **33**, 54–65.

Williamson, M. (1972). *The analysis of biological populations.* Edward Arnold, London and Crane, Russak & Co., New York.

Wilson, E. O. (1975). *Sociobiology. The new synthesis.* Belknap Press of Harvard University Press, Cambridge, Massachusetts.

Manuscript received by the editors July 1979.

8. Primary productivity

L. E. RODIN

Water input and transpiration

Water relations of plants in the Karakum Desert
(V. M. Sveshnikova, N.T. Nechaeva & A. P. Savinkin)

Diurnal and seasonal variations in transpiration intensity, water content in leaves, water deficiency, osmotic pressure and sucking force (Table 8.1) were evaluated in the dominant plant species of the Karakum Desert (The Repetek Reservation) during 1965 to 1969.

Plant species growing in the Karakum are characterized by varying intensities of water expenditure. Most representative values of transpiration activity prove, on the whole, to be rather small, while the limits of their variation are either wide for one group or species, or narrow for others. The maximum intensity of transpiration in tree and shrub species per gramme wet weight is never higher than 900 mg h^{-1}, while in herbaceous perennial plants these values are as high as 1750 mg h^{-1} and in ephemerals 2760 mg h^{-1}.

Correlation analysis was employed to estimate the dependence of transpiration intensity upon factors which govern this process. In spring, the most favourable season, one finds the most pronounced dependence of rate of water loss upon endogenous factors. In the spring–summer season, the high correlation between transpiration intensity and air temperature and humidity cannot be found. In summer, despite the increasing influence of environmental conditions, the role of endogeneous factors is still significant, and only in autumn do the effects of environmental conditions become prominent. At that time the transpiration intensity in all species increases with elevation of temperature and decreases in air humidity.

Comparison of the rate of water loss in summer with the rate of evaporation from a free water surface is representative of relative transpiration. The latter index reveals the degree of water retention by leaves: in various species the transpiration intensity proved to be 5.5–25% that of evaporation.

The amount of water in leaves or in the assimilating organs varied considerably between plant species (Table 8.1). The moisture content in herbaceous perennial plants is higher than in the tree and shrub species. Water content typically decreases during the growing season and in hot summer reaches its lowest level. Investigation of the water-retaining capacity – that is, the ratio of the amount of water at full saturation to the highest water content registered under natural conditions – showed that water storage in leaves of plants of the species studied is relatively large.

In defining the water relations of arid-zone plants, it is important to

Table 8.1. *Indicators of the water regime in dominant plants of the Karakum (Repetek)*

Life form, species	Transpiration intensity (mg g^{-1} wet wt h^{-1})		Water content (% wet wt)	Osmotic pressure (bar)		Suction force (bar)		Water deficit (% of saturation)		Degree of potential drying
	Maximum	Typical		Max.	Min.	Max.	Min.	Normal	Sub-lethal	
Trees, shrubs										
Haloxylon ammodendron	670	100–400	62–80	58	46	64	—	14	47–50	30
H. persicum	850	100–700	50–66	58	38	52	—	21	48–53	48
Calligonum caput-medusae	780	100–500	60–76	41	16	45	—	19	45–48	40
Salsola richteri	760	100–400	55–76	56	31	47	—	24	46–48	50
Ammodendron conollyi	810	100–600	51–76	50	24	47	—	19	43–46	65
Sub-shrubs										
Astragalus paucijugus	1540	680–700	63–83	—	4	49	29	23	45–49	
Smirnovia turkestana	2600	600	57–81	—	—	46	42	33	50–66	
Perennial herbs										
Aristida karelinii	930	240								
Heliotropium arguzioides	1750	780	52–89	—	—	44	25	33	50–67	
Carex physodes	2760	600								

elucidate the state of their water balance at various stages of growth. To this end, variations in the amount of water deficiency are considered. Studies of these variations in plants of the Karakum revealed that water deficiency is limited and, as a rule, does not exceed 33% (Table 8.1). In the case of arid-zone plants the span of observable variations in the water deficiency is extremely narrow: 14–33%. The lowest water-deficiency level was registered in the perennial tree and shrub species with a prolonged growing season, while the highest was in shrubs with a developed leaf plate and also in ephemeroids. The critical water-deficiency level has never reached the point of no return. The ratios of the values of the natural water deficiency to sub-lethal ones indicates the absence of notable alterations in the water supply of plants.

Indices of osmotic pressure and of sucking force in plant leaves are widely employed for analysing dependence of plant activity upon environmental factors. The maximum values for the osmotic pressure (41–58 bar) and for sucking force (47–64 bar) were estimated for the tree and shrub species (Table 8.1). (The sucking force in leaves of perennial herbaceous plants is also high.) Maximum levels of sucking force and osmotic pressure during the major portion of the studies proved to be of approximately the same order of magnitude specific for the stable state of the water balance. In the course of growth the osmotic pressure and sucking force values in plants of the Karakum display considerable variations from spring to summer.

Thus the results obtained revealed that trees and shrubs in the Karakum maintain their water balances, and these do not change noticeably even during periods of severe drought. This is reflected in the low level of water deficiency, in the range of variations in water content, and in the positive turgor pressure observed over almost the entire growing period. Every species is featured by its own complex of eco-physiological characteristics. Species-adaptive reactions and the pattern of water relations are adequately tailored to environmental conditions in the dominant species.

In the lowland Karakum (the Karrykul area), transpiration intensity was studied in the 42 most widely distributed species.

Differentiation of desert plants with respect to their transpiration intensity is shown in Table 8.2. Among the plants with low transpiration intensity, one finds *Ephedra strobilacea*, *Aellenia subaphylla*, *Salsola richteri*, *S. arbuscula* and *Haloxylon persicum* (100–250 mg g^{-1} wet wt h^{-1}); higher transpiration intensity (460–520 mg g^{-1} wet wt h^{-1}) is observed in *Ammodendron conollyi*, *Astragalus longipetiolatum* and *Calligonum* spp. High levels of transpiration intensity are found in the widely distributed *Artemisia kemrudica* and *Mausolea eriocarpa* (750–1020 mg g^{-1} wet wt h^{-1}). The transpiration intensity is exceptionally high (750 mg g^{-1} wet wt h^{-1}) in *Ammothamnus lehmannii*. Thus, the tree and shrub species and semi-shrublets are characterized by a wide range of values for water expenditure through transpiration.

Among the perennial herbaceous plants the species characteristic of the

Table 8.2. *Transpiration intensity of desert plants of the lowland Karakum (Karrykul area) (mg g⁻¹ wet wt h⁻¹)*

Perennials		Annuals	
Trees, shrubs		Winter–spring	
Aellenia subaphylla	100	*Alyssum desertorum*	690
Ephedra strobilacea	220	*Veronica biloba*	690
Haloxylon persicum	250	Early spring	
Salsola richteri	320	*Tetracme recurvata*	300
S. arbuscula	320	*Malcolmia grandiflora*	380
Calligonum setosum	460	*Hypecoum pendulum*	410
S. rubens	490	Later spring	
Ammodendron conollyi	500	*Senecio subdentatus*	660
Semi-shrubs, semi-shrublets		*Delphinium camptocarpum*	670
Salsola gemmascens	240	*Koelpinia linearis*	670
S. orientalis	270	*Lappula caspia*	1070
Astragalus longipetiolatus	520	Early summer	
Ammothamnus lehmannii	750	*Crucianella filifolia*	940
Artemisia kemrudica	750	*Euphorbia densa*	1020
Mausolea eriocorpa	1020	*Aphanopleura leptoclada*	1300
Herbs		Summer	
Astragalus hivensis	290	*Diarthron vesiculosum*	1220
Peganum harmala	360	*Chrozophora gracilis*	2550
Aristida pennata	370	Summer–autumn	
A. karelinii	560	*Climacoptera lanata*	190
Carex physodes	740	*Salsola sclerantha*	340
Haplophyllum pedicellatum	850	*Agriophyllum latifolium*	1750
Rheum turkestanicum	900	*Euphorbia cheirolepis*	1970
Schumannia karelinii	960		
Iris songarica	1100		
Alhagi persarum	1390		

winter–spring and spring–summer growing period (*Carex physodes*, *Haplophyllum pedicellatum*, *Tournefortia sogdiana*, *Alhagi persarum*) display high (700–1390 mg g^{-1} wet wt h^{-1}) levels of transpiration while species growing in the spring–autumn period (*Aristida pennata*, *A. karelinii*, *Astragalus chivensis*, *Peganum harmala*) have lower transpiration levels (300–500 mg g^{-1} wet wt h^{-1}). It is probable, therefore, that in the perennial herbaceous plants transpiration intensity is rather well correlated with humidity and temperature conditions.

Annual plants characteristic of winter–spring species (*Alyssum desertorum*, *Veronica biloba*) display transpiration intensities up to 690 mg g^{-1} wet wt h^{-1}. The early-spring species (*Malcolmia grandiflora*, *Tetracme recurvata*) have lower transpiration levels (300–410 mg g^{-1} wet wt h^{-1}) while the late-spring species (*Delphinium camptocarpum*, *Senecio subdentatus*) have somewhat higher levels (670–1070 mg g^{-1} wet wt h^{-1}). The early-summer species may also be grouped with the latter. Exceptionally high transpiration intensities are found in the summer and summer–autumn annual plants (up to 2000–2500 mg g^{-1} wet wt h^{-1}). Thus transpiration intensity is higher in plants growing in summer than in those growing in early spring.

In the majority of desert plants, maximum transpiration values are observed in spring when assimilation processes are most active. In a number of the Chenopodiaceae, the highest levels of transpiration and phytomass accumulation are registered in summer. The most widely distributed dominant species of desert vegetation display low transpiration intensity (100–250 mg g^{-1} wet wt h^{-1}) as in *Haloxylon ammodendron*, *H. persicum*, *Salsola richteri*, *S. arbuscula*, or a moderately low level (300–500 mg g^{-1} wet wt h^{-1}) as in *Ammodendron conollyi* and *Calligonum* spp. This property presumably served to promote the dominant role of these species in the vegetation cover.

Studies on transpiration intensity in plants of the Malye Barsuki area were performed on a number of species in the sagebrush communities. The diurnal pattern of transpiration in early spring, when available soil moisture is optimum, differs from the pattern registered in the summer months when moisture is insufficient. In spring, the transpiration intensity increases toward noon: the curve for transpiration intensity is essentially a one-peak curve and it follows the curve for solar radiation. At noon, transpiration rates for *Artemisia terrae-albae*, *Anabasis salsa* and *Rheum tataricum* are *c.* 1100, 700 and 800 mg g^{-1} wet wt h^{-1}, respectively.

At the beginning of summer (10–20 May), the parallel pattern of the solar radiation and transpiration curves can be observed only until 10.00–11.00 h. Later in the day when radiation increases, transpiration intensity declines. At noon *Artemisia terrae-albae* transpires 700 mg g^{-1} wet wt h^{-1} and *Anabasis salsa*, 400 mg g^{-1} wet wt h^{-1}.

In years when humidity is favourable, transpiration in the ephemerals and ephemeroids increases from the beginning of development to the period of maximum development (April). In the majority of perennial plants, this increase continues into May. In a dry spring period, transpiration of the long-lived plants decreases along with decreases in available soil water and increases after rainfall.

Water relations of plants in the northern Gobi Desert

(V. M. Sveshnikova & N. I. Bobrovskaya)

In 1970, which was a dry year in the Gobi Desert, variations in transpiration intensity were very high in *Reaumuria soongorica* (165–1450 mg g^{-1} wet wt h^{-1}, and only about half in *Zygophyllum xanthoxylon* (100–640) and *Salsola passerina* (60–760). In years of sufficient humidity, this span of values is lower, and for various species this parameter is rather similar (Table 8.3). Transpiration intensity is highest in *Reaumuria soongorica* and increases considerably with enhancement of drought effects. The water loss in *Zygophyllum xanthoxylon* is less than in other species.

Table 8.3. *Limits of variation of indices of the water regime in plants on brown saline soils in the Northern Gobi*

Species	Year of obser-vation	Intensity of transpiration ($mg\ g^{-1}$ wet wt h^{-1})				Water content of leaves (% of wet wt)				Water deficit (% of saturation)	Suction force (bar)	
		Mean	Maximum	Mini-mum	Range	Mean	Maxi-mum	Mini-mum	Range		Mini-mum	Maxi-mum
Reaumuria	1970	300–1045	470–1450	165	1285	60.0–48	72.0	41.0	31.0	8.0 24.0	—	100
soongorica	1971	190–250	210–450	80	370	63.0–48	71.0	48.0	23.0	11.1 26.1	24	30
Zygophyllum	1970	130–260	310–640	100	540	80.0–75.4	86.0	65.0	21.0	5.0 18.0	—	90
xanthoxylon	1971	110–180	140–320	40	280	83.0–80.1	89.0	78.0	11.0	2.8 5.8	25	37
Salsola passerina	1970	145–410	250–760	60	700	71.3–61.7	79.0	52.0	27.0	19.0 32.0	60	85
	1971	120–230	150–370	60	310	70.0–63.5	73.0	58.0	15.0	18.0 33.7	23	45
Brachanthemum	1970	200–490	290–680	120	560	59.5–51.6	68.0	48.0	20.0	15.0 27.0	70	100
gobicum	1971	150–300	180–380	40	360	70.1–64.2	73.0	62.0	11.0	12.3 17.7	24	42

Table 8.4. *Limits of variations in intensity of water regime for indicator species growing on sandy soils in the Northern Gobi*

Species	Years	Intensity of transpiration (mg g^{-1} wet wt h^{-1})				Water content of leaves (% of wet wt)				Water deficit (%)		Sucking force (bar)	
		Average	Maximum	Minimum	Difference	Average	Maximum	Minimum	Difference	Average	Maximum	Minimum	Maximum
Haloxylon ammodendron	1970	240–340	340–430	145	285	69–77	84	60	24	3.5	4.0	—	—
	1971	110–270	70–430	30–130	400	75–80	84	72	12	5.5	7.5	30	43
Nitraria sibirica	1970	260–620	450–960	160	800	75–82	89	60	29	9.5	14.0	20	80
	1971	140–460	370–740	60–260	680	76–82	89	73	16	8.2	11.0	25	30

Decrease in water content of leaves was in the following order: *Zygophyllum xanthoxylon*, *Salsola passerina*, *Reaumuria soongorica*, *Brachanthemum gobicum* (Table 8.3). In years with more humid conditions, variations in the water content in all species proved to be approximately half that in dry years.

Assessment of the sucking force in 1970 testifies to the exceptionally high tension in the water supply of plants (the absolute values reached 85–100 bar). These forces were quite different in years when humidity was normal: the highest values for the sucking force never exceeded 37–45 bar, the lowest were 23–25 bar.

The extent of water loss in leaves reflected by their water deficiency in the years under consideration was approximately the same. Water deficiency was as high as 18–32% in 1970 and 18–34% in 1971. The values reflect rather stable water balance in these species even under very different climatic conditions in these years. The highest water deficiency was found in leaves of *Salsola passerina*, the lowest in *Zygophyllum xanthoxylon*. Comparison of the water relations in plants growing on sandy soils in the Gobi lowlands with the species growing on brown desert soils revealed that in *Haloxylon ammodendron* and in *Nitraria sibirica* the water expenditure on transpiration is the most moderate due to the succulent leaf structure. *Haloxylon ammodendron* expends water most sparingly. The water content in *Haloxylon ammodendron* and in *Nitraria sibirica* leaves is high (Table 8.4).

The sucking force in plants was as high as 80 bar in 1970. Under favourable humidity conditions, its level never exceeded 43 bar. Water deficiency in *Haloxylon ammodendron* and in *Nitraria sibirica* was the lowest of all the other species studied: even the maximum values never exceeded 4–14% in 1970 and 7.5–11% in 1971.

Photosynthesis and respiration

Vegetation of the Kyzylkum, Karakum and Aralo-Caspian regions (L. N. Alexeeva, V. P. Bedenko, S. Fasylova, T. A. Glagoleva L. Kh. Naaber, O. A. Semikhatova, L. V. Shabanova, V. L. Voznesensky, I. L. Zakharjantz & O. V. Zalensky)

Studies on gas exchange, which is the basis for primary productivity of desert plants under natural environmental conditions, are of great interest with respect to the effects of high light and temperature intensities, high air dryness and frequently stringent water relations. Extensive investigations of the gas exchange in plants in the Malye Barsuki, the Kyzylkum desert and the Karakum desert have been started only recently (Zakharjantz *et al.*, 1971; Voznesensky, 1977; Zalensky, 1977). Formerly, only single determinations were made.

During the period of the International Biological Programme (1965–1972), we studied apparent photosynthetic rates with air flow techniques followed by either titration, or conductivity measurement of the alkali-absorbing solution. The potential photosynthetic rate was measured by manometric and radiometric techniques at saturation concentrations of CO_2 (Zalensky, 1963; Voznesensky, Zalensky & Semikhatova, 1965; Voznesensky, Zalensky & Austin, 1971). In the Malye Barsuki, photosynthetic rate was estimated by measuring accumulation of organic carbon every 3 h during a day. Of great significance are the investigations of plant respiration as a process having both its own physiological function and an influence on plant productivity. Respiration was estimated both by manometric technique (Semikhatova, Chulanovskaya & Metzner, 1971) and with air-flow. The effect of oxygen on photosynthesis and carbon metabolism was studied (Glagoleva, Mokronosov & Zalensky, 1978). Some data obtained in investigations, mostly on gas exchange in plants, are summarized in Table 8.5.

One of the significant results of this work is the discovery that in the majority of desert plants, active CO_2 assimilation during the growing period proceeds over the whole diurnal light period. Decrease or even cessation of photosynthesis was registered only in single cases of long and extremely hot weather, especially in ephemerals and mesoxerophytes of the Kyzylkum. The diurnal pattern of photosynthesis in all plants was characterized by gradually changing curves with the maximum values early in the day when water balance was normal. Any alteration of external factors (cloudiness, rain, dust storms, drought, etc.) produced abrupt shifts in the intensity of assimilation. Highest photosyntheic intensity was observed in all plants in spring which is the most favourable season in deserts, and in some species also in autumn. Generative stages of the majority of plants were observed in these seasons. The maximum photosynthetic rates in desert plants vary within rather wide limits: the potential photosynthesis, 30–115 mg CO_2 dm^{-2} h^{-1}; and the apparent one, 10–50 mg CO_2 dm^{-2} h^{-1}. The most active are the ephemerals and certain semi-shrub plants. Among families one can distinguish certain species of the Leguminoseae, Compositae, Gramineae and Cruciferae.

The species with long growing seasons display low, photosynthetic intensity. They are members of the Chenopodiaceae. At the same time there are species in every family that are characteristic of both low and high photosynthetic intensities. Within one species, however, the potential photosynthetic capacity remains constant under various ecological conditions of growth. This implies that the assimilating apparatus in plants is genetically determined.

According to the data obtained for desert plants in the Kyzylkum, the CO_2 concentration which saturates their photosynthetic capacity varies from that registered under natural conditions (in the case of halophytes) to 0.4%. The extent to which the maximum photosynthetic intensity increases during elevation of CO_2 concentration from the natural level to 1% allows one to follow the relationship of this parameter to the evolutionary age of the species under consideration, or the retention in these species of the evolutionarily fixed ability to withstand elevated CO_2 concentrations in the air. Among the plants studied, the highest increase of assimilation with elevated CO_2 concentration was registered in plants of the rather old genus *Ephedra*. All of the remaining species studied presumably evolved in the Neogene period or later in the Quaternary Era when CO_2 concentrations in the atmosphere were low.

Light saturation for potential photosynthetic intensity in the ephemerals is approximately half the light intensities observed at full solar illumination; in tree and shrub species it is two-thirds of full solar illumination. In some perennial herbaceous plants light saturation under natural conditions is never attained at elevated CO_2 concentrations.

Hot-desert plants are able to assimilate at temperatures within a wide range varying from approximately −5 to +55 °C. Evolution of CO_2 however, begins to exceed its assimilation at 40–45 °C and in some plants even at lower temperatures. The optimal temperature range for assimilation in the majority of plants appears to be 25–30 °C. In spring these values are somewhat lower. The most conspicuous feature of numerous desert plants is a wide range of the optimum temperatures for photosynthesis. *Haloxylon persicum*, in particular, is noted for its ability to assimilate at the same rate at temperatures from 10–45 °C. The lowest optimal temperatures were registered in the evergreen shrub *Ephedra strobilacea* which evidently accounts for its ability to assimilate actively, even in winter on relatively warm days.

The effect of different oxygen contents on $^{14}CO_2$ assimilation by dominant species of various ecological groups in the Karakum has been studied. The kinetic measurements were performed with exposures of 10–90 s with a CO_2 concentration of 0.04%. Use of pulse exposures favoured assessment of actual rate of this process. In order to decrease the oxygen content, the leaf chamber

Table 8.5. *Data on the gas exchange of desert plants in Central Asia*

Region and species	Life forms	Rate of photosynthesis (mg CO_2 g^{-1} dry wt h^{-1}): Apparent, Average	Rate of photosynthesis: Apparent, Maximum	Rate of photosynthesis: Potential maximum	Photosynthetic productivity (mg CO_2 day^{-1} g^{-1} dry wt): Average	Photosynthetic productivity: Maximum	Temperature optimum for photosynthesis (°C)	Light intensity saturating photosynthesis (potential intensity) (cal cm^{-2} min^{-1})	Rate of respiration: °C	Rate of respiration: mg CO_2 g^{-1} wet wt h^{-1}
Karakum										
Ammodendron conollyi	Tree	7.6	20	71	110	160	22–35	0.9	30	1.2
Haloxylon ammodendron	Tree	7.1	10	40	80	100	10–30	0.8	33	1.2
H. persicum	Tree	6.1	10	30	55	85	10–45	0.8	30	1.1
Salsola richteri	Shrub	6.0	18	35	120	160	20–30	0.9	30	0.6
Calligonum caput-medusae	Shrub	5.8	12	23	80	125	25	0.9	30	0.7
Ephedra strobilacea	Shrub	2.3	4	30	30	35	5–20	—	30	1.0
Astragalus paucijugus	Shrub	14	22	90	170	240	15–23	1.1	35	0.7
Aristida karelinii	Herbaceous perennial	42	52	110	250	460	30–40	1.4	32	1.6
Heliotropium arguzioides	Herbaceous perennial	8.6	24	77	170	215	15–30	1.4	35	1.1

Carex physodes	Ephemeroid	14	29	65	205	220	17–30	0.7	—	—
Malye Barsuki										
Artemisia terrae-albae	Semi-shrub	12	48	—	—	—	—	—	—	—
Kochia prostrata	Semi-shrub	12	33	—	—	—	—	—	—	—
Anabasis aphylla	Semi-shrub	11	31	—	—	—	—	—	—	—
A. salsa	Semi-shrub	14	42	—	—	—	—	—	—	—
Kyzylkum		(mg CO_2 dm^{-2} h^{-1})			(mg CO_2 dm^{-2} day^{-1})			(klx)		
Salsola orientalis	Shrub	10	23	22	—	200	15–35	70	22–25	1.1
Calligonum aphyllum	Shrub	25	32	34	—	300	—	30	20–25	1.3
Atraphaxis seravschanica	Shrub	—	24	42	—	—	20–35	40	20–25	1.6
Aellenia subaphylla	Shrub	20	22	31	—	270	15–35	50	20–25	0.5
Eurotia eversmanniana	Semi-shrub	—	40	50	—	—	15–35	60	20–25	0.9
Artemisia turanica	Semi-shrub	30	50	—	—	400	—	40	20–25	0.4
Alyssum desertorum	Ephemer	—	47	60	—	200	20–30	—	15	0.9
Malcolmia africana	Ephemer	—	27	57	—	230	25–30	—	15	0.6
Astragalus filicaulis	Ephemer	20	51	87	—	310	15–30	—	—	—
Ferula assa-foetida	Ephemeroid	28	30	34	—	340	—	—	22–25	0.8
Veronica campylopoda	Ephemer	—	25	58	—	—	30	—	15	0.3
Lamium amplexicaule	Ephemer	—	24	38	—	150	25	—	22–25	0.9

was flushed with helium for 120 s prior to infusing $^{14}CO_2$. The plants studied can be divided into three groups with respect to O_2. The C_3 plants display the usual Warburg effect (eight species). The enhanced rate of assimilation at 1% compared with 21% oxygen is, as a rule 30–50%. An analysis of the conditions under which the inhibiting action of oxygen is most pronounced has revealed that the maximum values of the Warburg effect are obtained at low illumination and low CO_2 concentration. The highest value for the Warburg effect was found in *Chenopodium murale*. A group of plants (Crassulaceae acid metabolism?) in which the photosynthetic activity is not changed with oxygen concentration is of great interest. Among these in the Karakum, one finds the aphyllous species widely distributed (*Haloxylon ammodendron*, *H. persicum*) and shrubs with cylindrical leaves (*Salsola richteri*) of the family Chenopodiaceae. These display a peculiar anatomical structure distinct from the typical Kranz-type, resembling succulent species by having a well-developed water-bearing tissue. With respect to carbon isotope discrimination $\delta^{13}C$, these species are close to C_4 plants. In contrast to the latter they display diurnal variations in their pH and organic acids. At night, under the conditions imposed, these plants assimilate only small amounts of CO_2. The stimulating effect of oxygen on $^{14}CO_2$ assimilation was observed in typical C_4 plants. Of great interest is the C_4 plant *Atriplex dimorphostegia* in which we observed both the inhibiting effects of oxygen on photosynthesis at low light intensity and its stimulating effect at saturating light intensity. In C_3 plants, oxygen inhibits processing of carbon through the Calvin cycle intermediates: the incorporation of ^{14}C into sugar phosphates and phosphoglycerate at 1% oxygen is substantially higher than at 21% oxygen. The involvement of carbon in the glycolate pathway fails to explain completely the inhibiting effect of oxygen on photosynthesis. In C_4 plants, oxygen stimulates the incorporation of ^{14}C into malate and aspartate. The incorporation of ^{14}C into the intermediates of the Calvin cycle in C_4 plants is inhibited much as in typical C_3 plants.

With respect to the productivity rate (amount of CO_2 assimilated per day) desert plants may be grouped into the tree and shrub species with long growing seasons which display low productivity rates and the herbaceous plants and semi-shrubs which grow actively for a short period until the beginning of summer and display high productivity rates. The considerable increment in the desert plants having short growth cycles is achieved with high photosynthetic intensities while in the long growth cycle species the relatively low photosynthetic intensities are counterbalanced by prolongation of their growth. As a result, one finds, for example, that the assimilating apparatus of various plants in the Karakum assimilates during a year approximately the same amount of CO_2 (*c.* $15{\cdot}0\,g\,CO_2\,g^{-1}\,yr^{-1}$) per unit dry wt of the assimilating organs. Annual productivity of plants in the Malye Barsuki area is *c.* $13\,g\,CO_2\,g^{-1}$ dry wt at the apparent photosynthetic intensity of $14–23\,mg\,CO_2\,g^{-1}\,h^{-1}$.

The principal indicator of dark gas exchange in desert plants is the respiration capacity assessed by the manometric technique at 20–25 °C in all plants at the period of florescence. The observed values of leaf respiration capacity in the 130 species of desert plants in the Kyzylkum and Karakum were found to range from 110–800 $m^3 O_2 g^{-1}$ wet wt h^{-1} (0.3–1.6 mg $CO_2 g^{-1}$ wet wt h^{-1}). Assessment of the respiration intensity of the Karakum plants with the air-flow technique under natural conditions revealed that at 30–35 °C the respiration intensity of the assimilating organs was 2–4.5 mg CO_2 g^{-1} dry wt h^{-1}. There are species displaying both high and low respiration intensity in every family. The highest respiration intensity is characteristic of plants of the Leguminosae. Among the plants of various life forms, highest respiration capacity was found in the ephemerals and ephemeroids.

The critical temperature above which leaf respiration rates were suppressed was 40–47 °C for the ephemerals and 45–54 °C for the shrubs and semi-shrubs. During the period of growth, the critical temperature increased with the onset of the hot period in some species, while in others it remained rather constant. The upper limit for the critical temperature for respiration in hot-desert plants is much higher than that for species found in the moderate zone, and is close to the values reported for tropical plants.

Calculation of the diurnal balance of dry matter for certain whole plants (excluding roots) among the ephemerals revealed that by the end of the growing period their expenditures on respiration had started to exceed those on photosynthesis which probably accounts for the death of these plants.

Thus, the majority of plants in the hot deserts of Central Asia have developed adaptive properties in the course of their evolution. Among these one finds certain aspects of the gas exchange such as maximum photosynthetic intensities in spring, somewhat elevated optimal temperatures for photosynthesis, a wide range of optimal temperatures for photosynthesis, high light intensities necessary for saturating potential photosynthesis, a shifting of the period of active growth to spring, high critical temperature for respiration and other features that allow plants to complete successfully the whole cycle of their development.

Vegetation of the Northern Gobi Desert (N. N. Slemnev & D. Bold)

Field studies on the photosynthetic capacity of principal plants of Mongolia were begun in 1970 (Slemnev & Bold, 1974, 1976) and are being continued at the present time. A comparison of plants from two communities in the Northern Gobi Desert with plants of Karakum indicates that they differ considerably in most photosynthetic parameters (Tables 8.5 and 8.6). For instance, the values of the photosynthetic indices of intensity and productivity

Table 8.6. *Some data on photosynthesis of plants in the Northern Gobi*

Plant	Rate of photosynthesis (mg CO_2 g^{-1} dry wt h^{-1})			Photosynthetic productivity (mg CO_2 g^{-1} dry wt day^{-1})		Temperature optimal for photosynthesis (°C)	Light intensity saturating photosynthesis (klx)
	Apparent		Potential maximum				
	Average	Maximum		Average	Maximum	Potential intensity	
Desert community							
Reaumuria songarica	3.1	11	35	42	73	25–35	> 100
Brachanthemum gobicum	3.0	9	48	41	76	20–35	80–> 100
Zygophyllum xanthoxylon	3.1	10	70	43	64	30–35	80–> 100
Salsola passerina	2.4	8	17	30	44	25–35	> 100
Nitraria sibirica	5.5	17	60	68	130	25–30	80–> 100
Haloxylon ammodendron	3.9	12	28	47	110	20–35	> 100
Desert-steppe community							
Stipa gobica	2.1	8	31	28	59	15–25	50
Cleistogenes songorica	3.4	16	28	43	91	25	50–80
Artemisia frigida	2.1	10	37	28	78	20–25	80–> 100
Allium polyrrhizum	2.0	8	57	26	61	25	50–80
Allium mongolicum	2.0	8	50	25	48	20–25	> 100

averaged for all Gobi species, are 27–71% of the corresponding values obtained in the Karakum. However, the maximum values of the intensity determined genetically for each species (Zalensky, 1977) also vary widely within the limits of the genera and families. A comparison of the mean and maximum values of photosynthetic intensity and productivity by seasons is of great interest. It makes it possible to determine the photosynthetic capacity both in communities and in individual species growing in different regions. The degree of determination in turn indicates the adaptability of the plants to the conditions of their habitat and the suitability of these habitats to the normal vital activity of the plants. At this level, the potential capabilities of the plants of the Northern Gobi for photosynthesis are limited to a greater degree than is the case with the plants of Karakum. This is seen particularly clearly in *Haloxylon ammodendron*. The ratio of the mean to the maximum photosynthetic intensity and productivity according to season in the Gobi plant is about 0.4 while that of the plant from the Karakum is about twice as great. An analysis of the daily variation in photosynthesis at different periods of vegetation, the dependence of photosynthesis on the temperature and illumination and also experiments with irrigation of desert-steppe plants, showed that the photosynthesis of plants under natural habitat conditions is limited mainly by temperature (higher than optimal for photosynthesis) and water deficiency (Slemnev, 1978). At the same time all the characteristics of water and temperature stress are less marked in the shrubs and sub-shrubs of the desert community than in herbaceous perennials of a desert-steppe community.

Cycles of mineral elements (N. I. Bazilevich, B. A. Bykov & L. J. Kurochkina)

Cycles of mineral elements, or the minor biological turnover of chemical elements between plants and soils, are governed both by productivity (annual increment, reserve of phytomass and litter fall, including the timber fall, the branch fall and the root residue), and by the particular features of selective uptake by plants, primarily the dominant species of phytocoenoses. Desert plants, as opposed to plants of the steppe and particularly of the forest zones, usually display a greater content of elements and nitrogen in all organs. Desert plants tend to have higher accumulations of chlorine, sometimes of sulphur and sodium (halogene elements) even in plants growing in non-salinated soils (Rodin & Bazilevich, 1967).

In *Artemisia terrae-albae*, which is the characteristic species of clay deserts similar to the situation in sagebrush species, potassium and calcium are the dominant elements; the sum of the elements is small, 3–6% (Table 8.7). The sum is markedly higher in plants growing on saline soils, for example *Anabasis aphylla*, whose green shoots accumulate such elements as sodium and chlorine

Table 8.7. *Content of nitrogen and ash elements in certain desert plants (% dry wt)*

Plant and organs	N	Si	Ca	K	Mg	Al	Fe	Mn	P	S	Na	Cl	Total –N
						M. Barsuki							
Artemisia terrae-albae													
Leaves and first year shoots	2.21	0.56	1.14	1.96	0.20	0.38	0.12	0.02	0.22	0.53	0.38	0.25	5.76
Stems	0.89	0.56	0.45	0.86	0.11	0.22	0.21	0.01	0.07	0.45	0.20	0.08	3.22
Roots	0.97	0.60	1.71	0.50	0.15	0.47	0.22	0.02	0.07	0.31	0.18	0.03	4.26
Anabasis salsa													
Green shoots	1.75	0.30	1.69	1.02	0.67	0.36	0.04	0.02	0.11	0.25	8.85	2.97	16.28
Stems	1.15	1.11	4.36	0.69	0.41	0.36	0.22	0.01	0.12	0.18	1.06	0.49	9·01
Roots	1.21	0.46	1.99	0.35	0.09	0.23	0.15	0.01	0.06	0.19	0.41	0.14	4.08
Rheum tataricum													
Flowering stems	1.90	0.17	1.48	2.13	0.38	0.26	0.06	0.01	0.28	0.34	0.92	0.29	16.32
Leaves	1.38	0.20	3.03	2.84	0.60	0.46	0.08	0.01	0.12	0.43	1.82	0.46	10.05
Roots	0.61	0.11	0.61	0.73	1.34	0.20	0.04	Traces	0.10	0.37	0.25	0.03	3.66
							Repetek						
Haloxylon ammodendron													
Green shoots	1.60	0.10	3.28	5.31	2.06	0.10	0.05	0.015	0.06	0.28	6.19	0.69	18.14
Branches	0.73	0.09	0.62	0.60	0.15	0.03	0.01	0.002	0.03	0.06	0.13	0.07	1.79
Trunk	0.87	0.08	0.73	0.34	0.05	0.004	0.002	0.003	0.004	0.08	0.09	0.02	1.40
Roots	1.39	0.11	1.87	1.10	0.95	0.10	0.02	0.005	0.05	0.32	0.73	0.56	5.82
Haloxylon persicum													
Green shoots	1.62	0.14	5.39	2.68	1.33	0.11	0.04	0.015	0.04	0.20	1.32	0.27	11.54
Branches	0.69	0.10	1.03	0.99	0.28	0.09	0.007	0.003	0.04	0.06	0.17	0.06	2.83
Trunk	0.79	0.08	1.44	0.40	0.11	0.08	0.004	0.007	0.03	0.08	0.11	0.02	2.36
Roots	1.09	0.13	2.48	1.42	1.10	0.13	0.01	0.005	0.04	0.15	0.39	0.06	5.92
Carex physodes													
Leaves, stems	1.42	0.60	1.15	1.63	0.26	0.04	0.05	Traces	0.21	0.17	0.06	0.18	4.35
Roots	0.92	0.35	1.22	0.48	0.15	0.13	0.07	Traces	0.05	0.10	0.14	0.16	2.85
Rhizomes	0.80	0.13	0.35	0.32	0.06	0.02	0.01	Traces	0.05	0.09	0.05	0.03	1.11

most vigorously in addition to calcium and potassium. Still higher amounts of halogene elements, especially sodium, are accumulated by *A. salsa* in its green shoots (the sum of the elements can be as great as 16%).

In sandy desert plants, for example, in *Ammodendron argenteum* (M. Barsuki), the sum of the elements is rather small, even in leaves it is only 3%. Potassium and calcium dominate as in *Artemisia* spp. Nitrogen content reaches high levels typical of the Leguminosae (2.5%). The ephemerals, and ephemeroids (e.g. *Rheum tataricum*) are characterized by a higher sum of elements than *Artemisia terrae-albae*, especially in the leaves. These high levels are mostly due to the high content of calcium and potassium, and frequently to sodium and chlorine as well. Similar generalities hold true also for the Taukum plants.

As for the sandy desert plants of the south-eastern Karakum (the Repetek Reservation) the biological peculiarities of *Haloxylon ammodendron* dominate in this area because of its prevalence. Since its roots extend down to the capillary layer of the saline ground waters, the accumulation of ash elements in annual shoots is higher than in plants on the highly saline soils of the clay desert (more than 18%). Sodium accumulation exceeds that of potassium and calcium (the latter two elements dominate in branches, stems and roots). In *H. persicum*, an ombrophyte plant, even the green shoots have a higher content of calcium and potassium than of sodium. *Carex physodes* constitutes the major population in the *Haloxylon ammodendron* association covering about 70% of the area. It is characterized by the predominant accumulation of potassium, calcium and silicon and by a low content of the ash elements as is usually the case for sedges and grasses.

Generally, reserves of ash elements and of nitrogen in the phytomass of the communities studied increases with increasing phytomass. This correlation, however, cannot be approximated by a straight-line dependence since the above-mentioned peculiarities in nutrient uptake by the dominant species affect the relationship (Table 8.8). In the saline soils of clay deserts (the *Anabasis salsa* community) and in sandy deserts (the *Haloxylon ammodendron* community) there are more ash elements per unit of phytomass than in the watershed-area communities.

Most nutrients accumulate in roots (66–81%). Despite the relatively small fraction of the phytomass which the annual shoots make up (less than 5%), the relative content of chemical elements in them is large (8–15%). Thus the composition of the dominant elements in the vegetation is determined essentially by the chemical composition of the underground sphere of the community.

The amount of chemical element uptake which is used in annual production and the return of these elements with the litter fall, including branch fall, timber fall and root residue, is dependent primarily upon the magnitude of the increment. Biological peculiarities of the dominant species also contribute

Table 8.8. *The amount of ash elements and nitrogen (kg ha^{-1}) which are used annually in production and returned to the soil surface*

Property	M. Barsuki			Taukumy			Repetek
	Ammodendron argenteum	*Artemisia terrae-albae* (1966)	*Anabasis salsa*	*Ephedra lomatolepis*	*Artemisia terrae-albae*	*Agropyron fragilis*	*Haloxylon ammodendron*
Ash elements in phytomass	1096.57	1267.18	481.87	481.96	293.99	227.78	1093.30
of which:							
in the green portion	144.04	103.64	53.87	55.99	37.01	18.53	165.01
in the green portion (%)	13	8	11	11	13	8	15
in wood	130.10	143.71	57.03	70.62	62.84	48.57	109.76
in wood (%)	12	11	12	15	21	21	10
in roots	822.43	1026.27	370.97	355.46	194.13	160.68	818.53
in roots (%)	75	81	77	74	66	71	75
Nitrogen in phytomass	345.61	303.70	91.30	150[a]	135.29[a]	60[a]	276.03
Used in new structural growth: ash elements	495.94	328.34	201.29	79.59	128.78	160.97	422.15
Used in new structural growth: nitrogen	155.46	75.98	37.93	25.00[a]	70.81[a]	49[a]	79.56
of which:							
green first-year shoots							
ash elements (%)	29	30	21	48	29	11	38
nitrogen (%)	37	32	13	56	22	18	25
wood, ash elements (%)	2	1	2	3	4	8	3
nitrogen (%)	3	2	6	4	2	2	6
roots, ash elements (%)	69	69	75	49	67	81	59
nitrogen (%)	60	66	85	40	76	80[a]	69

Returned in litter: ash elements	417.71	302.71	186.54	57.26	123.15	131.25	367.47
Returned in litter: nitrogen	135.65	69.70	35.56	18.00[a]	60.74[a]	48.00[a]	76.46
Relative ash in litter (%)	4.4	4.4	6.3	3.6	3.3	3.7	5.0
Relative content of nitrogen in litter (%)	1.1	1.0	1.2	1.2	1.6[a]	1.3[a]	1.0
Content in litter of biohalogens (Na + Cl + excess S) in the sum of ash elements (%)	7.5	4.5	24	8	Not determined	Not determined	29
Principal elements in phytomass	Ca, N, K, Si	Ca, N, K, Si	Ca, N, Na, Si	Ca, N, K, Mg	N, Ca, K, Si	Ca, N, Si, K	Ca, N, K, Na (Cl)
of which:							
in green portion	K, N, Ca, Si	Ca, N, K, Si	Na, Cl, N, Ca	Ca, N, K, S	N, K, Ca, Si	N, K, Ca, Si	K, Na, Ca, N (Cl)
in wood	Ca, N, K, Si	N, K, Ca, S	Na, Ca, N, Cl	N, Ca, K, Si	Ca, N, Si, K	Ca, Si, K, N	Ca, N, K, Mg
in roots	Ca, N, K, Si	Ca, N, Si, K	Ca, N, Si, Na	Ca, N, K, Mg	N, Ca, K, Si	Ca, N, Si, K	Ca, N, K, Na (Cl)
Principal elements in annual production	Ca, N, K, Si	Ca, N, K, Si	Ca, N, Na, Si	Ca, N, K, Mg	N, Ca, K, Si	Ca, N, Si, K	Ca, K, N, Na (Cl)
of which:							
in green portion	K, N, Ca	Ca, N, K, Si	Na, Ca, N, Cl	Ca, N, K, S	N, K, Ca, Si	N, K, Ca, Si	Na, K, Ca, N (Cl)
in roots	Ca, N, K, Si	Ca, N, Si, K	Ca, N, Si, Na	Ca, N, K, Mg	N, Ca, K, Si	Ca, N, Si, K	Ca, N, K, Cl (Na)
Principal elements in litter	Ca, N, K, Si	Ca, N, K, Si	Ca, N, Na, Si	Ca, N, K, Mg	N, Ca, K, Si	Ca, N, Si, K	Ca, N, Mg, K (Na, Cl)
of which:							
in green portion	K, N, Ca	Ca, N, K, Si	Na, Ca, N, Cl	Ca, N, K, S	N, K, Ca, Si	N, K, Ca, Si	Mg, Na, Ca, N (Cl)
in root residues	Ca, N, K	Ca, N, Si Na	Ca, N, K, Na	Ca, N, K, Mg	N, Ca, K, Si	Ca, N, Si, K	Ca, N, K, Cl (Na)

[a] Estimated.

to these relationships. The relative content of ash elements per unit of organic matter of annual increment or of litter fall is highest in the *Anabasis salsa* and *Haloxylon ammodendron* communities. This is seen from the average ash content of the litter fall. The absolute contribution of chemical elements in biological cycles is the greatest in the *Ammodendron argenteum* and *Artemisia terrae-albae* communities where the increment is highest.

The major portion of nutrient elements is used in the annual production of roots. Composition of the dominant elements, however, is conditioned to the same extent both by the uptake of elements by below-ground and above-ground (annual) organs. The elemental contribution of these latter organs is much higher (11–48%) than any of the other plant parts contributing to annual production or litter fall.

The most abundant elements in the chemical composition of the increment-litter fall in plant communities on non-saline soils are calcium, potassium and silicon. Increased soil salinity is accompanied by substitution of elements in the biological cycles, initially magnesium and sulphur (*Ephedra lomatolepis*) and later chlorine and sodium (*Anabaseta, Haloxyloneta ammodendroni*). The amount of halogene elements (sodium, chlorine and to some extent sulphur) increases from 5–7 to 8%, or even up to 25–30% of the ash elements in the litter fall (including timber fall, branch fall, and root residue).

The chemical elements released through mineralization of litter fall and the newly formed salts contributed to the formation of various soils differing in their salinity and in other properties. Thus, directly under the cover of *Haloxylon ammodendron* more than 40 g m^{-2} of the rather toxic carbonate salts of sodium can be released due to the enrichment of litter-fall ash elements with sodium. This results in significant alkalinization of soils which prevents their occupation by other plants (Bazilevich *et al.*, 1953). Similar phenomena, but on a smaller scale, are observed in the soils of the *Anabasis salsa* community. In contrast, decomposition of the litter fall in the *Artemisia terrae-albae* and *Ammodendron argenteum* communities is accompanied by accumulation of calcium carbonates (since the ash elements of the litter fall and other falls) are enriched in calcium. Essential elements such as nitrogen and potassium also accumulate there contributing to fertility of the soils.

A mathematic model of the population of *Haloxylon ammodendron*

(W. A. Vavilin & A. B. Georgievsky)

A model of the coenopopulation of *Haloxylon ammodendron* in the *Haloxylon ammodendron–Carex physodes* ecosystem is proposed. *Carex physodes* and *Haloxylon ammodendron* are the dominant species of the ecosystem. As a consequence of vigorous competition they never grow together. *Carex physodes* precludes restoration of *Haloxylon ammodendron.* It can, however,

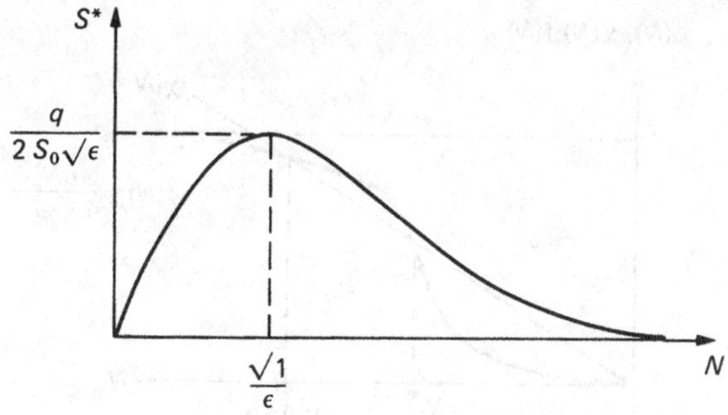

Fig. 8.1. Variations in the effective area (S*) (the favourable area of regeneration of *H. ammodendron*) under increase in the number of *H. ammodendron* trees.

be restored along the margins of the so-called 'undercrown spots' formed by the litter fall of *Haloxylon ammodendron* trees.

The leading variable is taken to be the quantity of *H. ammodendron* per unit area. The area occupied by *Carex physodes* is the variable determined by the area occupied by *Haloxylon ammodendron.* All of the area in the association is covered by *Carex physodes* and by *Haloxylon ammodendron.*

Variations in the quantity of trees over period Δt can be expressed by the equation:

$$\Delta N = Kp\frac{qN^2_{t-\tau}}{S_0(1+\epsilon N^2_{t-\tau})}\Delta t - lN_t\Delta t \qquad [8.1]$$

where N_t is the number of fruit-bearing *Haloxylon ammodendron* trees per area S_0 at the moment t; S_0 is the area covered by *Carex physodes* and *H. ammodendron*; $K\Delta t$ is the quantity of seeds produced by a tree over period Δt; $NK\Delta t$ is the total number of seeds on area S_0; $qN_t/1+\epsilon N^2_t$ is the effective area in the site suitable for the growth of *H. ammodendron* at time t; ϵ is a constant value that shows the influence of the number of trees N_t on the effective area under one tree; p is the probability that a seed found in the effective area will develop into an adult tree; τ is the time period required for *H. ammodendron* tree to become an adult fruit-bearing tree; $l\Delta t$ is the probability of death of the tree over period Δt (Fig. 8.1).

Assuming $\Delta t = 1$ yr, and designating $Kp(q/S_0) = \alpha$ we have:

$$N_{t+1} = \frac{\alpha N^2_{t-\tau}}{1+\epsilon N^2_{t-\tau}} - (l-1)N_t. \qquad [8.2]$$

Thus, we are considering a rather abstract coenopopulation of *Haloxylon ammodendron* for which we assume existence of a mean age that marks the beginning of the fruit-bearing state. We have taken into account the fact that the shoots can grow at the periphery of the canopy projections as well as the fact that the limiting factor was the decrease of the effective area after a

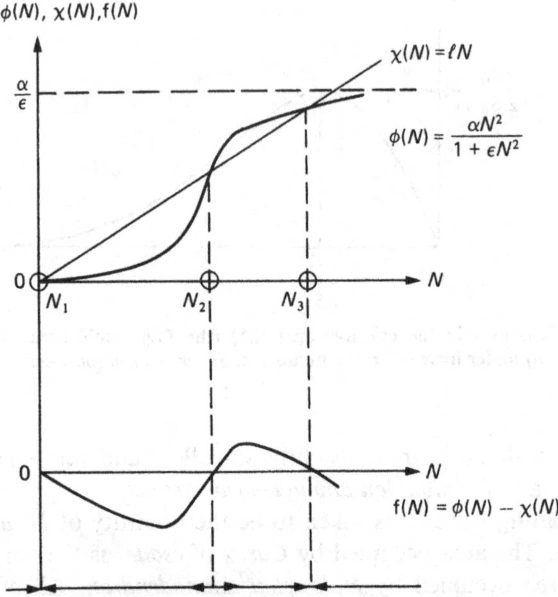

Fig. 8.2. The dependence of functions $f(N)$, $\phi(N)$, $\chi(N)$ upon variable N at $0 < (2l/\alpha)^2\epsilon < 1$. Arrows show the direction of system motion near special points N_1, N_2, N_3. It is clear that N_2 is an unstable special point, and N_1 and N_3 are stable.

certain optimal $N \doteqdot \sqrt{(1/\epsilon)}$ due to the worsening of the ecological parameters of the ecosystem with increase in the coenopopulation density.

For the analysis of equation [8.2], we assumed the following parameter variations:

α from 10^{-4} to 1; l from 10^{-3} to 1.10^{-1}; τ from 5 to 20. [8.3]

The indicated limits had been chosen according to the material obtained by other authors.

To estimate the range of parameter α, let K vary from 10^4 to 10^5 samples per year; probability p from 10^{-4} to 10^{-1}; area $q \sim 1\ \text{m}^2$; $S_0 = 10^4\ \text{m}^2$. Then at the extreme values of K and p we have $10^{-4} \leqslant \alpha \leqslant 1$. On the selection of the parameter ϵ see below.

Equation [8.2] has been calculated on a computer with concrete values of the parameters. When choosing various algorithms for the calculation the solutions derived from equation [8.2] have been considered, assuming $\tau = 0$, $\Delta t \to 0$.

$$\frac{dN}{dt} = \frac{\alpha N^2}{1+\epsilon N^2} - lN = f(N). \qquad [8.4]$$

The specific equation [8.4] points are: $N_1 = 0$ and

$$N_{2,3} = \frac{\alpha}{2l\epsilon}\left[1 \mp \sqrt{\left(1-\left(\frac{2l}{\alpha}\right)^2\epsilon\right)}\right] \qquad [8.5]$$

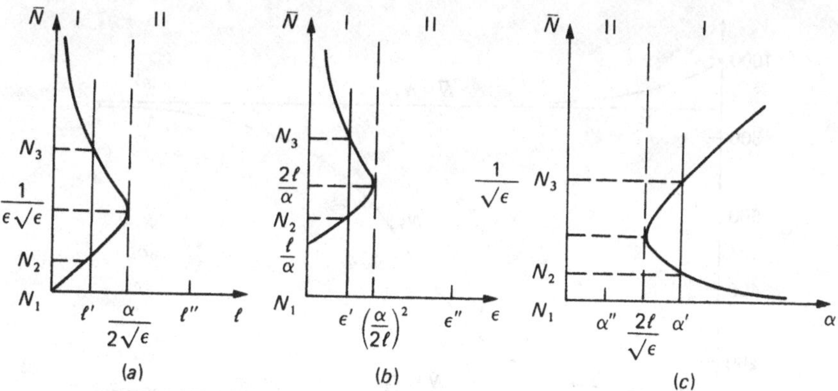

Fig. 8.3. Variations in balance values $\bar{N}$ for parameters $l(a)$, $\epsilon(b)$, $\alpha(c)$. I; the pattern at $0 < (2l/\alpha)^2\epsilon < 1$. The three balance states: N_1, N_2, N_3 correspond to any fixed value l' or ϵ' or α'. II; the pattern at $(2l/\alpha)^2\epsilon > 1$. The one balance state $N_1 = 0$ corresponds to any fixed value l'' or ϵ'' or α''.

The phase line for [8.4] and dependence of $\bar{N}$ values upon parameters are shown in Figs. 8.2 and 8.3. We limited our interests to the upper semisurface. At $0 < (2l/\alpha)^2\epsilon < 1$ the state of equilibrium N_1 and N_3 are stable, while N_2 is unstable. If $(2l/\alpha)^2\epsilon \ll 1$ then

$$N_2 \approx \frac{l}{\alpha}, \quad N_3 \approx \frac{\alpha}{l\epsilon}. \tag{8.6}$$

At $(2l/\alpha)^2\epsilon > 1$ the system is featured by one stable balance state $N_1 = 0$. Consequently, the generalized parameter $p = (2l/\alpha)^2\epsilon$ can give us information as to the qualitative aspects of the systems state. We have studied the pattern for the solution of equation [8.2] within the chosen range of the parameters.

Earlier (equation 8.6) the probable value for N_3 was presented. Here, assuming the real density of the coenopopulation being 10^2–10^3 samples* ha^{-1} we calculated the parameter ϵ. We found then that for equations [8.2] and [8.4], the stable state N_3 will ensue at $0 < (2l/\alpha)^2\epsilon \lesssim 1$ (pattern I), and at $p = (2l/\alpha)^2\epsilon \gtrsim 1$ (pattern II) such a state is not obtainable.

According to pattern I (Fig. 8.3) the system may be found in a stable state N_1 or N_3 depending on the initial conditions. Other states are the unstable ones. At chance alterations of the variable and the parameters the model may shift from one pattern into another (e.g. from N_3 to N_1). Hence, if we utilize the system (exclude part N) above the permissible level, the population will be extinguished under the pressure of severe environmental conditions. A rational utilization is imaginable which promotes the maximum increment of the population without the danger of destroying the system. This is possible at the point of inflection N_2 (Figs. 8.2 and 8.4), its specific location corresponding to the maximum values of the function $f(N)$. The unstable state N_2 under favourable environmental conditions (α is great, l is small) is

* Adult trees of *Haloxylon ammodendron.*

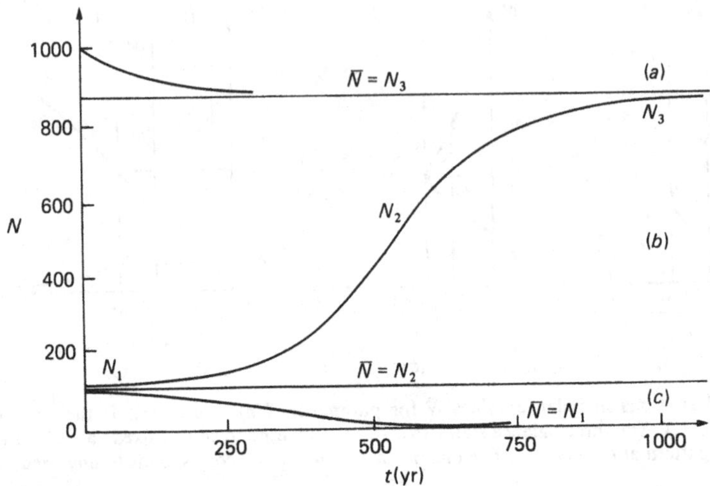

Fig. 8.4. Variations in the quantity of trees N with time according to the solution of equation 8.2 at the different initial numbers of trees. (*a*), $N_{-10}-N_{-9} = \ldots N_{-1} = N_0 = 1000$; (*b*), $N_{-10}-N_{-9} = \ldots N_{-1} = N_0 = 120$; (*c*), $N_{-10}-N_{-9} = \ldots N_{-1} = N_0 = 100$. The equation 8.2 has been calculated on a computer at $\tau = 10$, $\epsilon = 10^{-5}$, $l = 0.01$, $\alpha = 10^{-4}$. N_1, N_2, N_3, the balance state of system.

located near the stable state $N_1 = 0$. Therefore, the population will not be extinguished if only a single fruit-bearing plant is left.

Under extremely unfavourable conditions (poor fruiting, drought, competition) over a prolonged period of time $(2l/\alpha)^2\epsilon \approx 1$, any type of utilization may result in the destruction of the population since the unstable state N_2 is close to the stable one N_3. Overall analysis of the model yields two interesting consequences: duration of the period required for the population to reach a stable state, and occurrence of an ultimate density of the population on the site (see Fig. 8.4). The principal drawback of this model is that it does not incorporate the age composition of the coenopopulation. Such a model, however, requires a considerable number of parameters to be known which have not yet been estimated experimentally.

Conclusion (L. E. Rodin)

Seventeen ecosystems in deserts in the USSR and the Mongolian People's Republic were studied for ten years. The 17 ecosystems represented widespread desert associations of that region (Karakumy, Kyzylkumy, Malye Barsuki, Taukumy, Northern Gobi), and were typical of sub-boreal deserts characterized by a continental dry desert climate.

The majority of dominant species show signs of adaptation which permit them to complete the whole life cycle successfully (maximal intensity of photosynthesis in spring and at the beginning of summer, elevated optimal temperature of photosynthesis, high critical temperature of transpiration, etc.).

A few species play a dominant role in annual production and in mineral exchange. In the *Ammodendron argenteum* communities, five species (*Ammodendron argenteum, Artemisia tomentella, Kochia prostrata, Agropyron fragile, Astragalus ammodendron*) account for 89% of the annual production. In the *Haloxylon ammodendron* community only two species, *Haloxylon ammodendron* and *Carex physodes*, produce 83% of the annual production, and 89% of the ash content of the mineral elements.

Those communities with complex structure (several storeys both above ground and in the soil) have a higher annual production and accumulate more phytomass. In each of the above-mentioned communities several storeys are present above and below ground.

The animal life of deserts is unusual and is characterized by abundance of specialized highly endemic forms. The most peculiar is the psammophytic complex. In the psammophytic ecosystems the underground storey, the surface storey and the above-ground storey are sharply distinguished both in the structure of the vegetation and in the vertical distribution of the animal population. Together with the strictly stratified species populations there are many groups which migrate within limits of two or three storeys either daily or over long-term cycles.

The trophic specialization of animals is closely correlated with the phenophases of the vegetative cover and is correspondingly characterized by seasonal fluctuations.

Desert animals are characterized by a high degree of morphophysiological adaptation to the conditions of the arid environment. The behavioural reactions of the majority of the species allow them to avoid unfavourable coincidences of abiotic and biotic factors in the environment in daily and seasonal cycles. Variations in the degree of development, structure and productivity of the vegetation in the arid zone are well known. Fluctuations in the density of the animal populations in various years are equally characteristic. Decreases in numbers of one species is usually accompanied by an increase in numbers of another species but the population of some species may remain constant. The biomass of animals also varies greatly between years; the mass of invertebrates being ten times that of vertebrates.

Relationships between animals and vegetation are manifold; they are most clearly expressed in trophic chains. All living organisms, for example of psammophytic desert ecosystems, belong to one of eleven tropho-functional groups. Among desert vertebrates zoophages and carpophages are most common, less so phyllophages. Among Arthropoda, zoophages, phytoxylo-detritophages and phyllophages are most common.

In the desert ecosystem the role of the animal population in transforming matter and energy is exceptionally important and multi-faceted; it is, however, not yet fully understood because of insufficient information on several ecological parameters (spectra of food objects, level of energy exchange, annual life cycles, assimilation and dissimilation processes, etc.).

The negative influence of wild grass-eating animals on the vegetation is not so large as was first imagined (before we obtained accurate quantitative data): the quantity of vegetable mass consumed by them in a sandy desert is on average 3–4% of the annual production and only in years with large numbers of rodents will it reach 10%. Domestic animals, such as sheep, when imposing only a moderate 'load' on the pasture, browse ageing brushes and contribute to their 'rejuvenation' and ultimately to an increase in the mass of fodder.

Human interference which destroys (or only weakens) even one of the storeys leads, unavoidably, to a decrease in the biological (and economic) productivity. This leads to an important practical benefit from improvement of seral communities disturbed by a human interference – introduction (by means of seeding, sowing, strip-sowing) of such combinations of plants and various life forms which would best use various storeys of the aerial zone of the community, various soil horizons and also various seasons of the year.

The biological wisdom of different forms of protection (prohibition, rotation, reduction of exploitation norms of grazing, falling or pulling for firewood or for other reasons) lies in conserving or preserving the normal multi-storeyed structure of the community. This leads to an increase in economic productivity.

The retention of moisture is the most important condition for the successful functioning of desert ecosystems. Various measures need to be taken to conserve the sparse preciptation and ensure its full use by the vegetation (loosening of heavy, poorly permeable soils; prevention of surface run-off, improvement of mechanical properties of soils by adding sand; restriction of physical evaporation by mulching, etc.). Studies of the water regime of desert plants indicate that they are fully adapted to the harsh conditions of the arid climate and even catastrophic fluctuations do not destroy the whole ecosystem but only interfere with normal functioning of its individual structural elements. Root penetration to significant depth in many desert plants serves not only as adaptation to lack of moisture in the upper horizons but has also decreased competition with plants which have more shallow root systems.

In the assimilating organs, which are shed annually, the halogene elements (chlorine, sodium) are preferentially accumulated while in the roots of desert plants the biologically active elements (potassium, calcium, phosphorus) are accumulated. They also contain a large amount of energy. Therefore the dominance of root in the organic mass of desert ecosystems, and the concentration of elements and energy in the roots, constitutes a reserve of fertility in the soil. Fertility is assured by extended retention and circulation of materials produced by photosynthesis in the short-lived assimilatory organs.

The most likely way to improve and preserve productive, long-lived, self-perpetuating ecosystems in arid areas is the establishment of communities with deep, extensively rooted trees and shrubs together with other plant forms (semi-shrubs, ephemeral and perennial grasses and annual plants) with various periods of development to ensure multi-storeyed structure (both above and below ground). There are numerous examples of damage to desert ecosystems caused by various human factors (excessive grazing by domestic animals; felling trees and shrubs for building or fuel; exploiting plants for technical or medicinal raw materials; construction of railways and roads; construction of oil and gas pipelines, electric power lines; well sinking; establishment of permanent or temporary settlements, etc.). There are also many data on successions, both regressive and progressive. The existing information allows us to reach a general conclusion that desert ecosystems are highly vulnerable but this does not mean that they lack stability.

Certainly, the understanding of stability is always multi-faceted. Composition, structure, productivity and other properties of ecosystems are not stable; they suffer sharp fluctuations both in seasonal and annual aspect but they are highly stable in cycles covering many years or even centuries, if we consider only disturbances caused by the climatic factors. The same exosystem can show very little stability in respect to an influence of human and animal factors. In other words an ecosystem can be stable in one relation and unstable in another one.

In any case, the desert ecosystems require a careful approach to their exploitation. At the same time complete abandonment of their use (prohibition of grazing, cutting, etc.) also leads to unwanted consequences (degradation of vegetation and decrease in biological productivity of desert ecosystems). Wild grass-eating mammals have a significant effect on the structure and composition of the vegetative cover, However, as a result of its inherent resilience the vegetation could sustain removal by grazing of up to 75% of its phytomass. Animal numbers decrease when 50% of the phytomass is grazed off (Abaturov & Kuznetsov, 1975); in this way their 'pressure' is weakened.

The vegetation of all our deserts undoubtedly evolved under conditions of some optimal grazing regime affected both by the large nomadic ungulates and by the small 'resident' rodents. Therefore, moderate grazing ensures good condition of the desert vegetation (Nechaeva, Vasilevskaya & Antonova, 1970). Moderate grazing also has favourable effects on soil moisture and increasing water circulation (Zhambakin *et al.*, 1976). We can presume that a restricted and, let us add, rational (selection of dead trees and old trees to promote fast, new growth in those remaining) exploitation of wood (and other raw materials) also permits preservation of the desert assemblage without any danger of degradation as has been confirmed in forestry practice.

The stability problem is multi-faceted. Even though understanding of the interactions of factors increasing or reducing productivity with one another, and with a stable environment, is reasonably good (though not in all cases as good as is needed), we know relatively little about the importance of the controlling module in the trophofunctional structure of the desert biogeocoenosis.

A realization of the deficiencies in desert ecosystems has led to increases in their productivity (Rabotnov, 1960). New ecosystems with considerably higher productivity have been formed on large areas of poor desert pastures in Kazahkstan and Uzbekistan (Shamsutdinov, 1976). These experiments have worked well. However, we do not yet know how stable these artificial ecosystems, are, that is, if the biological circulation of matter and energy and circulation of heat, water, mineral material, gases and other factors of the environment is sufficiently balanced.

Despite the experimental data mentioned above on human and other disturbances of ecosystems we still do not possess convincing accurate information on regeneration of these ecosystems. There is evidence of a definite tendency towards regeneration but neither within the life-span of one generation nor even many generations spanning several centuries has a full restoration been factually established. On the contrary, there are any number of reports of non-restoration of former ecosystems and landscapes (often only one or two centuries ago). For example, tropical rain forests, mangroves, sub-tropical forests, boreal broad-leaf and coniferous forests which were converted to savannahs, solonchak marshes, thin xerophytic forests, small-leaf forests, meadows and other seral communities. Some of these were regarded by some investigators as terminal.

Also we do not have accurate data on ways to restore desert ecosystems after technological disturbances which have taken place with especial intensity during the last two or three decades. Nor do we have information on indirect influences (side effects) of construction of large canals, water reservoirs, various communication lines, etc. This is an extensive area of future studies, urgent and pressing.

The high vulnerability of desert ecosystems brings forward with a special urgency the problem of protection of desert landscapes – from virgin forests, wormwood or saltwort deserts to mobile sands and solonchaks – protection in the sense of their preservation in at least that stage of equilibrium in which they exist during the life of the present generation of people. In this connection the problem of protection of desert ecosystems must be understood as preservation of normal functioning of an ecosystem. And, as the evolution of the vegetation of desert ecosystems took place under the influence of wild grazing ungulates it is possible to permit grazing of domestic animals together with the resident wild grass-eaters (naturally within certain limits) in national parks in the desert zone and in the corresponding altitude zones in mountains. Of course, it does not follow that we should open the national parks for 'free' grazing but experiments should be carried out in some, sufficiently representative part of a national park. This will give us a 'clue' for management of the renewable resource of the arid zone – the vegetation cover, a 'clue' to the possibility of its rational use (and this means also improvement in cases of its disturbance) without exhaustion and deterioration.

Acknowledgements

The section on Water Input and Transpiration was edited by V. M. Sveshnikova. V. L. Voznesensky was editor for the section on Photosynthesis and Respiration. The section on Cycles of Mineral Elements was edited by N. I. Bazilevich. I wish to express my appreciation to these colleagues who helped me to prepare the Russian version of the paper and to Professor D. W. Goodall and Dr R. A. Perry whose attention and friendship gave me the opportunity to translate the manuscript and publish it in this book.

References

Abaturov, B. D. & Kuznetsov, G. V. (1975). Rol' mlekopitayushchikh v biosfere. In: *Biosfera i chelovek*, pp. 230–2. Nauka, Moscow.

Bazilevich, N. I., Hollerbach, M. M., Litvinov, M. A., Rodin, L. E. & Steinberg, D. M. (1953). O roli biologicheskikh faktorov v obrazovanii takyrov na trasse Glavnogo Turkmenskogo kanala. *Botanicheskiĭ Zhurnal* (*Leningrad*), **38**, 13–30.

Glagoleva, T. A., Mokronosov, A. T. & Zalensky, O. V. (1978). Oxygen effects on photosynthesis and ^{14}C metabolism in desert plants. *Plant Physiology*, **62**, 204–9.

Nechaeva, N. T., Vasilevskaya, V. K. & Antonova, K. G. (1970). Priznakĭ dlya ekologicheskoi klassifikatsii drevesnykh i poludrevesnykh rasteniĭ Karakumov. *Problemy osvoeniya pustyn*, **4**, 19–27.

Rabotnov, T. A. (1960). O floristicheskoĭ i tsenoticheskoĭ polnochlennosti tsenozov. Doklady Akademii Nauk SSSR. **130** (3), 671–3.

Rodin, L. E. & Bazilevich, N. I. (1967). *Production and mineral cycling in terrestrial vegetation*. Oliver & Boyd, Edinburgh & London.

Semikhatova, O. A., Chulanovskaya, M. V. & Metzner, H. (1971). Manometric method of plant photosynthesis determination. In: *Plant photosynthetic production. Manual of methods*, pp. 239–56. Dr W. Junk, The Hague.

Shamsutdinov, Z. Sh. (1976). *Sozdanie dolgoletnikh pastbishch v aridnoĭ zone Sredneĭ Azii*. Fan, Tashkent.

Slemnev, N. N. (1978). Vliyanie poliva na rasteniya pustynnostepnoi zony MNR (Mongol'skoi Narodnoi Respubliki). In: *Zakonomernosti geografii i dinamiki rastitel'nogo i zhivotnogo mira MNR*, pp. 55–9. Nauka, Moscow.

Slemnev, N. N. & Bold, D. (1974). O fotosinteze rastenii pustyni Gobi v Mongolii. *Botanicheskiĭ Zhurnal* (*Leningrad*), **59**, 1129–41.

Slemnev, N. N. & Bold, D. (1976). Sravnitel'naya kharakteristica fotosinteticheskoi deyatel'nosti pustynnykh i stepnykh rasteniĭ Mongolii. *Trudy Instituta botaniki Akademii nauk Mongol'skoĭ Narodnoĭ Respubliki*, **2**, 165–72.

Voznesensky, V. L. (1977). *Photosynthes pustynnykh rastenij*. Nauka, Leningrad.

Voznesensky, V. L., Zalensky, O. V. & Austin, R. B. (1971). Methods of measuring rates of photosynthesis using carbon-14 dioxide. In: *Plant photosynthetic production. Manual of methods*, pp. 276–93. Dr W. Junk, The Hague.

Voznesensky, V. J., Zalensky, O. V. & Semikhatova, O. A. (1965). Metody issledovaniya photosyntheza i dykhaniya rasteniĭ. Nauka, Leningrad.

Zakharjantz, I. L., Naaber, L. Kh., Fazylova, S., Alekseeva, L. N. & Oshanina, N. P. (1971). *Gazoobmen i obmen veshchestv pustynnykh rasteniĭ Kyzylkumov*. Phan, Tashkent.

Zalensky, O. V. (1963). Maximalnaya potentsialnaya intensivnost photosynteza rasteniĭ Pamira i drugikh klimaticheskikh oblasteĭ. *Trudy Pamirskoĭ biologicheskoĭ stantsii*, **1**, 53–60.

Zalensky, O. V. (1977). *Ekologo-phiziologischeskie aspekty izucheniya photosynteza.* Nauka, Leningrad.

Zhambakin, Zh. A., Aibasov, E. B. & Mollabekova, K. (1976). Vliyanie razlichnykh sposobov vypasa ovets na vodnyĭ balans pustynnykh pastbishch. In: *Problemy osvoeniya pustyn*, **1**, 55–9.

Manuscript received by the editors October 1974.

9. Production by desert animals

F. B. TURNER & R. M. CHEW

Introduction

The major component of the United States contribution to the International Biological Program has been the analysis of ecosystems, involving extensive work in such disparate environments as tundra, forests, grasslands and deserts. Much of this research has been devoted to understanding the principal processes involved in ecosystem function, and to creating mathematical models of these processes. The process of production, whether by plants or animals, is central to the maintenance of any ecosystem, and has a critical bearing on man's own existence in the biosphere. Indeed, the basic structure of the International Biological Program was strongly influenced by considerations of biological productivity. In this chapter we consider secondary production by animals of arid regions.

Two major aspects of secondary production are of ecological interest: the absolute rates of production (i.e., kcal ha^{-1} yr^{-1}) and the efficiency of production. Efficiency can be calculated in terms of the energy assimilated or ingested; or with respect to the total energy of net primary production, or to that limited portion of primary production available to the consumer in question. Efficiency is the more difficult aspect to evaluate. Because of the paucity of information on production efficiencies of desert animals, our principal emphasis will be on the rates of secondary production of new tissue – that material made available to other trophic levels by direct consumption and decay. Our basic intention is to define, as far as possible, the levels and efficiencies of secondary production by desert species, and to compare these with values observed in other environments. We will consider the relative importance of various kinds of consumers, as may be gauged by their respective productivities. Finally, we will attempt to explain observed variations in production by desert species, not only as expressed in differences between species but also by year-to-year differences in the same species.

We will confine ourselves to extremely arid regions, as defined by McGinnies, Goldman & Paylore (1968), and will not review research conducted in semi-arid grasslands or savannas. We will not consider domestic or semi-domestic animals, though we recognize that proper management and exploitation of such species are indispensable to man's continued occupancy and utilization of arid lands. This important problem area will be treated in another chapter. Evans (1967) has pointed out that secondary production includes the activities of all heterotrophic organisms, decomposers as well as primary and secondary consumers. The number of taxa of heterotrophic organisms prohibits analytical measurements of the secondary production of almost all ecosystems.

Production by desert animals has rarely been evaluated, as is evident from recent reviews. The compendium edited by Petrusewicz (1967) contained discussions of methodology, considerations of problems bearing on estimates of productivity, and a few papers giving actual production estimates. Neither this paper nor those edited by Lamotte & Bourlière (1967) and Petrusewicz & Ryszkowski (1969/70) concerned arid-land species. McNeill & Lawton (1970) reviewed 27 studies of animal production, including fish, freshwater and marine invertebrates, various terrestrial arthropods, and a dozen vertebrates. None of the animals involved was a desert species. Finally Noy-Meir (1974) reviewed the function of higher trophic levels in desert ecosystems, but his discussion did not include actual estimates of animal production. Because of the paucity of existing research, we include in our review new findings based on research in American deserts between 1963 and 1974. Our general approach is to examine pre-existing research and/or unpublished material relating to the dynamics of vertebrate populations and to derive production estimates from these data.

Procedures

With respect to absolute production by a consumer population one can pose two basic questions. First, how much new tissue does the population produce annually (and how may this amount vary in time)? Second, how much of this production is passed on to other trophic levels? The latter is the real output of the population to the rest of the ecosystem, by the irreversible processes of death and consumption. On the other hand, new tissue produced by an individual during one time interval may be catabolized later. In this case the energy involved is never transferred to other parts of the system. As the time interval of measurements becomes longer, the two measures become more similar.

Our procedures follow conventional practice in defining production as the energy equivalent of new tissue produced by a population per unit area per unit time (e.g., kcal $ha^{-1}\,yr^{-1}$). Our basic computational approach follows that given by Petrusewicz & Macfadyen (1970), so that production (P) may be expressed as:

$$P = E + \Delta B, \qquad [9.1]$$

with E the energy equivalent of all animals dying during the time interval in question, and ΔB the energy equivalent of the change in standing stock. R. M. Chew & A. E. Chew (1970) estimated production by mammals in Arizona in terms of the growth of all animals alive at the beginning of the time interval – either to the time of death or until the end of the interval – plus the growth registered by young individuals born during the interval. This method is equivalent to equation [9.1], for if, in energy equivalents: B_0 = initial standing stock; B_1 = final standing stock; W_1 = initial biomass of all animals that die; W_2 = initial biomass of all animals that survive; g_1 = weight change of all animals dying; g_2 = weight change of all animals surviving; then:

$$B_1 = W_2 + g_2$$

$$B_0 = W_1 + W_2$$

$$\Delta B = B_1 - B_0 = (W_2 + g_2) - (W_1 + W_2)$$

$$E = W_1 + g_1$$

and, by substitution in equation [9.1]:

$$P = W_1 + g_1 + [(W_2 + g_2) - (W_1 + W_2)]$$

$$P = g_1 + g_2. \qquad [9.2]$$

Further elaboration of this point is given by Petrusewicz & Macfadyen (1970, pp. 7–8).

Here it is worth commenting on Kozlovsky's (1968) use of terms in his discussion of ecological efficiencies. Kozlovsky defined 'net production' (*NP*) as the sum of four components. *P* (production) was defined as the output of a population to predators of higher trophic levels. *Dc* (decomposition) was the output to decomposers. *TA* represented tissue accumulation (or loss): the increase or decrease of standing stock. And *L* was defined as loss of energy or matter by physical export. Kozlovsky's formulation can be made congruent with equations [9.1] and [9.2] if *P* and *Dc* are interpreted as outputs of tissue produced within the time period being considered. For example, if an individual animal weighed 10 g at t_0 and 15 g at time of death due to predation, 5 g would be credited to *P* (not 15 g).

As defined for use in equation [9.1], *E* (elimination) is the energy equivalent of all animals dying during the time interval in question. Hence, the calculation of *E* requires information as to the rate at which animals die off and the manner in which their body weights change during the time period of reference. These computations may be handled by means of single-step approximations or by various continuous functions, the nature of which depends on the assumed mortality and growth functions. Further discussion of this problem is given by Petrusewicz & Macfadyen (1970, pp. 84–90).

Our new calculations of production by the lizard (*Cnemidophorus tigris*) assumed linear changes in body weights combined with exponential mortality rates. The continuous function for estimating elimination in this manner was derived by Hardin Strickland and used by Turner, Medica & Kowalesky (1976) in research on energy flow in Nevada populations of *Uta stansburiana*:

$$E(g) = N \cdot W(1 - e^{-\lambda t}) \pm \frac{kN}{\lambda}[1 - (\lambda t + 1)e^{-\lambda t}], \qquad [9.3]$$

where: e = 2.718; N = initial size of group; W = initial mean weight (g) of an individual; λ = daily death rate; t = number of days in interval; k = daily change in weight (g).

New estimates of production by pocket mice (*Perognathus formosus*) for the years 1972–74 were also based on exponential mortality rates and linear weight changes, but these analyses were done graphically.

The work of R. M. Chew & A. E. Chew (1970) on mammals in southeastern Arizona has been re-evaluated, and our review will present revised analyses pertaining to four of the 13 species discussed in the original paper. A recurring problem in analyses of mammalian population data is estimating recruitment and early survival of young, because animals are not trapped until after they have been weaned. For the Arizona data, these estimates were based on observed disappearance rates of animals of trappable age (λ, loss per head per day), and estimated ages (d, days) of rodents when first trapped. Thus, the recruitment of newborn (N_0) was estimated from the number of young trapped (N_1) as follows:

$$N_0 = N_1/e^{-\lambda d} \qquad [9.4]$$

Birth weights and rates of subsequent growth were based on observations in the field and laboratory, and from records in the literature (sometimes for closely related species). Production for various life history intervals was estimated by graphic analyses of survival and growth curves. Production of extra-fetal tissues was taken as 0.27 g g^{-1} of live birth weight (Kaczmarski, 1966). Unless specific caloric data were available, the energy equivalent of these tissues and those of neonate animals was taken as 1.03 kcal g^{-1}. The caloric equivalent of young mammals prior to weaning was taken as 1.43 kcal g^{-1} (Kaczmarski, 1966), and that of older animals 1.50 kcal g^{-1} (Gorecki, 1965).

Review of existing data

Mammals in southeastern Arizona

R. M. Chew & A. E. Chew (1970) estimated the density, biomass and energy flow of 13 species of mammals of a desert shrub ecosystem in southeastern Arizona. Their data are of particular interest because the measurements were made at the same time and place under the same conditions. Production was first estimated for one species (*Dipodomys merriami*) exhibiting limited fluctuations in monthly densities, and for another species (*Peromyscus eremicus*) with considerable variation in monthly densities. Production of the other species was then estimated by assuming that production efficiency was linearly proportional to the ratio of maximum:minimum monthly density (Chew, R. M. & Chew, A. E., 1970, pp. 12–13). This is a questionable

Table 9.1. *Selected data for four rodents of a desert shrub ecosystem in southeastern Arizona.* Adapted from R. M. Chew & A. E. Chew (1970)

Species	Food habit	Average body weight (g)	Biomass ($g\ ha^{-1}$)		Rate of disappearance, λ (per head per day)	Annual[a] maintenance energy ($kcal\ ha^{-1}$)	Annual[b] energy flow ($kcal\ ha^{-1}$)
			Maximum	Minimum			
Dipodomys merriami	Granivore	39.7	590	320	0.0025	54880	57900
Onychomys torridus	Carnivore	23.8	71	30	0.0116	6530	6888
Ammospermophilus harrisii	Omnivore	122	34	9	—	1860	2061
Peromyscus eremicus	Omnivore	20.2	67	4	0.0177	4340	4720

[a] Maintenance energy is based on weaned individuals that entered the trappable population.

[b] Energy flow estimates reflect maintenance energy values, energy equivalents of all growth (production) between conception and death, and metabolic costs of production. Energy flow values are revisions of those given by R. M. Chew & A. E. Chew (1970, Table 6).

Table 9.2. *Aggregate weight changes over a 1 yr period of individuals in eight classes of* Dipodomys merriami *in a shrub community in southeastern Arizona (8.34 ha)*

Body weight range (g)	Sex	Aggregate change in biomass (g)
29	m & f	241
29–34	m & f	228
34–38	m	276
	f	227
38–42	m	81
	f	266
42	m	−122
	f	−287
Total		910[a]

[a] 910 g × 1.5 kcal g^{-1} = 1365 kcal for an area of 8.34 ha.

assumption, and in the ensuing discussion we examine the Chews' data more rigorously.

In the Arizona study, rodents were live-trapped and registered for five consecutive nights at ten times during a year in a 30.5 m trap grid (6 × 11 trap stations). Effective trapping areas ranged from about 7–25 ha, depending on movement of each species. These ranges were estimated from the recapture loci of individual animals. Weights were obtained for most of the animals taken during each census. The quality of the data for estimating production varied with the completeness of the censuses, the frequency of recapture of individuals, and the turnover rate of each species. We concluded that enough information was available to permit good estimates of production by four species, exhibiting three dietary patterns (Table 9.1). These new estimates supersede the original values (Chew, R. M. & Chew, A. E., 1970) and differ slightly from derivatives given by Johnson & Schreiber (1979). Production estimates for the other nine species should be considered unsubstantiated.

Of the four species, the kangaroo rat (*Dipodomys merriami*) had the most stable biomass and the lowest rate of disappearance (Table 9.1). Individual rats were almost always captured when at risk. Weight changes were known for 322 rats captured and weighed at two successive trappings. Average weight changes among eight categories of animals were then computed for each of ten trapping periods. Growth of the entire sample population was reconstructed from these averages and information on the composition of the population at each census. Total weight changes among trappable rats are summarized in Table 9.2. The four categories with weights greater than 38 g lost weight during 19 of 40 cases, and classes greater than 42 g had a net loss of weight for the year. These rats maintained themselves partly on previously accumulated production. This method of estimating growth is satisfactory when individuals are free to enter or leave the trapping area. Only that growth accomplished during the known residency of an animal is credited to production of the study site.

Table 9.2 pertains only to kangaroo rats trapped, reflecting both the growth of animals present throughout the entire period of study (g_2) and of animals that disappeared (g_1). However, a major portion of growth occurs before weaning, when young rats are not trapped. Young kangaroo rats were first captured at an average weight of 28 g (estimated average age of 56 days). All rats weighing less than 34 g when first captured were assumed to have been born within the trapping area, rather than to have dispersed into it; 90 such young were registered during the year. From the average rate of disappearance ($\lambda = 0.0025$), we inferred a recruitment of 104 newborn rats. By considering the growth curve of the species, we estimated that six rats died before weaning at an average weight of 12 g, and that, of the 98 weaned, another eight died (at an average weight of 25.5 g) before they had an opportunity to be trapped. The various components of production by juvenile kangaroo rats prior to registration are summarized in Table 9.3. To this is added the energy represented by growth of trappable animals (1365 kcal) yielding a total annual production of 5358 kcal in an area of 8.34 ha.

Similar procedures were followed with data for three other species, modified according to the nature of the specific data available. The grasshopper mouse (*Onychomys torridus*) exhibited the second most stable population biomass (Table 9.1). Weights were obtained for nine adults and nine young, out of a total registry of 51 adults and 23 young. Only 31 mice were captured more than once. A growth curve was constructed based on observations of five mice in a litter raised in the laboratory and of mice trapped in the field. Mice not weighed during censusing were assumed to have the same initial weight as the average of those weighed.

Growth of animals in the field was estimated from the growth curve, assuming that an animal was present for half the period before its first capture and half the period after its last capture. From the number of young *Onychomys* trapped and the observed disappearance rate of trapped mice ($\lambda = 0.0116$), we estimated recruitment of 30 newborn mice. We estimated that seven mice died before weaning, at an average body weight of 7 g. Table 9.3 gives pre-weaning production. All weaned grasshopper mice were trapped and their aggregate post-weaning growth was 57.6 g. Growth of adult mice was estimated as 159 g. Total annual production was then 846 kcal in an effective trapping area of 11.45 ha.

Nineteen adult and 11 young ground squirrels (*Ammospermophilus harrisii*) were registered. We judged that nine of the young were born on the study area, the survivors of two average-sized litters totalling 12 young. Average litter size (6), birth weight (3.6 g) and growth to weaning size (44 g) were taken

Table 9.3. *Estimated production by juveniles of four species of rodents in southeastern Arizona. Data pertain to growth experienced before registration by live-trapping. Growth between conception and birth allows for the production of 0.27 g g^{-1} live weight of extra-fetal tissues.* Based on work of R. M. Chew & A. E. Chew (1970)

Events	Weight changes of survivors (g)	Number surviving	g_2 (g)	Number dying	g_1 (g)	Caloric equivalent (kcal g^{-1})	g_2 (kcal)	g_1 (kcal)	Production (kcal)
Dipodomys merriami									
Conception to birth	3.81	104	396	0	0	1.03	408	0	408
Birth to weaning	3–23	98	1960	6	54	1.43	2803	77	2880
Weaning to trappability	23–28	90	450	8	20	1.50	675	30	705
Totals		90	2806	14	74		3886	107	3993
Onychomys torridus									
Conception to birth	2.92	30	87	0	0	1.03	90	0	90
Birth to weaning	2.3–14	23	269	7	33	1.43	385	47	432
Totals		23	356	7	33		475	47	522
Ammospermophilus harrisii									
Conception to birth	4.57	12	55	0	0	1.03	57	0	57
Birth to weaning	3.6–44	9	364	3	44	1.43	521	63	584
Totals		9	419	3	44		578	63	641
Peromyscus eremicus									
Conception to birth	1.27	33	42	0	0	1.03	43	0	43
Birth to weaning	1.0–11	22	220	11	53	1.43	315	76	391
Weaning to trappability	11–15	17	68	5	11	1.50	102	16	118
Totals		17	330	16	64		460	92	552

Table 9.4. *Estimated annual production by four rodents in southeastern Arizona.* Based on re-evaluations of data of R. M. Chew & A. E. Chew (1970)

Species	Area of reference (ha)	Estimated production by juveniles prior to age of trappability (kcal)	Estimated production by juveniles of trappable age and adults (kcal)	Total annual production (kcal ha^{-1})
Dipodomys merriami	8.34	3993	1365	642.4
Onychomys torridus	11.45	522	324	73.9
Ammospermophilus harrisii	24.7	641	334	39.5
Peromyscus eremicus	9.4	552	214	81.5

from Neal (1964). Post-weaning growth was estimated from observed growth of animals recaptured in the field. Most adult animals in the field were of weights that Neal (1964) found to be seasonally stable, or the animals were captured only between August and October when body weights were unchanged. Hence, we estimated no growth for 16 of the 19 adults in the trapping area. Pre-weaning growth (Table 9.3) together with subsequent growth of young (150 g) and adults (73 g) gives a total production of 975 kcal in a trapping area of 24.7 ha.

Nineteen adults and 17 young cactus mice (*Peromyscus eremicus*) were registered, but only eight were captured more than once and none was weighed more than once. This species exhibited extreme variability in biomass (Table 9.1), and we estimated its production using data of marginal quality to permit comparisons with more stable species. A growth curve was constructed based on observations of two litters raised in the laboratory (R. M. Chew, unpublished), combined with growth data for the similarly sized *P. maniculatus* (Svihla, 1934). From the observed disappearance rate ($\lambda = 0.0177$), we estimated that the 17 young were survivors of 33 mice born in the trapping area. We also judged that 11 mice died before weaning (at an average body weight of 5.8 g), and that 22 weaned at 11 g, and that five more died before being trapped (at an average weight of 13.1 g). Table 9.3 summarizes growth of juvenile cactus mice before the age of trappability. The subsequent growth of trapped young and adults was estimated to be about 143 g. Total production was 766 kcal in an area 9.4 ha.

Table 9.4 summarizes the foregoing discussion, recapitulating early growth of non-trappable young rodents, as well as subsequent growth by young and growth of adults. Table 9.4 also gives areas of reference (based on capture-recapture loci) and estimates of annual production.

Table 9.5. *Estimated growth rates of* Dipodomys merriami *in Joshua Tree National Monument.* Figures were derived by Soholt (personal communication) based on data in Chew & Butterworth (1959)

Age interval (days)	Initial body weight (g)	Final body weight (g)	Growth during interval (g)	Growth rate, k (g^{-1} day^{-1})
0–12	3.0	9.0	6.0	0.0907
12–24	9.0	15.5	6.5	0.0450
24–48	15.5	21.8	6.3	0.0142
48–108	21.8	31.8	10.0	0.0063
108–204	31.8	36.1	4.3	0.0013

Dipodomys merriami *at Joshua Tree National Monument, California*

Soholt (1973) analyzed the demography and energetics of the population of *Dipodomys merriami* inhabiting a *Larrea tridentata* (D.C.) Cov (= *L. divaricata* Cav.)– *Cassia armata* community at the southern edge of the Mohave Desert in San Bernardino County, California. There were other rodents present in this area, but *D. merriami* made up 93.6% of the captures of nocturnally active species. Between September 1969 and September 1970 the average density of kangaroo rats was 16.1 ha^{-1}, which is higher than previously reported estimates for this area as well as for other populations of *D. merriami*. Hence, we judge that 1969–70 was an above-average year for this species. Soholt's analysis (1973) underestimated production by young-of-the-year because he did not evaluate the death of young prior to weaning and first capture. However, Soholt (personal communication) has provided further information so that this mortality can be estimated. Consequently, some of the production values herein are different from those in his paper in 1973.

The average weight of a newborn *D. merriami* is about 3.0 g (Chew & Butterworth, 1959). Chew & Butterworth analyzed the growth of this species in the laboratory, using animals from very near the area studied by Soholt. The growth of Soholt's animals is postulated to follow the same pattern observed by Chew & Butterworth (Table 9.5), but at a slower rate, so that the resulting growth curve fits actual body weights measured in the field. The full-grown animals studied by Soholt weighed less (36.1 g) than those in the laboratory and southeastern Arizona (> 42 g).

Survival rates of kangaroo rats in Soholt's study area can be determined directly from trapping registrations for those rats of trappable age, but must be estimated indirectly for earlier periods of life. Between September 1969 and September 1970, 86 lactating females were captured in the trapping area (5.27 ha). Autopsy showed that pregnant females from an adjacent area had an average of 2.2 fetuses, so we assumed the birth of 189 young. Soholt captured 63 juvenile rats, and at the time of first capture these animals were

Table 9.6. *Components of production by young-of-year* Dipodomys merriami *occupying 5.27 ha in Joshua Tree National Monument in southeastern California.* The analysis is based on data given by Soholt (1973) and other information acquired in personal correspondence

Period of life	Weight changes (g)	Number of individuals surviving	g_2 (g)	Number of individuals dying	g_1 (g)	Caloric equivalents (kcal g^{-1})	g_2 (kcal)	g_1 (kcal)	Production (kcal ha^{-1})
Conception to birth	0–3.81[a]	189	720	0	0	1.03	742	0	140.8
Birth to weaning	3–20	113	1921	76	656	1.44	2766	945	704.2
Weaning to trappability	20–28.5	63	536	50	213	1.44	772	307	204.7
Trappability to death	28.5–36.1	0	0	63	178	1.44	0	256	48.6
Totals			3177	189	1047		4280	1508	1098

[a] Includes 0.81 g of extra-fetal tissues plus 3 g live weight of newborn.

estimated, on average, to be 90 days old and weigh 28.5 g. If we assume a constant rate of mortality between birth and the age of 90 days, the daily loss rate per head (λ) is 0.0122. Then, at time of weaning (age 42 days), there would have been 113 surviving rats (of 189 born). The average weight of these animals is estimated as 20 g (Table 9.6). The trapped young disappeared from the study area at a higher rate ($\lambda = 0.037$); after about 210 days none of these animals would be present.

Production by animals born into the population is summarized in Table 9.6, based on growth by surviving animals (g_2) and growth by animals that died (g_1), as in equation [9.2]. We assumed production of extra-fetal tissues (0.27 g g^{-1} of live birth weight), and the caloric equivalent of neonate tissue (1.03 kcal g^{-1} live weight) as in our previous analyses. Soholt determined the caloric equivalent of juvenile and adult kangaroo rats in his population to be 1.44 kcal g^{-1} live weight. Total production by the 189 kangaroo rats born in the study area was 5788 kcal, or 1098 kcal ha^{-1}.

Relating this production estimate to other estimates of rodent production in this review entails two difficulties. To begin with, there is no estimate of production by adult *D. merriami*. However, our previous estimate (Table 9.2) showed that adult kangaroo rats contributed $\leqslant 10\%$ of the total growth $(g_1 + g_2)$.

A more important problem is to relate the calculated production to an appropriate time interval. Our analysis treated all young as one group born hypothetically at the same time. If this were true – say, with all births occurring in March – then the production estimate could be legitimately ascribed to one calendar year, and all young-of-the-year would have died within that time. Actually, however, young *D. merriami* were captured by Soholt at different times during the year and some were certainly alive at the end of the period of study (September 1970).

We think this source of error outweighs the lack of information on the adult production. Hence, the actual annual production by Soholt's kangaroo rats was something less than 1098 kcal ha^{-1}. Even making this allowance, the estimated productivity of the California population is much higher than that of the Arizona population of *D. merriami* (642 kcal ha^{-1}, see later). As we pointed out earlier, the year 1969–70 was apparently a good one for *D. merriami* in Soholt's study area, and modal production is almost certainly lower than that estimated here.

Uta stansburiana *in Nevada*

Turner *et al.* (1976) analyzed energy flow and production by populations of the lizard, *Uta stansburiana*, in southern Nevada. In contrast to the Chews' analysis of small mammals, this study concerned the energy dynamics of a single species population over a 3 yr period beginning in March 1965. Production during 1965–66 was estimated as 336 kcal ha^{-1}. For the next two years estimated production was 535 and 536 kcal ha^{-1}, but the dynamics of the 1966–67 and 1967–68 populations differed. The year 1966–67 was one of marked increase in the abundance of *Uta* following prodigious egg production during the spring of 1966 (see Turner, Hoddenbach, Medica & Lannom, 1970). Of the total 1966–67 production, 34% occurred between March 1 and July 31; production of young was estimated at around 300 ha^{-1} and the elimination of these individuals was roughly 288 kcal ha^{-1}. Spring production in 1967 was only 20% of the total; production of young was only 144 ha^{-1}, but the elimination from these lizards was approximately 263 kcal ha^{-1}. Turner *et al.* (1976) pointed out that 'In spite of reduced production during the spring of 1967, the rapid growth of young *Uta*. . .raised annual production to the same level observed in 1966–67'.

The events in 1967–68 resulted in a relatively high ratio of production to respiration energy (P/R). Whereas P/R values in 1965–66 and 1966–67 were 0.22–0.24, P/R in 1967–68 was 0.34. Turner *et. al.* (1976, Table 18) estimated monthly production by *Uta* and showed that production was minimal during July of each of the three years analyzed. Productivity was usually highest in September of each year, due to elimination of young lizards which had experienced several months of growth. Production was also high during April and May when adult lizards gained weight and females produced eggs.

Birds in Arizona

Russell, Gould & Smith (1973) and Russell & Gould (1974) estimated production by birds occupying a 20 ha site in the Avra Valley near Tucson, Arizona, during 1972 and 1973. The more recent paper includes some revisions of earlier data, and we have taken data given by Russell & Gould (1974) as definitive when differences occurred. Some further small corrections were made after personal discussions with Russell. These two papers pertained only to production of eggs and young birds, and were not concerned with changes in body weights of adults. Wiens (1973) pointed out that weights of most adult grassland birds changed little during the breeding season, being of the order of 8% or less. If we assume stable body weights of adults, production may be considered simply in terms of eggs laid and growth of young birds.

The data of Russell and coworkers must be considered in terms of the following problems. First, production estimates were based on eggs, nestlings and fledglings observed, but some nests were never located or could not be examined. For example, black-throated sparrows (*Amphispiza bilineata*) established 12 territories on the plot in 1972 but only four nests were found. Secondly, estimated production cannot always be properly ascribed to the area (20 ha) in question. For example, Russell *et al.* (1973) pointed out that mourning and white-winged doves nesting on the plot foraged in agricultural areas elsewhere and found only a portion of their food in nesting areas. At Russell's suggestion we ascribed 50% of the estimated production by these species to the 20 ha study area. Finally, Russell and his coworkers assumed weights of fledging birds to be equal to adult body weights, which was an overestimate of production to time of fledging. We recomputed production by fledglings in terms of more appropriate body weights. Fledging weights for the mourning dove, verdin, cactus wren, curve-billed thrasher, brown towhee, rufous-winged sparrow, and black-throated sparrow were taken from Russell & Gould (1974, Table 2). For other birds we assumed an average fledging weight of 65% of the adult body weight.

The energy content of eggs of altricial birds was estimated from information in A. L. Romanoff & A. J. Romanoff (1949, p. 114). The shell, albumen, and yolk were assumed to compose 7.0, 73.2, and 19.8% of the total egg weight. The solid contents of albumen and yolk were taken as 10.5 and 42.9%, respectively (Romanov, A. L. & Romanov, A. J., p. 322), and we used a caloric equivalent of 6.15 kcal g^{-1} for the dry weight of egg contents. Energy in egg shells was not considered. Thus, a 1 g egg of an altricial bird would contain about 1 kcal:

$$[(0.732 \times 0.105) + (0.198 \times 0.429)] \times 6.15 = 0.995. \qquad [9.5]$$

This is close to the value used by Pinowski (1967) for the eggs of tree sparrows

Table 9.7. *Estimated minimal annual production by 20 species of birds in a 20 ha area near Tucson, Arizona*

Data are taken from Russell *et al.* (1973, Table 4) and Russell & Gould (1974, Table 5). The bases for converting weights to kilocalories are given in the text. Production by doves is half that estimated by Russell and coworkers.

Species	Energy content of unhatched eggs (kcal)		Energy content of unfledged nestlings (kcal)		Energy content of fledged birds (kcal)		Minimum annual production (kcal ha^{-1})	
	1972	1973	1972	1973	1972	1973	1972	1973
Lophortyx gambelii	—	—	—	—	1287.6	6625.2	64.4	331.3
Zenaida asiatica	6.8	17.0	96.9	5.8	158.3	79.2	13.1	5.1
Zenaida macroura	34.7	34.7	53.4	—	—	574.5	4.4	30.5
Geococcyx californianus	—	19.0	—	—	—	1560.1	—	79.0
Micropallas whitneyi	—	—	—	—	86.8	—	4.3	—
Colaptes chrysoides	—	—	—	—	564.5	564.5	28.2	28.2
Centurus uropygialis	—	—	—	—	759.9	607.9	38.0	30.4
Myiarchus tyrannulus	—	—	—	—	199.7	349.5	10.0	17.5
Myiarchus cinerascens	9.9	—	—	—	—	91.2	0.5	4.6
Auriparus flaviceps	8.0	15.0	36.7	126.9	182.9	219.4	11.4	18.1
Campylorhynchus brunneicapillus	28.8	25.2	307.3	247.2	250.5	801.6	29.3	53.7
Mimus polyglottos	—	16.0	—	—	—	—	—	0.8
Toxostoma curvirostre	47.2	70.8	544.4	327.3	1028.7	1122.2	81.0	76.0
Polioptila minuta	8.1	7.2	—	—	24.0	18.0	1.6	1.3
Molothrus ater	18.4	11.5	—	—	—	—	0.9	0.6
Pyrrhuloxia sinuata	19.0	11.4	51.8	26.7	192.9	154.3	13.2	9.6
Carpodacus mexicanus	3.6	18.0	60.1	—	—	186.4	3.2	10.2
Pipilo fuscus	8.6	8.6	80.2	—	53.4	320.6	7.1	16.5
Aimophila carpalis	—	9.0	—	38.4	39.7	139.1	2.0	9.3
Amphispiza bilineata	6.0	—	—	—	172.8	288.1	8.9	14.4
Totals	199	263	1231	772	5002	13702	322	738

in Poland (1.02 kcal g^{-1}). To estimate the energy content of hatchlings which died and of birds which successfully fledged we assumed a water content of 70% and a caloric equivalent of 1.67 kcal g^{-1} live weight. This last value was based on data of Odum, Marshall & Marples (1965) for non-migrant birds of eastern United States.

Table 9.7 gives minimal production estimates for 20 species of Arizona birds during 1972 and 1973. Estimated annual production by 65% of these species never exceeded 20 kcal ha^{-1}, and only the estimate for quail in 1973 exceeded 100 kcal ha^{-1}. The lack of information relating to unhatched eggs and unfledged nestlings of some species probably does not have much effect on the general magnitude of these estimates, since by far the majority of production is represented by fledged birds. Among Arizona birds the proportion was about 78% in 1972 and 93% in 1973. Pinowski (1967) estimated that about 92% of production by young tree sparrows was embodied in surviving fledglings.

The lack of data on subsequent growth by fledglings is a more important deficiency, for fledgling biomass may represent only 60–85% of the growth ultimately realized by these birds (see Wiens, 1973, p. 263). We estimated the amount of production which would have occurred in 1972 and 1973 had all fledglings survived and grown to adult weights. For all 20 species, this was about 122 kcal ha^{-1} in 1972 and 354 kcal ha^{-1} in 1973. Of the 1973 estimate, roughly half was growth by 36 fledgling quail. The potential production by growing fledglings was about 38% of the 1972 minimum production estimate and about 48% of the 1973 estimate.

It is clear that calculations of avian production will depend in an important way on estimates of survival rates of fledgling birds and their body weights at time of death. However, estimates of production were still modest even when we assumed complete survival of fledglings. Production in excess of 20 kcal ha^{-1} would have been achieved by only five species in 1972 (quail, gilded flicker, Gila woodpecker, cactus wren and thrasher), and only one of these estimates would have exceeded 100 kcal ha^{-1} (thrasher). Production was greater in 1973, and ten species would have produced more than 20 kcal ha^{-1} had all fledglings survived. Production by Gambel's quail would have been about 510 kcal ha^{-1}, and the roadrunner and thrasher would have produced more than 100 kcal ha^{-1}.

From the foregoing data we estimate the range of potential production by 20 species of Arizona birds to be from around 322–443 kcal ha^{-1} in 1972 and from 737–1091 kcal ha^{-1} in 1973. The most productive birds were Gambel's quail (*Lophortyx gambelii*), the curve-billed thrasher (*Toxostoma curvirostre*), the cactus wren (*Campylorhynchus brunneicapillus*), the roadrunner (*Geococcyx californianus*), and the Gila woodpecker (*Centurus uropygialis*).

The cactus wren has been studied in detail by A. H. Anderson & A. Anderson (1973). This wren is fairly common in the Sonoran Desert where cholla cacti (*Opuntia* spp.), palo verde (*Cercidium* spp.) and saguaro cacti (*Carnegiea gigantea*) provide nesting sites. The data of Russell and coworkers (Table 9.7) showed minimum annual production of 29 and 54 kcal ha^{-1} in 1972 and 1973, respectively. These estimates are increased to 33 and 65 kcal ha^{-1} by assuming complete survival and growth of all birds fledged. Male wrens maintain territories throughout the year, so transient birds are excluded unless a vacant area exists. Juveniles are eventually forced out by the adults. In a given area the number of territories and the density of adult birds are stable.

A. H. Anderson & A. Anderson (1973) reported numbers of breeding pairs, numbers of eggs laid and hatched, and numbers of hatchlings fledged over the period 1963–68 for a site in the Saguaro National Monument, east of

Table 9.8. *Estimated annual production by cactus wrens* (Campylorhynchus brunneicapillus) *on a 19.6 ha site in Saguaro National Monument, Arizona.* Based on data in A. H. Anderson & A. Anderson (1973)

Year	Energy content of unhatched eggs (kcal)	Energy content of unfledged nestlings (kcal)	Energy content of fledged birds (kcal)	Energy invested in growth after fledging (kcal)	Total annual production (kcal ha^{-1})
1963	36	101	2145	169	125.0
1964	65	171	1940	154	118.8
1965	54	1076	3524	275	251.4
1966	29	591	2911	231	191.9
1967	11	307	2554	198	156.6
1968	130	855	3881	309	263.9
Averages	54	517	2826	223	184.6

Tucson. They measured growth rates of hatchlings and survival of fledglings at another location just north of Tucson. We combined information from the two study areas and assumed that growth and survival were the same at both localities.

Our analysis of production was similar to that used for the data of Russell and coworkers. The average weight of eggs was taken as 3.6 g (Russell & Gould, 1974), and caloric equivalents were assumed to be 1 kcal g^{-1} live weight for eggs, and 1.67 kcal g^{-1} for hatchlings and fledglings. The weights of hatchlings at the time of death were estimated from a growth curve given by A. H. Anderson & A. Anderson (1973, p. 132). Elimination of hatchlings was estimated by assuming a constant death rate between the time of hatching and fledging (20 days). Hatchlings fledged at an average weight of 31.3 g, and fledglings reached adult weight (38.9 g) after 38 days. Corresponding figures given by Russell & Gould (1974) were 30.0 g and 38.3 g. Growth and elimination of young birds before their disappearance were estimated from survivorship and growth curves. Only 5% of the young were still present at the end of any one calendar year.

The estimates for six years summarized in Table 9.8 indicate a fairly uniform level of annual production by cactus wrens. The highest estimate (1968) is about 2.2 times as high as the lowest (1964). Wren production in the area studied by the Andersons is four times higher than that observed by Russell and coworkers in the Avra Valley. The maximum estimate based on Russell's observations, 65 kcal ha^{-1} is only one-third of the 6-yr mean based on the Andersons' work. These findings presumably reflect some substantial difference in the quality of wren habitats in the two areas investigated.

Production by Bootettix punctatus *in southern Nevada*

Mispagel (1978) estimated energy flow in populations of *Bootettix punctatus* in southern Nevada for five years (1971–75). This grasshopper is host-specific on creosotebush (*Larrea tridentata*). Numbers of grasshoppers of all ages were estimated from samples taken by vacuuming shrubs at two-week intervals. Growth rates of immature grasshoppers were inferred from changes in mean dry weights of individuals in D-Vac samples.

Production was estimated in terms of growth, exuviae, and eggs laid. From observations of captive grasshoppers, Mispagel estimated that adult females laid one clutch of eggs per season. Growth of young grasshoppers accounted for most of the estimated production (67–85%) in all years. For each sampling interval, growth was calculated as $1/2(N_0+N_1)(W_1-W_0)$, with N_0 and N_1 equal to numbers at the beginning and end of the interval and W_0 and W_1 the respective mean weights of grasshoppers at these times. Annual production was taken as the sum of all bi-weekly estimates. Mispagel's estimates of annual production ranged from 50 kcal ha^{-1} (1974) to 250 kcal ha^{-1} (1972), with a 5-yr mean of 170 kcal ha^{-1}. The most productive years (1972, 1975) were not the years of highest primary production. Mispagel suggested that parasitism of egg masses by beeflies in 1973 was responsible for the low numbers of grasshoppers in 1974.

The largest source of error in Mispagel's analysis was the assumption that D-Vac samples from 4–12 shrubs were reliable indexes of the size and composition of grasshopper populations (the density of *Larrea* in the study area was about 1000 ha^{-1}). Aside from purely statistical vagaries, it was acknowledged that around 15% of the early instars and up to 29% of adults avoided capture.

The grasshopper densities reported by Mispagel generally showed an increase to a maximum in late June (or early July) followed by a decline through September. In 1972, maximum numbers were reported in late March. In 1974, a few grasshoppers were captured in March, but no others were collected until September. Production between March and September was taken as zero, though the sampling data imply that grasshoppers were present through the summer. The production estimate for the last interval in 1974 (5.1 g) does not appear to be consistent with the reported increase in numbers (60–330 ha^{-1}) and growth (from a mean weight of 27 mg to 54 mg). In spite of these problems Mispagel's analyses are important as the only estimates of production by a desert arthropod and we know of no other predicated on sampling of total populations.

Production by soil nematodes

Production by nematodes in Rock Valley during 1974 was estimated by Freckman (1978). This was done indirectly by first estimating respiration of soil populations for each month of the year. Soil samples (200 cm^3) were taken three times each month at three depths and three distances in the vicinity of four shrub species. One-third of the samples were taken in open areas; the others from beneath shrubs. Nematodes recovered from samples were assigned to one of four trophic groups: fungal feeders, microbial feeders, omnivore-predators and root parasites (Freckman, Mankau & Ferris, 1975). Numbers of nematodes varied according to season, depth, distance from shrubs, and shrub species, but not shrub size (Freckman & Mankau, 1977). Total estimated numbers of nematodes of all types (to a depth of 30 cm) ranged from around 100000–156000 m^{-2}, depending on the time of year. To estimate monthly respiration 100–150 nematodes of each trophic group were used to determine mean body weights (W) of individuals of each group. These ranged from about 0.15–0.30 μg and did not differ seasonally. The hourly respiration (R_h) of an average individual at 20 °C (nl O_2 h^{-1}) was computed after Klekowski, Wasilewska & Paplinska (1972):

$$R_h = 1.4W^{0.72}. \quad [9.6]$$

Respiration rates were corrected for soil temperatures, and aggregate monthly respiration obtained by summing over individuals, hours and trophic groups. Refinements of extraction procedures (Freckman, 1977) showed that nematodes were anhydrobiotic (dormant) when soil moisture content was $< 2.7\%$. These adverse moisture conditions occurred at all depths in June 1974 and at certain depths during some other months. Because respiration is essentially suspended during anhydrobiosis, monthly respiration estimates were based only on subsamples taken from sufficiently moist soils.

Annual production was estimated using a transposition of an equation relating annual respiration (R) and production (P) of poikilothermic animals (Engelmann, 1966):

$$\log R = 0.62 + 0.86 \log P. \quad [9.7]$$

Annual nematode respiration in 1974 was estimated to be 1.157 kcal m^{-2}, and production 0.23 kcal m^{-2} (Freckman, 1978).

We question the final calculation of production on two grounds. First, to estimate production from respiration by an algebraic transposition of equation 9.7 is statistically improper. Engelmann should have regressed his estimates of production on those of respiration, in the manner followed by McNeill & Lawton (1970). If this were the only problem it would be simple enough to compute the appropriate regression equation. However, McNeill & Lawton (1970) examined far more data than were available to Engelmann. Their estimate of annual production (for short-lived poikilotherms) was given as:

$$\log P = 0.8262 \log R - 0.0948. \qquad [9.8]$$

When Freckman's estimate of population metabolism is used in this equation annual production by Rock Valley nematodes is computed as 0.91 kcal m^{-2}. There is an important distinction between the older model and that of McNeill & Lawton: the slope of Engelmann's regression exceeds one while that of McNeill & Lawton is less than one. McNeill & Lawton (1970, Table 2) discussed the implications of this point, namely, that as population respiration decreases the relative amount of production increases. At low values of R (< 10 kcal m^{-2}) the discrepancy between the two models is pronounced. Because the McNeill & Lawton model was based on many more data we favor it over the older formulation.

New estimates of secondary production

Production by Perognathus formosus

Production during 1963 and 1966

Pocket mice were studied in Rock Valley from 1961–68 by French and associates. Populations occupying three, 9 ha enclosures and an unfenced 9 ha plot were censused by live-trapping. Remarkably complete registries of the resident populations were maintained. Details of trapping, marking and data management were given by French (1964) and French *et al.* (1974). As work continued the age distributions of these populations became progressively better defined and by 1967 the enclosures contained few mice of unknown age (French *et al.*, 1974, p. 55). Pocket mice moved in and out of the unfenced area, and one of the enclosures was subjected to continuous γ-irradiation, so we restrict ourselves to data from the two control enclosures (Plots A and C of French *et al.*, 1974). Work began in Plot A in the summer of 1962 and in Plot C in January 1964. Animals in the plots were never intentionally killed and no body weights were taken.

Pocket mice were kill-trapped in areas adjoining the enclosures between 1961 and 1968. These animals were weighed and females used to assess reproductive activity. The ages of these animals were estimated from tooth wear (French *et al.*, 1974, p. 49). Over the 8-yr period, 1601 mice were trapped: 806 females and 795 males. Data from the enclosures also indicated a sex ratio of 1:1, although males and females exhibited seasonal differences in susceptibility to trapping. The largest kill-trapping samples were taken in 1961, 1963 and 1966. Demographic data were available only from 1963 to 1968. Hence, our analyses of production by *Perognathus formosus* were limited to Plot A in 1963 and to Plots A and C in 1966.

Table 9.9. *Numbers and mortality of adult pocket mice* (Perognathus formosus) *in 9-ha enclosures in Rock Valley during 1963 and 1966*

Month	Plot A 1963	Deaths in 1963	Plot A 1966	Plot C 1966	Totals 1966	Deaths in 1966
January	126	0	69	98	167	0
February	126	0	69	98	167	0
March	126	0	66	98	164	3
April	121	5	61	95	156	8
May	75	46	55	84	139	17
June	68	7	54	84	138	1
July	62	6	54	84	138	0
August	57	5	54	82	136	2
September	53	4	54	72	126	10
October	50	3	51	72	123	3
November	49	1	45	65	110	13
December	47	2	45	59	104	6

Pocket mice were relatively dense (*c.* 14 ha^{-1}) in Plot A in the spring of 1963. A greater density (*c.* 17 ha^{-1}) was observed only in 1967 (French *et al.*, 1974, p. 53). Reproduction was negligible in Plot A during 1963; no young were trapped in the plot that year (French *et al.*, 1974, p. 61), although there was some reproduction by female pocket mice kill-trapped nearby (French *et al.*, 1974, Table 12). Densities in Plot A and Plot C in 1966 were about 7.5 and 11 ha^{-1}, respectively. Whereas the 1963 population was mainly composed of adult year-classes, 1966 populations were predominantly mice born in 1965.

Reproduction and survival of pocket mice in 1963 and 1966

Table 9.9 gives numbers of adult pocket mice estimated to be alive at the end of each month in Plot A (1963 and 1966) and Plot C (1966). We assumed no mortality during January.

No reproduction occurred in 1963. In 1966, juvenile pocket mice were first captured in mid-May, when mice were about one month old. Hence, we assumed the young were born about mid-April. Assuming a sex ratio of 1:1, we estimated the number of adult females alive on April 15 as 29 in Plot A and 44.8 in Plot C (Table 9.9). About 90% of the females reproduced in 1966, and mean litter size was 5.9 (French *et al.*, 1974, Table 12), so we estimated total births as 392 (154 in Plot A and 238 in Plot C). The total of young mice actually trapped in these plots was 314. French *et al.* (1974, p. 57) estimated that mortality during the first month of life, before young mice were trappable, was about 17%. Hence, the number of young trapped at an age of one month or older is reasonably related to our estimate of total births.

We used a survival rate of 0.83 for the first month of life (April 15–May

Table 9.10. *Survival of young* Perognathus formosus *in two, 9 ha plots in Rock Valley during 1966*

Survival during the first month of life was taken as 0.83. Survival between May 15 and July 1 based on survivorship curves given by French *et al.* (1974, Fig. 11). Subsequent survival was based on actual trapping records.

Date	Plot A	Plot C	Totals	Total deaths
April 15	154	238	392	—
May 1	140	217	357	35
June 1	118	184	302	55
July 1	115	159	274	28
August 1	112	154	266	8
September 1	107	128	235	31
October 1	106	128	234	1
November 1	100	96	196	38
December 1	98	87	185	11
January 1	98	87	185	0

15), and the survivorship curves given by French *et al.* (1974, Fig. 11) for the remainder of 1966. Table 9.10 gives estimated numbers of pocket mice alive in Plots A and C during 1966 and the combined number of deaths during each month.

Body weights of pocket mice

Monthly mean weights of pocket mice in Plots A and C during 1963 and 1966 were calculated from weights of kill-trapped individuals from outside the plots. All weights for 1963 pertain to adults; during 1966 weights were determined for both adults and juveniles.

Body weights of adults during 1963. Table 9.11 gives mean body weights for male and female mice during the months of 1963 when data were available. Analysis of variance showed a significant effect of season (January–April, May–June, and August–October) and of sex on body weight. However, the differences between males and females were generally slight – less than 1 g – between May and October. Since males and females were equally abundant and apparently died at the same rate, we simplified the weight data by using overall averages for each month.

Table 9.11 shows the apparent seasonal pattern of weight change among these mice; a rise to maximum in late spring or early summer and a decline during the hot and dry period of late summer. We also believe mice lose weight during November and December, when feeding activity is reduced and mice are underground. In extending the data in Table 9.11, we made several simplifications and assumptions. We assumed a monthly loss during November and December of 2.5% of the October body weight. We estimated a common

Table 9.11. *Mean body weights of adult male and female* Perognathus formosus *trapped in Rock Valley during 1963.* The bases for estimating modified overall mean body weights are given in the text

Month	Males		Females		Overall mean body weight (modified) (g)
	n	Mean body weight (g)	*n*	Mean body weight (g)	
January	4	13.5	1	13.7	13.70
February	19	16.8	3	14.3	15.30
March	—	—	1	12.7	15.30
April	38	16.6	25	14.3	15.50
May	26	16.9	33	16.0	16.45
June	18	17.5	29	16.8	17.15
July	14	17.3	24	16.4	16.85
August	37	16.1	40	15.6	15.90
September	16	15.7	19	16.0	15.90
October	17	16.0	22	16.0	15.90
November	—	—	—	—	15.50
December	—	—	—	—	15.10

mean weight (13.9 g) for five females collected during January, February and March; estimated a common mean weight (16.7 g) for 57 males collected during February, March and April; and estimated common mean weights for both sexes during August–October (15.8 g for females, 16.0 g for males). These modifications are the basis for the overall mean body weights in Table 9.11.

Body weights of juveniles during 1966. The birth weight of pocket mice was estimated as 1.6 g, from weights of three newborn mice (1.5, 1.5, and 1.7 g). Early growth is rapid, for when young mice were first trapped at an age of about one month, they weighed almost as much (~ 15 g) as adults. We do not know how mortality was distributed over the first month of life, but assumed an average weight at death of 7.5 g for mice dying during this interval. Table 9.12 summarizes mean body weights based on weights of 96 mice born during the spring of 1966. Males were heavier than females, and we estimated missing weights accordingly. The basic data imply an increase in body weight at the end of the year, but this is based solely on the mean weight of three females collected in October. We chose a more conservative course, taking an overall mean weight of females for September and October (16.3 g). We assumed 16.7 g as the average weight of males in October. We reduced November and December weights by 2.5% and 5% of October weights, as with 1963 adults. Table 9.12 gives overall mean body weights for the last nine months of 1966, modified according to these assumptions.

Table 9.12. *Mean body weights of juvenile* Perognathus formosus *trapped in Rock Valley during 1966.* The bases for estimating modified overall mean body weights are given in the text

Month	Males		Females		Overall mean body weight (modified) (g)
	n	Mean body weight (g)	*n*	Mean body weight (g)	
April	—	—	—	—	7.5
May	2	14.9	—	—	14.7
June	26	17.1	23	16.9	17.0
July	1	20.5	—	—	20.0
August	4	19.6	3	18.0	18.8
September	15	16.7	19	15.9	16.5
October	—	—	3	18.6	16.5
November	—	—	—	—	16.1
December	—	—	—	—	15.7

Table 9.13. *Mean body weights of adult* Perognathus formosus *trapped in Rock Valley during 1966.* Pregnant females are not included. Bases for modification of basic data are given in the text

Month	*n*	Mean body weights (g)	Modified mean body weights (g)
January	—	—	18.0
February	2	18.20	17.6
March	55	17.58	17.6
April	36	19.15	19.2
May	2	19.55	19.6
June	25	18.90	18.9
July	—	—	18.3
August	—	—	17.6
September	3	17.00	17.0
October	—	—	17.0
November	—	—	16.6
December	—	—	16.2

Body weights of adults during 1966. Available weights of adult mice trapped in 1966 are given in Table 9.13; 39 obviously pregnant females were omitted from this tabulation. Analysis of variance showed no significant effect of sex on body weight, so sexes were combined to estimate the mean weights in Table 9.13. Season had a highly significant effect on weights. With respect to missing weights, we first calculated an overall mean weight of 17.6 g for February and March; then we assumed a January weight 2.5% greater than this mean. We interpolated linearly between 18.9 and 17.0 g to estimate means for July and August. We assumed a mean weight of 17.0 for October, and reduced

Table 9.14. *Elimination* (*E*) *of adult* Perognathus formosus *in Rock Valley during 1963.* Data pertain to one, 9 ha enclosure

Month	Deaths	Mean body weight (modified data) (g)	*E* (g)
April	5	15.50	77.50
May	46	16.45	756.70
June	7	17.15	120.05
July	6	16.85	101.10
August	5	15.90	79.50
September	4	15.90	63.60
October	3	15.90	47.70
November	1	15.50	15.50
December	2	15.10	30.20
Totals	79		1291.85

November and December weights as with previous data. These modifications are given in Table 9.13. The same seasonal weight changes occurred in this year with reproduction as in 1963 when no reproduction occurred in the study plots.

Elimination (*E*) *of pocket mice*

Losses, or elimination, from the 1963 and 1966 populations were estimated by combining mortality data from Tables 9.9 and 9.10 with weight data.

Table 9.14 recapitulates monthly mortality data from Table 9.9 and corresponding mean monthly weights from Table 9.11. It also gives resulting monthly and annual estimates of elimination in Plot A (9 ha) during 1963.

Table 9.15 recapitulates monthly mortality data from Table 9.10, corresponding monthly mean weights from Table 9.12, and gives resulting monthly and annual estimates of elimination of juveniles in Plots A and C combined (18 ha) during 1966. Table 9.15 also gives monthly mortality data from Table 9.9, corresponding monthly mean weights from Table 9.13, and resulting monthly and annual estimates of elimination by adults in Plots A and C (18 ha) during 1966.

Changes in standing stock (ΔB)

The change in standing stock for 1963 was simply the difference between the adult standing stock at the beginning of 1963 and that at the end of the year, because no young were born. The mean body weight of 126 mice in January was estimated to be 13.7 g and that of 47 surviving mice in December was 15.1 g. Then $\Delta B = (47 \times 15.1) - (126 \times 13.7)$ g, or -1016.5 g.

Table 9.15. *Elimination (E) of juvenile and adult* Perognathus formosus *in Rock Valley during 1966*. Data pertain to Plot A and Plot C combined (18 ha)

Month	Juvenile deaths	Mean body weights (modified data) (g)	Juvenile elimination (g)	Adult deaths	Mean body weights (modified data) (g)	Adult elimination (g)
March	—	—	—	3	17.6	52.8
April	35	7.5	262.5	8	19.2	153.6
May	55	14.7	808.5	17	19.6	333.2
June	28	17.0	476.0	1	18.9	18.9
July	8	20.0	160.0	—	—	—
August	31	18.8	582.8	2	17.6	35.2
September	1	16.5	16.5	10	17.0	170.0
October	38	16.5	627.0	3	17.0	51.0
November	11	16.1	177.1	13	16.6	215.8
December	—	—	—	6	16.2	97.2
Totals	207		3110.4	63		1127.7

Table 9.16. *Recapitulation of estimated elimination, change in standing stock, and secondary production by populations of pocket mice* (Perognathus formosus) *in Rock Valley during 1963 and 1966*

Estimate	Year 1963	1966
Area of reference (ha)	9	18
Adult elmination (g)	1292	1128
Juvenile elimination (g)	0	3110
Total elimination (g)	1292	4238
Adult ΔB (g)	−1017	−1320
Juvenile ΔB (g)	0	2904
Total ΔB (g)	−1017	1583
Production (g) $E+\Delta B$	275	5821
Production (kcal ha^{-1})	45.8	485.2

The standing stock on January 1, 1966 was the number of adult mice alive in Plots A and C (167) multiplied by a mean body weight of 18.0 g (3006.0 g). At the end of the year the adult biomass was reduced to 104×16.2 g, or 1684.8 g. However, owing to reproduction during the year, the population also included 185 juveniles with a mean body weight of 15.7 g (2904.5 g). Total change in standing stock was thus $1684.8 + 2904.5 - 3006.0 = 1583.3$ g (Table 9.16).

Table 9.17. *Monthly density and biomass values of four populations of* Perognathus formosus *in Rock Valley*

Month	Low density 1972–73 (n ha^{-1})	Low density 1972–73 (g ha^{-1})	Medium density 1972–73 (n ha^{-1})	Medium density 1972–73 (g ha^{-1})	High density 1972–73 (n ha^{-1})	High density 1972–73 (g ha^{-1})	Extremely high density 1973–74 (n ha^{-1})	Extremely high density 1973–74 (g ha^{-1})
March	1.10	19.2	2.93	56.9	4.29	70.0	2.1	38.4
April	1.10	21.3	2.93	63.2	9.03	154.0	2.5	37.6
May	1.32	17.5	6.77	129.8	8.80	174.3	13.9	203.5
June	2.30	30.0	16.25	286.5	15.12	278.1	63.1	680.5
July	3.07	53.4	22.80	393.9	26.64	465.7	99.1	1283
August	5.26	82.1	19.64	363.0	20.09	364.6	99.7	1573
September	4.39	69.3	15.35	279.9	13.54	225.7	93.1	1628
October	3.84	57.1	11.51	214.4	9.26	168.2	77.5	1367
November	3.29	49.1	9.26	164.8	6.32	114.0	63.5	1106
December	2.74	41.3	7.00	126.4	4.29	77.9	52.4	898.2
January	2.41	34.8	5.64	97.1	2.93	52.8	43.8	739.1
February	2.08	29.9	4.29	76.8	2.03	37.5	36.7	610.8
March	1.75	25.1	3.39	59.4	1.35	25.3	30.7	538.6
Means	2.67	40.8	9.83	177.9	9.51	169.9	52.2	823.4
Total (mouse-days or kg-days)	1016	15.55	3813	68.98	3701	66.15	20236	320.92

Production of pocket mice

The foregoing estimates were combined to determine secondary production by pocket mice during the two years in question (Table 9.16). We assumed a caloric equivalent of 1.5 kcal g^{-1} live weight, which is equivalent to a water content of 70% and 5.0 kcal g^{-1} dry weight.

Production during 1972–73 and 1973–74

In 1972, densities of *Perognathus formosus* were manipulated in the same enclosures described in the previous section, in order to observe effects of density on demographic function (Chew *et al.*, 1973). In Plot A (9 ha) the natural density of *P. formosus* was only 1.1 ha^{-1} on April 1 (Table 9.17). Plot C (9 ha), was divided into two halves. On April 1, the naturally occurring density of *P. formosus* on one side was reduced to 2.93 ha^{-1}, and the natural density of mice on the other side was artificially increased to 9.03 ha^{-1}. Thus, at the beginning of the period of above-ground activity, populations of low, medium and high density existed simultaneously on the same food resource base and under the same weather conditions. Pocket mice were live-trapped from March through to August, at monthly intervals in Plot A and biweekly in the other areas. Mice were weighed at each census, and were almost always captured if present. Production for the period March 1 to August 1, 1972 was calculated directly from weight changes of individuals, i.e., $P = g_1 + g_2$ (equation 9.2).

For mice that were resident in March, the weight change before death (g_1) was the difference between the weight at last capture and the initial weight in March. The weight change of mice that survived (g_2) was the difference between August and March weights. For mice that were introduced into the population, the initial weight was taken as that determined during the first census after introduction. Individuals that disappeared before the first census did not enter into the calculations. For mice born into the population, g_1 was the weight at last capture and g_2 was the weight at last census in August.

The average size of litters of *P. formosus* has been given as 5.6 (French *et al.*, 1974); hence, we estimated the number of young mice that did not survive long enough to be registered (N) as

$$N = 5.6\,Pr - Nw, \qquad [9.9]$$

with Pr the number of successful pregnancies observed and Nw the number of young mice registered. The mean weight of unregistered young was estimated as half that of the average weight of registered young at their first capture: 13.5, 14.2 and 15.5 g for low, medium and high density populations, respectively.

Weight changes had to be estimated for the interval between September and March 1, 1973, because mice were not active above ground for most of this time and were not censused. We assumed that (*a*) deaths occurred at a constant rate from the number present in August 1972 to the number censused in March 1973, (*b*) body weights declined linearly at such a rate that adults surviving to March 1 were 90% of their weight in August, and (*c*) juveniles lost weight in the same way, but emerged earlier than adults and regained their August 1972 weight by the time of the first census in March 1973.

Table 9.17 summarizes densities and biomasses of three 1972 populations. Figures for March through to August are based on direct measurements; values for September to March 1, 1973 were estimated as described above. Table 9.18 gives estimates of production by the three populations.

The two manipulated populations (medium and high density) converged in density and biomass in August 1972. Survival thereafter was better in the population that was initially of medium density. This group emerged in March 1973 with a density 2.5 times greater and a biomass 2.3 times greater than the population of initially high density.

In 1973, the two halves of Plot C (4.5 ha) were experimentally manipulated. Equal numbers of pocket mice were added progressively to both areas until densities were about 27 ha^{-1}. One area was structurally modified with 60 sheet-metal barriers 15.2 m long, placed in a repetitive pattern. The experiment was a test of the hypothesis that carrying capacity was limited by density-

Table 9.18. *Annual production* (*P*) *by* Perognathus formosus *populations of different densities*. Values are g live wt ha^{-1} (March 1 to March 1). g_1, weight change by mice dying during year; g_2, weight change by mice surviving to end of year

	Low density 1972–73			Medium density 1972–73			High density 1972–73			Extremely high density 1973–74		
Categories	g_1	g_2	P	g_1	g_2	P	g_1	g_2	P	g_1	g_2	P
Adults												
Resident	2.0	−0.6	1.4	−0.8	0.0	−0.8	−0.2	0.0	−0.2	24.6	0.0	24.6
Introduced	—	—	—	2.0	−1.6	0.4	34.3	4.1	38.4	34.9	16.0	50.9
Young-of-year	55.0	23.1	78.1	275.3	50.9	326.2	309.6	25.3	334.9	986.9	409.8	1396.7
Total g ha^{-1}			79.5			325.8			373.1			1472.2
Total kcal ha^{-1}			119			489			560			2208

Table 9.19. *Weights* (g) *of pocket mice introduced into Plots A and C in Rock Valley during 1972*

Category	Average initial weight	Average weight at time of first census (*a*)	Average peak weight (June–July) (*b*)	Average August weight (*c*)	Weighted mean body weight $\frac{(a+2b+c)}{4}$	Average March weight
Adult females	17.4	18.9	22.7	19.3	20.9	17.4
Juvenile females	11.8	14.5	18.0	17.3	17.0	17.3
Adult males	19.1	20.6	22.6	21.4	21.8	19.3
Juvenile males	11.2	14.0	18.0	17.4	16.9	17.4

dependent interactions and that barriers would reduce such interactions and allow a high carrying capacity. Details are described in Chew & Turner (1974). Both mouse populations in Plot C increased to peak densities, exceeding previous observations in Rock Valley (Table 9.18). The two areas have been combined for the present analysis of production because the barriers had little effect on density and survival. Both areas were censused every two weeks, but mice were not weighed. Consequently, weight changes have been estimated using assumptions based on 1972 weight data.

Mice introduced into the enclosures in 1973 were assumed to have weighed the same as mice introduced in 1972, and to have grown at the same rate (Table 9.19). Mice captured only once or twice after their first release were assumed to have died at their initial weight. Mice that died later, but before August, were assumed to have died at a weighted mean based on the body weight at first capture, the peak body weight later attained, and the projected body weight in August (Table 9.19). Mice surviving until August 1973 were assumed to have grown as in 1972.

Growth between March and August of 1973 was assumed to have occurred at the same rate as in the preceding year. August 1972 body weights were related to March 1972 body weight as follows:

$$W_{Aug} = 10.08 - 0.435 W_{Mch}. \quad [9.10]$$

This regression was used to estimate August 1973 weights from those in March of that year. Mice that survived until March 1974 were assumed to then weigh 90% of their estimated August 1973 weight. Mice not recaptured after August 1973 were assumed to have died at 90% of their August weight. Mice failing to live through August of 1973 were assumed to have died at half their potential weight increase between March and August. Mice dying before March 31, 1973, were assumed to have died at their March 1 weight. Most of these assumptions are conservative.

In order to obtain a reasonable fit between the time and number of observed pregnancies and the time of registry and number of young in each group of new juveniles, it was assumed that the first group of young entered the trappable population at 50 days of age. We further assumed that successive groups entered at 48, 46, 44, and 42 days, and all later cohorts at 40 days of age. The initial weights of young were estimated from the growth curve developed for the medium density population in 1972. This population exhibited the highest rate of growth in 1972.

Weight gains of young mice dying before September 1, 1973 were estimated by a month-by-month integration of the general growth curve and the survivorship curve for each group of juveniles. Mortality before the first census was assumed to have been 5.7%. This is the difference between the number of young registered and the number expected, assuming six mice per successful pregnancy in this particularly favorable habitat (rather than the average of 5.6 used previously). The low mortality is another indication of the favorable conditions for pocket mice in 1973. Mice dying before registration were assumed to have died at 7 g, which is intermediate to the birth weight (1.2 g) and the average initial weight of young at first capture in 1972 (15.2 g). For young surviving until September 1 or March 1, 1974, we assumed the same death rates and weight losses as before.

Table 9.17 gives densities and estimated biomasses of the extremely dense population of 1973–74, and Table 9.18 shows net secondary production. The four populations show an 18.6-fold range of annual production, i.e., 119–2208 kcal ha^{-1}.

Production by Cnemidophorus tigris

This analysis is predicated largely on data reported by Turner, Medica, Lannom & Hoddenbach (1969*b*). The year 1965–66 was one of conspicuous increase by whiptail lizards. Female whiptail lizards are normally reproductive at an age of about 22 months, and lay one clutch of eggs. However, during 1965 some 2 yr and older females laid two clutches. As a result of this unusually good reproduction, densities increased almost two-fold between the spring of 1965 and the spring of 1966. Naturally, the age distributions of the 1965 and 1966 populations differed; the latter was composed principally of eight-month old lizards surviving from the summer of 1965.

Overall annual mortality among age groups of lizards alive in the spring of 1965 could be estimated from recoveries of marked lizards during 1966 and subsequent years. However, Turner *et al.* (1969*b*) gave no information on how total annual mortality was distributed throughout the year. The only information given on body weights was associated with estimates of total population biomass (Turner *et al.*, 1969*b*, p. 192), and this was not broken down on an age-specific basis. Although survivors of the 1965 reproduction were registered in 1966 and later years, Turner *et al.* did not census 1965 hatchlings. Hence, the total number of young produced could only be estimated indirectly, using minimal survival observed among 60 marked hatchlings (52%). The data published by Turner *et al.* (1969*b*) were used to estimate production by using unpublished body weights and making further assumptions about reproduction and survival.

Changes in numbers of adults and age-specific survival

Hence we are concerned with survival of lizards between April 20, 1965, and the same date in 1966. Although counts of whiptails were reported from four areas, each of 9 ha, in Rock Valley (Turner *et al.*, 1969*b*, p. 192), we will restrict ourselves to those lizards occupying three fenced areas. Plots 1 and 3 are Plots A and C of French *et al.* (1974). Turner *et al.* (1969*b*) reported both the total whiptails registered in each area, and estimates based on capture–recapture data. For 1965, we used capture–recapture estimates of density in plots 1, 2 and 3; for 1966 we used either the total registry or the capture–recapture estimate, whichever was larger. Age distributions of the 1965 and 1966 populations were given by Turner *et al.* (1969, p. 193), and we used these distributions and associated densities to estimate states of the 1965 and 1966 populations (Table 9.20).

From this table we estimated annual survival rates of 0.73 between the ages of eight and 20 months, 0.60 between the ages of 20 and 32 months; and 0.67 for older lizards. Because sex ratios of the whiptail lizard populations were

Table 9.20. *Estimated numbers of* Cnemidophorus tigris *of various ages occupying three, 9 ha enclosures in Rock Valley during the springs of 1965 and 1966*

Year	Plot	Age (months) 8	20	32	44+	Plot Totals
1965	1	35	20	19	9	83
	2	19	11	35	—	65
	3	25	34	29	1	89
Totals		79	65	83	10	237
1966	1	100	22	10	15	147
	2	74	18	11	27	130
	3	103	18	18	20	159
Totals		277	58	39	62	436

Table 9.21. *Estimated age-specific daily death rates for* Cnemidophorus tigris *in Rock Valley (1965–66)*

Age group (months)	λ, first 106 days of year	λ, last 259 days of year
8	0.00244	0.000195
20	0.0039	0.000378
32+	0.00312	0.0003

1:1 (Turner *et al.*, 1969*b*, p. 192), we assumed the same mortality rates for males and females.

Field observations indicated that adult whiptail lizards went into aestivation on August 4, 1965, after 106 days of activity, and were dormant for the remainder of the year. In the absence of specific information, we assumed that 85% of the annual mortality occurred during the 106 days of above-ground activity. The daily death rates (λ) for the various age groups are given in Table 9.21.

Changes in body weights of adults

Weight changes could not be well inferred from an examination of individual weight records. Individual variation was appreciable, and body weights often went up and down during the summer, possibly depending on states of nutrition and/or hydration at time of measurements. Females showed distinct oscillations in body weight associated with production of eggs. Our approach was to compare the mean body weights of various age groups at the beginning and end of the annual period in question, and to assume that the changes

Table 9.22. *Mean body weights of* Cnemidophorus tigris *in Rock Valley in 1965 and 1966*

Sex and initial age (months)	n	Mean body weight in 1965 (g)	n	Mean body weight in 1966 (g)
Males (8)	25	6.46	45	19.36
Females (8)	19	6.55	36	18.61
Males (20)	44	15.99	23	22.06
Females (20)	34	14.80	15	19.31
Males (32)	41	18.96	21	21.74
Females (32)	55	17.10	31	20.84
Males (44)	4	19.50	6	23.81
Females (44)	3	16.31	1	20.00

in group means were acceptable estimators of net annual changes in weight. Because males were heavier than females, the analysis of growth was based not only on different age groups, but also on sex. Table 9.22 summarizes mean weights of four age groups in 1965 and mean weights of these same groups in 1966. For later calculations we assumed initial mean weights of 8 month old lizards to be 6.5 g, and all other means were rounded to the nearest 0.1 g.

We assumed that no weight gains occurred after August 4, when aestivation began. Some weight was lost during the ensuing 259 days of inactivity. Nagy (1972) estimated weight losses of two overwintering chuckwallas as 0.045 g per 100 g day^{-1}. Hence our growth model assumed an overwintering weight loss of 0.05 g per 100 g day^{-1}. Total annual weight changes were adjusted to compensate for presumed losses during the 259 days of hibernation. For example, 8 month old males averaged 6.5 g on April 20, 1965. A year later the survivors of this cohort averaged 19.4 g (Table 9.22). Then, if x is the weight at beginning of hibernation:

$$x-[259(0.05x/100)] = 19.4 \qquad [9.11]$$

and x equals 22.3 g. Weight increments and decrements for periods of activity and below-ground inactivity are given in Table 9.23.

Egg production by females during 1965

The basic data to be considered here are the mean clutch sizes for two age groups of female whiptail lizards reported by Turner *et al.* 1969*b*, p. 199). They assumed that every female survived for the entire reproductive season and laid two clutches. We re-evaluated egg production in 1965 using more realistic assumptions. We assumed that some female mortality occurred before the laying of the first clutch and additional mortality before the second clutch. From field observations we judged that the first clutch was laid on June 15 and the second on July 18. We assumed that all females laid one clutch and

Table 9.23. *Estimated weight increments and decrements in* Cnemidophorus tigris *in Rock Valley (1965–66)*

		During 106 days of growth		During 259 days of inactivity	
Sex	Age (months)	ΔW (g)	g day^{-1}	ΔW (g)	g day^{-1}
Male	8	15.8	0.149	−2.9	−0.0112
	20	9.3	0.088	−3.3	−0.0127
	32	5.9	0.056	−3.2	−0.0124
	44	7.8	0.074	−3.5	−0.0135
Female	8	14.9	0.141	−2.8	−0.0108
	20	7.4	0.070	−2.9	−0.0112
	32	6.8	0.064	−3.1	−0.0120
	44	6.7	0.063	−3.0	−0.0116

Table 9.24. *Estimated egg production by* Cnemidophorus tigris *in Rock Valley in 1965*

	20 month old females	Older females	All females
Initial numbers April 20	32.5	46.5	79
Alive June 15	26.1	39.0	65.1
Mean size of first clutch	2.46	3.73	—
Eggs laid on June 15	64.3	145.5	210
Alive July 18	23.0	35.2	58.2
Mean size of second clutch	2.49	3.33	—
Eggs laid on July 18[a]	42.9	88.0	131
Total eggs laid	107.2	233.5	341

[a] Adjusted for only ¾ of females producing a second clutch.

that three-fourths of those surviving until July 18 laid a second (Table 9.24). Because we assumed no egg mortality it follows that 341 young were added to the population in the summer of 1965.

Juvenile survival

We used estimated numbers of 8-month old whiptails in the three enclosures (277) as measures of survivors of the 1965 reproductive effort (Table 9.20). We assumed that members of both clutches survived at the same rate, and that hatchlings were active until October 15, 1965. Field observations indicated that whiptail lizards hatched from the first clutches around August 15, and were active for 61 days before hibernation. Lizards hatched from second clutches about September 18 and were active for only 26 days before hibernation. Both groups hibernated 186 days. As with adults, we assumed

Table 9.25. *Mean body weights of young* Cnemidophorus tigris *in Rock Valley during the autumn of 1965 and the spring of 1966*

Sex	Clutch	Date	n	Mean body weight (g)	Range (g)
Male	First	September–October, 1965	27	4.22	1.62–6.82
		April–May 1966	43	8.72	4.34–14.73
	Second	Autumn 1965	2	1.55	1.48–1.61
		Spring 1966	9	5.35	3.25–7.14
Female	First	Autumn 1965	27	3.81	1.53–7.65
		Spring 1966	36	8.92	5.14–13.69
	Second	Autumn 1965	8	1.96	1.35–2.65
		Spring 1966	8	5.33	2.83–6.72

that 85% of the mortality between hatching and the spring of 1966 was sustained during the period of above-ground activity. The daily mortality rate for the period of activity was 0.0037; for the 186 day period of hibernation 0.00018.

Weight changes among hatchling lizards

The smallest lizards weighed in the field were about 1 g. On August 2, 1973, a female 30 mm long weighed 0.86 g. We adopted a hatching weight of 0.9 g and assumed that male and female hatchlings grew at the same rate until the spring of 1966. How much did hatchlings grow before hibernation in mid-October? We attacked this problem by examining body weights taken towards the end of the 1965 season, and by looking at body weights of seven to eight month old lizards in the spring of 1966 (Table 9.25). All but one of the autumn measurements occurred in September and all but two of the spring measurements in May. The discrepancies reflect growth in late September and early October 1965, and growth after emergence in the spring of 1966. Although the picture is unclear, we assumed that hatchlings of the first clutch weighed 7 g on emergence in the spring of 1966 and that hatchlings of the second clutch weighed 4.5 g. Because of weight losses during hibernation, body weights on October 15, 1965, would have been 7.72 g and 4.96 g, respectively. Table 9.26 summarizes our assumptions regarding weight changes among hatchlings.

Estimating elimination (E) and change in standing stock (ΔB)

Elimination for all sex and age groups was estimated using equation 9.3. Table 9.27 summarizes estimates of elimination for adult and hatchling lizards for populations occupying a total area of 27 ha. The standing stocks of the

Table 9.26. *Weight increments and decrements of young* Cnemidophorus tigris *in Rock Valley between hatching in 1965 and the spring of 1966*

Clutch	Initial body weight (g)	Body weight October 15, 1965 (g)	ΔW (g)	Daily weight change (g)	Body weight April 20, 1966 (g)	ΔW (g)	Daily weight change (g)
1	0.9	7.72	6.82	0.112	7.0	−0.72	−0.0039
2	0.9	4.96	4.06	0.156	4.5	−0.46	−0.0025

Table 9.27. *Elimination (E) of* Cnemidophorus tigris *in Rock Valley in 1965–1966*

	Elimination (g)
Adult males	775.0
Adult females	715.5
Hatchlings from first clutch	218.0
Hatchlings from second clutch	53.3
Total	1762

Table 9.28. *Estimated standing stock of* Cnemidophorus tigris *in three, 9 ha enclosures in Rock Valley as of April 20, 1965*

Sex	Age (months)	*n*	Mean body weight (g)	Biomass (g)
Male	8	39.5	6.5	256.8
	20	32.5	16.0	520.0
	32	41.5	19.0	788.5
	44+	5.0	19.5	97.5
Female	8	39.5	6.5	256.7
	20	32.5	14.8	481.0
	32	41.5	17.1	709.6
	44+	5.0	16.3	81.5
Totals		237		3192

populations occupying the three, 9 ha areas in the spring of 1965 were derived from information given in Tables 9.20 and 9.22. Table 9.28 gives the estimated composition and biomass of these populations as of April 20, 1965. Similarly, the standing stock composed of lizards 20+ months of age on April 20, 1966, was estimated as in Table 9.29. Although the original population in 1965 of 237 lizards was reduced by mortality to 159 individuals, the growth of the

Table 9.29. *Estimated standing stock of* Cnemidophorus tigris *20+ months of age in three, 9 ha enclosures in Rock Valley as of April 20, 1966*

Sex	Age (months)	*n*	Mean body weight (g)	Biomass (g)
Male	20	29	19.4	562.6
	32	19.5	22.0	429.0
	44	28	21.7	607.6
	56+	3	23.8	71.4
Female	20	29	18.6	539.4
	32	19.5	19.3	376.4
	44	28	20.8	582.4
	56+	3	20.0	60.0
Totals		159		3229

survivors was such that biomass remained essentially unchanged, with ΔB equal to +37 g. However, the 1966 population also included 277 young lizards surviving from 1965 reproduction. According to earlier assumptions, young whiptail lizards surviving from the first clutches laid in 1965 weighed 7 g in the spring of 1966, while those from the second clutches weighed only 4.5 g. The 162 young surviving from the first clutch represented a biomass increment of 1134 g, while the 115 survivors from the second clutch contributed 518 g. Total biomass of young lizards in 1966 was, then, 1652 g. Adding the slight annual increase in standing stock of adult lizards (37 g) gives ΔB of 1689 g for an area of 27 ha.

Estimating annual production by Cnemidophorus tigris

We are now able to estimate production during 1965–66 by whiptail lizards. Calorimetric determinations of whiptail lizards collected in Rock Valley during May and June of 1973 were done under the direction of James MacMahon at Utah State University. The mean caloric content of these lizards was about 4800 cal g^{-1} dry weight, or (assuming a water content of 70%) 1.45 kcal g^{-1} live weight. Table 9.30 gives estimates of annual elimination of adult and hatchling whiptails, annual change in standing stock, and annual production for the three study areas (27 ha). Table 9.30 also gives annual production in terms of kcal ha^{-1}.

Production by Phrynosoma platyrhinos

Medica, Turner & Smith (1973) reported on reproduction and survival of horned lizards (*Phrynosoma platyrhinos*) occupying two enclosures, each of 0.4 ha, near Mercury, Nevada. Although no information on growth was

Table 9.30. *Estimated annual production (kcal) by* Cnemidophorus tigris *(1965–66) in three, 9 ha enclosures in Rock Valley*

Adult elimination	2161
Hatchling elimination	393
Total E	2554
ΔB	2449
Production for 27 ha	5003
Annual production per ha	185

Table 9.31. *Composition and biomass of populations of* Phrynosoma platyrhinos *occupying two, 0.4 ha enclosures in Rock Valley as of March 1, 1969*

Groups	n	Approximate mean body weight (g)	Biomass (g)
8 Months old			
males	4	2.5	10
females	2	2.5	5
20 Months old			
males	3	19.0	57
females	0	—	—
Older			
males	0	—	—
females	3	24.0	72
Totals	12		144

reported in this paper, live body-weight measurements of these lizards were available and were used to estimate production between the spring of 1969 and the spring of 1970. The composition of the aggregate population occupying the two enclosures as of March 1, 1969, is given in Table 9.31. Of these 12 adults, three died during the ensuing year. Two, 8 month old lizards died at estimated body weights of 13 and 14 g, and an older female died at a weight of 28 g. Total adult elimination during the year was, then, 55 g.

The three older females all laid two clutches of eggs during the late spring of 1969 (Medica *et al.*, 1973, p. 81). The first clutches were laid sometime between May 12 and early June, and the second clutches between mid-June and mid-July. We assumed that all hatchlings from first clutches emerged on July 12 and that all hatchlings from the second clutches appeared on August 15. Thirty-four hatchlings (18 males, 16 females) were registered in the two enclosures during the late summer of 1969. This number is reasonably related to the mean clutch size of horned lizards at the Nevada Test Site (6.7) reported by Tanner & Krogh (1973).

At least 13 (seven males, six females) survived until the following spring

Table 9.32. *Weight changes among yearling* Phrynosoma platyrhinos *in two, 0.4 ha enclosures during the spring and summer of 1969*

Sex	Number	Date	Live body weight (g)
Male	1112	March 25	2.45
		May 13	6.70
		June 24	13.30
	1124	May 8	6.14
	1417	March 27	2.28
		June 5	11.14
		June 24	13.12
	3222	April 22	4.01
		May 13	5.03
		July 14	13.57
Female	1127	May 12	8.53
		June 24	18.48
		August 1	22.89
	3212	March 25	2.97
		May 12	6.78
		July 14	17.38
		August 19	21.01

(Medica *et al.*, 1973, Table 2). The smallest hatchling horned lizard captured in the Nevada enclosures weighed 1.1 g, so we assumed an average hatching weight of 1.0 g. Body weights of 8 month old horned lizards in the spring of 1970 suggested that lizards from first clutches weighed about 2.8 g, while those from second clutches weighed about 1.7 g. If we assume the same pattern of weight loss during winter (September 30–March 10) as postulated for *Cnemidophorus*, these lizards would have weighed about 3 g and 1.8 g just before hibernation.

Between the summer of 1969 and the spring of 1970 21 juvenile horned lizards died, but we do not know how these deaths were distributed in time. If lizards from the two clutches died in equal numbers the maximum elimination by juveniles would be about 51 g (i.e., $11 \times 3 + 10 \times 1.8$). The smallest estimate would obtain if all hatchlings died right after birth (*c.* 21 g). We assumed an overall average weight, at death, of 1.8 g for all hatchlings, and estimated elimination by juveniles as 38 g. Total elimination from the population (E) was 93 g (55 g + 38 g).

Change in standing stock over a year's time was estimated from the biomass of the spring populations of 1969 and 1970. As given above, the youngest lizards in the spring of 1970 weighed around 2.8 g or 1.7 g, depending on time of hatching during the summer of 1969. We used an overall mean weight of 2.2 g for these yearlings. Growth of yearling lizards during 1969 was inferred from data in Table 9.32. It is difficult to determine exactly what happened

Table 9.33. *Composition and biomass of populations of* Phrynosoma platyrhinos *occupying two, 0.4 ha enclosures in Rock Valley as of March 1, 1970*

Groups	*n*	Approximate mean body weight (g)	Biomass (g)
8 Months old			
males	7	2.2	15.4
females	6	2.2	13.2
20 Months old			
males	2	19.0	38
females	2	19.0	38
Older			
males	3	24.0	72
females	2	24.0	48
Totals	22		225

towards the end of the season, but making allowance for loss of weight during the winter, we judged that a mean weight of 19 g in March of 1970 (as estimated for 1969) was reasonable.

Data for older lizards indicated some weight gain by the three, 20 month old males during 1969, and we assumed an increase to 24 g. Weight changes of the three, 32 month old females during 1969 were erratic because of the development of laying of two clutches of eggs (Medica *et al.*, 1973, Table 3). All three females attained weights of 30 g or more during April and May and again in June. However, two of the females declined to less than 20 g during July. We assumed no net change in their weight (24 g) over the course of the year. Table 9.33 summarizes the estimated state of the March 1970 population. The change in standing stock (ΔB) between March 1969 (Table 9.31) and March 1970 was +81 g (225 g − 144 g). Annual production (P) in the 0.4 ha plots was then 81 g + 93 g, or 174 g. Assuming a live-weight equivalent of 1.4 kcal g^{-1} implies an annual production of 305 kcal ha^{-1}.

Of this production, about half was represented by the elimination of three adults and 21 hatchlings, and the other half by an increase in standing stock. Growth of yearlings was the biggest component of the latter. The six yearlings present in the spring of 1969 (with an aggregate weight of 15 g) grew so rapidly that the four surviving until 1970 weighed 76 g. No other age group demonstrated anything comparable to this growth. The time of greatest weight increase of horned lizards is apparently during the spring following hatching, whereas hatchling *Uta stansburiana* put on most weight during the summer of birth and enter the next spring at nearly adult weight (Turner, Hoddenbach & Lannom, 1964; Turner *et al.*, 1976).

The pattern of weight increase of horned lizards is not clearly correlated with increases in body length. Tanner & Krogh (1973, pp. 335–6) estimated that hatchling horned lizards increased in snout-vent length at around 0.25 mm day^{-1} during the summer, but growth in the ensuing spring was only about 0.13 mm day^{-1}.

The foregoing estimate of annual production by horned lizards (~ 300 kcal ha^{-1}) is high relative to estimates for two more abundant lizards (*c.* 185 kcal ha^{-1} for *Cnemidophorus tigris* and 335–535 kcal ha^{-1} for *Uta stansburiana*). However, several points need to be kept in mind. First, the area surveyed was small and second, reproduction by horned lizards in 1969 was better than usual. These lizards ordinarily produce but one clutch of eggs, and in particularly bad years (e.g., 1970) may not reproduce at all (Medica *et al.*, 1973, p. 81). Third, the densities of horned lizards in enclosures in 1969 and 1970 (10–32 ha^{-1}) were higher than the densities recorded in 9 ha enclosures in Rock Valley (1–7 ha^{-1}) and those reported by Tanner & Krogh (1973) on Frenchman Flat (5 ha^{-1}). Medica *et al.* discussed the possibility that fencing the 0.4 ha enclosure may have enforced local increases in density which would not normally have occurred. The fencing of plots may also have had some favorable influence on survival, for minimal survival of adults in 9 ha plots was around 50% (Medica *et al.*, 1973, p. 82).

Production by birds in southern Nevada

Observations of breeding birds were made by Hill in Rock Valley between 1971 and 1973, and some of these data have been recapitulated in annual summaries of research associated with the Desert Biome section of the United States contribution to the International Biological Program (Turner, 1972, 1973). The best information was acquired during 1973, and from it we can estimate production by two species which breed regularly in Rock Valley: the black-throated sparrow (*Amphispiza bilineata*) and LeConte's thrasher (*Toxostoma lecontei*). Hill's observations extended over several plots totalling 43 ha. Spring breeding densities were estimated by conventional spot-map censuses and observations were made along parallel lines 50 m apart. Three to eight censuses were taken each month from February to November. Mean live body weights were based on breeding birds taken by mist nets and weighed at their nests. Dry weights were assumed to be 30% of live weights. The estimated density of *Amphispiza* was 26 pairs ha^{-1}; that of *Toxostoma* was 1.5 pairs ha^{-1}.

The mean clutch size (C) of *Amphispiza* (based on nine nests) was 4.0; that of *Toxostoma* (based on four nests) was 3.8. Whereas the sparrows nested only once in 1973, Hill's observations indicated that the thrashers bred more often. Breeding by these birds normally begins in February and may be continued through to June. Hill followed three pairs of thrashers during the 1973 breeding season. One of these pairs nested four times, but only two nests were successful. One of the other pairs nested twice. We assumed that each pair of thrashers laid two clutches of eggs in 1973. Hatching success (*H*) was 78.1% for sparrow eggs and 73.3% for thrasher eggs. Nestling survival (*S*) was 68% for sparrows and 72.7% for thrashers (Hill & Burr, 1974).

To estimate production by these birds we made the same assumption regarding the stability of adult body weights as we did in discussing production by Arizona birds (see p. 211). Furthermore, we had to estimate egg weights, weights of nestlings, and weights of fledgling thrashers, since these were not measured in the field. The live weights of eggs were estimated using the following formula for the eggs of altricial birds (Schoenwetter, 1924; Romanoff, A. L. & Romanoff, A. J., 1949, p. 107):

$$\text{egg weight (g)} = 0.5463\, LB^2, \tag{9.12}$$

where *L* and *B* are the length and breadth of the eggs (cm). Estimated egg weights (including shells), W_e, were 2.04 g for *Amphispiza* and 5.61 g for *Toxostoma*. The energy content of unhatched eggs was taken as 1 kcal g^{-1} live weight (see equation 9.5). Of 25 nestling sparrows under observation in 1973, eight died at an average age of about three days. Work on white-crowned sparrows (Banks, 1959; Morton & Orejuela, 1972) has shown that 3 day old nestlings attained about 45% of fledging weight. The live weight of fledging *Amphispiza* (W_f) in the Rock Valley was 12 g. Therefore, we took 5.4 g as the weight of nestlings which died in the nest (W_n). Nestling thrashers which died in 1973 did so at an age of about two days. We assumed that these birds attained 20% of fledging weight. Russell *et al.* (1973, Table 10) reported fledging weights of curve-billed thrashers (*Toxostoma curvirostre*) in Arizona to be 68% of adult weight. The adult weight of *T. lecontei* in Rock Valley is about 65 g. Hence, we took the fledging weight (W_f) as 44.2 g and the weight of dying nestlings (W_n) as 8.8 g.

Production by one pair of birds was computed as the energy content of (*a*) unhatched eggs, *U*, (*b*) dying hatchlings, *D*, and (*c*) fledged birds, *F*. Then:

$$U = 1.0 \text{ kcal g}^{-1}\, [CW_e(1-H)] \tag{9.13}$$

$$D = 1.5 \text{ kcal g}^{-1}\, [CW_n H(1-S)] \tag{9.14}$$

$$F = 1.5 \text{ kcal g}^{-1}\, (CW_f HS). \tag{9.15}$$

Production by a pair of sparrows was about 48.1 kcal, but the density of breeding pairs was 0.6 ha^{-1}, so annual production was 28.9 kcal ha^{-1}. Production by one pair of thrashers was about 300.4 kcal (recalling that the average number of clutches laid was two). Breeding-pair density was 0.035 ha^{-1}, and annual production was about 10.5 kcal ha^{-1}.

The above calculations assume that every resident pair with recorded territories built nests and laid eggs. This was true of thrashers observed. Observations of sparrows indicated that at least 15 pairs reproduced. The most conservative estimate of production by sparrows would be, then,

15/26 × 28.9 kcal ha^{-1} (16.7 kcal ha^{-1}). On the other hand, as with the Arizona data of Russell *et al.* (1973), we have not made allowances for growth of fledglings. Hence, production is underestimated by whatever growth was realized by fledged birds. We can roughly estimate the maximum extent of this production if we assume that all fledglings survived and grew to adult size. One pair of sparrows produced 2.12 fledglings (each 12 g) and one pair of thrashers produced 4.05 fledglings (each 44.2 g). Growth of these birds to adult size (about 14 g and 65 g, respectively) would have resulted in further production of about 4 kcal ha^{-1} for each species (assuming 1.5 kcal g^{-1} live weight).

Discussion

Net primary production in deserts and its limitation of aggregate secondary production

It is unlikely that total secondary production of any desert community can ever be directly evaluated. However, we can draw some general inferences as to aggregate secondary production by considering levels of net primary production (NPP) and general efficiencies of consumers.

Net primary production of deserts is at the low end of the spectrum of vegetation types. Whittaker (1970, p. 83) gave the normal range of annual above-ground NPP of desert scrub vegetation as 100–2500 kg ha^{-1} and Noy-Meir (1973, pp. 44–5) gave estimates of 300–2000 kg ha^{-1} for above-ground NPP and 1000–4000 kg ha^{-1} for total NPP. Because so little is known of the fate of below-ground production, we will consider here only NPP above the ground.

In the Mohave desert, where most of the work discussed in this paper was carried out, above-ground NPP tends toward the minimum levels given by Whittaker and Noy-Meir. Bamberg and his colleagues, using harvest methods, estimated that annual above-ground NPP in Rock Valley averaged about 530 kg ha^{-1} in 1971–73, with a range of 162 to 1247 kg (Norton, 1974; Bamberg *et al.*, 1974; Ackerman, Bamberg, Hill & Kaaz, 1974). These estimates are only about 35% of estimates based on net carbon dioxide exchange rates (Bamberg *et al.*, 1973) and we judge that true NPP is higher than harvest-based estimates reported by Bamberg *et al.* (1974).

We assume that in deserts most of the NPP is not consumed but decays, leading to as yet unassessed production by reducers (bacteria and fungi). A small portion is decomposed abiotically. Whittaker (1970, p. 99) stated that, 'In terrestrial communities, as much as 90% of the net primary production remains unharvested as living plant tissue and must be utilized as dead tissue by saprobes and soil animals', but specific estimates for desert communities are lacking. Applying 10% to the upper limit of annual above-ground NPP

(2000–2500 kg ha^{-1}) suggests that up to several hundred kilograms per hectare are ingested annually by above-ground consumers.

The efficiency with which this material is converted to new tissue varies depending on the types of consumers and their relative abundances (see Turner, 1970). At present we will assume a conversion of 10% for the total mix of consumer efficiencies, which range from ~ 1% for endotherms to 20% or more for heterotherms and ectotherms. There is no evidence yet that desert-inhabiting consumers have significantly different efficiencies to other consumers (see Chew, R. M. & Chew, A. E., 1970; Soholt, 1973).

If we assume the general values given above, maximum annual, aggregate secondary production from above-ground NPP in deserts may be estimated as: 2500 (kg ha^{-1}) × 4 × 10^{3} (kcal kg^{-1}) × 0.10 (fraction consumed) × 0.10 (fraction into production), or 10^{5} kcal ha^{-1}. This is only 1% of above-ground NPP. As pointed out by Whittaker (1970, p. 99) secondary production by reducers should exceed that of consumers; we expect the excess to be considerable in desert ecosystems.

Production by desert animals contrasted with that of animals in other environments

As shown in the previous section, little can be inferred about the aggregate secondary production in desert ecosystems. It may be more useful to consider how the energy of primary production is subdivided among various consumer species. However, progress in this direction is limited by the paucity of data for arthropods. Production by these diverse and abundant consumers may far exceed that of vertebrates. We have commented briefly on production estimates for grasshoppers (50–250 kcal ha^{-1}) in southern Nevada (Mispagel, 1979). Nutting, Haverty & LeFage (1975, Table 17) estimated that 10 species of termites in desert grassland of the Santa Rita Experimental Range in Arizona assimilated about 247 kg of detritus ha^{-1} yr^{-1}. Assuming conservatively a production:assimilation ratio of 0.1, and a caloric equivalent of 4 kcal g^{-1} dry wt of termite, the assimilation of 247 kg ha^{-1} implies a termite production of 24.7 kg ha^{-1} or 98 000 kcal ha^{-1} yr^{-1}. Turner *et al.* (1976) estimated annual energy flow in southern Nevada populations of *Uta stansburiana* to range from 1831–2794 kcal ha^{-1}. If one assumes an assimilation efficiency of 0.8 this would imply the consumption of about 2300–3450 kcal ha^{-1} of arthropod prey. Similarly, annual production of 74 kcal ha^{-1} by the insectivorous rodent, *Onychomys torridus* (Table 9.4) implies a minimum annual production of 9250 kcal ha^{-1} by its arthropod prey, if one assumes a production: assimilation ratio of 0.01 (Table 9.39) and an assimilation: consumption ratio of 0.8 for this rodent.

Our general conclusion with regard to desert vertebrates is that annual production by any one species is ordinarily some tens to hundreds of kcal ha^{-1}. In particularly unproductive environments, or during exceedingly dry years (e.g., 1963, 1970 in Rock Valley), annual production by some species may fall to near zero. In particularly good years (e.g., 1973 in Rock Valley) some species might produce of a few thousands of kcal ha^{-1}.

Among Rock Valley vertebrates, and this may be true of deserts in general, the bulk of production is by mammals and reptiles, with birds contributing little to the total. This was observed in desert grassland by Wiens (1973). With regard to a site at the Jornada Experimental Range in New Mexico, he stated (p. 264): 'Here bird density and biomass were extremely low, but small mammal density and biomass were much higher than elsewhere [among six other grassland sites].' In addition, reptiles 'were noticeably more diverse and abundant at Jornada than elsewhere. It seems likely that the ecological roles played by birds in grassland ecosystems proper are assumed by reptiles and small mammals...at Jornada.'

In general, maximum production among desert vertebrates of southwestern United States appears to be accomplished by one or two species of abundant mammals. We would expect production by the most common insectivorous lizards to exceed that of less numerous herbivorous mammals. The lowest productivity would be exhibited by predatory and herbivorous lizards, snakes, large carnivorous mammals, and birds. Oliver Pearson (personal communication) pointed out that mammals may rank high in production in southwestern deserts because of the exceptional way heteromyids (*Dipodomys*, *Perognathus*) exploit seed resources. Ecological equivalents are absent among rodents of the Peruvian Desert. On the other hand, rodents would rank even higher if fossorial forms were included. Where pocket gophers or equivalents (South American tucu tucu) are reasonably abundant, they outweigh all other rodents (Dingman & Byers, 1974; O. Pearson, personal communication).

Comparisons between production of desert vertebrates and those of more mesic ecosystems are tenuous. Amounts of energy utilized by consumer populations are influenced by various factors: body size and metabolic rate, relative abundance, and type of food eaten. Hence, comparisons of production by individual species occupying different environments are meaningless unless viewed in terms of all relevant variables. Time variations in production have to be considered also; we have already commented on conspicuous variations in annual production by *Perognathus formosus*.

Annual production estimates for herbivorous small mammals occupying grasslands or old fields range from 1200 kcal ha^{-1} by white-footed mice in Georgia (Odum, Connell & Davenport, 1962) to 5200 kcal ha^{-1} by voles in Michigan (Golley, 1960). Annual production by sparrows is around 400–500 kcal ha^{-1} in Georgia and Michigan old fields (Odum *et al.*, 1962; Wiegert & Evans, 1967). These values are higher than those for desert vertebrates of the same general size and trophic function. Production by desert rodents is more similar to that reported for small mammals of temperate forest and taiga.

Table 9.34. *Small mammal production per rodent-year and per gram-year.* A rodent-year is 365 rodent-days lived and a gram-year is the maintenance of 1 g of live biomass for a 1 yr period

Species	Habitat type	Mean body weight (g)	Production (kcal rodent-year^{-1})	Production (kcal gram-year^{-1})	Source
Dipodomys merriami	Desert	39.7	50.4	1.27	R. M. Chew & A. E. Chew (1970); Tables 9.3 and 9.4
Onychomys torridus	Desert	23.8	40.4	1.70	
Ammospermophilus harrisii	Desert	122	180	1.87	
Perognathus formosus	Desert				
Low density		15.3	42.9	2.80	Tables 9.17 and 9.18
Medium density		18.1	46.8	2.59	
High density		17.9	55.2	3.08	
Extremely high density		15.9	39.8	2.50	
Microtus ochrogaster	Grassland	28.5	115	4.04	Golley, 1960
Clethrionomys glareolus	Beech forest	17.3	151	8.73	Grodziński *et al.* (1970)
Apodemus flavicollis	Beech forest	23.9	106	4.44	
Clethrionomys rutilis	Taiga	22.1	92	4.16	Grodziński (1971)
Tamiasciurus hudsonicus	Taiga	229	320	1.40	

Grodziński, Bobek, Drożdż & Gorecki (1970) reported annual productions ranging from 300–2660 kcal ha^{-1} for voles and yellow-backed mice in beech forest in Poland; Grodziński (1971) reported values of 100–1100 kcal ha^{-1} for rodents of Alaskan taiga. In this connection, the annual production ascribed to these Alaskan species (0.2–2.4 kcal ha^{-1}) by Ryszkowski & Petrusewicz (1967, p. 145) were in error by several orders of magnitude. The erroneous values were also repeated by McNeill & Lawton (1970).

Comparisons of production by various desert and non-desert small mammals are given in Table 9.34 (in terms of rodent-years and gram-years). In terms of kcal gram-year^{-1}, production by desert species is less than that by four of the other five species. This may indicate that production by individual animals in deserts is limited by a relatively impoverished environment.

Intraspecific variations in production

Noy-Meir (1973, p. 26) defined deserts as 'water-controlled ecosystems with infrequent, discrete, and largely unpredictable water inputs'. Net primary production may be highly variable from one year to the next and the productivities of populations of desert animals are inevitably influenced by year-to-year differences in available food and energy. Noy-Meir discussed what has been termed the 'pulse and reserve' model of desert function, according to which some of each 'pulse' of energy is diverted into various reserves, which are relatively undepleted during periods of no growth. The reserve states of desert annuals are seeds, while perennial reserves may be stored in roots or old stems. According to Noy-Meir, the ability of a population to oscillate between active (producing) and inactive or dormant states is highly adaptive in intermittently favorable environments, and 'the prevalence of this pattern among desert organisms explains the long-term stability of the system despite its extreme short-term variability'. Animals must accommodate to the pulse–reserve dynamics of plants in various ways. Noy-Meir points out that pulse–reserve patterns have evolved in some animals (e.g., many insects), whereas other species utilize reserves of other organisms or have extremely labile feeding habits. Riechert (1979, p. 798 *et seq.*) has summarized many instances in which year-to-year differences in rainfall and temperature influence reproduction (and productivity) of desert animals.

Some lizards evidently have the capacity to 'turn off' during particularly unfavorable years. Under these conditions there is no reproduction and little above-ground activity. For example, Turner, Lannom, Medica & Hoddenbach (1969*a*) reported the apparent lack of reproduction by leopard lizards (*Crotaphytus wislizenii*) during 1964, and this evidently happened again in 1970.

Similarly, horned lizards did not reproduce in 1970 (Medica *et al.*, 1973). Evidently the unfavorable conditions of 1970 existed over much of the Mohave desert, for Nagy (1973) reported a marked decline in above-ground activity and pronounced losses of body weight among chuckwallas in San Bernardino County, California, during the last half of the year. These lizards apparently did not reproduce in this area during 1970, and Nagy suggested that such behavior may be a regular adaptive response to exceedingly dry years. In these species, and under such unfavorable conditions, we would expect very low production or perhaps none.

Uta stansburiana, on the other hand, reproduces to some extent during even the most unfavorable years and is apparently adapted to subsist on whatever arthropods are present under both good and bad conditions. Annual production by *U. stansburiana* varied from 336 kcal ha^{-1} in 1965–66 to around 535 kcal ha^{-1} in 1966–67 and 1967–68, and these data indicated a fair degree of year-to-year stability in productivity. What variation did occur was primarily associated with variations in reproductive success, but also reflected annual differences in growth rates of young lizards.

Many desert birds apparently function in a similar manner. Keast & Marshall (1954) pointed out that some birds of the Australian desert nest and lay eggs within weeks after a significant rain. Breeding may be highly facultative and related to the occurrence of rainfall, regardless of time of year (see also Serventy & Marshall, 1957). During periods of drought, breeding ceases and some species may go for several years without reproducing. The number of breeding species also varies in response to amounts and distribution of rainfall, with much reduced numbers of reproducing species in areas which have experienced prolonged drought. Winterbottom & Rowan (1962) reported similar observations of birds in the Kalahari desert of southwest Africa.

As with lizards, we infer that production by some species of Australian and African desert birds is zero during times of severe drought. In Rock Valley, Nevada, black-throated sparrows (*Amphispiza bilineata*) and thrashers (*Toxostoma lecontei*) reproduced each year between 1971 and 1973. However, during the exceptionally favorable year of 1973 three other species – *Mimus polyglottos, Spizella breweri* and *Amphispiza belli* – which had not reproduced during 1971 and 1972 laid eggs and reared young (Hill & Burr, 1974).

At least some species of *Perognathus* are adapted to pulse–reserve economies. Storage of seeds allows more effective use of available food, torpidity and/or inactivity during late autumn and winter reduce demands on energy reserves, and long life spans allow populations to survive one or more years without reproduction (French *et al.*, 1974; Chew, 1975). Our earlier analyses of *Perognathus formosus* showed enormous annual variations in production, from an estimated low of 46 kcal ha^{-1} in 1963 to over 2000 kcal ha^{-1} in 1973–74. These variations are not easy to interpret. Some years (1966, 1973–74) followed particularly rainy periods that enhanced mouse reproduction and led to increased productivity. This was particularly true of 1973–74, when levels of plant production far exceeded any previously observed in Rock Valley (Ackerman *et al.*, 1974, p. 29; Bamberg *et al.*, 1974; Nelson & Chew, 1977).

The prodigious *Perognathus* production in 1973–74 was, in part, promoted by artificial augmentation of pocket mouse numbers in the experimental plots. Whereas artificially increased densities led to impairments of reproduction and weaning in 1972–73 (a year of modest net primary production), the artificial increase of numbers in the spring of 1973, as well as reproduction by these mice, was supported and sustained by enormous primary production that spring. In 1972–73, production varied according to the number of mice present and their density-dependent interactions. In 1973–74 the abundance of food mitigated density effects.

Productivity depends in part on the stock with which a population begins its active season, and is also determined by the age structure and extent and timing of natality and mortality of the population. An adult animal can contribute little to production. Only its growth during a given interval can be considered production, and growth of adults is often negative (Table 9.35). The preponderance of production is by young animals (Table 9.35).

An adult contributes most to production by dying after making its reproductive contribution. Food resources otherwise required for maintenance of post-reproductive adults are then potentially available for growth of young animals. Surplus adult males are a particular drain on production.

Young animals contribute most to production by surviving until they reach

Table 9.35. *Percentage of total net production (P) by categories of* Perognathus formosus *in four populations of different densities.* Figures are derived from data in Table 9.18

	Low density			Medium density			High density			Extremely high density		
Categories	g_1	g_2	P	g_1	g_2	P	g_1	g_2	P	g_1	g_2	P
Adults												
Resident	2.5	−0.7	1.8	−0.3	0.0	−0.2	−0.1	0.0	−0.1	1.7	0.0	1.7
Introduced	—	—	—	0.6	−0.5	0.1	9.2	1.1	10.3	2.3	1.1	3.4
Young	69.1	29.1	98.2	84.5	15.6	100.1	83.0	6.8	89.8	67.0	27.8	94.9

Table 9.36. *Production by cohorts of young-of-the-year produced by an extremely dense population of* Perognathus formosus *during 1973–74*

Time of first registration	n	Proportion alive September 1, 1973 (%)	Proportion alive March 1, 1974 (%)	Annual production (g, live weight)	Annual production (g, per-mouse)
May 16	6	0.50	0.17	114	19.0
May 30	2	0.50	0.00	38	19.0
June 11	41	0.73	0.22	747	18.2
June 27	145	0.85	0.41	2461	17.0
July 10	195	0.85	0.35	3268	16.8
July 25	123	0.87	0.32	1989	16.1
August 8	151	0.85	0.12	2229	14.8
August 21	67	0.91	0.16	1067	15.9
September 11	20	0.95	0.05	259	13.0
September 19	14	1.0	0.43	202	14.4
Total or mean	764	0.80	0.22	12374	16.4

adult weight, and the magnitude of this production depends on birth date and survival rate. Early birth has two potential effects: it allows more time for completion of growth before the end of the growing season (about September 1), but in poor or average years it also increases the risk of death before full growth has occurred. In a particularly favorable year, (e.g., 1973–74) the latter effect may be limited or unexpressed. All birth groups in the 1973–74 population showed good survival (Table 9.36). The earlier the mice were born, the more growth they achieved before September 1, and the more they contributed to production.

Table 9.37 compares age structure, natality and mortality for four pocket mouse populations. Here, the population with the lowest productivity per individual was the one with the lowest ratio of young to adults. In this

Table 9.37. *Relative production by* Perognathus formosus *in four populations of differing densities*

Population characteristic	Relative density: Low	Medium	High	Extremely high
Density (n ha^{-1})				
Adult	1.1	7.2	19.9	30.4
Juvenile	5.4	19.9	16.5	81.6
Production (g ha^{-1})				
Adult	1.5	1.1	38.1	50.9
Juvenile	78.1	326.2	334.8	1396.7
Production (g per mouse)				
Adult	1.3	0.2	1.9	1.7
Juvenile	14.5	16.5	20.3	17.1
Total	12.3	12.1	10.3	12.9
Young: adult ratio	4.9	2.8	0.8	2.7
Average birth date	May 19	April 18	April 6	June 1
Survival to September 1	0.78	0.61	0.41	0.86

high-density population the relative paucity of young counteracted their superior productivity. Production per mouse in the other three populations ranked in the same order as average birth date and survival of young to September 1. In the average year of 1972–73, the greater mortality associated with early birth was apparently the principal consequence of birth date. The low-density population had the highest young: adult ratio of the three, but the young were born late and achieved the least growth per animal. The extremely dense population of 1973–74 had the lowest young: adult ratio, but this was counterbalanced by exceptional survival of young.

Although annual production by pocket mice is extremely variable, production by some other rodents may be less so. For example, *Dipodomys merriami* showed much less variation of biomass in a year's time (see Table 9.1) than other herbivorous rodents in the community studied by R. M. Chew & A. E. Chew (1970). This stability might also persist over a period of several years.

Efficiency of secondary production

The efficiency of production by an animal population can be considered from several viewpoints. To measure the performance of the population as it contributes to energy transfers of its community, one can calculate the ratio of production to assimilation, P/A. Upon death, P is output of the population to the next trophic level. Assimilation (A) is the input of matter-energy into the population and is equivalent to energy flow (EF), or production (P) plus respiration (R). Less exactly, one can use the ratio of P/R as an approximation of output/input efficiency.

As an index of performance relative to food supply, one can compute the ratio of P to the total food supply, for example, the production by an herbivore relative to net primary production. To evaluate the performance of the population more precisely, one can estimate the ratio of P to the fraction of the total food that is physically available to the population and of positive value when eaten (Chew & Woodman, 1974). This approach presupposes that one knows the food habits of the species in question and can reasonably assess availability. The ratio of production to available food varies with the kind of food considered, the density of the population, and the species' preferences for different available foods (Chew, R. H. & Chew, A. E., 1970, p. 18).

The new data on *Perognathus formosus* in 1972–74 (see above) are particularly interesting because they allow us to assess efficiency in relation to population density within the same time period and between time periods. To calculate P/A for *P. formosus* we estimated energy flow of populations following the general procedures of R. M. Chew & A. E. Chew (1970):

$$EF = P/p + M, \qquad [9.16]$$

where p is the efficiency of growth or production (which varies with life history stage) and M is the energy cost of maintenance. Maintenance (M) was calculated following procedures of Mullen & Chew (1973), where respiration for a month of d days was:

$$M\,(\text{cal month}^{-1}) = \bar{B}\cdot d\cdot(R_b H_n + R_b H_s + 13.92\, H_s), \qquad [9.17]$$

with $\bar{B}$ the average daily biomass (g), H_n the number of hours spent in the nest each day, and H_s the hours active on the surface. The hourly cost of activity on the surface is 13.92 cal g^{-1} and R_b is the hourly resting respiratory rate:

$$R_b\,(\text{cal g}^{-1}) = 39.95 - 1.037\, T, \qquad [9.18]$$

with T the nest or surface temperature in °C. Nest temperatures were estimated from soil temperatures, and surface temperatures during hours of activity were estimated as the means of air temperatures measured at 18.00, 24.00, and 06.00 hours at a height of 15 cm in open spaces between shrubs. The hours spent by *P. formosus* each day in their nests and on the surface were derived from information in French, Maza & Aschwanden (1966) and French *et al.* (1974).

The major determinants of efficiency of production are (*a*) cost of maintenance, i.e., cal g^{-1} day^{-1}, (*b*) proportion of females involved in reproduction, (*c*) number of offspring weaned per pregnant female (parity), and (*d*) rate of turnover of numbers or biomass (θ). We define θ as 365 days divided by the average life span of a mouse (in days). Maintenance costs are much higher

Table 9.38. *Mean density, annual energy flow and production, and related attributes of four populations of* Perognathus formosus *in Rock Valley (1972–74)*

Values of θ are weighted by the numbers of individuals in three categories: residents at start of year, introduced animals (mostly adults), and young-of-the-year. Parity is expressed as the number of young weaned per resident and introduced female.

Population	Mean density (n ha^{-1})	Production (kcal ha^{-1})	Energy flow (kcal ha^{-1})	*P/EF*	θ	Parity
Low density, 1972	2.7	119	7545	0.016	2.42	6.43
Medium density, 1972	9.8	489	25919	0.019	2.74	3.59
High density, 1972	9.5	559	23969	0.023	5.85	1.38
Extremely high density, 1973	52.2	2208	148454	0.015	1.91	5.16

for endothermic animals than for ectothermic ones, and are inversely and exponentially related to body size. Production can only be realized from reproducing females, and continued maintenance of non-reproductive individuals is inefficient (except as adults are needed to assure production during the next season). Consequently, we expect the degree of ectothermy, the size of the individual animal, parity of females, and rate of population turnover to each have, *per se*, a direct effect on the efficiency of production (*P/EF*).

Table 9.38 summarizes estimates of annual energy flow (derived using equation 9.16) for four Rock Valley populations of *Perognathus formosus.* This table also gives associated average densities, estimates of production, measures of efficiency (*P/EF*), turnover rates (θ) and estimates of female parity. Turnover rate is the dominant determinant of production efficiencies of the four populations in Table 9.38. The ranking of populations by efficiency is the same as that by turnover rates. However, the relationship between efficiency and parity is inverse. This is not in accord with previously derived expectations, possibly because the relationship was confounded by simultaneous density-dependent responses of the three 1972 populations. In 1972, both the number of young weaned per successful pregnancy and survival were inversely related to population density. Interactions among individuals apparently acted to reduce reproductive success and increase turnover (Chew *et al.*, 1973).

Table 9.39 gives additional data for other desert vertebrates. The Arizona rodents (data of R. M. Chew & A. E. Chew, 1970) may exhibit a direct relationship between efficiency and turnover. *D. merriami*, which had the lowest turnover, had the lowest efficiency, and *P. eremicus* had the highest turnover and the highest efficiency.

Among mammals for which we have data (Tables 9.38 and 9.39), the

Table 9.39. *Efficiency of production (P/EF) among desert vertebrates*

Species	Annual energy flow (kcal ha^{-1})	Annual production (kcal ha^{-1})	P/EF	θ	Sources
Dipodomys merriami	88134	1068	0.012	—	Soholt (1973)[a]
Dipodomys merriami	57900	642	0.011	1.9	R. M. Chew & A. E. Chew (1970)'
Onychomys torridus	6888	74	0.011	8.5	R. M. Chew & A. E. Chew (1970)'
Peromyscus eremicus	4720	82	0.017	12.9	R. M. Chew & A. E. Chew (1970)'
Ammospermophilus harrisii	2061	39	0.019	—	R. M. Chew & A. E. Chew (1970)'
Perognathus formosus			0.015–0.023		Table 9.38
Uta stansburiana					
1965–66, 1966–67			~0.19	—	Turner *et al.* (1976)
1967–68			0.25	—	

[a] Revised from original data (e.g. see Tables 9.4 and 9.5).

absolute level of production (kcal ha^{-1}) is predominantly a function of population biomass or density. The small differences in efficiency (1.1–2.3%) have little effect on production when weighed against the effect of different biomasses. It is not until efficiency is an order of magnitude larger, as with the ectothermic *Uta stansburiana* with *P/EF* 19–25% (Table 9.39), that we can see the impact of efficiency on production. The annual production of this lizard in Rock Valley (*c.* 336–536 kcal ha^{-1}) equalled or surpassed that of most desert rodents we discussed. In the energy budget of the individuals of a population, energy not used for thermoregulation can enter into production.

Small variations in efficiency can be expected among mammals, depending upon species differences in metabolic rate related to body size and life style. For example, three species of shrew (*Sorex* spp.), which are carnivores noted for their high metabolic rates, had an average *P/EF* ratio of 0.0069. This may be compared to an average of 0.021 for ten species of herbivorous and granivorous rodents (Grodziński & French, 1974). However, the predominantly carnivorous grasshopper mouse, *Onychomys torridus*, does not have a low efficiency compared to the granivorous *D. merriami* and *P. formosus* (Table 9.39).

The average efficiency of the nine rodent populations given in Tables 9.38 and 9.39 is 0.0158. Among 22 populations of herbivorous and granivorous mammals of temperate grassland and forest, and taiga, selected from the summary of Grodziński & French (1974) the average *P/EF* is 0.0023 (0.0015 – 0.0037). It will be necessary for authors to re-evaluate their data carefully before potential differences between desert and non-desert species can be tested. Turnover rates of the two groups do not differ.

Models of production

As pointed out by McNeill & Lawton (1970), 'There is...need in ecology for short cut methods for analysing the trophic-dynamics of communities'. We have already commented on Engelmann's (1966) analysis and those of McNeill & Lawton (1970) in which quantitative relationships between production and respiration by different kinds of animals were explored (see equations 9.7 and 9.8). Another interspecific model relating production (*P*) by small rodents to population respiration (*R*) has been developed. Grodziński & Wunder (1975) examined this relationship among 38 small mammal populations:

$$P\,(\text{kcal ha}^{-1}\,\text{yr}^{-1}) = 0.00546\,R^{1.1365}. \qquad [9.19]$$

French, Grant, Grodziński & Swift (1976) gave a slightly different version (based on a few more data):

$$P = 0.00643\,R^{1.116}. \qquad [9.20]$$

The McNeill & Lawton model for endotherms expresses production as an essentially constant 1.8% of respiration, while the small mammal model predicts annual production to be relatively lower (< 1%) when annual respiration is around 5–40 kcal ha^{-1}, about 1% when *R* is 100 kcal ha^{-1}, and around 1.4% when *R* is 1000 kcal ha^{-1}.

One may also consider models seeking to explain annual differences in productivity by particular species. Turner *et al.* (1976) developed several such models for *Uta stansburiana*. The simplest was a linear regression model based on population state variables (i.e., density, biomass) at various times of year. Over a 3-yr period respiration was positively correlated with density, but production was correlated only with biomass. Correlations of production with monthly biomass estimates and with combined biomass of various pairs of months were also examined. The best estimator ($r = > 0.99$) of production by *U. stansburiana* was:

$$P\,(\text{kcal ha}^{-1}) = 2.2\,(\text{April} + \text{November biomass}) - 220. \qquad [9.21]$$

This makes some sense biologically, for the November biomass reflects numbers of young born during the summer and their subsequent growth and survival. However, the equation has an extremely large negative intercept. We re-analyzed the data for *U. stansburiana* using the three annual production estimates as dependent variables and 12 monthly biomass estimates as independent variables. With a floating intercept the best predictor (multiple $R^2 = 1.000$) of production was:

$$P\,(\text{kcal ha}^{-1}) = 40.93(\text{August biomass}) - 4.76(\text{April biomass}) - 3060.1. \qquad [9.22]$$

Table 9.40. *Annual production by* Perognathus formosus *in enclosures in Rock Valley, Nevada.* All annual periods begin on March 1

Annual period	Plot	Production (kcal ha^{-1})
1963–64	A	9
1966–67	A	398
1966–67	C	632
1972–73	A	119
1972–73	C_s	489
1972–73	C_n	559
1973–74	C	2208

When we forced a zero intercept the best predictor (multiple $R^2 = 0.998$) of production was:

$$P\ (\text{kcal ha}^{-1}) = 4.79(\text{September biomass}) - 2.02(\text{August biomass}). \quad [9.23]$$

With only three data points involved, all the above models rendered good agreement between predictions and actual observations. All three models reflected the importance of summer recruitment. In our view equations [9.21] and [9.22] are most apt to reflect biological reality, and we are inclined to favor equation 9.21 because of its smaller negative intercept.

Can we examine data pertaining to *Perognathus formosus* in a similar way? Annual production by pocket mice in enclosures in Rock Valley was estimated for calendar 1963 (Plot A), calender 1966 (Plots A and C), between March 1972 and March 1973 (Plots A and C), and between March 1973 and March 1974 (Plot C).

The 12 month intervals analyzed were chosen in terms of the best density and body-weight data available. However, for analytical purposes these estimates were adjusted to the same 12 month interval. Hence, the 1963 and 1966 analyses were modified to pertain to annual intervals beginning on March 1. Table 9.40 gives revised production estimates for annual intervals beginning March 1. The production estimate for calendar 1963 (46 kcal ha^{-1}) is greater than the estimate for March 1, 1963, to February 28, 1964 (9 kcal ha^{-1}), because the new interval involved a decrease in ΔB from -1017 g to -1242 g. The number of animals dying in the new interval was the same as before, but the initial and final body weights differed. This resulted in the decrease in ΔB and the corresponding decline in estimated production.

Can the production estimates given in Table 9.40 be correlated with the states of the pocket mouse populations at any time during the annual intervals in question? Monthly densities (n ha^{-1}) and standing stocks (g ha^{-1}) of the 1972 and 1973 populations were given in Table 9.17. Table 9.41 gives

Table 9.41. *Monthly densities (n ha^{-1}) and standing stocks (g ha^{-1}) of 1963 and 1966 populations of* Perognathus formosus *in Rock Valley*. For each plot the first column gives densities and the second column standing stocks

	Plot A 1963–64		Plot A 1966–67		Plot C 1966–67	
March	13.9	212.1	7.3	127.8	11.1	194.6
April	13.3	206.3	23.6	255.9	37.6	407.3
May	8.3	135.7	21.5	345.0	34.0	545.8
June	7.5	128.3	18.9	322.9	30.3	532.2
July	6.8	114.9	18.6	361.7	27.4	532.4
August	6.3	99.7	18.3	336.2	26.6	489.6
September	5.8	92.7	17.7	295.2	22.6	376.5
October	5.5	87.5	17.3	287.8	22.6	376.5
November	5.4	85.7	16.0	265.7	18.2	303.5
December	5.2	82.2	15.7	262.0	16.5	275.2
January	5.2	78.6	15.7	260.5	16.5	273.2
February	5.2	75.5	15.6	257.7	16.4	270.0

corresponding data for the 1963 and 1966 populations. These variables were used in stepwise multiple regression analyses with production as the dependent variable. Production (P) exhibited high positive correlations (> 0.95) with all monthly densities between June and December and with all monthly biomass values between July and December. When production was analyzed in terms of densities the first two variables entered were July density (d_J) and March density (d_M):

$$P\ (\text{kcal ha}^{-1}) = 22.213\ d_J - 8.278\ d_M + 32.439. \qquad [9.24]$$

When production was analyzed in terms of biomass the first two variables entered were August biomass (b_A) and March biomass (b_M):

$$P\ (\text{kcal ha}^{-1}) = 1.410\ b_A - 0.616\ b_M + 27.718. \qquad [9.25]$$

Both models involving densities and biomass (equations 9.24 and 9.25) yielded multiple R^2 values $\geqslant 0.995$, so there is little basis for choosing one over the other in terms of the amount of explained variation.

However, we believe measures of biomass are better predictors of production than measures of density. This was clearly true of *Uta stansburiana*, where production was not well correlated with numbers but was highly correlated with biomass variables. Pocket mouse production can be estimated using both density and biomass state variables because densities and standing stocks are so highly correlated. Even when populations were composed of high proportions of juveniles (e.g., April 1966), young mice attained adult body weight so quickly that numbers and biomass showed a generally consistent relationship. For the populations represented in Tables 9.17 and 9.41 the relationship between biomass, B (g), and density ha^{-1} (n) was calculated (with a forced zero intercept) as $B = 15.81\ n$ ($r = 0.990$). On the other hand, high

Table 9.42. *Comparisons of estimates of secondary production (kcal ha^{-1}) by* Perognathus formosus *and predictions by models based on density and biomass state variables.* All annual intervals began on March 1. Multiple R^2 values are given in parentheses

Plot	Year	Estimated annual production	Annual production predicted by equation 9.24 (0.998)	Annual production predicted by equation 9.25 (0.999)
A	1963–64	9	68	38
A	1966–67	398	385	423
C	1966–67	632	549	598
A	1972–73	119	92	132
C_s	1972–73	489	515	504
C_n	1972–73	559	588	509
C	1973–74	2208	2216	2222

summer densities of *U. stansburiana* (following the hatching of young) were not associated with maximum standing stocks because the newly hatched lizards were so small. Peak biomass levels were not reached until November, after young animals had experienced some growth (Turner *et al.*, 1976, Table 8).

Table 9.42 compares estimates of production by seven pocket mouse populations with predictions of equations [9.24] and [9.25]. Both equations [9.21] and [9.25] involve a measure of summer or autumn standing stock, reflecting the success of reproduction. Intuition clearly supports the importance of recruitment in production, so it is logical that these variables should enter as positive and dominant influences. However, it is more difficult to interpret the biological significance of other variables in the biomass models. Both equation [9.21] and equation [9.25] include a measure of the initial state of the breeding population (March for *Perognathus* and April for *Uta*).

In the pocket mouse model the March biomass functions as a decrement. It is possible that this reflects a real density-dependent response, for Chew *et al.* (1973) showed that reproductive success of *Perognathus* was inversely related to spring densities. When densities were artificially lowered, weaning and survival of young were improved. At augmented densities these processes were impaired. In equation [9.21] April biomass of *Uta* acts positively. In experiments involving artificial manipulations of *Uta* densities, Turner *et al.* (1974) reported no density effect on egg production. In our view, the biomass models of production by *Uta* and *Perognathus* can be clearly interpreted only in terms of the post-reproductive standing stocks. We have speculated as to biological implications of the other variables, but these remain only ideas requiring further testing.

Another aspect of modeling production is concerned with the sensitivity of models to changes in various input parameters. The general approach is illustrated by tests performed by Turner *et al.* (1976) with a computer simulation pertaining to a population of *Uta stansburiana.* This simulation was used to compute energy of respiration (R), but some of the findings may be applicable to production estimates as well. For example, errors in estimates of density were more important than erroneous assumptions as to mortality rates. It was also reported that 'discrepancies introduced by rather coarse assumptions as to adult body weights were small compared to other sources of error' (Turner *et al.*, 1976).

We made tests of the analysis of production by whiptail lizards (pp. 228–34), and how changes in some of the assumptions influenced production estimates. In the original analysis we assumed that 85% of mortality occurred during above-ground activity. But one might also assume that overall annual mortality was distributed equally throughout the year. We originally assumed loss of weight during hibernation, but an alternative would be to assume no change in body weight during inactivity. Finally, we assumed that three-fourths of the females laid a second clutch of eggs, while a contrasting assumption would be that all females laid two clutches. Annual production estimates were computed for all eight combinations of three sets of two contrasting assumptions. The estimates ranged from as low as 178 kcal ha^{-1} to as high as 227 kcal ha^{-1}. These calculations served as a form of sensitivity analysis and revealed the following points.

Whether we assumed that all females or only three-fourths of them laid two clutches was an unimportant distinction, leading to an average difference of about 3–4% in alternative estimates. However, the number of young surviving until the spring of 1966 was always taken as 277 and if egg production was low, assumed survival was necessarily better. Hence, the failure of differences in assumed fecundity to obviously influence production estimates must be gauged in this light. Neither was it important whether we assumed no weight changes during hibernation or a slight loss of weight. Differences were on the order of 3 – 5%. However, the contrasting assumptions as to how mortality was distributed throughout the year had clearly observable effects on production estimates. The highest estimates (averaging around 215 kcal ha^{-1}) were obtained when we assumed age-constant mortality. When we assumed that 85% of annual mortality occurred during above-ground activity, estimates were around 15% less (*c.* 183 kcal ha^{-1}). If lizards die off more rapidly during the late spring and summer (i.e., during the time of growth) then fewer individuals attain maximal weights. It is more productive (in terms of elimination calories) to experience greater mortality during hibernation because the dying animals have achieved maximal body weights prior to death.

Acknowledgments

We thank Frederic Wagner for critical advice during the planning and organization of this review and acknowledge support by the Desert Biome of the United States International Biological Program. Stephen Russell and Lars Soholt made available useful unpublished data related to their research on birds and kangaroo rats. Bernardo Maza gave important assistance in estimating production by *Perognathus formosus* in 1963 and 1966. We are particularly grateful to Deborah Garrison for invaluable secretarial and editorial assistance. The preparation of this review was supported, in part, by Contract AT(04-1)GEN-12 between the United States Atomic Energy Commission and the University of California and Contract E(04-1)GEN-12 between the Energy Research and Development Administration and the University of California.

References

Ackerman, T., Bamberg, S. A., Hill, H. O. & Kaaz, H. W. (1974). Annual plant populations. In: *Rock Valley Validation Site Report* (ed. F. B. Turner & J. F. McBrayer), pp. 25–9. US/IBP Desert Biome Research Memorandum RM 74-2. Utah State University, Logan.

Anderson, A. H. & Anderson, A. (1973). *The cactus wren*. University of Arizona Press, Tucson.

Bamberg, S. A., Wallace, A., Kleinkopf, G. E., Vollmer, A. & Ausmus, B. S. (1973). *Plant production and its utilization in Mojave desert shrubs*. US/IBP Desert Biome Research Memorandum RM 74-10. Utah State University, Logan.

Bamberg, S. A., Ackerman, T., Hill, H. O., Kaaz, H. W. & Vollmer, A. (1974). Perennial plant populations. In: *Rock Valley Validation Site Report* (ed. F. B. Turner & J. F. McBrayer), pp. 30–5. US/IBP Desert Biome Research Memorandum RM 74-2. Utah State University, Logan.

Banks, R. (1959). Development of nestling white-crowned sparrows in central coastal California. *Condor*, **61**, 96–109..

Chew, R. M. (1975). *Effect of density on the population dynamics of* Perognathus formosus *and its relationships within a desert ecosystem*. US/IBP Desert Biome Research Memorandum RM 75-19. Utah State University, Logan.

Chew, R. M. & Butterworth, B. B. (1959). Growth and development of Merriam's kangaroo rat, *Dipodomys merriami*. *Growth*, **23**, 75–95.

Chew, R. M. & Chew, A. E. (1970). Energy relationships of the mammals of a desert shrub (*Larrea tridentata*) community. *Ecological Monographs*, **40**, 1–21.

Chew, R. M., Turner, F. B., August, P., Maza, B. & Nelson, J. (1973). *Effect of density on the population dynamics of* Perognathus formosus *and its relationships within a desert ecosystem*. US/IBP Desert Biome Research Memorandum RM 73-18. Utah State University, Logan.

Chew, R. M. & Turner, F. B. (1974). *Effects of density on the population dynamics of* Perognathus formosus *and its relationships within a desert ecosystem*. US/IBP Desert Biome Research Memorandum RM 74-20. Utah State University, Logan.

Chew, R. M. & Woodman, J. C. (1974). Nutritive value of three common chaparral plants for the woodrats, *Neotoma fuscipes* and *N. lepida*. *Bulletin of the Southern California Academy of Science*, **73**, 115–16.

Dingman, R. E. & Byers, L. (1974). *Interaction between a fossorial rodent (the pocket gopher,* Thomomys bottae) *and a desert plant community.* US/IBP Desert Biome Research Memorandum RM 74-22. Utah State University, Logan.

Engelmann, M. D. (1966). Energetics, terrestrial field studies, and animal productivity. *Advances in Ecological Research*, **3**, 73–115.

Evans, F. C. (1967). The significance of investigations in secondary terrestrial productivity. In: *Secondary productivity in terrestrial ecosystems*, vol. 1 (ed. K. Petrusewicz), pp. 3–15. Państowe Wydawnictwo Naukowe, Warsaw.

Freckman, D. W. (1977). A comparison of techniques for extraction and study of anhydrobiotic nematodes from dry soils. *Journal of Nematology*, **9**, 176–81.

Freckman, D. W. (1978). Ecology of anhydrobiotic soil nematodes. In: *Dry biological systems* (ed. J. H. Crowe & J. S. Clegg), pp. 345–57. Academic Press, New York.

Freckman, D. W. & Mankau, R. (1977). Distribution and trophic structure of nematodes in desert soils. In: *Soil organisms as components of ecosystems* (ed. U. Lohm & T. Persson). *Ecological Bulletin, (Stockholm)*, **25**, 511–14.

Freckman, D. W., Mankau, R. & Ferris, H. (1975). Nematode community structure in desert soils: nematode recovery. *Journal of Nematology*, **7**, 343–6.

French, N. R. (1964). Description of a study of ecological effects on a desert area from chronic exposure to low-level ionizing radiation. *United States Atomic Energy Commission Report, UCLA*, 12–532. Los Angeles.

French, N. R., Grant, W. E., Grodziński, W. & Swift, D. M. (1976). Small mammal energetics in grassland ecosystems. *Ecological Monographs*, **46**, 201–20.

French, N. R., Maza, B. G. & Aschwanden, A. P. (1966). Periodicity of desert rodent activity. *Science*, **154**, 1194–5.

French, N. R., Maza, B. G., Hill, H. O., Aschwanden, A. P. & Kaaz, H. W. (1974). A population study of irradiated desert rodents. *Ecological Monographs*, **44**, 45–72.

Golley, F. B. (1960). Energy dynamics of a food chain of an old-field community. *Ecological Monographs*, **30**, 187–206.

Gorecki, A. (1965). Energy values of body in small mammals. *Acta Theriologica*, **10**, 333–65.

Grodziński, W. (1971). Energy flow through populations of small mammals in the Alaskan taiga forest. *Acta Theriologica*, **16**, 231–75

Grodziński, W., Bobek, B., Drożdż, A. & Gorecki, A. (1970). Energy flow through small rodent populations in a beech forest. In: *Energy flow through small mammal populations* (ed. K. Petrusewicz & L. Ryszkowski), pp. 291–8. Państowe Wydawnictwo Naukowe, Warsaw.

Grodziński, W. & French, N. R. (1974). Production and respiration in populations of small mammals. *Transactions of the First International Theriological Congress*, **1**, 206.

Grodziński, W. & Wunder, B. A. (1975). Ecological energetics of small mammals. In: *Small mammals: their productivity and population dynamics* (ed. F. B. Golley, K. Petrusewicz & L. Ryszkowski), pp. 173–204. Cambridge University Press, Cambridge.

Hill, H. O. & Burr, T. (1974). Birds. In: *Rock Valley Validation Site Report* (ed. F. B. Turner & J. F. McBrayer), pp. 51–5. US/IBP Desert Biome Research Memorandum RM 74-2. Utah State University, Logan.

Johnson, D. R. & Schreiber, R. K. (1979). Assimilation, respiration and production: (b) vertebrates. In: *Arid land ecosystems: structure, functioning and management*, vol. 1 (ed. D. W. Goodall & R. A. Perry), pp. 731–42. Cambridge University Press, Cambridge.

Kaczmarski, F. (1966). Bioenergetics of pregnancy and lactation in the bank vole. *Acta Theriologica*, **11**, 408–17.

Keast, J. A. & Marshall, A. J. (1954). The influence of drought and rainfall on reproduction in Australian desert birds. *Proceedings of the Zoological Society of London, Series C*, **124**, 493–9.

Klekowski, R. Z., Wasilewska, L. & Paplinska, E. (1972). Oxygen consumption by soil-inhabiting nematodes. *Nematologica*, **18**, 391–403.

Kozlovsky, D. C. (1968). A critical evaluation of the trophic level concept. I. Ecological efficiencies. *Ecology*, **49**, 48–59.

Lamotte, M. & Bourlière, F. (eds.) (1967). *Problèmes de productivité biologique.* Masson et Cie., Paris.

McGinnies, W. G., Goldman, B. J. & Paylore, P. (eds.) (1968). *Deserts of the World.* University of Arizona Press, Tucson.

McNeill, S. & Lawton, J. H. (1970). Annual production and respiration in animal populations. *Nature, London*, **225**, 472–4.

Medica, P. A., Turner, F. B. & Smith, D. D. (1973). Effects of radiation on a fenced population of horned lizards (*Phrynosoma platyrhinos*) in southern Nevada. *Journal of Herpetology* , **7**, 79–85.

Mispagel, M. E. (1978). The ecology and bioenergetics of the acridid grasshopper, *Bootettix punctatus*, on creosotebush, *Larrea tridentata*, in the northern Mojave Desert. *Ecology*, **59**, 779–88.

Morton, M. & Orejuela, J. (1972). The biology of immature mountain white-crowned sparrows (*Zonotrichia leucophrys oriantha*) on the breeding ground. *Condor*, **74**, 423–30.

Mullen, R. K. & Chew, R. M. (1973). Estimating the energy metabolism of free-living *Perognathus formosus*: a comparison of direct and indirect methods. *Ecology*, **54**, 633–7.

Nagy, K. A. (1972). Water and electrolyte budgets of a free-living desert lizard, *Sauromalus obesus*. *Journal of Comparative Physiology*, **79**, 39–62.

Nagy, K. A. (1973). Behaviour, diet and reproduction in a desert lizard, *Sauromalus obesus*. *Copeia*, 1973, 93–102.

Neal, B. J. (1964). Comparative biology of two southwestern ground squirrels: *Citellus harrisii* and *C. tereticaudus*. PhD Thesis. University of Arizona, Tucson.

Nelson, J. F. & Chew, R. M. (1977). Factors affecting seed reserves in the soil of a Mojave Desert ecosystem, Rock Valley, Nye County, Nevada. *American Midland Naturalist*, 97, 300–20.

Norton, B. E. (1974). IBP studies in the desert biome. *Bulletin of the Ecological Society of America*, **55**, 6–10.

Noy-Meir, I. (1973). Desert ecosystems: environment and producers. *Annual Review of Ecology and Systematics*, **4**, 25–51.

Noy-Meir, I. (1974). Desert ecosystems: higher trophic levels. *Annual Review of Ecology and Systematics*, **5**, 195–214.

Nutting, W. L., Haverty, M. I. & LaFage, J. P. (1975). *Demography of termite colonies as related to various environmental factors: population dynamics and role in the detritus cycle.* US/IBP Desert Biome Research Memorandum RM 75-32. Utah State University, Logan.

Odum, E. P., Connell, C. E. & Davenport, L. B. (1962). Population energy flow of three primary consumer components of old-field ecosystems. *Ecology*, **43**, 88–96.

Odum, E. P., Marshall, S. G. & Marples, T. G. (1965). The caloric content of migrating birds. *Ecology*, **46**, 901–4.

Petrusewicz, K. (ed.) (1967). *Secondary productivity of terrestrial ecosystems* (*Principles and methods*), vols. 1 & 2. Państowe Wydawnictwo Naukowe, Warsaw.

Petrusewicz, K. & Macfadyen, A. (1970). *Productivity of terrestrial animals.* IBP Handbook No. 13, F. A. Davis Co., Philadelphia.

Petrusewicz, K. & Ryszkowski, L. (eds.) (1969/1970). *Energy flow through small mammal populations.* Państowe Wydawnictwo Naukowe, Warsaw.

Pinowski, J. (1967). Estimation of the biomass produced by a tree sparrow (*Passer m. montanus* L.) population during the breeding season. In: *Secondary productivity in terrestrial ecosystems*, vol. 1 (ed. K. Petrusewicz), pp. 357–67. Państowe Wydawnictwo Naukowe, Warsaw.

Riechert, S. E. (1979). Development and reproduction in desert animals. In: *Arid land ecosystems: structure, functioning and management*, vol. 1 (ed. D. W. Goodall & R. A. Perry), pp. 797–822. Cambridge University Press, Cambridge.

Romanoff, A. L. & Romanoff, A. J. (1949). *The avian egg.* Wiley, New York.

Russell, S. M., Gould, P. J. & Smith, E. L. (1973). *Population structure, foraging behavior and daily movements of certain Sonoran Desert birds.* US/IBP Desert Biome Research Memorandum RM 73-27. Utah State University, Logan.

Russell, S. M. & Gould, P. J. (1974). *Population structure, foraging behavior and daily movements of certain Sonoran Desert birds.* US/IBP Desert Biome Research Memorandum RM 74-27. Utah State University, Logan.

Ryszkowski, L. & Petrusewicz, K. (1967). Estimation of energy flow through small rodent populations. In: *Secondary productivity in terrestrial ecosystems*, vol. 1 (ed. K. Petrusewicz), pp. 125–46. Państowe Wydawnictwo Naukowe, Warsaw.

Schoenwetter, M. (1924). Relatives Schalengewicht inbesondere bei Spar-und Doppeleiern. *Beiträge zur Fortpflanzungsbiologie der Vögel mit Berücksichtigung der Oologie*, **1**, 49–52.

Serventy, D. L. & Marshall, A. J. (1957). Breeding periodicity in western Australian birds: with an account of unseasonal nestings in 1953 and 1955. *Emu*, **57**, 99–126.

Soholt, L. F. (1973). Consumption of primary production by a population of kangaroo rats (*Dipodomys merriami*) in the Mojave Desert. *Ecological Monographs*, **43**, 357–76.

Svihla, A. (1934). Development and growth of deermice (*Peromyscus maniculatus artemisiae*). *Journal of Mammalogy*, **15**, 99–104.

Tanner, W. W. & Krogh, J. E. (1973). Ecology of *Phrynosoma platyrhinos* ȧt the Nevada Test Site, Nye County, Nevada. *Herpetologica*, **29**, 327–42.

Turner, F. B. (1970). The ecological efficiency of consumer populations. *Ecology*, **51**, 741–2.

Turner, F. B. (ed.) (1972). *Rock Valley validation site report.* US/IBP Desert Biome Research Memorandum RM 72-2. Utah State University, Logan.

Turner, F. B. (ed.) (1973). *Rock Valley validation site report.* US/IBP Desert Biome Research Memorandum RM 73-2. Utah State University, Logan.

Turner, F. B., Hoddenbach, G. A. & Lannom, J. R. Jr. (1964). Growth of lizards in natural populations exposed to gamma irradiation. *Health Physics*, **11**, 1585–93.

Turner, F. B., Hoddenbach, G. A., Medica, P. A. & Lannom, J. R. (1970). The demography of the lizard, *Uta stansburiana* Baird and Girard, in southern Nevada. *Journal of Animal Ecology*, **39**, 505–19.

Turner, F. B., Lannom, J. R. Jr, Medica, P. A. & Hoddenbach, G. A. (1969*a*). Density and composition of fenced populations of leopard lizards (*Crotaphytus wislizenii*) in southern Nevada. *Herpetologica*, **25**, 247–57.

Turner, F. B., Medica, P. A., Lannom, J. R. Jr & Hoddenbach, G. A. (1969*b*). A demographic analysis of fenced populations of the whiptail lizard, *Cnemidophorus tigris*, in southern Nevada. *Southwestern Naturalist*, **14**, 189–201.

Turner, F. B., Medica, P. A. & Kowalewsky, B. W. (1976). *Energy utilization by a desert lizard,* Uta stansburiana. US/IBP Desert Biome Monograph No. 1. Utah State University Press, Logan.

Turner, F. B., Medica, P. A. & Smith, D. D. (1974). *Reproduction and survivorship of the lizard,* Uta stansburiana, *and the effects of winter rainfall, density and predation on these processes.* US/IBP Desert Biome Research Memorandum RM 74-26. Utah State University, Logan.

Whittaker, R. H. (1970). *Communities and ecosystems.* Macmillan, New York.

Wiegert, R. G. & Evans, F. C. (1967). Investigations of secondary productivity in grasslands. In: *Secondary productivity in terrestrial ecosystems,* vol. 2 (ed. K. Petrusewicz), pp. 499–518. Państowe Wydawnictwo Naukowe, Warsaw.

Wiens, J. A. (1973). Pattern and process in grassland bird communities. *Ecological Monographs,* **43**, 237–70.

Winterbottom, J. M. & Rowan, M. K. (1962). Effect of rainfall on breeding birds in arid areas. *The Ostrich,* **33**, 77–8.

Manuscript received by the editors July 1975

PART II

Ecosystem dynamics

10. Introduction

J. A. MACMAHON

This part of the volume emphasizes the effects of the temporal and spatial variability, so characteristic of the world's arid-lands, on ecological processes. As Noy-Meir (Chapter 18) rightly points out, deserts are often considered as examples of simple ecosystems because ultimately the organisms occurring there are limited by a single factor – water; little feedback exists between the sparse, open vegetation and certain microenvironmental conditions (e.g. temperature) and there is generally thought to be a low species richness.

The common stereotype of 'simple deserts' ignores the high degree of variation in both abiotic and biotic components that deserts exhibit over time and space. These variations, particularly with regard to water and its concomitant effects on the biota, are well represented by the intensive studies of a Chihuahuan Desert site by Ludwig & Whitford (Chapter 11). Additionally, West (Chapter 12) and Binet (Chapter 13) detail the effects of spatial and temporal variation in deserts on various aspects of nutrient cycles.

Such environmental vagaries as the above chapters describe create problems for those attempting to model desert ecosystems and their dynamic properties. If water availability is controlled and other simplifying assumptions are made, highly predictive models of attributes such as the growth of a crop species can be developed (van Keulen & de Wit, Chapter 17). However, when all aspects of the natural system are to be modeled, there is still considerable controversy even as to what form models of desert might take. Goodall (Chapter 15) presents details of several modeling philosophies and gives examples of their application to desert systems. Despite his extensive work, he concludes 'So far, this [arid-land ecosystem modelling activity] has taken the form of a proliferation of ideas rather than firm achievement'. One part of the problem in desert models (or other ecosystem models for that matter) is their failure to include the great spatial variability of desert landscapes. This specific point forms the focus of Noy-Meir's creative discussion (Chapter 16).

When we view desert ecosystems on time scales greater than a few years, the possible role of temporal and spatial variations can be shown to be significant to ecological phenomena such as community evolution and succession (Le Houérou, Chapter 14).

I should like to underline some of the points made by the authors of this section, and take the advantage of overviewing their contributions and those of other workers to make a few additional points.

First, there is no question that precipitation, in terms of both intensity and frequency, in deserts is unpredictable compared to other ecosystems, with the

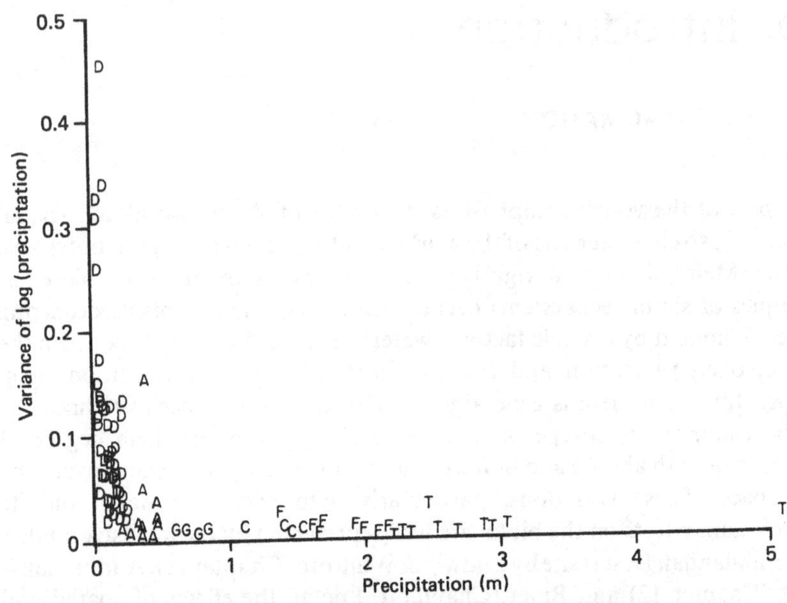

Fig. 10.1. A plot of the variance of the log of mean annual precipitation versus mean annual precipitation. Weather data are taken from Clayton (1944) and Clayton & Clayton (1947) for various world localities. D, desert; F, deciduous forest; T, tropical rainforest; G, grassland; C, coniferous forest; A, tundra. (From MacMahon, 1980).

exception of some arctic tundras (Fig. 10.1). Note that since precipitation has a fixed lower boundary (0 cm yr^{-1}), standard deviations and means of precipitation can be expected to be correlated with one another. Log transformation of precipitation values reduces this problem (Lewontin, 1966).

Colwell (1974) points out that predictability is actually composed of two components: constancy and contingency, both fostering predictability, but each having potentially different ecological implications. Constancy measures the degree to which a parameter exists in a given state throughout the period of interest. Contingency measures the degree to which a parameter exists in a given state, at a given, repeated interval within the period of interest. The unpredictability of desert rainfall comes from both of these attributes.

A result of the temporal desert rainfall pattern is that events directly driven by water availability (e.g., plant growth and reproduction), and those ecosystem processes indirectly tied to such events (e.g. animal reproduction) exist in a 'feast–famine' context with the interval-length of feast or famine being unpredictable.

For desert organisms, the response to these scenarios can include an increase in growth or a change in chemical composition for individuals, (Bamberg, Vollmer, Kleinkoff & Ackerman, 1976; Schreiber & Johnson, 1975), an increase in density for populations (Ludwig & Whitford, Chapter

11), and an increase in species richness for communities during periods of water surfeit. The converse for each of these trends might obtain in times of insufficiency. Another group of desert species occur, however, in which there is a damping of a species' response to surfeit by not changing in a strong positive manner to abundance nor negatively to insufficiency of precipitation. There are many examples of species showing only moderate response to rainfall significantly greater than the long-term mean. The kangaroo rat (*Dipodomys*) data of Ludwig & Whitford (Chapter 11) are of this type, while their cricetid rodent data exemplify the relatively undamped situation; perennials versus ephemerals exemplify the broad differences among plants to the same conditions. Noy-Meir (1973, 1974) has developed an approach to desert ecosystems, termed the pulse–reserve paradigm, that details the feast–famine aspects of the adaptational suites of desert organisms in a modeling context.

The responses and adaptations of individuals and populations to the unpredictable nature of deserts are well known and often mentioned. I would like to emphasize here a series of correlations, inferred to be cause–effect in nature, which involve the levels of biological organization from the population through the ecosystem. This is not meant to be an exhaustive presentation, merely a 'potpourri' which might suggest that deserts are not, by any criteron, 'simple' or 'uninteresting' systems. In fact they are so polar relative to the spectrum of world ecosystems along many biotic and abiotic axes that we need to understand their dynamics to place other ecosystems into their proper context in general ecological theory. I shall separately address a series of ecological topics emphasizing how the temporal variability in water availability, the main driving variable of most deserts, relates to some topics of contemporary ecological interest.

Competition, coexistence, and species richness

The popular feeling that deserts have depauperate floras and faunas is untrue in both a relative and absolute sense, except by comparison to other extreme ecosystems such as the wet tropical forests. For example, characteristic North American hot desert sites, usually a few hectares in extent, regularly contain *c.* 10–20 woody perennial plant species, perhaps 40–70 annuals, up to 10–14 species each of small mammals and reptiles, 25 breeding bird species, 1000 macroarthropods, and an unknown number of nematodes and soil microarthropods, probably totalling 200 or so species (MacMahon, 1979; MacMahon & Wagner, 1980). These numbers compare well with, and in many cases exceed, forest and grassland values from areas of similar extent. For example, a paper comparing aspen, spruce and fir forest stands in northern Utah (Schimpf, Henderson & MacMahon, 1980) lists as maximum species values for any of these three forest types the following: 12 woody perennials, 11 small mammals, 1 reptile, 21 breeding bird species and about 700 macroarthropods.

The question we might ask then, is not why is the desert simple in terms of species richness, but rather why is the species richness relatively so great in view of the apparent 'harshness' of the physical environment? I believe, as do others, that a partial answer lies in the environmental variation discussed above. Unpredictable variation in the physical environment is constantly changing the competitive milieu of each species. Thus, no one species retains a competitive advantage long enough to exclude others because of variation in the resonses shown by competitors to a changing environment.

The unpredictable resources in desert communities are often scarce. The result is that many species vie for the same limiting resource and the outcome is nearly classical diffuse competition. The demonstrably competitive interactions of desert birds, small mammals and ants, all using seeds of annuals, are such a case (Brown & Davidson, 1977).

Recently, the effect of disturbance on ecological systems has been a topic of interest. A variety of models and other analyses from various ecosystems suggest that disturbance of moderate intensity fosters species richness (e.g. Barclay, 1975; Jacobs, 1977*a*, *b*; Huston, 1979; Paine, 1979). While these specific models involve competitive interactions, this does not imply that competition is the only driving variable important in determining species richness in deserts. Extending the arguments of Wiens (1974, 1977) based on birds in grasslands of varying climatic stability to deserts, the climatic instability itself may organize community components, without invoking competition as a mechanism.

Life history strategies

One reason that deserts superficially seem so simple is that many desert species are small, in addition to being inconspicuous during unfavorable periods e.g. annuals 'hide' as seeds, amphibians as deep soil inhabitants and many invertebrates as various resting stages. The small size of many desert organisms is partly a consequence of the, by now, well known *r*- versus *K*-selection continuum. In frequently disturbed areas, species with high maximum rates of natural increase, large litter or clutch size, small body size and short life spans are often favored. I hasten to point out that on theoretical grounds this is not always the case, as shown by Schaffer's (1974) model (see also Stearns, 1976).

While *r*- versus *K*-strategies are, I believe, an oversimplification of a more complex suite of life history characteristics (e.g. Wilbur, Tinkle & Collins, 1974; Grime, 1979; Whittaker & Goodman, 1979), there is some semblance of *r*-strategy selectiveness fostered by desert environments. In addition, the feast–famine nature of deserts also permits existence of species that can temporally vary between *r–K* characteristics (Nichols, Conley, Batt & Tipton, 1976) as well as *K*-strategists who are physiologically more resilient in the face of environmental extremes and can survive between *r*-strategist favorable periods, waiting for *K*-strategist conditions to obtain. Thus, many life-history tactics may work equally well in deserts, a factor potentially affecting species richness.

Factors other than the temporal variation in water

I have emphasized above the temporal variability in the desert milieu. Desert storm cell sizes are often small, creating a mosaic of wet versus dry sites within close (a few tenths of kilometer) proximity to one another. This spatial variation in precipitation, along with variation in the effectiveness of rain water getting into soil zones where plants can use the water – often caused by geomorphic drainage patterns, soil particle-size sorting, and the presence of hardpans and surface crusts – creates immense spatial hetereogeneity in the landscape, and concomitantly in the status of local species assemblages. This varying array can by itself enhance ecosystem stability (i.e. persistence via redundancy) and thus maintain species richness see the model of Crowley, 1978).

Although I have emphasized water, other resources having high spatial and temporal diversity would also enhance species richness. Thus, the stark mosaic of soil organic matter, nitrogen and phosphate (West, Chapter 12; Binet, Chapter 13) can have influences on species richness. Note that none of these resources need necessarily be independent of another to have the effect postulated – any independence among them only further increases environmental heterogeneity.

Succession

It has been argued that succession does not occur in deserts (or tundra for that matter) (see reviews in MacMahon, 1979; MacMahon & Wagner, 1980). The reason for this apparent lack of an 'intermediate' biota following a disturbance until the reappearance of the 'climax' communities may be due to unpredictable rainfall. I have recently argued (MacMahon, 1980) that low desert and tundra precipitation and the linked extremes of precipitation variability may limit the pool of colonizers to those already present in the mature assemblages. The result is that at least the life forms, and often the species that occupy a 'climax' desert community, are the only forms that can recolonize and survive in a disturbed desert habitat. Climax species are already pre-adapted to disturbance by the natural environmental variations and these species are the only forms with proximate propagules and the potential to survive. They thus succeed themselves – autosuccession.

This does not negate species or life-form compositional changes when there is a change in the essential character of the environment either over the short or long term (MacMahon, 1980; Le Houérou, Chapter 14). But the altered environments are, by definition, different from those preceding them and they support new communities, even though such communities contain species from the old assemblage. In fact this is how the assemblages of species now characteristic of deserts probably evolved (see Axelrod, 1979).

The perspective which, I believe, derives from the above discussion and a careful consideration of the chapters in this section is that deserts are interesting, highly variable, complex communities worthy of serious study – not simple, uninteresting, desolate expanses.

Acknowledgments

Douglas Andersen, Peter Landres and Kimberly Smith liberally commented on a draft of this manuscript. Linda Finchum interpreted my own form of crypsis. Robert Bayn and Bette Peitersen helped with manuscript preparation.

References

Axelrod, D. I. (1979). Age and origin of Sonoran desert vegetation. *Occasional Papers of the California Academy of Sciences*, **132**, 1–74.

Bamberg, S. A., Vollmer, A. T., Kleinkopf, G. E. & Ackerman, T. L. (1976). A comparison of seasonal primary production of Mojave Desert shrubs during wet and dry years. *American Midland Naturalist*, **95**, 398–405.

Barclay, H. (1975). Population strategies and random environments. *Canadian Journal of Zoology*, **53**, 160–5.

Brown, J. H. & Davidson, D. W. (1977). Competition between seed-eating rodents and ants in desert ecosystems. *Science*, **196**, 880–2.

Clayton, H. H. (1944). World weather records. *Smithsonian Miscellaneous Collection*, **79**, 1–1199.

Clayton, H. H. & Clayton, F. L. (1947). World weather records 1931–1940. *Smithsonian Miscellaneous Collection*, **105**, 1–646.

Colwell, R. K. (1974). Predictability, constancy, and contingency of periodic phenomena. *Ecology*, **55**, 1148–53.

Crowley, P. H. (1978). Effective size and the persistence of ecosystems. *Oecologia*, **35**, 185–95.

Grime, J. P. (1979). *Plant strategies and vegetation processes*. Wiley, New York.

Huston, M. (1979). A general hypothesis of species diversity. *American Naturalist*, **113**, 81–101.

Jacobs, J. (1977*a*). Coexistence of similar zooplankton species by differential adaptation to reproduction and escape in an environment with fluctuating food and enemy densities. I. A model. *Oecologia*, **29**, 233–47.

Jacobs, J. (1977*b*). Coexistence of similar zooplankton species by differential adaptation to reproduction and escape in an environment with fluctuating food and enemy densities. II. Field data analysis of *Daphnia*. *Oecologia*, **30**, 313–29.

Lewontin, R. C. (1966). On the measurement of relative variability. *Systematic Zoology*, **15**, 141–2.

MacMahon, J. A. (1979). North American deserts: their floral and faunal components. In: *Arid-land ecosystems: structure, functioning and management*, vol. 1 (ed. D. W. Goodall & R. A. Perry), pp. 21–82. Cambridge University Press.

MacMahon, J. A. (1980). Ecosystems over time: Succession and other types of change. In: *Forests: fresh perspectives from ecosystem analyses* (ed. R. Waring). Proceedings of the Oregon State University Biology Colloquium, 26–27 April, 1979, Corvallis, in press.

MacMahon, J. A. & Wagner, F. H. (1980). The Mojave, Chihuahuan, and Sonoran deserts of North America. In: *Warm desert ecosystems*, ed. I. Noy-Meir. Ecosystems of the World, vol. 12. Elsevier, New York, in press.

Nichols, J. C., Conley, W., Batt, B. & Tipton, A. R. (1976). Temporally dynamic reproductive strategies and the concept of *r*- and *K*-selection. *American Naturalist*, **110**, 995–1005.

Noy-Meir, I. (1973). Desert ecosystems: Environment and producers. *Annual Review of Ecology and Systematics*, **4**, 25–51.

Noy-Meir, I. (1974). Desert ecosystems: higher trophic levels. *Annual Review of Ecology and Systematics*, **5**, 195–214.

Paine, R. T. (1979). Disaster, catastrophe, and local persistence of the sea palm *Postelsia palmaeformis*. *Science*, 205, 685–7.

Schaffer, W. M. (1974). Optimal reproductive effort in fluctuating environments. *American Naturalist*, **108**, 783–90.

Schimpf, D. J., Henderson, J. A. & MacMahon, J. A. (1980). Some aspects of succession in the spruce-fir forest zone of northern Utah. *Great Basin Naturalist*, **40**, in press.

Schreiber, R. K. & Johnson, D. R. (1975). Seasonal changes in body composition and caloric content of Great Basin rodents. *Acta Theriologica*, **20**, 343–64.

Stearns, S. C. (1976). Life-history tactics: a review of the ideas. *Quarterly Review of Biology*, **51**, 3–47.

Whittaker, R. H. & Goodman, D. (1979). Classifying species according to their demographic strategy. I. Population fluctuations and environmental heterogeneity. *American Naturalist*, **113**, 185–200.

Wiens, J. A. (1974). Climatic instability and the 'ecological saturation' of bird communities in North American grasslands. *Condor*, **76**, 385–400.

Wiens, J. A. (1977). On competition and variable environments. *American Scientist*, **65**, 590–7.

Wilbur, H. M., Tinkle, D. W. & Collins, J. P. (1974). Environmental certainty, trophic level, and resource availability in hife history evolution. *American Naturalist*, **108**, 805–17.

11. Short-term water and energy flow in arid ecosystems

J. A. LUDWIG & W. G. WHITFORD

Introduction

In desert ecosystems precipitation is low and largely unpredictable in its spatial and temporal distribution (Noy-Meir, 1973). The evaporation potential is high. Water redistribution on the surface of the landscape after a precipitation event involves complicated processes (infiltration, run-off, storage, drainage). Each arid landform has different surface and soil characteristics which affects water availability, hence the type of plant community occupying it. The dominant animal species of each plant community are fairly distinctive with respect to their responses to available water and plant biomass availability.

The objective of this chapter is to discuss the dynamic behavior of the biotic components of arid ecosystems in relation to water and energy flow between and within seasons. The growth responses of the different plant species occupying adjacent ecosystems in a watershed will be related to water availability and to heat energy levels. Growth and behavioral responses of groups of animals with contrasting adaptations to changing conditions of food and water availability will be discussed. Plant and animal litter redistribution and decomposition patterns will be discussed with respect to patterns of energy flow and climatic variation. Finally, general conclusions about desert ecosystem dynamics with respect to short-term water and energy flow will be stated.

Approach and methods

The major emphasis in this chapter will be on data and discussion of short-term water and energy flow in a northern Chihuahuan Desert watershed under study on the Jornada Experiment Station of New Mexico State University, which is 40 km north-northeast of Las Cruces, New Mexico. This watershed and the data it has produced on environmental driving forces (precipitation, soil water, radiant energy, and temperature) and biotic responses (plant productivity, animal population changes and microbial activity) for time intervals of weeks during the growing season are most familiar to the authors. However, data from other arid lands in North America (e.g. Wallace & Romney, 1972; Romney *et al.*, 1973) and other parts of the world (e.g. Evenari, Shannan & Tadmor, 1971) will be used in a comparative way and will be needed to draw general conclusions about short time-scale ecosystem dynamics of water and energy flow in deserts.

Water availability in the soil will be indicated by the level of water stored by soil volume. Plant water content will be an important variable for the animals. Energy in desert ecosystems will be considered in two basic ways: first, the radiant energy available and its capture and conversion to chemical or potential energy by the plants; second, the radiant energy which is absorbed, reflected, and conducted by the soil and air and converted to heat energy. The flow of chemical energy to animals and litter will be of considerable importance to the system. The importance of temperature effects on the dynamics of the ecosystem can be considered as it interacts strongly with water flow.

Results and discussion

Water dynamics and growth responses of producers

A Chihuahuan Desert system

To illustrate the dynamics of water and the response of plants to water in the different ecosystems of a Chihuahuan desert watershed, we will show data for precipitation, soil water storage, plant response and temperature. Precipitation totals for periods of consecutive days of precipitation and the water stored in the soil profile to a given depth will be shown. The changes in energy content (calories) of the leaves and fruits produced by the plants through the different seasons will be given. Minimum daily air temperatures are used as an expression of the energy (heat) budget of the system through time.

Precipitation was recorded using a weighing bucket rain gauge. Soil moisture storage was monitored using electrical resistant gypsum blocks and the bulk densities of the soils. Air temperatures were measured using hygrothermographs housed in standard weather instrument shelters. Data from two weather stations will be used, one located on the alluvial fans and the other in the basin.

Productivity estimates were based on harvest methods (Milner & Hughes, 1968) and dimensional analysis (Newbould, 1967). Details of methodology of measuring productivity of species is beyond the scope here, but are described in Ludwig, Reynolds & Whitson (1975) and Whitford (1974).

Alluvial fans (bajadas)

The upper alluvial fans are characterized by *Larrea tridentata*. The precipitation events are highly seasonal (Fig. 11.1). Storage of water in the profile on these alluvial fans follows the precipitation patterns. Response of *L. tridentata* follows the water storage patterns. When minimum air temperatures drop

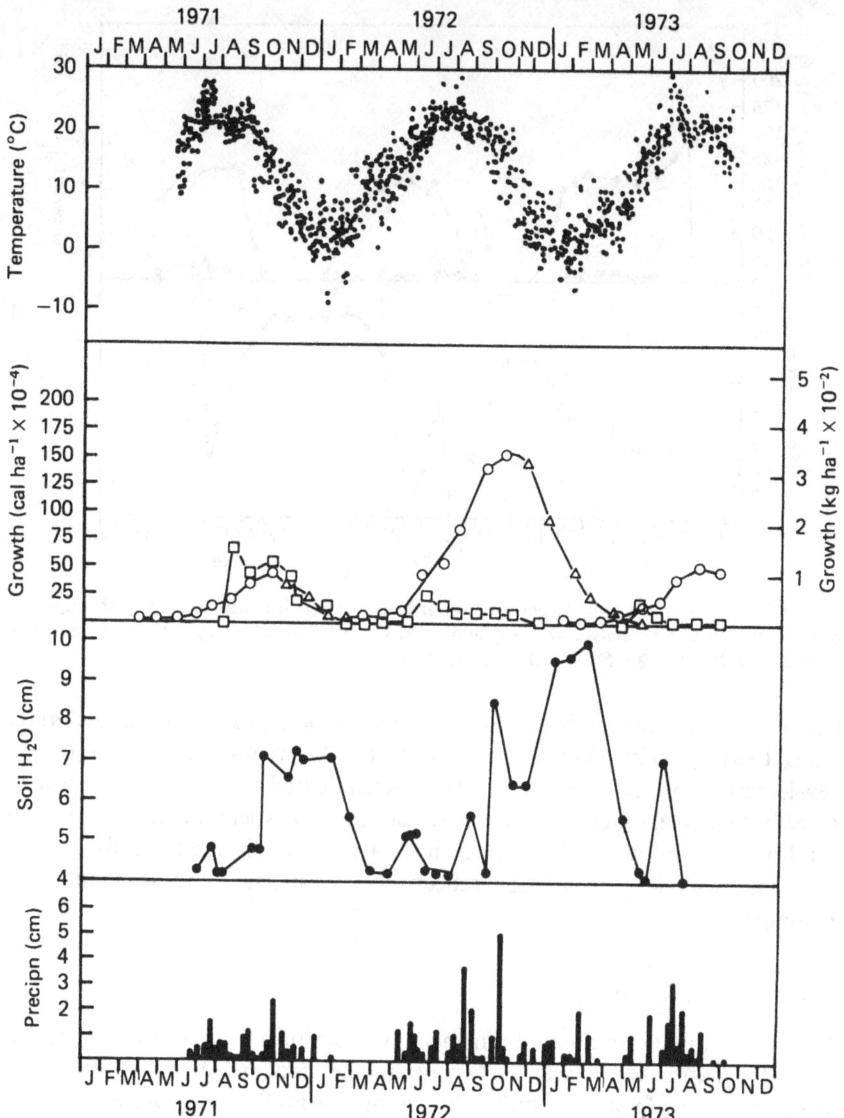

Fig. 11.1. Chihuahuan Desert alluvial fans (bajadas). Precipitation, soil water storage, growth responses of creosotebush (*Larrea tridentata*) and daily minimum air temperatures for 1971–73. △—△, Old leaves; ○—○, new leaves; □—□, fruits.

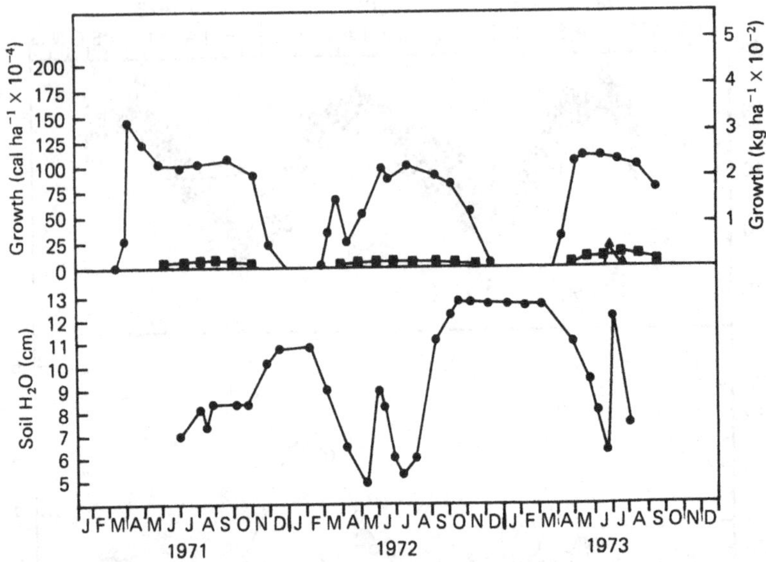

Fig. 11.2. Chihuahuan Desert larger water courses (arroyos). Soil and water growth responses of mesquite (*Prosopis glandulosa*). Daily minimum air temperatures and precipitation as in Fig. 11.1 ●—●, leaves; ■—■, shoots; ▲—▲, fruits.

below 0 °C, usually in November and December, growth of *L. tridentata* stops. Leaves produced in the previous year will have mostly fallen off before new leaves are produced in the next year. As indicated by its pattern of growth, *L. tridentata* is relatively slow in its production of new leaf material with good conditions of soil water. It seems to have a relatively low but steady growth rate. The peak leaf production reflects the consistency and seasonality of available water.

Water courses (arroyos)

The water courses on alluvial fans occupy a position where they will receive water from run-off. *Yucca elata* occurs in the smaller water courses and it produces new leaf material during the summer rainfall periods of July, August and September (Smith & Ludwig, 1976). Reproduction is highly variable. When production of fruits is high, production of new leaves remains very low. Thus *Y. elata* shows a different strategy in water utilization and energy flow relative to new leaf production and reproduction than *Larrea tridentata*.

In the large arroyos, the soils are deeper with the profile extending to an average depth of 120 cm. Storage of water exceeded 12 cm during many of the seasons (Fig. 11.2). The growth data for *Prosopis glandulosa*, mesquite, shows that the leaf biomass produced at old nodes on old stems of this species

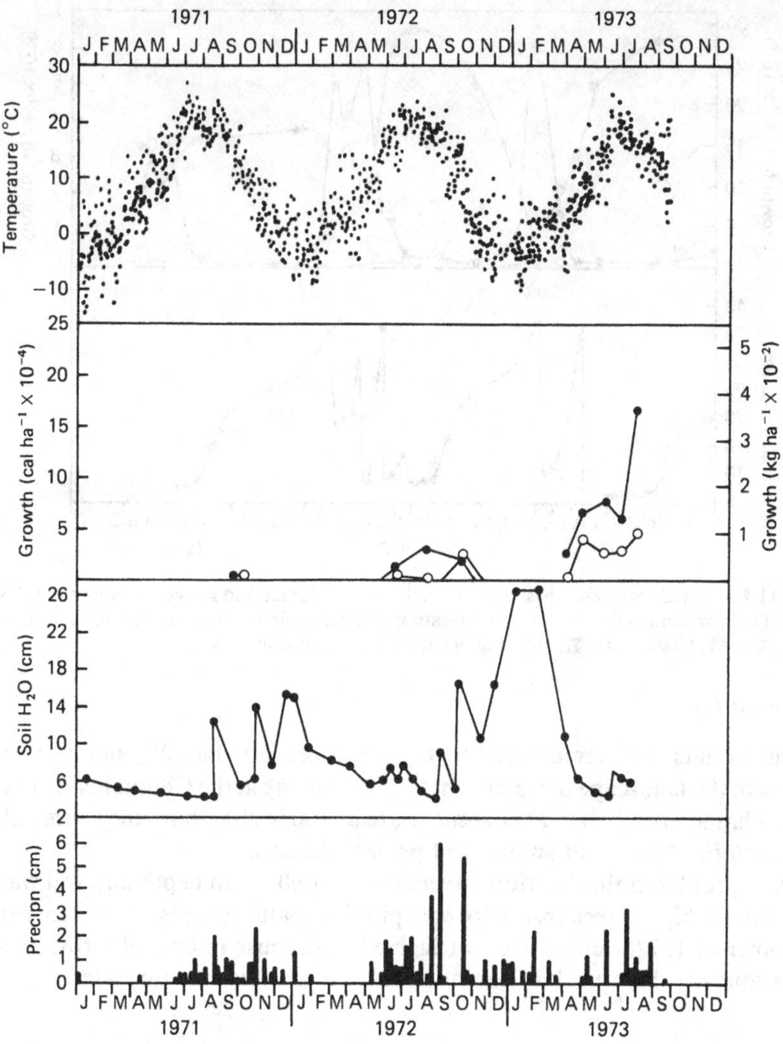

Fig. 11.3. Chihuahuan Desert alluvial flats. Precipitation, soil water storage and growth responses of small annual and perennial forbs, and daily minimum air temperatures for 1971–73. ●—●, Annual forbs; ○—○, perennial forbs.

is generally constant from year to year, but there may be some die-back in the months of May and June. In early April 1972, there was a frost and leaf biomass did not recover until June, since water storage was low until late May. Reproduction in *P. glandulosa* was essentially zero in 1971 and 1972, but a large quantity of fruits were produced in 1973. This same pattern is evident for the production of new shoots. In November, when minimum daily air temperatures drop below freezing, *P. glandulosa* loses its leaves. The reproductive response in *P. glandulosa* is similar to that of *Y. elata*.

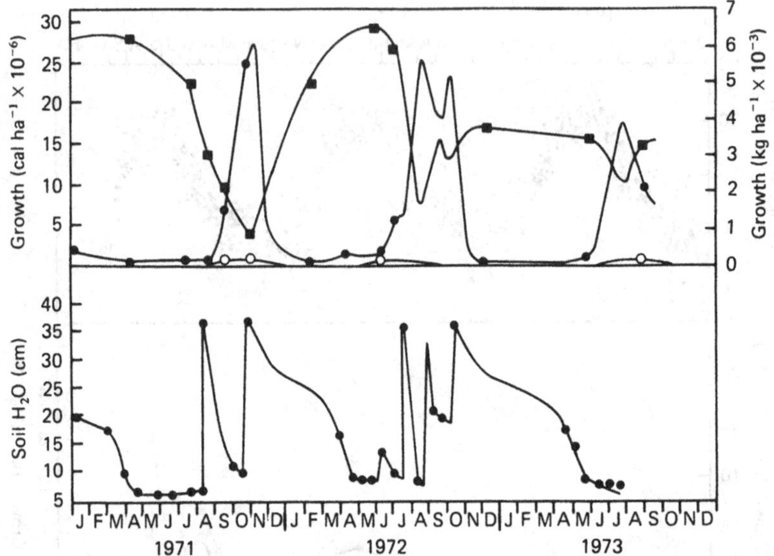

Fig. 11.4. Chihuahuan Desert basins (swales). Soil water storage and growth responses of tobosa grass (*Hilaria mutica*) for 1971–73. Precipitation and daily minimum air temperatures as in Fig. 11.3. ●—●, Green; ■—■, standing dead; ○—○, reproductions.

Alluvial flats

Alluvial flats are characterized by having slopes less than 2% and represent areas in the landscape of water run-on. These areas in the Chihuahuan Desert are characterized by *Flourensia cernua* (tarbush), but they are also characterized by small annual and perennial forbs.

Soil profiles in this landform average about 60 cm in depth and soil water storage is highly responsive to precipitation patterns (Fig. 11.3). In early summer of 1972, rainfall events triggered a response in annual forbs, which was followed by a peak in perennial forbs about two months later.

Basins (swales)

As one proceeds down the alluvial flat to areas with less than 1% slope, the soil texture changes to a clay loam. Swales in southern New Mexico are characterized by *Hilaria mutica* (tobosa grass). This large perennial grass has a deep and diffuse root system. Soils are deeper than on the alluvial flats (about 90 cm). Since these areas are lower on the watershed and have deep soils, they represent the system with maximum soil water storage. *H. mutica* is highly responsive to both temperature and the soil water. The dynamics of water utilization and energy accumulation by this species (Fig. 11.4) shows that its production is considerable. It reached a peak calorie content of over 25 Mcal ha^{-1} in both 1971 and 1972.

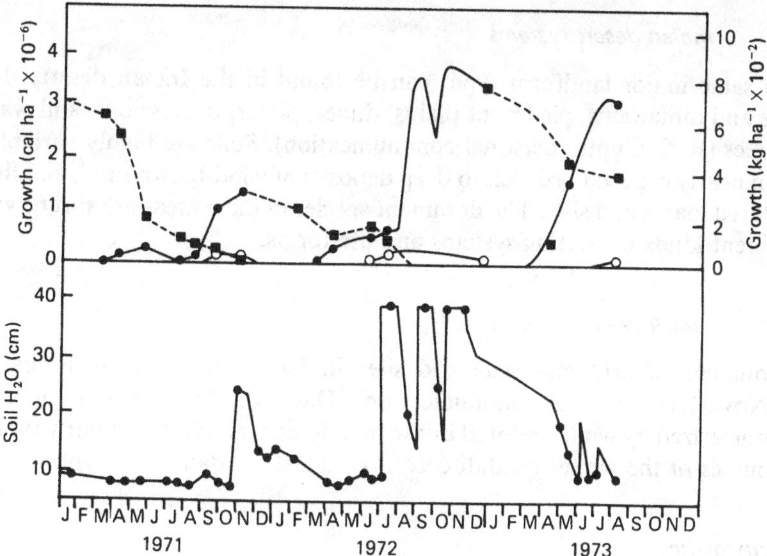

Fig. 11.5. Chihuahuan Desert sinks (playas). Soil water storage and growth responses of vine mesquite grass (*Panicum obtusum*) for 1971–73. Precipitation and daily minimum air temperatures as in Fig. 11.3. ●—●, Green; ■—■, standing dead; ○—○, reproductions.

Sinks (playas)

As one proceeds down the alluvial flats into a sink (basin or playa) the area is generally level and the soil texture is a clay with a hard pan at about 60 cm. The playa in our system is not saline and is characterized by a plant cover of *Panicum obtusum* (vine-mesquite grass). This perennial grass reproduces vegetatively by stolons and also sexually by seeds. It has diffuse root systems which remain relatively shallow due to the clay pan. Significant growth occurs when a series of large rainfall events keeps this basin under water for periods of up to a month (Fig. 11.5). Production of seeds by *P. obtusum* will follow good vegetative growth.

Australian desert systems

A desert region very similar to our Chihuahuan Desert is the area around Alice Springs, Australia. It is characterized by summer rainfall with an average yearly total of about 25 cm (Slatyer, 1962). It also is a region characterized by landforms similar to those found in the Chihuahuan Desert (M. A. Ross, personal communication). In the communities receiving run-on, such as mulga groves, productivity within the groves exceeded by five times that outside of the groves (900 : 180 kg ha^{-1}) following periods of heavy rainfall (Ross & Lendon, 1973).

Indian desert systems

The same major landform types can be found in the Indian deserts: low hills and mountains, piedmont plains, dunes, plains, depressions, and water courses (R. S. Gupta, personal communication). Soils are highly variable – from nearly exposed bedrock to deep deposits of wind-blown sand, and finer textured loams and silts. The dominant species in these areas are shrubs with different kinds of rooting systems and life forms.

Israeli desert systems

A number of arid and semi-arid sites in Israel are under investigation (I. Noy-Meir, personal communication). Data for three different sites, all characterized by winter rainfall but with different soil types, are fairly typical examples of the winter rainfall deserts of Israel, Jordan and Egypt.

Sarayia site

This site, located on the north slope of a limestone hill southeast of Hebron, is at an altitude of 700 m. The average rainfall is about 250 mm, with most occurring between November and April. The soil is shallow (10–25 cm) and is over a hard fissured limestone material. The vegetation is characterized by summer deciduous shrubs and herbaceous plants. The shrubs cover about 3–4%, and are low in stature (20–35 cm).

The seasonal dynamics of growth for two shrubs (*Sarcopoterium spinosum* and *Artemisia herba-alba*) are shown in Fig. 11.6. *S. spinosum* green material (total minus woody) reaches a peak at about 135 kg ha^{-1} in April. *Artemisia herba-alba* green biomass (total minus woody) reaches peak biomass in July at about 80 kg ha^{-1}. *Poa bulbosa* peaks at about 300 kg ha^{-1} in March. Annual forbs and grasses peak rapidly in April at about 500 kg ha^{-1}.

Sde-Boqer site

This site is located on a plain with loess soils, which are increasingly saline below 50 cm. Average rainfall on this site is about 75 mm, occurring from October to April. Minimum air temperatures, about 2 °C, occur in January. The vegetation is characterized by a shrub cover of about 3–5%, the major shrub being *Hammada scoparia*. Other plant groups are annual and perennial forbs.

Maximum soil moisture storage is reached in the winter and spring months. However, maximum growth does not occur until temperatures reach the favorable growth levels in March and April.

The growth response of *Erodium hirtum* to precipitation inputs and soil

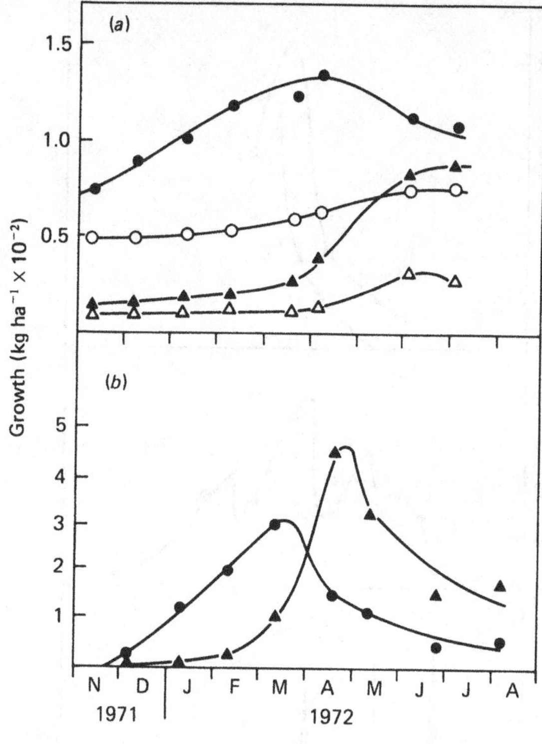

Fig. 11.6. Negev Desert Sarayia site. Seasonal growth dynamics for two shrubs (*Sarcopoterium spinosum* and *Artemisia herba-alba*), a perennial grass (*Poa bulbosa*) and annual forbs and grasses. Rainfall periods are indicated. Data provided by I. Noy-Meir (personal communication). (*a*) *Sarcopoterium*: ●, total; ○, woody. *Artemisia*: ▲, total; △, woody. (*b*) ●, *Poa bulbosa*; ▲, annuals.

moisture (% by weight) at 30–45 cm is shown in Fig. 11.7. This species is characterized by a maximum growth peak in April at about 320 kg ha^{-1}. The growth of annuals follows very closely that of *E. hirtum*, peaking at about 250 kg ha^{-1} in April.

Migda site

This site is located on a plain of deep loess soils. Vegetation is dominated by annual grasses and forbs due to past grazing. The average rainfall for this area is 250 mm, with essentially all of it occurring between October and April. Mean January air temperature minima are about 6 °C. The peak biomass is reached in April for annual grasses and forbs at about 6000 kg ha^{-1} (Fig. 11.8). After this there was no rain and the green biomass of annual grasses and forbs decreased rapidly to virtually zero by June 1.

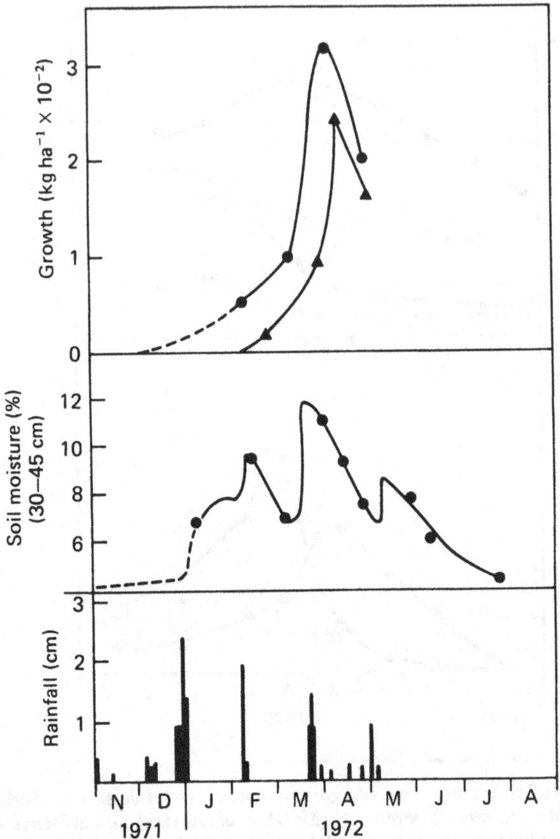

Fig. 11.7. Negev Desert Sde-Boqer site. Precipitation, percent soil moisture by weight and seasonal growth dynamics for *Erodium hirtum* and annual forbs and grasses. Data provided by I. Noy-Meir (personal communication). ● *Erodium hirtum*; ▲, annuals.

These three Israeli sites also illustrate the importance of soil depth on different landscape areas relative to the amount of soil moisture that is stored after precipitation inputs. The sites with the deepest soil had the maximum amount of biomass. Most species or species groups showed maximum biomass occurring in April, after soil temperature increased to allow maximum growth in March. This contrasts with the Chihuahuan Desert system where water inputs occur during the warm season. Here temperature regimes are adequate and the controlling factor of temporal biomass or energy flow dynamics is water input. In the Israeli deserts, the factor controlling temporal dynamics is largely temperature.

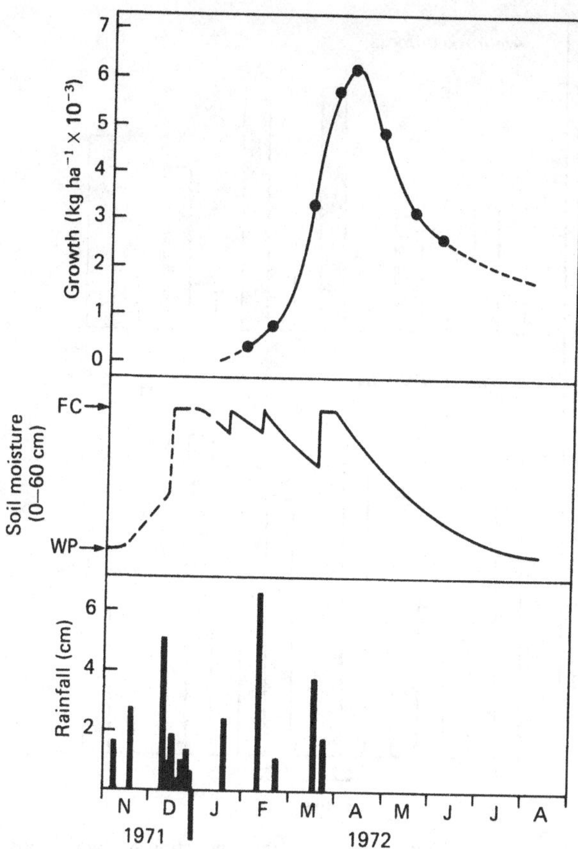

Fig. 11.8. Negev Desert Migda site. Precipitation, estimated soil water potential and seasonal growth dynamics for annual forbs and grasses. Data provided by I. Noy-Meir (personal communication). FC, field capacity; WP, wilting point.

Mohave desert systems

Data describing Mohave desert dynamics of water and energy flow are available from IBP studies at a Rock Valley site in Nevada. The stem production of two Mohave desert shrubs (*Ambrosia dumosa* and *Krameria parvifolia*) are given in Fig. 11.9, for six different zones in 1971 and 1972 (Romney *et al.*, 1973). *A. dumosa* reached a peak biomass of new stems in Zone 22 of only about 3 kg ha^{-1} in 1971. This contrasts sharply with 1972 when it reached a peak stem production of about 10 kg ha^{-1} in Zone 22.

The biomass of these two shrubs indicates that for this winter rainfall desert, temperature is also the critical factor in determining peak biomass. Biomass peaks in the spring of the year were very similar to the Israeli sites.

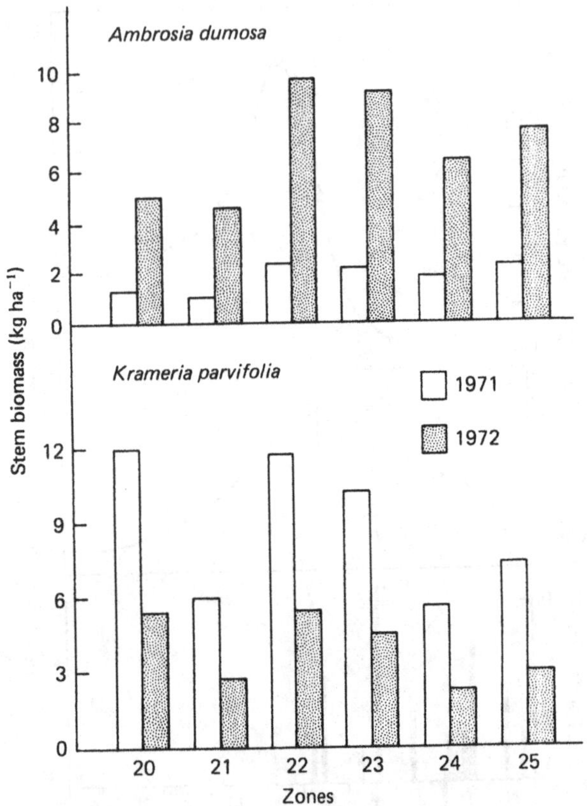

Fig. 11.9. Mohave Desert Rock Valley site. Dry stem production by two shrubs (*Ambrosia dumosa* and *Krameria parvifolia*) in six homogeneous zones. Data provided by S. A. Bamberg and F. B. Turner (personal communication).

Tunisian desert systems

In an area called the Pre-Saharan Zone of Tunisia, there are similarities of different landforms in relationship to other desert ecosystems (C. Floret, personal communication). Their plains can be divided depending on whether the substrate is calcareous, or gypsum. Dune areas may be stable or mobile and various kinds of depressions may occur. These areas may be flooded with salt or fresh water. Plant species with quite different characteristics occur on these different geomorphic landscapes.

Comparison of producers in desert ecosystems

Peaks in new biomass production

The peaks of new biomass produced in different years and seasons for selected species or species groups from different desert landscapes and regions across the world are shown in Table 11.1. Comparing the southwestern United States deserts on equivalent alluvial fan landscapes and comparing *Larrea tridentata* with *Krameria parvifolia*, we see that in *L. tridentata* peak production occurs in the fall whereas in *K. parvifolia* peak production occurs in the spring. Peak production of new shoots in the Chihuahuan Desert is about 200 kg ha^{-1} in 1971, whereas in *K. parvifolia* in 1971 it was 12 kg ha^{-1}, indicating the difference in production. Of course, it would have been better to compare the same species in these two deserts. Both species occur in both deserts, but comparable growth data were not available.

In comparing annual forb production in the three different deserts (the Chihuahuan and Mohave in North America, and an Israeli site), we can see that in a year of good growth such as 1972 (precipitation above average in all three regions), that peak production in the Chihuahuan Desert was about 55 kg ha^{-1}. In the Mohave Desert, a maximum production occurred in the spring at about 7.5 kg ha^{-1}. This compares with the peak production in the Israeli Sde-Boqer site of 250 kg ha^{-1}. In comparing these results, the Israeli site was recently grazed (I. Noy-Meir, personal communication) and shrub cover is low (3–4% v. 15–20% in the United States). The response of annuals in a disturbed area is likely to be greater.

Rates of productivity

Another way of comparing different ecosystems is to compare their rates of productivity during periods of maximum growth when water storage (availability) is at a maximum.

In order to calculate rates of production with reasonable accuracy, estimates of biomass must be made at relatively short intervals and based on the time interval of maximum (near-linear) growth.

The relative rates of production for the two shrubs and the annuals in our Chihuahuan Desert and Noy-Meir's Sde-Boqer site are given in Table 11.2. Since caloric data is not available from both sites, product rates are in terms of biomass, with rates over the time interval indicated.

The production rates of *Larrea tridentata* exceeds *Hammada scoparia*, however, it must be stressed that the rates are given on an area basis, thus if the density and biomass of *L. tridentata* is greater than *H. scoparia*, this could account for the difference. However, using standing biomass estimates to calculate productivity rates also assumes that turnover rate (death) and

Table 11.1. *Production peaks of new biomass in categories for selected species or species groups in different years and seasons from certain desert landscapes and regions*

Desert	Precipitation (mm)	Landscape	Species or group	Category	Year	Season	Biomass (kg ha^{-1})
Chihuahuan, USA	197	Alluvial fans	*Larrea tridentata*	Shoots	1971	Autumn	200
	395				1972	Autumn	707
	235				1973	Summer	265
				Fruits	1971	Summer	150
					1972	Spring	93
					1973	Spring	120
		Alluvial flats	Annual forbs and grasses	Shoots	1971	Autumn	2
					1972	Summer	55
					1973	Spring	550
Mohave, USA	114 (1972)	Alluvial fans	*Krameria parvifolia*	Stems	1971	Spring	12
					1972	Spring	6
			Annual forbs and grasses	Shoots	1971	Spring	6
					1972	Spring	7
Sde-Boqer, Israel	160	Alluvial fans	*Hammada scoparia*	Green shoots	1972	Spring	350
			Erodium hirtum	Shoots	1972	Spring	320
			Annual forbs and grasses	Shoots	1972	Spring	250

Table 11.2. *Rates of production for shrubs and annuals in two desert sites in 1972*

Desert	Species or group	Component	Production rates (kg ha^{-1} day^{-1})	Interval
Chihuahuan				
	Larrea tridentata	Green shoots	4·6	June–September
	Annuals	Above-ground	2·3	July–August
			6·9	
Sde-Boqer	*Hammada scoparia*	Green shoots	1·2	April–June
	Erodium hirtum	Above-ground	4·5	February–April
	Annuals	Above-ground	4·2	February–April
			9.9	

consumption (by all animals) are not significant effects (over the time interval considered).

The rates given in Table 11.2 cannot be considered as estimates of net primary production, since they only consider the rates of biomass increase of certain plant components. Root biomass is not considered. In the shrubs, biomass increments of the older woody parts was not used. Further, these rates in terms of kg ha^{-1} day^{-1} cannot be multiplied by the number of days in the year to obtain total annual production since these rates are averaged only over the time of maximum growth.

Growth and behavioural responses of animals

Factors affecting short-term responses

Energy flow through consumer populations in a desert ecosystem may best be summarized by examining the major groups of organisms consuming specific portions of the producer species rather than considering individual species. The relative amounts of seed production and herbage varies from one year to the next depending on the timing and intensity of rainfall events as has been shown in the previous sections. In addition, the availability of seeds of a specific type and rate of production of fresh green vegetation may vary greatly even within one season depending on the timing and intensity of rainfall. The time of year when seeds and/or herbage of a particular type are available to consumers largely determines the behavioural and/or physiological responses of these populations. Thus, years with identical total rainfall and total primary productivity may result in very different productivities of animal species.

Irrespective of the taxon, the rate of energy flow through consumers is largely a function of availability of suitable sources of energy and therefore ultimately a function of the past climate.

In periods of water stress, feeding preferences of consumers may be shifted to provide adequate water for maintenance. If adequate water is unavailable, the viability of the population is reduced. Therefore, in order to understand the flow of energy and water through consumers over short periods, we suggest that only when the minimum physiological requirements for water are met is a population capable of acquiring the extra energy needed for reproduction. If there are insufficient water sources, increased mortality serves to reduce energy flow through a taxon.

Factors affecting water and energy flow in consumers

Seed consumers

Most species of heteromyid rodents are capable of maintaining water balance on a seed diet without resorting to free water or green vegetation (Schmidt-Nielsen, B. & Schmidt-Nielsen, K., 1952; Chew, 1965). Although some authors (Beatley, 1969; Bradley & Mauer, 1971) claim that the reproductive success of heteromyids is a function of the availability of vegetation with good moisture content, Whitford (1976) suggests that survivorship and recruitment is a function of caloric content of forage, not its succulence. Non-heteromyids require varying amounts of succulent vegetation or insect food to meet their water requirements (MacMillen, 1964). Energy flow through desert rodent populations appears to be a function of precipitation resulting in abundant food supply which has been suggested by French *et al.* (1974) as the primary factor limiting population growth.

Whitford (1976) summarizes four years of intensive study on rodent communities of the Chihuahuan Desert, which showed that density and biomass of cricetid rodents responded rapidly to rainfall that exceeded the long-term average, but that heteromyid rodent biomass showed significant change only following a drought period when seed supplies were scarce, thus reducing the availability of energy sources and metabolic water (Fig. 11.10). As this model shows, variation in rainfall above the minimum physiological needs for 'water independent' (heteromyids or other species with similar physiological traits) species has little effect on their density and biomass. Establishment and survival of 'non-water independent' species depends on patches of favorable habitat which provides their water needs during drought periods. The amount of precipitation to meet their minimum physiological needs is greater than that of heteromyids and exceeds the long-term average rainfall of the most arid deserts. For example, Turner (1972, 1974) reported only one species of *Peromyscus* at the Rock Valley Site in Nevada, which

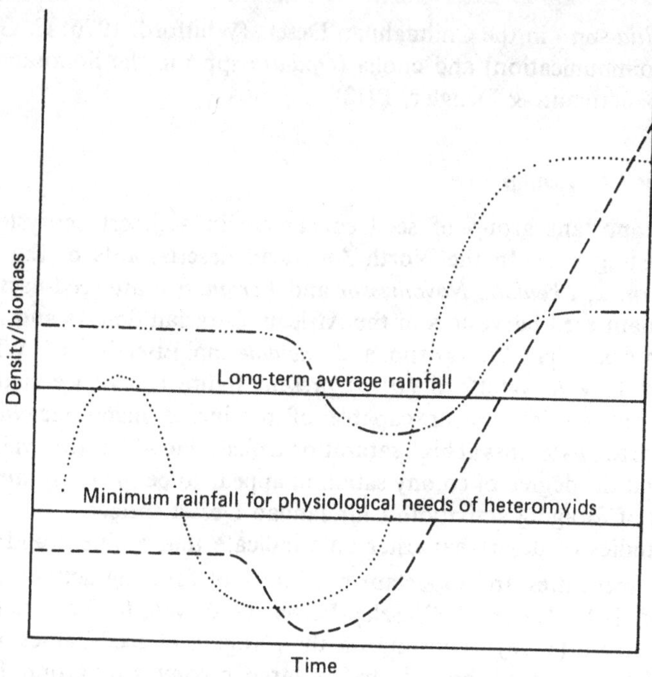

Fig. 11.10. A generalized scheme showing the relationship between rainfall and density or biomass of Chihuahuan Desert rodents, based on data in Whitford (1976). —· · ·—, Heteromyids; ---, cricetids; . . ., annual rainfall.

occurred at extremely low density (< 0.1 ha^{-1}) even during a year which exceeded average rainfall (< 100 mm yr^{-1}) by a factor of two. We submit that this model may be generally applicable to rodent populations in other deserts of the world. The dominant rodents in the African and Eurasian deserts, jerboas (*Jaculus* spp.) and gerbils (*Taterillas* and *Tatera* spp.) and in the Australian deserts, *Notomys* spp, possess physiological adaptations similar to kangaroo rats (Kirmiz, 1962; MacMillen & Lee, 1967; Chew, 1965).

In these deserts, as in the North American deserts, stability of rodent biomass will be a function of the amount of rainfall. This model suggests that fluctuations in numbers and biomass will increase as rainfall amounts approach the minimum physiological requirements for these 'water independent' species. 'Non-water independent' rodent species will make up significant parts of the rodent biomass only in those areas that are marginal deserts or that have recently undergone desertification.

The density and biomass of some species like woodrats (*Neotoma* spp.) is limited by habitat. *Neotoma* spp. build nests in areas which supply ample succulent food and nest building materials such as *Yucca* spp. and prickley pear (*Opuntia* spp.) in the Chihuahuan Desert (Whitford, 1976; C. Grenot, personal communication) and cholla (*Opuntia* spp.) in the Sonoran desert (Brown, Liebermann & Dengler, 1972):

Seed-harvesting ants

Another important group of seed consumers in a desert ecosystem are seed-harvesting ants. In the North American deserts, ants of the genera *Pogonomyrmex*, *Pheidole*, *Novomessor* and *Veromessor* are seed-harvesters. The dominant seed-harvesters in the African–Eurasian deserts and *Messor* spp. and in Australia *Chelaner* spp. and *Pheidole* spp. Ettershank & Whitford (1973) and Kay & Whitford (1975) present indirect evidence that some harvester ant species appear capable of producing metabolic water to compensate for water loss at high saturation deficits. However, the availability of seeds and the degree of colony satiation appear to be the most important regulators of foraging activity in Chihuahuan Desert species.

Most studies of desert harvester ants indicate that ambient and/or soil surface temperatures are important regulators of foraging activity (Shaeta & Kaschef, 1971; Delye, 1967; Szlep-Fessel, 1970; Whitford & Ettershank, 1975). Whitford (1978), reporting on the foraging of seed-harvesting ants (*Pogonomyrmex* spp.), showed that a large colony size group forager, *Pogonomyrmex rugosus* (> 1000 foragers per colony) harvested intensively in a year with high annual production following a drought but was nearly inactive during the subsequent year in which annual plant production was also high. Species that were individual foragers with small colony size (< 1000 foragers per colony) foraged with equal intensity during both years. Thus, precipitation appears to affect seed removal by harvester ants in two ways: production of a seed resource and, if excess seeds are available, group foraging colonies may store sufficient seeds to eliminate the necessity for surface activity in the following year.

When we estimated the energy flow through the various seed consumers in the Chihuahuan Desert, the rodents and ants exhibited reciprocal behavior in terms of energy flow: for example, in 1971, rodents, 1241 kcal ha^{-1}; ants, 0.9 kcal ha^{-1}; birds, 3.1 kcal ha^{-1}; and in 1972, rodents, 270 kcal ha^{-1}; ants, 12 kcal ha^{-1}; birds, 4.0 kcal ha^{-1}. These estimates show that in terms of energy flow, ants and birds are much less important than rodents in a Chihuahuan Desert ecosystem, which is probably true for other desert ecosystems as well. Seed-harvesting ants are not affected by short-term drought as are rodents, hence are able to respond quickly to abundant seeds following a drought. Although harvester ants required only an estimated 12 kcal ha^{-1} for maintenance and growth in 1972, they harvested approximately 71 kcal of seeds.

Table 11.3. *Arthropod biomass (g dry wt ha^{-1}) for time periods specified and for arthropod groups indicated.* (Data are from Whitford, 1971, 1972, 1973, 1974). Blank spaces indicate data not collected or samples not processed

Arthropod	1971		1972		1973		1974	
	May	August	May	August	May	August	May	August
Playa								
Shrub dwelling	17·2	54·6	3·1	175·7	8·9	57·7	—	—
Lepidoptera and Hymenoptera	—	—	—	3·2	5·6	1·1	2·0	0·2
Orthoptera	—	—	—	48·0	98·0	51·0	0·0	0·8
Bajada								
Shrub dwelling	19·9	21·9	46·6	7·6	68·8	12·3	—	—
Lepidoptera and Hymenoptera	—	—	—	0·1	0·3	0·3	0·9	0·5
Orthoptera	—	—	—	10·0	25·0	100·0	0·0	0·0

Seed-eating birds

The avifauna of a desert region must be considered in two parts: (*a*) the breeding or resident population and (*b*) migrants or transient populations. While the resident populations may be important as rate regulators in the system, the transients are potentially of greater importance in energy flow and as exploiters of excess resources.

In the Chihuahuan Desert, resident birds are primarily insectivores and even the gramnivorous black-throated sparrow consumes some insects during the breeding season and at times of water stress (R. J. Raitt, personal communication). The productivity of resident birds is directly a function of moisture availability and insect productivity is the ultimate factor (Raitt & Pimm, 1976). The generalized model proposed for rodents (Fig. 11.10) is also applicable to birds. Resident birds behave like heteromyid rodents in that insufficient moisture during the breeding season results in reproductive failure but the adults exhibit normal survivorship. Rainfall close to the long-term average allows reproduction and maintenance of a relatively stable bird population. Excess moisture produces large blooms of annuals, and hence seeds in excess of those that can be exploited by rodents and ants allowing migrant seed-feeding birds such as lark buntings and horned larks to enter the system to exploit this resource during the non-growing season (Raitt & Pimm, 1976). Hence their impact on the system is like that of cricetid rodents.

Plant-dwelling arthropods

Population growth in plant dwelling arthropods appears to vary as a function of plant water potential and phenology of the host plant (Table 11.3). Changes in insect biomass on shrubs like *Prosopis glandulosa* and *Larrea tridentata*, follow periods of growth of new leaves. Most of the shrub-dwelling insect biomass is made up of sucking insects (Psyllidae, Membracidae, Miridae) which depend upon the quality of phloem sap and cell sap for their maintenance and growth. Shrub productivity varies directly with rainfall (Table 11.1, Figs. 11.1–11.5). Thus, herbivorous insect biomass varies with rainfall.

Plant exudate and honeydew feeders

Energy flow through animal populations which feed on plant exudates and/or honeydew is obviously tied to plant water status. Most of these consumers are either social insects which have a variety of ways to integrate short-term fluxes in energy availability and/or water stress, or insects like the dipterans, which avoid stress in the egg stage and are capable of rapid development to the adult stage given favorable environmental conditions.

Colony densities of exudate/honeydew feeding ants reach 77 ha^{-1} and many of these species forage most of the year (Schumacher & Whitford, 1974). The high colony densities suggest that exudate/honeydew feeders are important but as yet unevaluated consumers.

Domestic stock

Cattle and similar consumers respond directly to moisture availability, and thus, to new growth per unit time. Thus, energy flow through these populations does not exhibit the time lags evident in the seed consumers. Cattle grazing probably has an effect on energy flow through insect foliage consumers. Insects exhibited peak activity with a lag of about three weeks after initial growth response of grasses and forbs to rainfall. A similar response was noted in the Sahara by Cloudsley-Thompson (1964). Reduction of standing live grasses and forbs by cattle grazing probably limited energy flow through foliage-consuming insects to some degree. The intensity of grazing on our study area was not as severe as that reported by Cloudsley-Thompson (1964) where grazing by domestic stock denuded the area, thus effectively eliminating herbivorous insects.

Predators

Data on energy flow through predators in any ecosystem is difficult to obtain. During the peak of the drought cycle in a Chihuahuan Desert ecosystem arthropod predators exceeded the biomass of herbivorous insects on *Prosopis glandulosa* and *Larrea tridentata*. However, the biomass of arthropod predators did not respond as rapidly to increased productivity following the onset of rains as did the herbivorous insects (Whitford, 1971, 1972, 1973, 1974).

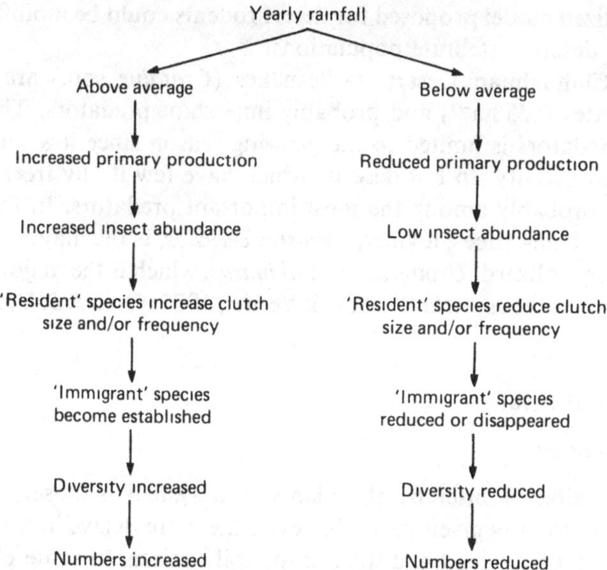

Fig. 11.11. A scheme showing the relationship between rainfall, density and diversity of desert lizard communities, based on data in Whitford & Creusere (1977).

In discussing the interrelationships of arthropods in the Sahara, Cloudsley-Thompson (1968) suggested that a high proportion of carnivores is characteristic of desert fauna. He suggests that seasonal cycles of abundance are related to rain events and that in the non-rainy season these animals exhibited reduced activity.

In our studies (Whitford, 1971, 1972, 1973, 1974) mid-summer predatory species accounted for nearly 50% of the arthropod biomass. Many of the predatory arthropod species are known to be long-lived. The low turnover of many species of desert predatory arthropods and high turnover of most herbivorous insects probably accounts for this apparent paradox in relative biomass.

Whitford & Creusere (1977) discuss the short-term changes in numbers and biomass of a desert lizard community. Seasonal activity periods of adult and hatchling lizards were longer during dry years than wet years. During wet years adults and juveniles of most species exhibited allochronic activity. Densities and biomass of most resident species varied directly with changes in productivity and relative abundance and activity of arthropods.

Whitford & Creusere (1977) present the following model for the relationship of lizard productivity to rainfall (Fig. 11.11). In lizards, as in insectivorous birds, rainfall below a certain critical level results in insufficient insects for normal reproduction, hence a drop in numbers and biomass with a time lag of one to two years after the drought event. These relationships suggest that the generalized model proposed for desert rodents could be modified slightly to fit most desert vertebrate populations.

In the Chihuahuan Desert, rattlesnakes (*Crotalus* spp.) are abundant (approximatey 0.25 ha^{-1}) and probably important predators. Their importance as predators is limited to the growing season since low temperatures restrict their activity. In hot deserts, which have few if any freezing nights, snakes are probably among the most important predators. In the Algerian Sahara near Beni-Abbes, a viper, *Cerastes cerastes*, is the major predator on the herbivorous lizard, *Uromastix acanthinurus*, which is the major vertebrate herbivore in that ecosystem (Grenot & Vernet, 1972; Grenot & Vernet, 1973).

Dynamics of litter

Production of litter

The production of litter by the plants in a system is closely tied to the productivity for any given year. However, there are delays in turnover from growth to standing dead and then to the soil surface. In some plant groups (e.g., annuals) this turnover may be very rapid. In contrast, many of the wood stems of shrubs may persist in the canopy as standing dead for many years.

We used litter collectors under selected shrubs to obtain data on the dynamics of litter production by *Larrea tridentata* (Fig. 11.1). The changes in biomass of the stems and leaves produced as litter follow very distinct patterns. After growth is reduced due to either cool autumn temperatures (1972) or summer drought (1973), older stems and leaves are shed. This differs from previous reports on litter production in this species (Burk & Dick-Peddie, 1973), which indicate that the abscission of stems occurs during periods of rapid growth, rather than after major growth. Slight phasing was also found in the loss of stems and leaves by Oechel, Strain & Odening (1972).

The dynamics of litter produced by the two perennial grasses, *Hilaria mutica* and *Panicum obtusum*, which occur within the swale and playa ecosystems, respectively, are shown in Figs. 11.4 and 11.5. The patterns in 1971 are quite similar for both species, based on data obtained by area harvesting of litter biomass. The peaks in July reflect a decrease in standing dead material produced by growth in 1970. With the summer rains, there was decomposition of some of the litter material, but an increase occurred again in the autumn, following late summer growth. In the winter, there was some decomposition again since precipitation did occur during this period.

Role of detritivores and decomposers

Of all of the sources of energy available to consumers, detritus, of all categories, is probably the most constant since it certainly represents the greatest percentage of the productivity for any given year. There is little known about the litter consumers in North American desert ecosystems. The information which is available is based on studies in very different deserts (Wallwork, 1972*a*, *b*; Wooten & Crawford, 1974, 1975; Haverty & Nutting, 1974*a*, *b*; 1975*a*, *b*, 1976; LaFage, Nutting & Haverty, 1973; Haverty, LaFage & Nutting, 1974; Haverty, Nutting & La Fage, 1975; McBrayer, Mamolito & Franco, 1975; Johnson & Whitford, 1975; Crawford, 1976). Wallwork (1976) summarizes the available literature. The studies on the activities of detritivores in other arid lands of the world have been largely limited to studies of termites (Lee & Wood, 1971*a*, *b*; Boullion, 1970) and studies on breakdown of animal dung (Anderson & Coe, 1974; Ferrar & Watson, 1970).

Wallwork (1972*a*) found that microarthropods were extremely scarce even in moist desert microhabitats and was forced to limit his study to the microarthropods of the litter of juniper trees, the only site where they occurred in significant numbers in Joshua Tree Monument. This has been substantiated by the studies of Edney, McBrayer, Franco & Phillips (1974, 1975). In survey studies in the Chihuahuan Desert we found microarthropods only in the soil and buried leaf litter of a large arroyo (Whitford, 1973). The micro-arthropods were predominately mites varying in density from zero to 31 individuals per 100 g soil between June and August. Further, Wallwork (1976) concludes from his studies in 1972 and those of Wood (1971) that microarthropods reach their highest densities in winter, and that in the Mohave Desert, a smaller population peak is observed in spring. These findings and those of Wood (1971) and Wallwork (1972*a*) strongly suggest that microarthropods are much less important in litter breakdown in deserts than are other arthropod groups.

Noy-Meir (1974) has stated that microbial decomposition of litter and wood at or near the soil surface must be limited to the short periods when this layer is moist (after rain or dew), and O'Brien (in Whitford, 1974) has shown that active microbial decomposition occurs only when soils are moist and temperatures moderate.

Since microbial decomposition is limited by moisture near the soil surface and microarthropods and the other litter consumers appear to be considerably affected by moisture, termites, which are largely independent of soil moisture, appear to be the most important litter consumers in hot desert ecosystems. Johnson & Whitford (1975) showed that subterranean termites consumed 7.9 ± 2.8 and 1.2 ± 0.4 kg ha^{-1} during July and August in two Chihuahuan Desert communities. They estimated the input of detritus in one system at 10.3 ± 10^6 cal ha^{-1} and termite consumption at 3.4 ± 10^6 cal ha^{-1}. This one group of detritivores accounted for more than 50% of the net primary production of these Chihuahuan Desert ecosystems. These estimates of energy flow through a consumer group that appears to be relatively independent of rainfall (as long as litter is buried below the soil surface) suggest that termites may be the most important consumers in hot desert ecosystems.

Conclusions

In setting out our original objective to discuss the dynamic behavior of the biotic components of desert ecosystems in relation to short-term water and energy flow, we had hoped some underlying principles about the short-term dynamics of deserts would emerge. We knew that we would have to rely heavily on our own data which limits the scope. Even when other data were kindly provided for study, that deep intuition associated with a thorough knowledge of your own desert site was missing. It will need the data and intuitions of many more desert ecologists before a real integration of the principles of short-term dynamics in deserts can emerge.

In spite of the above dissatisfaction, our data and discussion does quantify and test, in a sense, some hypotheses put forth earlier by desert ecologists. Our conclusions regarding some of these hypotheses are as follows.

(1) The quantity of available water is not only affected by precipitation inputs, but by the soil characteristics and the position of an ecosystem in the landscape. Our data on water storage shows that, of the ecosystems considered, the water courses on the alluvial fans have the greatest total and temporal storage, followed by swales. The lowest storage was found for the alluvial fans themselves (see Figs. 11.1–11.5).

(2) The season of precipitation inputs, hence water availability, greatly affects the growth responses of the different species occupying the various ecosystems. Winter or late summer rainfall will not trigger as large a pulse of production as spring or early summer rainfall, when temperature conditions are nearer to optimum. Also, the reproductive response of some species is related to this seasonality in complex ways.

(3) The sequencing of favorable and unfavorable moisture conditions also strongly influences productivity of certain desert plants. For example, our data show that annuals will attain a much higher level of production if a series of favorable growth seasons occur in sequence. The first production peaks may represent the maximum that can be obtained because of limiting seed reserves. If this is followed by a second favorable growth season, the production peaks will be orders of magnitude higher due to a larger, new seed reserve being available (see Bridges *et al.*, 1972; and Noy-Meir, 1973, for a discussion of the pulse–reserve strategy in desert plants).

(4) The response strategy of annual plants (perhaps ephemerals in general whether annual or perennial) appears to be characterized by the saying, 'Hurry before it's too late'. The data for annuals in two sites show relatively high rates of productivity following significant water inputs. Maximum growth and production of annuals occurs within one to two months after initiation of growth.

(5) The response strategy of evergreen, perennial shrubs appears to be

characterized by the saying, 'Slow and steady wins'. The data for two shrubs in two sites shows the interval of maximum production to be about four months. The total (peak) production of new shoot material may be equivalent to, or exceed, the peak of the annuals because of this longer growth period. These values for shrub production do not account for the large amounts of energy going into secondary growth of old stems and roots.

(6) The past disturbance of desert regions by domestic animals makes present comparisons of short-term energy flow difficult. Present 'openness' with respect to shrub cover may in some areas be due to past grazing. Production of annuals appears to be greater in open areas.

(7) The ecological efficiencies of two desert sites characterized by two shrubs and annual grasses and forbs is less than 0.5%. This follows previous estimates of ecological efficiency for deserts (Whittaker, 1975). Since the values reported are minimum estimates, the total (Lindemann) ecological efficiency would probably be about double that given.

(8) The seed consumer community in a desert ecosystem is composed of gramnivorous rodents and birds and harvester ants. The impact of each component of the seed consumer community is dependent on the immediate past climatic history of the system. Variation in rainfall amounts above the minimum physiological requirements of 'resident' vertebrates results in only small changes in biomass and numbers but rainfall amounts below that minimum result in significant reduction in numbers.

Migratory gramnivorous birds exploit seed sources following periods of greater than average precipitation when seed production exceeds the amounts used and stored by resident gramnivores.

(9) Arthropods that feed on live vegetation or on plant honeydew/exudates account for a small portion of the energy flow through the ecosystem and are directly dependent on the water status of the host plants.

(10) In hot desert regions, we suggest that termites are the single most important consumer group, processing more than 50% of the primary productivity. Since termites of a variety of families inhabit the hot desert areas of the world (Krishna & Weesner, 1970) it is probable that this is generally true. However, physiological differences in termites in other deserts may restrict foraging to periods of enhanced soil moisture.

(11) Decomposition by microorganisms is soil moisture dependent and thus considerably variable. Microbial decomposition and mineralization of fecal material in and around subterranean galleries of social insects may be more constant and thus of greater importance.

Acknowledgements

We are greatly indebted to those individuals who provided us with their data on short-term growth dynamics and to those individuals who provided us with descriptive information on their desert regions. Without these contributors the scope of this chapter would have been severely limited. We also want to express our appreciation to all our colleagues who helped collect data on our site and have provided much of the stimulation necessary for an undertaking such as this. Discussions of these data with Gary Cunningham, Jeff Delson, Carol Kay, Fenton Kay, Richard Mishaga, Robert O'Brien, Stuart Pimm and Ralph Raitt have contributed greatly to the development of ideas expressed in this chapter. However, we accept full responsibility for any errors which may be in the text.

The work on this chapter was carried out as part of the US/IBP Desert Biome Programme and was supported by the US National Science Foundation Grant Number GB15886.

References

Anderson, J. M & Coe, M. J. (1974). Decomposition of elephant dung in an arid, tropical environment. *Oecologia*, **14**, 111–25

Beatley, J. C. (1969). Dependence of desert rodents on winter annuals and precipitation. *Ecology*, **50**, 721–4

Bouillon, A. (1970). Termites of the Ethiopian region. In: *Biology of termites*, vol. 2 (ed. K. Krishna & F. M. Weesner), pp. 153–280. Academic Press, New York.

Bradley, W. G. & Mauer, R. A. (1971). Reproduction and food habits of Merriam's kangaroo rat, *Dipodomys merriami. Journal of Mammalogy*, **52**, 479–507.

Bridges, K. W., Willcott, C., Westoby, M., Kickert, R. & Wilken, D. (1972). *Nature: a guide to ecosystem modelling*. Minneapolis: IBP Ecosystems Modelling Symposium American Institute of Biological Sciences Meeting (presented paper, available from authors).

Brown, J. H., Lieberman, G. A. & Dengler, W. G. (1972). Woodrats and cholla: dependence of a small mammal population on the density of cacti. *Ecology*, **53**, 310–13.

Burk, J. H. & Dick-Peddie, W. A. (1973). Comparative production of *Larrea divaricata* Cav. on three geomorphic surfaces in southern New Mexico. *Ecology*, **54**, 1094–102.

Chew, R. M. (1965). The water metabolism of mammals. In: *Physiological mammalogy – mammalian reactions to stressful environments*, vol. 2 (ed. W. V. Mayer & R. A. Van Gelder), pp. 43–178. Academic Press, New York.

Cloudsley-Thompson, J. L. (1964). The insect fauna of the desert near Khartoum: seasonal fluctuation and the effect of grazing. *Proceedings of the Royal Entomological Society, London*, **A39**, 41–6.

Cloudsley-Thompson, J. L. (1968). The Merkhiyat jerbels: a desert community. In: *Desert biology* (ed. G. W. Brown), pp. 1–20. Academic Press, New York.

Crawford, C. S. (1976). Feeding-season production in the desert millipede *Orthoporus ornatus* (Girard) (Diplopoda). *Oecologia*, **24**, 265–76.

Delye, G. (1967). Recherches sur l'ecologie, la physiologie, et l'ethologie des fourmis du Sahara. PhD Thesis, Université d'Aix-Marseille.

Edney, E. V., McBrayer, J. F., Franco, P. J. & Phillips, A. W. (1974). *Distribution of soil arthropods in Rock Valley, Nevada.* US/IBP Desert Biome Research Memorandum RM 74-32. Utah State University, Logan.

Edney, E. B., McBrayer, J. F., Franco, P. J. & Phillips, A. W. (1975). *Abundance and distribution of soil microarthropods in Rock Valley, Nevada.* US/IBP Desert Biome Research Memorandum RM 75-29. Utah State University, Logan.

Ettershank, G. & Whitford, W. G. (1973). Oxygen consumption of two species of *Pogonomyrmex* harvester ants (Hymenoptera: Formicidae). *Comparative Biochemistry and Physiology*, **46**, 605–11.

Evenari, M., Shannan, L. & Tadmor, N. H. (1971). *The Negev: the challenge of a desert.* Harvard University Press, Cambridge, Massachusetts.

Ferrar, P. & Watson, J. A. L. Termites associated with dung in Australia. *Journal of the Australian Entomological Society*, **9**, 100–2.

French, H. R., Maza, B. G., Hill, H. O., Aschwander, A. P. & Kaaz, H. W. (1974). A population study of irradiated desert rodents. *Ecological Monographs*, **44**, 45–72.

Grenot, C. & Vernet, R. (1972). Les reptiles dans l'ecosysteme au Sahara occidental. *Compte Rendu des Séances de la Sociètè de Biogeographie*, **433**, 96–112.

Grenot, C. & Vernet, R. (1973). Ecologie animale – sur une population *d'Uromastix acanthinurus* Bell isolée au milieu de Grand Erg Occidental (Sahara algerien). *Comptes Rendus hebdomadaires des Séances de l'Acadèmiè des Sciences, Serie D*, **276**, 1349–52.

Haverty, M. I. & Nutting, W. L. (1974*a*). Density, dispersion, and composition of desert termite foraging populations and their relationship to superficial dead wood. *Environmental Entomology*, **45**, 480–6.

Haverty, M. I. & Nutting, W. L. (1974*b*). Natural wood consumption rates and survival of a dry wood and a subterranean termite at constant temperatures. *Annals of the Entomological Society of America*, **77**, 153–7.

Haverty, M. I. & Nutting, W. L. (1975*a*). Natural wood preferences of desert termites. *Annals of the Entomological Society of America*, **68**, 533–6.

Haverty, M. I. & Nutting, W. L. (1975*b*). A simulation of wood consumption by the subterranean termites, *Heterotermes aureus* (Snyder), in an Arizona desert grassland. *Insectes Sociaux*, **22**, 93–102.

Haverty, M. I. & Nutting, W. L. (1976). Environmental factors affecting the geographical distribution of two ecologically equivalent termite species in Arizona. *American Midland Naturalist*, **95**, 20–7.

Haverty, M. I., LaFage, J. P. & Nutting, W. L. (1974). Seasonal activity and environmental control of foraging of the subterranean termite, *Heterotermes aureus* (Snyder), in a desert grassland. *Life Sciences*, **15**, 1091–101.

Haverty, M. I., Nutting, W. L. & LaFage, J. P. (1975). Density of colonies and spatial distribution of foraging territories of the desert subterranean termite, *Heterotermes aureus* (Snyder). *Environmental Entomology*, **4**, 105–9

Johnson, K. A & Whitford, W. G. (1975). Foraging ecology and relative importance of subterranean termites in Chihuahuan Desert ecosystems. *Environmental Entomology*, **4**, 66–70.

Kay, C. A & Whitford, W. G. (1975). Influences of temperature and humidity on oxygen consumption of Chihuahuan Desert ants. *Comparative Biochemistry and Physiology*, **52**, 281–6.

Kirmiz, J. P. (1962). *Adaptation to desert environments: a study on the jerboa rat and man.* Butterworth, London.

Krishna, K. & Weesner, F. M. (1970) (eds.). *Biology of termites*, vol. 2. Academic Press, New York.
LaFage, J. P., Nutting, W. L & Haverty, M. I. (1973). Desert subterranean termites: a method of studying foraging behaviour. *Environmental Entomology*, **2**, 954–6.
Lee, K. E. & Wood, T. G. (1971*a*). *Termites and soil*. Academic Press, New York.
Lee, K. E. & Wood, T. G. (1971*b*). Physical and chemical effects of soils of some Australian termites and their pedological significance. *Pedobiologia*, **11**, 376–409.
Ludwig, J. A., Reynolds, J. F. & Whitson, P. D. (1975). Size-biomass relationships of several Chihuahuan Desert shrubs. *American Midland Naturalist*, **94**, 451–61.
MacMillen, R. E. (1964). Population ecology, water relations, and social behaviour of a southern California semi-desert rodent fauna. *University of California Publications in Zoology*, **71**, 1–59.
MacMillen, R. E. & Lee, A. K. (1967). Australian desert mice: independence and exogenous water. *Science*, **518**, 383–5.
McBrayer, J. F., Mamolito, G. E & Franco, P. J. (1975). *The functional relationships among organisms comprising detritus-based food chains at Rock Valley Site*. US/IBP Desert Biome Research Memorandum RM 75-30. Utah State University, Logan.
Milner, C. & Hughes, R. E. (1968). *Methods for the measurement of the primary production of grasslands*. IBP Handbook No. 6. Blackwell, Oxford.
Newbould, P. J. (1967). *Methods for estimating the primary productions of forests*. IBP Handbook No. 2. Blackwell, Oxford.
Noy-Meir, I. (1973). Desert ecosystems: environment and producers. *Annual Review of Ecology and Systematics*, **4**, 25–51.
Noy-Meir, I. (1974). Desert ecosystems: higher trophic levels. *Annual Review of Ecology and Systematics*, **5**, 195–214.
Oechel, W. C., Strain, B. R. & Odening, W. R. (1972). Tissue water potential, photosynthesis, ^{14}C-labelled photosynthate utilization and growth in the desert shrub *Larrea divaricata* Cav. *Ecological Monographs*, **42**, 127–41.
Raitt, R. J. & Pimm, S. L. (1976). Dynamics of bird communities in the Chihuahuan Desert, New Mexico. *Condor*, **78**, 427–42.
Romney, E. M., Hale, V. Q., Wallace, A., Lunt, O. R., Childress, J. D., Haaz, H., Alexander, G. V., Kinnear, J. E. & Ackerman, T. L. (1973). *Some characteristics of soil and perennial vegetation in northern Mojave desert areas of the Nevada Test Site. Los-Angeles: UC-38 Biomedical and Environmental Research TID-4500*. Clearinghouse for Scientific & Technical Information, Springfield, Virginia.
Ross, M. A. & Lendon, C. (1973). Productivity of *Eragrostis eriopoda* in a mulga community. *Tropical Grasslands*, **7**, 111–16.
Schmidt-Nielsen, B. & Schmidt-Nielson, K. (1952). Water metabolism of desert rodents. *Physiological Review*, **32**, 135–66.
Schumacher, A. & Whitford, W. G. (1974). The foraging ecology of two species of Chihuahuan Desert ants: *Formica perpilosa* and *Trachymyrmex smithii neomexicanus* (Hymenoptera: Formicidae). *Insectes Sociaux*, **21**, 317–30.
Shaeta, M. N. & Kaschef, A. H. (1971). Foraging activities of *Messor aegyptiacus* Emery (Hymenoptera: Formicidae). *Insectes Sociaux*, **18**, 215–25.
Slatyer, R. O. (1962). Climate of the Alice Springs Area. *CSIRO Land Research Series No. 6*, pp. 109–28.
Smith, S. D & Ludwig, J. A. (1976). Reproductive and vegetative growth patterns in *Yucca elata* Engelm. (*Liliaceae*). *Southwestern Naturalist*, **21**, 177–84.
Szlep-Fessel, R. (1970). The regulatory mechanism in mass foraging and the recruitment of soldiers in *Pheidole*. *Insectes Sociaux*, **17**, 232–44.

Turner, F. B. (1972). *Rock Valley Validation Site Report*. US/IBP Desert Biome Research Memorandum RM 72-2. Utah State University, Logan.

Turner, F. B. (1974). *Rock Valley Validation Site Report*. US/IBP Desert Biome Research Memorandum RM 74-2. Utah State University, Logan.

Wallace, A. & Romney, E. M. (1972). *Radioecology and ecophysiology of desert plants at the Nevada Test Site*. United States Atomic Energy Commission Monograph, TID-2594.

Wallwork J. A. (1972*a*). Distribution patterns and population dynamics of the microarthropods of a desert soil in southern California. *Journal of Animal Ecology*, **41**, 291–310.

Wallwork, J. A. (1972*b*). Mites and other microarthropods from the Joshua Tree National Monument, California, *Journal of Zoology*, **163**, 91–105.

Wallwork, J. A. (1976). *The distribution and diversity of soil fauna*. Academic Press, New York.

Whitford, W. G. (1971). *Jornada validation site report*. US/IBP Desert Biome Research Memorandum RM 71-4. Utah State University, Logan.

Whitford, W. G. (1972). *Jornada validation site report*. US/IBP Desert Biome Research Memorandum RM 72-4. Utah State University, Logan.

Whitford, W. G. (1973). *Jornada validation site report*. US/IBP Desert Biome Research Memorandum RM 73-4. Utah State University, Logan.

Whitford, W. G. (1974). *Jornada validation site report*. US/IBP Desert Biome Research Memorandum RM 74-4. Utah State University, Logan.

Whitford, W. G. (1976). Temporal fluctuations in density and diversity of desert rodent populations. *Journal of Mammalogy*, **57**, 351-69.

Whitford, W. G. (1978). Structure and seasonal activity of Chihuahuan Desert ant communities. *Insectes Sociaux*, **25**, 79–88.

Whitford, W. G. & Creusere, F. M. (1977). Seasonal and yearly fluctuations in Chihuahuan Desert lizard communities. *Herpetologica*, **33**, 54–65.

Whitford, W. G. & Ettershank, G. (1975). Factors affecting foraging activity in Chihuahuan Desert harvester ants. *Environmental Entomology*, **4**, 689–96.

Whittaker, R. H. (1975). *Communities and ecosystems*, 2nd edn. Macmillan Publishing Co., New York.

Wood, T. G. (1971). The distribution and abundance of *Folsomides deserticola* (Collembola: Isotomidae) and other microarthropods in arid and semi-arid soils in southern Australia. *Pedobiologica*, **11**, 446–68.

Wooten, R. C. Jr & Crawford, C. S. (1974). Respiratory metabolism of the desert millipede *Orthoporus ornatus* (Girard) (Diplopoda). *Oecologia*, **17**, 179–86.

Wooten, R. C. Jr & Crawford, C. S. (1975). Food, ingestion rates and assimilation in the desert millipede *Orthoporus ornatus* (Girard) (Diplopoda). *Oecologia*, **20**, 231–6.

Manuscript received by the editors June 1977

12. Nutrient cycling in desert ecosystems

N. E. WEST

Some general features of nutrient cycling in arid ecosystems

Nutrient or mineral cycling is an essential process in the production of organic matter (Likens & Bormann, 1972). All ecosystems tend to conserve and concentrate essential elements (Odum, 1968), since mechanisms have evolved that promote recycling within ecosystems. Succession can be viewed as a process through which the biota accumulate enough nutrients to allow the rise of succeeding populations. Climax or mature communities perpetuate their stability, in part, by conserving the essential nutrients. Although energy flow is not the highest, nutrient utilization is highest and most efficient in climax communities (Woodwell & Whittaker, 1967).

Energy flow is useful in describing many ecosystem processes; however, radiant energy is not known to be a limiting factor to desert ecosystem productivity. The functioning of arid ecosystems is limited more by the availability of water and mineral nutrients. The predominant effect of the aridity factor on desert life has attracted most ecologists' attention to water stress phenomena, and thus the more normative mineral nutrient cycling processes in desert ecosystems have received relatively little study. The effects of the accumulation of excess elements through salinization of these habitats has, however, been much researched and consequently reached a high level of understanding, particularly where agronomic alterations are made. Most of such accumulations of salts occur in the alluvial soils of arid regions. Because of their present or potential arability, these areas have attracted disproportionate attention compared to their relative extent. Aeolian and skeletal soils are much more common in arid regions.

Influence of precipitation variability

The flush of plant growth commonly observed when soil moisture is favorable in deserts has given the impression that soils of arid areas are unusually fertile. This impression is strengthened by the reasonable hypothesis that in arid climates leaching losses are negligible and nutrients may accumulate from weathering of plant materials. It is only in recent years that ecologists have begun to appreciate that minimal amounts of the vital elements could be secondary controlling factors of desert productivity. This realization came about as more detailed analyses of desert productivity were performed. For

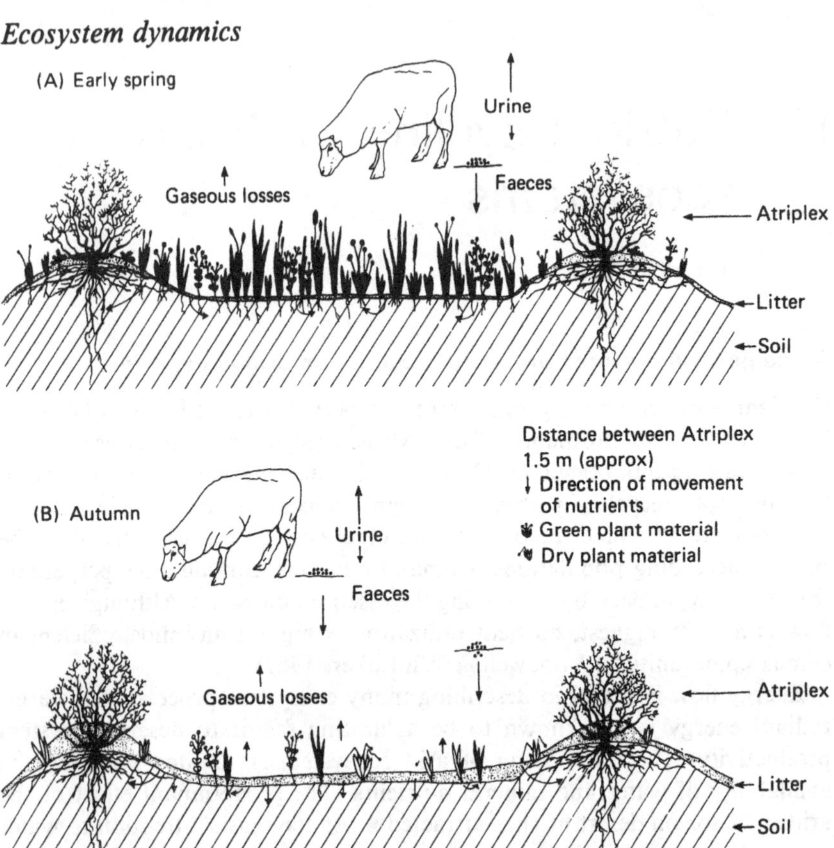

Fig. 12.1. Organic matter and nutrient flow in an *Atriplex vesicaria*-dominated ecosystem of the Riverine Plain, western New South Wales, Australia in early spring and autumn (Rixon, 1970).

example, Trumble & Woodroffe (1954) observed that plant, and consequently sheep, productivity was not as great as expected in a wetter than average year which immediately followed another above average year in South Australia. They attributed this phenomenon to a deficiency of available mineral nutrients. Charley & Cowling (1968) observed similar phenomena in New South Wales. Cowling (1969) experimentally examined this type of phenomenon and concluded that nitrogen and phosphorus could indeed be limiting to the growth of *Atriplex vesicaria*. Cline & Rickard (1973) found nitrogen to be limiting when better than average soil moisture was present on rangelands in arid southeastern Washington.

Decomposition and mineralization processes can be rapid in desert soils during wet periods (Cowling, 1969; Garcia-Moya & McKell, 1970). The overall response is a generally higher net yearly rate of detritiviore and microbial activity in desert ecosystems yielding faster litter breakdown and mineralization than was previously assumed for arid regions (Charley, 1972).

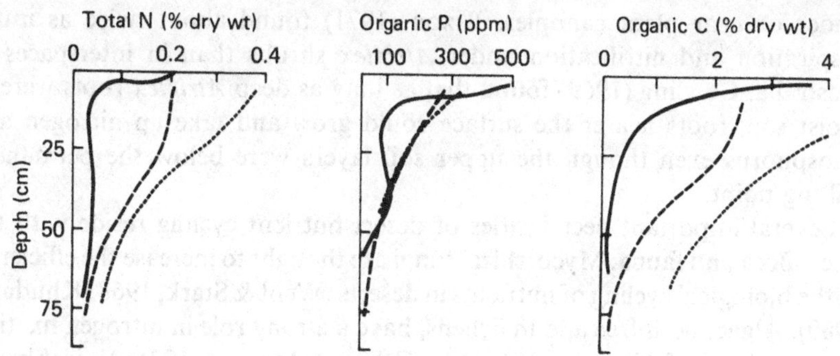

Fig. 12.2. Vertical distribution of total nitrogen, organic phosphorus and organic carbon in arid, sub-humid, and humid regions. Data from: Fuller & McGeorge (1951); Nicholls & Turner (1956); Pearson & Simonson (1939); Walker & Adams (1958). —, Arid; --, sub-humid; --, humid.

Structural variability

Another distinctive feature of nutrient cycling processes in deserts is the way in which loci of biological activity and importance are concentrated. Both vertical and horizontal structural heterogeneity are stronger than in most other kinds of ecosystems. Numerous recent papers have shown the 'islands of fertility' that exist around desert trees and shrubs (Garcia-Moya & McKell, 1970; Bowns & West, 1976; Cline, 1973; Wells, 1967; Tiedemann & Klemmedson, 1973*a*, *b*; Rickard, Cline & Gilbert, 1973; Charley & West, 1975). This horizontal structure is illustrated in Fig. 12.1, a depiction of organic matter and thus nutrient concentration in an *Atriplex vesicaria* community on the Riverine Plain of New South Wales, Australia (Rixon, 1970). Stark (1973) has demonstrated how water infiltrates eight to ten times faster under shrubs than in the open on desert soils high in clay. This pattern of infiltration influences rates of decomposition and downward movement of nutrients. Single shrubs and surrounding bare interspaces constitute a 'cell' of a desert ecosystem (Charley, 1972).

Micro-relief can also be highly important in patterning nutrient distribution (Charley & McGarity, 1964). Vertical gradients are especially strong with the more volatile, non-ash elements found in surface crusts. Ash elements, especially the salts, generally accumulate with increasing depth forming petrocalcic (caliche) or natric (saline–sodic) horizons or pans common under desert landscapes. Generalizations on vertical distribution of nitrogen, phosphorus and organic carbon in arid soils compared with more mesic soils are shown in Fig. 12.2.

As the vital elements are highly localized, so one would expect, and finds, that rates of biological activity are similarly structured. For instance, Skujins (1973) found that processes such as respiration, nitrification, proteolysis and dehydrogenase activity are largely concentrated in the surface 3 cm layer under or near plant canopies. Rixon (1971) found about twice as much respiration and nitrification under *Atriplex* shrubs than in interspaces in Australia. Cowling (1969) found that as long as deep *Atriplex* roots were in moist soil, roots nearer the surface could grow and take up nitrogen and phosphorus even though the upper soil layers were below the permanent wilting point.

Several important peculiarities of desert nutrient cycling reside with the microflora and fauna. Mycorrhizal fungi are thought to increase the efficiency of the biological cycling of nutrients in deserts (Went & Stark, 1968; Khudairi, 1969). Algae, both free and in lichens, have a strong role in nitrogen fixation (Rogers, Lange & Nicholas, 1966; MacGregor & Johnson, 1971; Novichkova-Ivanova, 1972; Rogers & Lange, 1972; Snyder & Wullstein, 1973; West & Skujins, 1977). Cryptogamic crusts also stabilize desert soils against erosion (Fletcher & Martin, 1948).

The probable importance of below-ground dwelling invertebrates in nutrient cycling is probably greater than now generally realized (Wood, 1970). The little work that has been done indicates that termites probably redistribute large amounts of essential elements (Dregne, 1968; Watson & Gay, 1970). They have also been found to be nitrogen fixers (Benemann, 1973). Ants are suspected to be of similar, if not greater importance in nutrient redistribution (Went, Wheeler & Wheeler, 1972; Burrows, 1972). The observation of calcium concentrations in krotovinas (burrows), especially those of cicadas (Hugie & Passey, 1963), leads us to believe their importance may well have been overlooked. Burrows (1972) also mentions the probable importance of a ground cockroach in litter comminution on an arid shrub site in southwestern Queensland.

Although relatively little has been done on the role of large desert herbivores on nutrient cycles, Odum (1971) postulates that since microbial decomposers will be sharply limited by dryness, the seemingly large population of rodents probably plays an important part in nutrient redistribution. The work of Rotschil'd (1968) seems to bear this out for nitrogen.

Temporal variability

The cycling of minerals in desert ecosystems occurs as a function of time. Although the annual cycle has been most intensively studied, daily and seasonal cycles and successional changes are probably important. What little extant data there are on other than annual bases will be covered following a consideration of the major pathways of recycling.

Recycling in ecosystems can take place through four pathways (Odum, 1971); microbial decomposition of detritus, direct (internal) transfer (such as between mycorrhizae and higher plants), animal excretion, and autolysis (direct, non-biological chemical release). These processes may be further categorized as biological or non-biological.

The biological ('closed' or internal) cycle involves the more or less tight cyclic circulation of nutrients between the organic and inorganic state in the soil and biotic community. This cycle includes the phenomena of uptake, retention (or growth increment) and restitution (or losses). Retention is measured by estimates of annual biomass increment and the chemical content of tissues comprising that net production. Restitution losses are similarly evaluated from data on weights and chemical content of plant litter and other detritus.

Uptake should not be interpreted as physiological demand for function since 'luxury uptake' is often mixed with 'limited uptake', as determined by the time course of availability for each separate element. The retained elements accumulate in the biomass and when measured at any given time constitute the mineral standing crop or 'mineralomass' (Duvigneaud & Denaeyer-De Smet, 1970). This total content of different mineral elements in a desert ecosystem varies with site, species composition, season, and successional stage.

The actual flux rates are very difficult to measure in deserts. The hydrologic-based input–output balance widely worked out for forests (Likens & Bormann, 1972) is not as feasible in drier areas that are without permanent streams. Mineral dynamics of deserts are usually inferred from standing crop biomass, growth rate (productivity) and litter turnover data, and presented as annual budgets.

Using these productivity data, either of two strategies can be used to estimate mineral cycling rates in ecosystems. First, if community equilibrium (i.e., climax or mature status) is assumed, then annual increases in elemental biomass should be zero and uptake should equal restitution; or input equals output (Witkamp, 1971). New growth (measured before significant litter fall and nutrient translocation) should also approximate uptake.

These techniques work fairly well in deserts with a distinct season of growth and utilizationa of vegetation. Under such circumstances, standing crop values and new growth: old growth ratios might estimate increment uptake and turnover rates fairly closely (Bjerregaard, 1971). Theoretically, if an ecosystem is in equilibrium, one year's data represents one year's cycle. In practice, variable rainfall influences the system's response. Bjerregaard (1971) used an average of three years data to smooth out fluctuations due to the climatic variations. In other deserts where precipitation, and therefore production, litter fall, and mineral throughput is more erratic, this approach has little merit.

Rodin & Basilevich's (1965) data give little evidence of equilibria existing. Annual increments, apparently in woody components, are indicated from all their desert data. No data on mortality are given. Perhaps the small increments are balanced by mortality and creation of standing dead material. Another possibility is that their samples are not from climax stands. Even if these are not the problems, Rodin & Basilevich (1965) state 'In reality, however, the mineral elements are retained by plant communities do not equal the difference between uptake and return, owing to the existence of lesser cycles within the annual cycles. These lesser cycles include those associated with the leaching of mineral elements by atmospheric precipitation, and their translocation from the leaves to the branches, roots, etc. Compilation of the complete annual balance of the biological cycle of matter is therefore provisional at the present stage'. There are probably no true nutrient cycles, except on a global basis.

An alternative technique to estimating nutrient budgets involves the use of growth rings and dimensional analysis of the vegetation. This has had great vogue in forest ecology, but can be also applied to shrubs (Whittaker, 1961). Either annual growth increments and their mineral contents are measured or mineral content of various aged stands are sampled with adjustments for litter lost through time. Essentially, biomass is divided by age to get average annual increment. However, in deserts we cannot rely as well on growth ring information due to anomalous growth resulting in many false or missing rings and stem-splitting. It is also difficult to estimate the volume of the wood cylinders between two xylem layers on irregularly shaped species. No techniques exist for estimating age or growth rates of many herbaceous or succulent species common to deserts. The importance of soil cryptogamic crusts in nutrient cycling processes of deserts also complicates the picture. Therefore, it is not surprising that relatively few, even inferential studies of nutrient dynamics in deserts exist.

Radio-tracer technology promises, however, some improvements in direct studies of flux rates (Olson, 1968; Wallace & Romney, 1972) and we look forward to progress in defining the biological side of nutrient cycling through this means in coming decades. Computer-assisted system analysis and simulation are readily available if more data on short term dynamics become available.

The geochemical, external or 'open' cycle involves the input and output from the system through physical processes. This includes such things as the transfers due to precipitation, leaching, erosion, harvests, amendments and weathering of parent rock. I agree with Bormann (1969) that these aspects have been greatly overlooked by ecologists.

Physical processes seem to have larger relative roles in nutrient input and outputs in deserts compared to other ecosystems. For example, West & Skujins (1977) found that approximately one-fifth of the biologically active nitrogen enters a cool desert system as wet precipitation and dry fallout. As Drover & Barrett-Lennard (1953) earlier found the greater part of this input is organic nitrogen associated with dust. If the net import of dust is negligible, then this source could be less important. In any case, the amounts involved are comparatively larger than more mesic systems (Bobritskaya, 1962).

Pauli (1964) points to the direct role of the sun in physio-chemical degradation of humus. Physical processes also dominate chemical ones in mineral weathering. When primary minerals are weathered kaolinitic clays result. These clays are the least retentive of nutrients. Combined with low humic content, this makes the mineral exchange complex the least effective of any terrestrial ecosystem.

Gifford & Busby (1973) and Porcella *et al.* (1973) have pointed out the importance of overland flow during high intensity rainstorms in redistributing organic matter and its attendant elements. Erosional truncation of soil profiles with loss of the nutrient-rich surfaces can almost permanently lower the productivity of desert ecosystems (Charley & Cowling, 1968). Other catastrophic events such as wind and hail storms, fires, or insect defoliation may cause sudden changes in nutrient contents of certain compartments. Where small total elemental standing crops of an element exist in deserts, the effect may be more dramatic and of longer duration than it would for more mesic ecosystems with larger components.

Accumulation of nutrients

Parallel and partially dependent on the mineral cycles is the accumulation of nutrients in the soil despite the demands for plant growth. Accumulation takes place in or near the rooting zone and may lead to the formation of physiologically rich or indurated horizons. This is so because of the 'pumping' action created by plant growth. All tissues of desert plants, even those growing in non-salinized soil, have notably higher content of nitrogen and the ash elements than do the plants of more mesic environments (Rodin & Basilevich, 1965).

For instance, Rodin & Basilevich (1965) list the following vertical stratification from the surface of elemental accumulation for wormwood associations or gray-brown desert soils in central Asian deserts (Ustyurt): N > Ca > Na, K > Cl, Mg. Where more annuals exist in light sierozem soils, silicon replaces calcium. In the reddish soils in the warmer deserts of Syria the sequence is: Ca > Si > N > Al, Fe > K, Mg. Since leaching is seldom sufficient in desert climates to remove the more abundant ions such as Na^{+}, Ca^{2+} and Cl^{-}, we see the phenomena of salinization, alkalization, and calcification, particularly when supplemental moisture is supplied or water tables raised in irrigation (Basilevich *et al.* 1972). Evaporation of water carries many elements to the soil surface. Accumulations of lime ($CaCO_3$) gypsum ($CaSO_4$), and silicates commonly form the matrix of indurated (petrocalcic horizons which restrict water penetration and root distribution. These phenomena along with vertical mixing (gilgai) are more developed the older the land form. Australian and African deserts exhibit the extremes. Whereas more mesic, temperate soils accumulate increasing humus as succession advances, there seem to be no such trends in desert soils. Either organic matter content is so low that trends are undetectable or successional changes have negligible effects on desert soils.

Intra-seasonal variation

Although some data on seasonal variation of nutritional and elemental content of desert plants exists (Cook, Stoddart & Harris, 1954; Chatterton *et al.*, 1971; Wallace & Romney, 1972; Caldwell, West & Goodman, 1971; West, 1972; Sharma, Tunny & Tongway, 1972; Moore, Breckle & Caldwell, 1973), we can only infer at this time how fluxes, such as plant uptake, vary. Even before they die, senescing organs have lost minerals through translocation, foliage drip and stem flow (Stenlid, 1958). Therefore, flushes of growth and litter fall undoubtedly mean considerable redistribution. For instance, J. C. Turner (in Wilcox, 1974) and Beale (1971) found that there was considerable withdrawal of nitrogen and phosphorus before phyllode shedding of *Acacia aneura* in Australia. Mack (1971) found a steady loss of phosphorus and potassium in leaves of *Artemisia tridentata* in varying degrees of senescence. Dina & Klikoff (1973) found that this species also has declining leaf nitrogen as water stress develops and the nitrogen is transferred to twigs. Charley (1959) and Cowling (1969) observed similar translocations for leaves and fruits of *Atriplex vesicaria* in Australia. Such redistribution of critical elements may be a conservation mechanism allowing the plants to re-use these minerals and overcome the need for additional uptake when the soil is drier.

Excess elements, on the other hand, are often concentrated in plant parts to be shed. For instance, the vesicular hairs on *Atriplex* concentrate sodium which is either lost by surface leaching or shedding (Mozafar & Goodwin, 1970). Perhaps the salt pumping phenomena such as described by Roberts (1950), Rickard (1965*a*, *b*), Rickard & Keough (1968), Sharma & Tongway (1973), Sharma (1973), Denaeyer-De Smet (1970), Wallace, Romney & Hale (1973) and Basilevich *et al.* (1972) will be found to be prominent only during a part of the year.

Skujins & West (1973) have found that simulated rainfall can leach elements from cool desert shrubs. Much of this increment is as dust rather than direct leaching. Romney *et al.* (1963) have abundantly illustrated the ability of desert shrubs to catch all sorts of particulate fallout. Wallace & Romney's (1972) work also indicates that dust rather than direct leaching from clean plants is implicated in any nutrient recharge to the soil. On the other hand, Charley (1959) found significant amounts of salts could be removed from the leaves of *Atriplex vesicaria* bushes grown in dust-free conditions in the glasshouse. The salinity of the rooting medium had a pronounced influence on the amount of salt which could be removed by leaching. Whether plant leaching phenomena have much of a net effect over a large desert area is not known, but doubtful.

Of more practical significance are the redistributions of nutrients caused by livestock grazing and human activity and their effects on plant community structure and composition. The grazing animal, by removing plant parts containing nutrients, decreases vegetation retention times. Erosion may prevail with sparse plant cover and soil micro-relief may be altered in turn to end up with redistributed nutrients and lowered productivity (Charley & Cowling, 1968; Rixon, 1970) or enhanced (Trumble & Woodroofe, 1954), depending on the circumstances.

Another short-term trend is the increase in total mineral elements in relation to dry weight which occur as litter ages. Rodin & Basilevich (1965) have noted increased calcium, silicon, magnesium, aluminium and sometimes, sulfur in Russian shrub litter. An impoverishment of potassium was also noted with increasing mineralization. Caldwell *et al.* (1971) and West (1972) have noted similar trends in standing dead *Artemisia* material in the western USA. Rixon (1970) found increases in the nitrogen content of the fecal pellets of sheep as they aged.

Distribution of minerals between ecosystem components

Soviet workers (Rodin & Basilevich, 1965; Rodin, Basilevich & Miroshnichenko, 1972; Chepurko, Basilevich, Rodin & Miroshnechenko, 1972; Basilevich, Rodin & Gorina, 1972; Tatlyanova, 1972; Rodin, 1977) have done much to summarize the available data on elemental content and standing crops in desert vegetation of Eurasia, the Near East and North Africa. Wallace & Romney (1972), Caldwell *et al.* (1971), West (1972), Moore, *et al.* (1973), Wiebe & Walter (1972) and Klemmedson & Smith (1972, 1973) have added data for North American deserts, Sharma (1968) for Indian trees, Ayyad & El-Ghareeb (1972) for Egyptian desert plants, Al-Ani, Habib, Abdulaziz & Ouda (1971) for native desert plants in Iraq, and Cowling (1969), Beadle, Whalley & Gilson (1957), Keay & Bettenay (1969), Lange (1967), Burrows (1972), and Siebert, Newman & Nelson (1968) for Australian shrubs. From their mineralomass data, estimates of turnover rates can be made, if stability is assumed.

Less is known, however, about the role of groups of organisms other than plants and how their nutrient cycling functions compare with more mesic ecosystems. The total amount of biomass of terrestrial heterotrophic organisms is less than 0.1% of the earth's total living matter (Rodin & Basilevich, 1968). The bulk of minerals in living matter are therefore found in plants. The grazing animal returns as feces and urine most of the nutrient elements eaten in herbage (Cowling, 1977) and usually only small proportions of plant nutrients are immobilized in animal tissues and exported from the system. Animals, however, exert nutrient cycling control functions out of proportion to their low biomass and energy flow (Noy-Meir, 1974).

Periodic population explosions of such species as hares and locusts may seriously deplete the producer compartment. Whether this depletion acts to the extent of nutritionally regulating animal numbers, such as Schultz (1969) has proposed for the tundra, is unknown (Jordan & Kline, 1972). Desert rodents and lagomorphs can pick out the most nutritious forage (Westoby, 1973) and exise or void undesirable material (Kenagy, 1972).

Trumble & Woodroffe (1954) observed increased plant and sheep productivity on moderately to heavily grazed compared to ungrazed *Maireana* (previously *Kochia*) and *Atriplex* ranges in northwestern South Australia. They concluded that sheep grazing increases the content of nitrogen in the new shoots produced and fecal deposits enhance nutrient status. The browsed plants were also thought to be more efficent users of water and other nutrients. It is believed that grazing animals generally shorten and accelerate nutrient cycling, provided soil moisture is favorable (Cowling, 1977). Animals increase both the amount of plant material returned to the soil and its decomposition rate. The rate of turnover of nitrogen in particular is accelerated because some 50–75% of the amount ingested by sheep on arid ranges is returned as urine. A further 20% is returned in dung, and undergoes more rapid decomposition than non-ingested plant material (Cowling, 1977).

Although desert microflora are small in mass, they respond quickly to accumulations of soil moisture. Charley & Cowling (1968) have found that mineralization of nitrogen occurs after showers insufficient to cause response by higher plants. Their study area has light, erratic rainfall, low above-ground biomass but rapid cycling of carbon, nitrogen and phosphorus.

Comparison of turnover rates with other ecosystems

Although Charley (1972) maintains that the proper approach to nutrient cycling should be through studies of soil metabolism, few such data exist for desert ecosystems. Most studies of this type involve disturbed soil samples which exhibit activity rates two to three times that of *in situ* samples.

The compilation of Rodin & Basilevich (1965) currently offers the best comparisons with other ecosystems through plant nutrient accumulation and return data. Regrettably, soil pools and flux rates have not yet been similarly compiled (Charley, 1972). Cycling times for elements in ecosystems are influenced by rates of element uptake, release by plants, and comminution and decomposition of litter. The general concept in the literature seems to be of a slow fertility cycle under arid conditions because of low biomass production and inference of slow litter decomposition (Rodin & Basilevich, 1965). High calcium content of soils is also thought to slow cycling rates (Jordan & Kline, 1972). The estimates of Holmgren & Brewster (1972) lend credence to these views. They calculate average turnover times of 14 years and six years for above-ground and below-ground plant detritus, respectively, in a salt-desert shrub ecosystem in western Utah, USA. Bjerregaard (1971) gives an overall estimate of 10 years for the turnover time in similar ecosystems.

Recent work (Cowling, 1969; Garcia-Moya & McKell, 1971; Burrows,

1972) indicates that, based on the percentage of total above-ground levels, the annual turnover rate of nitrogen and phosphorus in desert shrub communities can be greater than some more humid areas. Cowling (1969) found the respectively 46, 36 and 65% of the above-ground plant organic carbon, nitrogen and phosphorus is annually returned to the soil as *Atriplex vesicaria* litter. The proportion of annuals in the community accelerates the turnover rate.

Cowling (1969) believes that three to four year total turnover times for deserts are due to higher nitrogen contents and pH of litter promoting microbial activity. Comanor & Prusso's (1973) data allow calculation of a two-year turnover time for litter of *Artemisia tridentata* in northwestern Nevada. Burrows (1972) computes a 20 month decomposition time for *Eremophila gilesii* litter in southwestern Queensland.

Woody plants are believed to rely to a great extent upon the recycling of litter nutrients on soils of low nutrient status (Carlisle, Brown & White, 1966). The inference of rapidity of decomposition is further confirmed by the absence of litter mats in most desert contexts. This may be confounded, however, by the redistribution of litter by wind and overland flow of water.

The above rates compare to grassland where Dahlman and Kucera (1965) found a 4.1 yr mean residence time for litter in a tall-grass prairie, and deciduous forest where Cromack (1973) found a 2.5 yr turnover time for everything in the litter compartment except the phenols and stable carbon products.

Although there are exceptions (Noy-Meir, 1973), usually more than 50% of a desert community's biomass is in its root system. Root:shoot ratios of desert species between 1 and 20 have been reported (Noy-Meir, 1973). Although no one has measured it, the bulk of mineral accumulation and litter production probably occurs below ground. Defining litter to include below-ground debris explains the high Soviet values, compared to literature from elsewhere. Absolute values of mineral standing crop in the USSR vary widely from 6–25 kg ha^{-1} on salt flats to 600 kg ha^{-1} on herbaceous *Artemisia* (wormwood) with annuals in sierozem soil. Because of the proportion of biomass return (45–60%) as litter, the annual nutrient return in wormwood desert litter is comparable to that of coniferous forests and that of wormwood deserts with annuals can exceed even broad-leaved forest and approximate grasslands, in this respect.

A characteristic feature of the litter fall in desert communities is its high nitrogen content, averaging 1.5%, compared to 1.2% for grasslands and only 0.6% for forests (Rodin & Basilevich, 1965).

Ash elements in desert litter fall varies from 3.5% dry wt of glycophytes to over 10% dry wt of halophytes (Rodin & Basilevich, 1965). Most of the ash content of glycophytes is made up of calcium, potassium, phosphorus and organogenic sulfur. The share of saline elements (sodium, chlorine and excess sulfur) in the ash of halophytes and succulents goes up to 80%. Silicon is higher where annual plants are more abundant (Rodin & Basilevich, 1965).

Whereas most ecosystems lose the bulk of their nutrients through leaching, desert systems lose minerals largely through surface erosion or animal export. Gaseous losses of nitrogen and other non-ash elements appear high. The overall cycling of nutrients in deserts seems generally more open or loose than in more mesic systems where a higher frequency of biological conservation mechanisms operate at higher densities. Weathering of rock is generally slight because of lack of moisture for chemical and biological activity. Large buffering compartments are lacking, increasing the likelihood of large pulses of nutrient loss. Presence of large compartments in an ecosystem may generally diminish the effects of large variations in the physical environment or of irregularities in the cycles and flows.

Because of the lower levels of microbial activities under arid conditions, much mineralization can take place by autolytic or direct chemical means. Pauli (1964) has already pointed out the major photolytic alteration of humus found in desert environments. Volatilization of ammonia and rates of biological denitrification are higher in desert soils (MacGregor, 1972). Chemo-denitrification is also common in deserts, being catalyzed by the abundance alkaline metal ions (Wullstein, 1969).

Mineral recycling rates in deserts are high relative to input and output rates. They, like forests, have low relative stability, that is, there is a long recovery time following perturbation (Jordan, Kline & Sasscer, 1972).

Special problems of specific minerals

Vital elements may be categorized in several ways. A common distinction is made on how they behave for chemical analysis. The volatilizable elements (N, H, O, C, S) are lost when plants or soils are ashed. The ash elements are those that remain. A more meaningful classification is commonly based on the quantity necessary for plant growth. These are commonly grouped, as follows:

(*a*) macronutrients (C, H, O, N, K, Ca, Mg, P, S);
(*b*) micronutrients (Mo, Cu, Zn, Mn, Fe, B, Na, Cl);
(*c*) non-essential elements (Al, Ba, Sr, Rb, Pb, and others).

In the following discussion we have space for only some of the more specialized problems peculiar to desert ecosystems. We also lack information on how some elements may function in a desert context. We have organized our discussion into two ecologically meaningful categories: deficient and excessive elements.

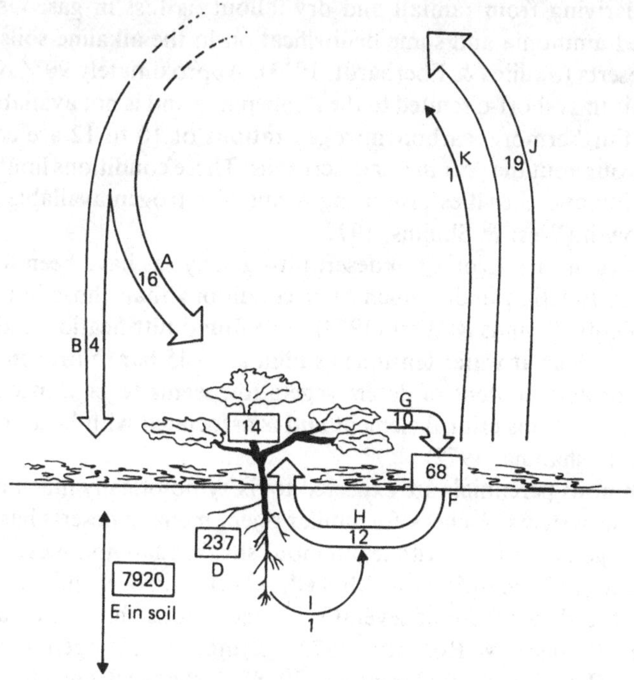

Fig. 12.3. Summary of the main pathways and annual budget of the nitrogen cycle in cool desert ecosystems in the Western USA, under assumed equilibrium conditions and no net erosion. Derived from West & Skujins (1977). Arrows represent annual fluxes and boxes represent standing crops, with the amounts of nitrogen expressed in kg ha^{-1}. A, biological fixation; B, wet and dry (dust) precipitation; C, above-ground phytomass; D, below-ground biomass, to 90 cm depth; E, soil, profile depth to 90 cm (limit of moisture recharge and rooting); F, litter; G, above-ground litter fall; H, plant uptake; I, in below-ground litter fall; J, denitrification; K, volatilization.

Deficient elements

Nitrogen

Dregne (1968) and Charley (1972) have surveyed the situation previously and conclude that nitrogen is probably the most deficient vital element in the world's arid regions. Knowing this, the US/IBP Desert Biome has invested considerable effort into understanding nitrogen pathways, pools, and flux rates. An annual budget summarizing our present collective knowledge on the nitrogen cycle in North American cool desert ecosystems is shown in Fig. 12.3 (West, 1975; West & Skujins, 1978).

Our cool desert ecosystems have markedly high rates of nitrogen fixation, primarily due to the surface-incrusting blue-green algae and lichens (Rychert & Skujins, 1973). However, much of the biologically-fixed nitrogen, as well as that deriving from rainfall and dry-fallout, is lost in gaseous forms as volatilized ammonia and some denitrification in the alkaline soils prevalent in our deserts (Skujins & Eberhardt, 1973). Approximately 90% of the fixed nitrogen is thus short-circuited to the atmosphere and is not available to plant growth. Furthermore, carbon:nitrogen rations of 10 to 12 are consistently found throughout the year in our desert soils. These conditions limit microbial assimilation and activities producing forms of nitrogen available for higher plant growth (West & Skujins, 1977).

Microorganisms involved in desert nitrogen cycles have been found to be adapted to function under much drier conditions than those in more mesic environments. Skujins & West (1973) have found nitrification and ammonification to occur at water tensions as high as −45 bar. Nitrogen storage in the high protein content of desert vegetation seems to be a mechanism for conservation of this critical element and is associated with heat and drought tolerance (Oshanina, 1972).

Leguminous perennials are expected to fix symbiotically much nitrogen in more mesic systems. A check for similar phenomena in deserts has shown no more nitrogen associated with leguminous shrubs than non-leguminous ones (Wells, 1967; Garcia-Moya & McKell, 1970). Nitrogen fixation has been found in the rhizosphere of several non-leguminous shrub genera of North America (Wallace & Romney, 1972). Symbiotic nitrogen fixation is a possibility (Farnsworth & Hammond, 1968). Perhaps nitrogen fixing annuals are a more important source of additional nitrogen than perennials. Beadle (1959) estimated that a good cover of *Swainsona sweinsonioides* (purple pea) may add up to 280 kg of nitrogen ha^{-1} yr^{-1} to desert soils in western New South Wales, however, such large flushes of growth are rare.

Although heterotrophic fixation is rare in deserts because of high carbon requirements the requisite organisms, rhizospheric associations of *Azobacter* and *Clostridium* are important loci of fixation in some desert contexts (Mahmoud, Abou El-Fadl & Elmofty, 1964; Wallace & Romney, 1972).

Repeated drying and wetting of the soil accentuates the Birch Effect to yield higher mineralization rates (Charley, 1972). Nitrogen is mineralized even after light showers insufficient to start higher plant growth. When adequate rain occurs, a pool of available nitrogen is present to allow rapid plant growth. Nearly all of the nitrogen returned in animal feces and urine is lost in gaseous forms (Rixon, 1970). This is the reason why animal activity has been ignored in Fig. 12.3.

Phosphorus

Phosphorus has also been singled out as an element limiting plant productivity in deserts (Pauli, 1964). Many examples of insufficient phosphorus exist from the Australian continent (Beadle, 1962). More recently, Jurinak & Griffin

(1972, 1973) have noted deficiencies of available phosphorus, as have Wallace & Romney (1972), for maximum growth of North American desert shrubs and grasses.

Phosphorus has a further importance in that the possibility exists that its concentration may influence the degree of nitrogen accumulation through limitation of biological fixation mechanisms (Walker, 1962; Jackson, 1957) and that deficiency can retard transformation of nitrogen from the organic to the inorganic pool (Cowling, 1969).

In the phosphorus cycle there is an unidirectional movement of the element out of the system (Williams, 1964) suggesting this cycle is less perfect than the nitrogen cycle, where the irrecoverable losses are balanced by atmospheric and biological fixation.

Excessive elements

Sodium

Primarily because of low precipitation, many cations and anions normally leached from soil profiles in more mesic environments, tend to accumulate in soils of arid regions. Sodium is foremost among those that create problems by their excess and where it is predominant 'sodic' soils form.

Sodium has been found to be essential, in the parts per billion range, for certain desert halophytes such as *Atriplex* (Brownell, 1968; Epstein, 1969; Waisel, 1972). Natural concentrations, even in the most mesic environments, would meet these requirements. Numerous examples of selective uptake and tolerance of sodium have been noted (Wallace, Romney & Hale, 1973; Wallace, Mueller & Romney, 1973).

Chlorine

Chloride build-up is the main causal agent for the development of saline soils. If sodium is also present then saline–sodic soils create a doubly difficult environment for non-tolerant species. A whole spectrum of salt tolerant plants and animals have evolved. These adaptations are believed not to be out of need for high salt environments but usually due to exclusion of more successful competitors prevailing in non-halomorphic soils (Barbour, 1970).

Some halophytes may actually have their growth stimulated by high chloride concentrations (Greenway, 1968). Halophytes commonly 'pump' NaCl to the plant surface (Fireman & Hayward, 1952) where it is variously excreted by specialized glands or in excisable tissue. Both active uptake and transpirational gradients are involved. The canopies under such plants can thus become more salinized than interspaces (Jessup, 1969; Sharma & Tongway, 1973).

Boron

Many desert soils have unusually high boron content. Desert species have evolved tolerances for this element above those that most mesic species will endure (Khudairi, 1960; Chatterton, McKell, Goodin & Bingham, 1969; Chatterton, McKell, Bingham & Clawson, 1970; Wallace & Romney, 1972).

Selenium

Selenium is also often found to be unusually abundant in some desert ecosystems. Various groups of plants such as *Astragalus* have been found to accumulate this element (Beath, 1941; Davis, 1972).

Ecologists are coming to realize that is is probably less common for a single element to impose a definite limit on an ecosystem than for more complex interactions to occur (Pomeroy, 1970). The development of the systems point of view has given new impetus to nutrient cycling studies.

Conclusions

Recent research is providing abundant illustrations of the second order limitations that nutrient factors play in desert ecosystems. Key elements in short supply are nitrogen and phosphorus. Water remains the dominant factor; however, plants cannot fully capitalize on soil moisture unless nutrients are available, and vice versa. Efficiency of water use declines under nutrient stress.

The overall rates for nutrient cycles in natural desert ecosystems are not much slower than more mesic systems, such as grasslands. However, as with other systems, a large portion of the available supply of the critical elements is locked up at any given time in either living or dead plant material and partly decomposed remains. All desert plant tissues have higher concentrations of nitrogen and ash elements than plants of more mesic environments. A large percentage of the critical nutrients in this detritus are mineralized each year permitting limited production when soil moisture conditions allow plant growth. Labile nutrients may accumulate in the soil during a cycle of dry days, weeks, or years when virtually fallow conditions prevail; then with the occurrence of favorable moisture a relatively lush growth occurs which gives a misleading impression of fertility. But should another favorable season follow the first, the soils are usually incapable of sustaining production and very little growth is made.

Nutrient distribution and transformation activities are highly structured in desert ecosystems. Most of the more critical elements are located either in younger plant tissues or in the surface soil. Islands of fertility exist around the larger perennial plants where litter buildup is accentuated and soil moisture and temperature relationships are moderated. However, accumulation of the more toxic minerals such as sodium and boron, and organics such as phenols from plant litter may also result in reduced growth or exclusion of associated species.

Cryptogamic crusts on the soil surface can be of great importance in preventing erosion and fixation of nitrogen, through the blue-green algal component. The more usual forms of decomposer organisms such as fungi and actinomycetes are less diverse, less abundant, and active for shorter periods than where moisture is less limiting.

Physical processes of nutrient transformation reach greater relative importance in desert than more mesic habitats. Nutrient input through precipitation can account for up to 20% of the annual increment of nitrogen. Gaseous losses of nitrogen through volatilization of ammonia and denitrification are promoted by high temperatures, high pH and catalytic metal ions. Catastrophic losses of nutrients can be pronounced because of small pools of tightly structured, surface-concentrated compartments and slow recovery rates. Cycles are therefore less buffered and more susceptible to environmental extremes than in more mesic ecosystems. They are, however, not necessarily less organized biologically.

Nitrogen is probably the most deficient element in most desert ecosystems. Nitrogen transformations may be limited because of the low carbon : nitrogen ratios. Phosphorus is deficient on occasion, particularly in Australia. Most cations are adequate or in excess. Salinity and/or alkalinity problems are typically a feature of agricultural areas in desert regions. Other problems such as selenium and boron toxicity are found mostly in deserts.

There is much of interest for ecologists investigating nutrient cycling processes in desert ecosystems. Relatively little research has been done in arid wild-land ecosystems compared to more mesic habitats and agronomic contexts. Although considerable data for nutrient contents at one time and place exist, present generalizations are derived from relatively few reports of direct research on nutrient dynamics. Much more definitive conclusions and new relationships are bound to emerge as research progresses on this topic.

Acknowledgements

The author wishes to thank his colleagues, J. Skujins, J. L. Charley and L. E. Rodin for helpful reviews of this manuscript.

The work on which this report is based was carried out as a part of the US/IBP Desert Biome, and was supported in part by National Science Foundation Grant No. GB-15886.

References

Al-Ani, T. A., Habib, I. R., Abdulaziz, A. I. & Ouda, N. A. (1971). Plant indicators in Iraq. Part 2. Mineral composition of native plants in relation to soils and selective absorption. *Plant and Soil*, **35**, 29–36.

Ayyad, M. A. & El-Gareeb, R. (1972). Microvariations in edaphic factors and species distribution in a Mediterranean salt desert. *Oikos*, **123**, 125–31.

Barbour, M. G. (1970). Is any angiosperm an obligate halophyte? *American Midland Naturalist*, **84**, 105–20.

Basilevich, N. I., Rodin, L. E. & Gorina, A. I. (1972). Productivity and biogeochemistry of succulent communities on solonachaks. In: *Ecophysiological foundations of ecosystems productivity in arid zone* (ed. L. E. Rodin), pp. 203–7. Nauka, Leningrad.

Basilevich, N. I., Hollerbach, M. M., Litnov, M. A., Rodin, L. E. & Steinberg, D. M. (1972). Biological factors in takyr formation at the Main Turkmenian Canal (translated title). *Botanicheskii Zhurnal*, **38**, 3–30.

Beadle, N. C. W. (1959). Some aspects of ecological research in semi-arid Australia. In: *Biogeography and ecology in Australia. Monographiae Biologicae, 3* (ed. A. T. Keast, R. L. Crocker & C. S. Christian) pp. 452–60, Dr W. Junk, The Hague.

Beadle, N. C. W. (1962). Soil phosphate and the delimitation of plant communities in eastern Australia. *Ecology*, **43**, 281–8.

Beadle, N. C. W., Whalley, R. D. B. & Gilson, J. B. (1957). Studies in halophytes. II. Analytic data on the mineral constituents of three species of *Atriplex* and their accompanying soils in Australia. *Ecology*, **38**, 340–4.

Beale, I. F. (1971). The productivity of two mulga (*Acacia aneura* F. Muell) communities after thinning in south-west Queensland. M. Ag. Sc. thesis University of Queensland, Brisbane.

Beath, O. A. (1941). The use of indicator plants in locating seleniferous areas in the Western US. IV. Progress report. *American Journal of Botany*, **28**, 887–900.

Benemann, J. R. (1973). Nitrogen fixation in termites. *Science*, **181**, 164–5.

Bobritskaya, M. A. (1962). Nitrogen uptake in soil from atmospheric precipitation in various zones of the European USSR. *Soviet Soil Science*, **12**, 1363–8.

Bormann, F. J. (1969). A holistic approach to nutrient cycling problems in plant communities. In: *Essays in plant geography and ecology* (ed. K. N. H. Greenidge), pp. 149–65. Nova Scotia Museum, Halifax.

Bowns, J. E. & West, N. E. (1976). *Blackbrush* (Coleogyne ramosissima *Torr.*) *on southwestern Utah rangelands*. Utah Agricultural Experiment Station Research Report, 27. Logan.

Brownell, P. F. (1968). Sodium as an essential micro-nutrient element for some higher plants. *Plant and Soil*, **18**, 161–4.

Burrows, W. H. (1972). Productivity of an arid zone shrub (*Eremophila gilesii*) community in south-western Queensland. *Australian Journal of Botany*, **20**, 317–29.

Caldwell, M. M., West, N. E. & Goodman, P. M. (1971). *Autecological studies of* Atriplex confertifolia *and* Eurotia lanata. US/IBP Desert Biome Research Memorandum RM 71-4. Utah State University, Logan.

Carlisle, A., Brown, A. F. & White, E. J. (1966). Litter fall, leaf production and the effects of defoliation by *Tortrix viridana* in a small oak (*Quercus petraea*) woodland. *Journal of Ecology*, **54**, 65–85.

Charley, J. L. (1959). Soil salinity–vegetation patterns in western New South Wales and their modification by overgrazing. PhD thesis, University of New England, Armidale, New South Wales.

Charley, J. L. (1972). The role of shrubs in nutrient cycling. In: *Wildland shrubs – their biology and utilization* (ed. C. M. McKell, J. P. Blaisdell & J. R. Goodwin), pp. 182–203. USDA Forest Service General Technical Report INT-1.

Charley, J. L. & McGarity, J. W. (1964). High soil nitrate levels in patterned saltbrush communities. *Nature, London*, **201**, 1351–2.

Charley, J. L. & Cowling, S. W. (1968). Changes in soil nutrient status resulting from overgrazing and their consequences in plant communities of semi-arid zones. *Proceedings of the Ecological Society of Australia*, **3**, 28–38.

Charley, J. L. & West, N. E. (1975). Plant-induced soil chemical patterns in some shrub-dominated semi-desert ecosystems of Utah. *Journal of Ecology*, **63**, 945–63.

Chatterton, M. J., McKell, C. M., Goodin, J. R. & Bingham, F. T. (1969). *Atriplex polycarpa*. II. Germination and growth in water cultures containing high levels of boron. *Agronomy Journal*, **61**, 451–3.

Chatterton, N. J., McKell, C. M., Bingham, F. T. & Clawson, W. J. (1970). Absorption of Na, Cl and B by desert saltbush in relation to composition of nutrient solution culture. *Agronomy Journal*, **62**, 351–2.

Chatterton, N. J., Goodin, J. R., McKell, C. M., Parker, R. W. & Rible, J. M. (1971). Monthly variation in the chemical composition of desert saltbush. *Journal of Range Management*, **24**, 37–40.

Chepurko, N. L., Basilevich, N. I., Rodin, L. E. & Miroshnichenko, Yu. (1972). Biogeochemistry and productivity of *Haloxyloneta ammodendroni* in south-eastern Karakum Desert. In: *Ecophysiological foundations of ecosystems productivity in arid zone* (ed. L. E. Rodin), pp. 198–203. Nauka, Leningrad.

Cline, J. F. & Rickard, W. H. (1973). Herbage yields in relation to soil water and assimilated nitrogen. *Journal of Range Management*, **26**, 296–8.

Cline, L. G. (1973). Bromegrass productivity in relation to precipitation, shrub canopy and soil nitrogen content. MS thesis, Utah State University, Logan.

Comanor, P. L. & Prusso, D. C. (1973). *Decomposition and mineralization in an* Artemisia tridentata *community in northern Nevada*. US/IBP Desert Biome Research Memorandum RM 73-79.

Cook, C. W., Stoddart, L. A. & Harris, L. E. (1954). The nutritive value of winter range plants in the Great Basin. *Utah Agricultural Experiment Station Bulletin* No. 372.

Cowling, S. W. (1969). A study of vegetation activity patterns in a semi-arid environment. PhD dissertation. University of New England, Armidale, New South Wales.

Cowling, S. W. (1977). Effects of herbivores on nutrient cycling and distribution. In: *Impact of herbivores on arid and semi-arid rangelands*, pp. 277–98. Australian Rangeland Society, Perth.

Cromack, K. (1973). Litter production and decomposition in a mixed hardwood watershed and a white pine watershed at Coweeta Hydrologic Station, N.C. PhD dissertation, University of Georgia, Athens.

Dahlman, R. C. & Kucera, C. L. (1965). Root productivity and turnover in native prairie. *Ecology*, **46**, 84–9.

Davis, A. M,. (1972). Selenium accumulation in a collection of *Atriplex* spp. *Agronomy Journal*, **64**, 823–4.

Denaeyer-De Smet, S. (1970). Note on the chemical composition of salts secreted by various gypso-halophytic species of Spain. *Bulletin de la Société Botanique de Belgique*, **103**, 273–8.

Dina, S. J. & Klikoff, L. G. (1973). Effect of plant moisture stress on carbohydrate and nitrogen content of big sagebrush. *Journal of Range Management*, **26**, 207–9.

Dregne, H. E. (1968). Appraisal of research on surface materials of desert environments. In: *Deserts of the world: an appraisal of research into their physical and biological environments* (ed. W. G. McGinnies, B. J. Goldman & P. Paylore), pp. 287–377. University of Arizona Press, Tucson.

Drover, D. P. & Barrett-Lennard, I. P. (1953). The amount of nitrate and ammonium ion in rainwater for six West Australian centres. *Proceedings of the Australian Conference on Soil Science, Adelaide*, 2. 6. 13, 3.

Duvigneaud, P. & Denaeyer-De Smet, S. (1970). Biological cycling of minerals in temperate deciduous forests. In: *Analysis of temperate forest ecosystems* (ed. D. E. Reichle), pp. 199–225. Springer-Verlag, New York.

Epstein, E. (1969). Mineral metabolism of halophytes. In: *Ecological aspects of the mineral nutrition of plants* (ed. I. H. Rorison), pp. 345–55. Blackwell, London.

Farnsworth, R. B. & Hammond, M. W. (1968). Root nodules and isolation of endophyte on *Artemisia ludoviciana. Proceedings of the Utah Academy of Science, Arts and Letters*, **45**, 182–8.

Fireman, M. & Hayward, H. E. (1952). Indicator significance of some shrubs in the Escalante Desert. *Botanical Gazette*, **114**, 143–55.

Fletcher, J. E. & Martin, W. P. (1948). Some effects of algae and molds in the raincrust of desert soils. *Ecology*, **29**, 95–100.

Fuller, W. H. & McGeorge, W. T. (1951). Phosphates in calcareous Arizona soils III. Distribution in some representative profiles. *soil Science*, **71**, 315–23.

Garcia-Moya, E. & McKell, C. M. (1970). Contribution of shrubs to the nitrogen economy of a desert-wash plant community. *Ecology*, **29**, 95–100.

Gifford, G. F. & Busby, F. E. (1973). Loss of particulate organic materials from semi-arid watersheds as a result of extreme hydrologic events. *Water Resources Research*, **9**, 1443–9.

Greenway, H. (1968). Growth stimulation by high chloride concentrations in halophytes. *Israel Journal of Botany*, **17**, 169–77.

Holmgren, R. C. & Brewster, S. F. Jr (1972). Distribution of organic matter reserve in a desert shrub community. *USDA Forest Service Research Paper* INT-30.

Hugie, V. K. & Passey, H. B. (1963). Cicadas and their effect upon soil genesis in certain soils in southern Idaho, northern Utah and north-eastern Nevada. *Proceedings of the Soil Science Society of America*, **27**, 78–82.

Jackson, E. A. (1957). Soil features in arid regions with particular reference to Australia. *Journal of the Australian Institute of Agricultural Science*, **23**, 196–208.

Jessup, R. W. (1969). Soil sâlinity in saltbush country of north-eastern South Australia. *Transactions of the Royal Society of South Australia*, **93**, 69–78.

Jordan, C. F. & Kline, J. R. (1972). Mineral cycling: some basic concepts and their application in a tropical rain forest. *Annual Review of Ecology and Systematics*, **3**, 33–50.

Jordan, C. F., Kline, J. F. & Sasscer, D. S. (1972). Relative stability of mineral cycles in forest ecosystems. *American Naturalist*, **106**, 237–53.

Jurinak, J. J. & Griffin, R. A. (1972). *Factors affecting the movement and distribution of anions in desert soils*. US/IBP Desert Biome Research Memorandum RM 72-38.

Jurinak J. J. & Griffin, R. A. (1973). *Soil as a factor in modelling the phosphorus cycle in the desert ecosystem*. US/IBP Desert Biome Research Memorandum RM 73-46. Utah State University, Logan.

Keay, J. Bettenay, E. (1969). Concentration of major nutrient elements in vegetation and soils from a portion of the Western Australian arid zone. *Journal of the Royal Society of Western Australia*, **52**, 109–18.

Kenagy, G. J. (1972). Saltbush leaves: excision of hyper-saline tissue by a kangaroo rat. *Science*, **178**, 1094–6.

Khudairi, A. K. (1960). Boron toxicity and plant growth. In: *Proceedings of the Tehran symposium on salinity in the arid zones*, pp. 175–9. UNESCO, Paris.

Khudairi, A. K. (1969). Mycorrhiza in desert soils. *BioScience*, **19**, 598–9.

Klemmedson, J. O. & Smith, E. L. (1972). *Distribution and balance of biomass and nutrients in desert shrub ecosystems.* US/IBP Desert Biome Research Memorandum RM 72-14. Utah State University, Logan.

Klemmedson, J. O. & Smith, E. L. (1973). *Biomass and nutrients in desert shrub ecosystems.* US/IBP Desert Biome Research Memorandum RM 73-8. Utah State University, Logan.

Lange, R. T. (1967). Nitrogen, sodium and potassium in foliage from some arid and temperate zone shrubs. *Australian Journal of Biological Sciences*, **20**, 1029–32.

Likens, G. E. & Bormann, F. H. (1972). Nutrient cycling in ecosystems. In: *Ecosystem structure and function* (ed. J. A. Wiens), pp. 25–68. Oregon State University Press, Corvallis.

Mack, R. N. (1971). Mineral cycling in *Artemisia tridentata*. PhD dissertation, Washington State University, Pullman.

MacGregor, A. N. (1972). Gaseous losses of nitrogen from freshly wetted desert soils. *Proceedings of the Soil Science Society of America*, **36**, 594–6.

MacGregor, A. N. & Johnson, D. E. (1971). Capacity of desert algal crusts to fix atmospheric nitrogen. *Proceedings of the Soil Science Society of America*, **35**, 843–4.

Mahmoud, S. A. Z., Abou El-Fadl, M. & Elmofty, M. K. (1964). Studies on the rhizosphere microflora of a desert plant. *Folia Microbiologia*, **9**, 1–8.

Moore, R. T., Breckle, S. W. & Caldwell, M. M. (1973). Mineral ion composition and osmotic relations of *Atriplex confertifolia* and *Eurotia lanata*. *Oecologia*, **11**, 67–78.

Mozafar, A. & Goodwin, J. R. (1970). Vesiculated hairs: a mechanism for salt tolerances in *Atriplex halimus*. *Plant Physiology*, **46**, 62–5.

Nicholls, K. D. & Turner, B. M. (1956). Pedology and chemistry of the basaltic soils of the Lismore District, New South Wales. *CSIRO Soil Publication No. 7.* Canberra.

Novichkova-Evanova, L. N. (1972). Soil algae of Middle Asian deserts. In: *Ecophysiological foundations of ecosystems productivity in arid zone* (ed. L. E. Rodin), pp. 180–2. Nauka, Leningrad.

Noy-Meir, I. (1973). Desert ecosystems. I. Environment and producers. *Annual Review of Ecology and Systematics*, **4**, 25–51.

Noy-Meir, I. (1974). Desert ecosystems. II. Higher trophic levels. *Annual Review of Ecology and Systematics*, **5**, 195–214.

Odum, H. T. (1968). Work circuits and systems stress. In: *Proceedings of a symposium on primary productivity and mineral cycling in natural ecosystems* (ed. A. E. Young), pp. 81–138. University of Maine Press, Orono.

Odum, E. P. (1971). *Fundamentals of Ecology*, 3rd edn. Saunders, Philadelphia.

Olson, J. S. (1968). Use of tracer techniques for the study of biogeochemical cycles. In: *Functioning of terrestrial ecosystems at the primary production level* (ed. F. E. Eckardt), pp. 271–88. UNESCO, Paris.

Oshanina, N. P. (1972). Nitrogen exchange of plants in the south-western Kyzlkum Desert. In: *Ecophysiological foundations of ecosystems productivity in arid zone* (ed. L. E. Rodin). pp. 214–15. Nauka, Leningrad.

Pauli, F. (1964). Soil fertility problems in arid and semi-arid lands. *Nature, London*, **204**, 1286–8.

Pearson, R. W. & Simonson, R. W. (1939). Organic phosphorus in seven Iowa soil profiles: distribution and amounts as compared to organic carbon and nitrogen. *Proceedings of the Soil Science Society of America*, **4**, 162–7.

Pomeroy, L. R. (1970). The strategy of mineral cycling. *Annual Review of Ecology and Systematics*, **1**, 171–90.

Porcella, D. B., Fletcher, J. E., Sorenson, D. L., Pidge, G. C. & Dogan, A. (1973). *Nitrogen and carbon flux in a soil–vegetation complex in the Desert Biome*. US/IBP Desert Biome Research Memorandum RM 73-76. Utah State University, Logan.

Rickard, W. H. (1965*a*). The influence of greasewood on soil-moisture penetration and soil chemistry. *Northwest Science*, **39**, 36–42.

Rickard, W. H. (1965*b*). Sodium and potassium accumulation by greasewood and hopsage leaves. *Botanical Gazette*, **126**, 116–19.

Rickard, W. H. & Keough, R. F. (1968). Soil–plant relationships of two steppe desert shrubs. *Plant and Soil*, **29**, 205–113.

Rickard, W. H., Cline, J. F. & Gilbert, R. O. (1973). Soil beneath shrub halophytes and its influence upon the growth of cheatgrass. *Northwest Science*, **47**, 213–17.

Rixon, A. J. (1970). Cycling of nutrients in a grazed *Atriplex vesicaria* community. In: *The biology of* Atriplex (ed. R. Jones), pp. 87–95. CSIRO Division of Plant Industry, Canberra.

Rixon, A. J. (1971). Oxygen uptake and nitrification by soil within a grazed *Atriplex vesicaria* community in semi-arid rangeland. *Journal of Range Management*, **24**, 435–9.

Roberts, R. C. (1950). Chemical effects of salt tolerant shrubs on soils. *Transactions of the 4th International Congress of Soil Science*, **1**, 404–6.

Rodin L. E. (1977) (ed.). *Productivity of vegetation in the arid zone of Asia*. (*Synthesis of the Soviet Studies for the International Biological Programme, 1965–1974*). Nauka, Leningrad (in Russian).

Rodin, L. E. & Basilevich, N. I. (1965). *Production and mineral cycling in terrestrial vegetation*. (English translation by G. E. Fogg, 1967). Oliver & Boyd, Edinburgh.

Rodin, L. E. & Basilevich, N. I. (1968). World distribution of plant biomass. In: *Functioning of terrestrial ecosystems at the primary production level* (ed. F. E. Eckardt), pp. 45–50. UNESCO, Paris.

Rodin, L. E., Basilevich, N. I. & Miroshnichenko, Y. M. (1972). Productivity and biogeochemistry of *Artemisieta* in the mediterranean area. In: *Ecophysiological foundations of ecosystems productivity in arid zone* (ed. L. E. Rodin), pp. 193–8. Nauka, Leningrad.

Rogers, R. W. & Lange, R. T. (1972). Soil surface lichens in arid and semi-arid south-eastern Australia. *Oikos*, **22**, 93–100.

Rogers, R. W., Lange, R. T. & Nicholas, D. J. D. (1966). Nitrogen fixation by lichens of arid soil crusts. *Nature, London*, **209**, 96–7.

Romney, E. M., Lindberg, R. G., Hawthorne, H. A., Brystrom, G. B. & Larson, K. H. (1963). Contamination of plant foliage with radioactive fallout debris. *Ecology*, **44**, 343–9.

Rotshil'd, E. V. (1968). *Desert nitrophilous vegetation and animals*. Moscow University.

Schultz, A. M. (1969). The study of an ecosystem: the arctic tundra. In: *The ecosystem concept in natural resource management* (ed. G. M. Van Dyne), pp. 77–93. Academic Press, New York.

Sharma, B. J. (1968). Chemical analysis of some desert trees. In: *Symposium on recent advances in tropical ecology* (ed. R. Misra & B. Gopal), part 1, pp. 248–51. Banaras Hindu University.

Sharma, M. L. (1973). Soil physical and physico-chemical variability induced by *Atriplex nummularia*. *Journal of Range Management*, **26**, 426–30.

Sharma, M. L. & Tongway, D. G. (1973). Plant induced soil salinity patterns in two saltbush (*Atriplex* spp.) communities. *Journal of Range Management*, **26**, 121–5.

Sharma, M. L., Tunny, J. & Tongway, D. J. (1972). Seasonal changes in sodium and chloride concentration of saltbush (*Atriplex* spp.) leaves as related to soil and plant water potential. *Australian Journal of Agricultural Research*, **23**, 1007–19.

Siebert, B. D., Newman, D. M. R. & Nelson, D. J. (1968). The chemical composition of some arid zone pasture species. *Tropical Grasslands*, **2**, 31–40.

Skujins, J. (1973). Dehydrogenase: an indicator of biological activities in arid soils. *Bulletin of the Ecological Research Committee, Stockholm*, **17**, 235–41.

Snyder, J. M. & Wullstein, L. H. (1973). The role of desert cryptogams in nitrogen fixation. *American Midland Naturalist*, **90**, 257–65.

Stark, N. (1973). Nutrient cycling in a desert ecosystem. *Bulletin of the Ecological Society of America*, **54**, 21.

Stenlid, G. (1958). Salt losses and redistribution of salts in higher plants. In: *Encyclopaedia of plant physiology* (ed. W. Richland), vol. 4, pp. 615–37. Springer-Verlag, Berlin.

Tatlyanova, A. A. (1972). Elementary chemical composition of some *Artemisia* species. *Botanicheskii Zhurnal, Leningrad*, **57**, 469–81.

Tiedemann, A. R. & Klemmedson, J. O. (1973*a*). Effect of mesquite on physical and chemical properties of the soil. *Journal of Range Management*, **26**, 27–9.

Tiedemann, A. R. & Klemmedson, J. O. (1973*b*). Nutrient availability in desert grassland soils under mesquite (*Prosopis guliflora*) trees and adjacent open areas. *Proceedings of the Soil Science Society of America*, **37**, 107–11.

Trumble, H. C. & Woodroffe, K. (1954). The influence of climatic factors on the reaction of desert shrubs to grazing by sheep. In: *Biology of deserts* (ed. J. L. Cloudsley-Thompson), pp. 129–47. Institute of Biology, London.

Waisel, Y. (1972). *Biology of halophytes*. Academic Press, New York.

Walker, T. W. (1962). Problems of soil fertility in a grass–animal regime. *Transactions of the International Society of Soil Science Commissions IV and V*, pp. 704–14.

Walker, T. W. & Adams, A. F. R. (1958). Studies on soil organic matter. I. Influence of phosphorus content of parent materials on accumulation of carbon nitrogen, sulfur and organic phosphorus in grassland soils. *Soil Science*, **85**, 307–18.

Wallace, A. & Romney, E. M. (1972). *Radioecology and ecophysiology of desert plants at the Nevada Test Site*. United States Atomic Energy Commission TID-25954.

Wallace, A., Mueller, R. T. & Romney, E. M. (1973). Sodium relations in desert plants. 2. Distribution of cations in plant parts of three different species of *Atriplex*. *Soil Science*, **115**, 390–4.

Wallace, A., Romney, E. M. & Hale, V. G. (1973). Sodium relations in desert plants. I. Cation contents of some plant species from the Mojave and Great Basin deserts. *Soil Science*, **115**, 284–7.

Watson, J. A. L. & Gay, F. J. (1970). The role of grass-eating termites in the degradation of a mulga ecosystem. *Search*, **1**, 43.

Wells, K. F. (1967). Aspects of shrub–herb productivity in an arid environment. MS thesis, University of California, Berkeley.

Went, F. W. & Stark, N. (1968). Mycorrhiza. *BioScience*, **18**, 1035–9.

Went, F. W., Wheeler, J. & Wheeler, G. C. (1972). Feeding and digestion in some ants (*Vermessor* and *Manica*). *BioScience*, **22**, 82–8.
West, N. E. (1972). *Biomass and nutrient dynamics of some major cold desert shrubs.* US/IBP Desert Biome Research Memorandum RM 72-15. Utah State University, Logan.
West, N. E. (1975). The nitrogen cycle in North American cool desert ecosystems. *XII International Botanical Congress, Moscow*. Abstracts vol. 1, p. 174.
West, N. E. & Skujins, J. (1977). The nitrogen cycle in North American cold-winter semi-desert ecosystems. *Oecologia Plant* (*Paris*), **12**, 45–53.
West, N. E. & Skujins, J. (eds) (1978). *Nitrogen in Desert Ecosystems*. US/IBP Synthesis Series 9. Dowden, Hutchinson & Ross, Stroudsburg, Pennsylvania.
Westoby, M. (1973). The impact of black-tailed jackrabbits (*Lepus californicus*) on vegetation in Curlew Valley, northern Utah. PhD dissertation, Utah State University, Logan.
Whittaker, R. H. (1961). Estimation of net primary production of forest and shrub communities. *Ecology*, **42**, 177–80.
Wiebe, H. H. & Walter, H. (1972). Mineral ion composition of halophytic species from northern Utah. *American Midland Naturalist*, **87**, 241–5.
Wilcox, D. G. (1974). Morphogenesis and management of woody perennials in Australia. In: *Plant morphogenesis as the basis for scientific management of range resources*, pp. 60–71. United States Department of Agriculture Miscellaneous Publication No. 1271.
Williams, O. B. (1964). Energy flow and nutrient cycling in ecosystems. *Proceedings of the Australian Society for Animal Production*, **5**, 291–300.
Witkamp, M. (1971). Soil as a compartment of ecosystems. *Annual Review of Ecology and Systematics*, **3**, 85–110.
Wood, T. G. (1970). Micro-arthropods from soils of the arid zone in southern Australia. *Search*, **1**, 75–6.
Woodwell, G. M. & Whittaker, R. H. (1967). Primary production and the cation budget of the Brookhaven Forest. In: *Proceedings of a symposium on primary productivity and mineral cycling in natural ecosystems* (ed. H. E. Young), pp. 52–80. University of Maine Press, Orono.
Wullstein, L. H. (1969). Reduction of nitrate deficits by alkaline metal carbonates. *Soil Science*, **108**, 222–6.

Manuscript received by the editors September 1975

13. Short-term dynamics of minerals in arid ecosystems

P. BINET

Introduction

In an ecosystem, the dynamics of mineral elements must include a microbial phase which takes place in the soil. This phase can be appreciated as a part of mineral dynamics when, for example, the molecular nitrogen fixation by free soil microorganisms is followed by decomposition, ammonification, nitrification of their residues and, finally, by denitrification. Most often, however, higher plants and animals are involved in these dynamics.

In arid environments, this rule remains valid for the researcher trying to study particular processes; a major difficulty being to define what the soil represents. Most often, it is a question of a simple 'substrate' for vegetation in which it is impossible to define a profile (Leredde, 1957). According to Aubert (1960), desert soils are soils of inorganic minerals which have undergone very little evolution, including: (*a*) soils of ablation or 'regs', poor in fine elements which have been blown away by the wind; (*b*) soils of transport, consisting of fine elements accumulated by the wind in the form of overlays, crusts, nebkas, barkhans or sand dunes in ribbons, hills or mountains; (*c*) soils of skeletal material formed by rocks, more or less broken up, in which the debris remains in place (e.g. soils of hammadas).

It is especially the physical characteristics of such soils, particularly depth and water retention characteristics, that determine the occurrence of living organisms (Kassas & Imam, 1954; Migahid, Abd el Rahman, el Shafei Ali & Hammouda 1955; Quezel & Simonneau, Kassas & Girgis, 1970).

These soils are enriched *in situ* in organic matter, which is utilized by the microorganisms to continue development, thereby permitting mineralization of elements in the organic material.

In order to study general characteristics of mineral dynamics in arid ecosystems, we will begin with the organic matter contained in these soils and the microorganisms associated with them, and will then examine how higher plants and animals participate in these processes. Finally, we will examine the cycles of nitrogen, phosphorus, potassium, sodium and the chloride ion.

General characteristics of the dynamics of minerals in arid ecosystems

Organic matter in arid soils

Whether observing the light color of moist arid soils (Krause, 1958) or measuring the percentage of organic carbon in these soils (Kellogg, 1953; Kassas & Imam, 1954, for the soils of *Zilletum spinosae* in Egypt; Leredde, 1957, in the Tassili n'Ajjer; Quezel & Simonneau, 1963, in the Guir hammada; Dubost & Hethener, 1966, in the oasis of Tassili n'Ajjer; Rougieux, 1966, in the Sahara), or humus content (Killian & Feher, 1935, 1938, in the Sahara; Montasir & Shafey, 1951, among the *Fagonia arabica*; Montasir & Foda, 1956, in Egypt regarding *Zygophyllum album*), the poverty of organic material is clearly evident (Fig. 13.1). Usually a soil of this kind is reported to contain a fraction of a gram to several grams of organic carbon per kilogram. However, this amount is undoubtedly underestimated as these analyses measure the ground soil passing through a 2-mm sieve while not taking into account organic debris of larger particle size (Leredde, 1957). Vegetation evidently plays an important role in the accumulation of such debris. Soil is enriched by contact with underground plant organs which die and produce various exudates (Thornton, 1953). The roots of certain species are particularly apt at invading fresh soil. *Psoralea lanceolata* forms slender exploratory roots in mobile sand dunes, assuring fixation and enrichment in organic matter (Kearney *et al.*, 1914). Perennial species participate in the enrichment of the soil in organic matter in several ways: (*a*) lower branches may create obstacles to wind-carried particles which, in mingling with dead plant organs, constitute a soil richer in organic matter than that of the immediate environment (e.g. *Ambrosia dumosa*, formerly *Franseria dumosa*, and *Larrea tridentata* in the Colorado desert, Muller, 1953; and *Anabasis aretioides* in the Sahara, Vargues, 1953; Hethener, 1967); (*b*) decayed organs accumulate at the base of the plant; under *Tamarix*, a true litter forms which blends eolian sand deposits and in which humification is very slow when the substrate is rich in NaCl (Killian, 1974*b*; Litav, 1957; Leredde, 1957; Fig. 13.1*b*). In other cases (e.g. *Calligonum*), the litter remains intact for a long time and does not mix with the sand (Killian, 1947*b*). However, accumulation of organic matter at the base of perennial species does not always occur. For example, this accumulation does not take place under *Encelia farinosa* (Muller, 1953) or *Acacia cyanophylla* (Litav, 1957) whose litter is removed by the wind and whose simple stem is not an obstacle to drifting debris; (*c*) they offer shelter to other perennial or annual species which, in turn, share in the organic matter enrichment of the soil (Osborn, Wood & Peltridge, 1932; Shreve, 1942; Garcia-Moya & McKell, 1970).

The input of litter to the soil system exhibits both spatial and temporal

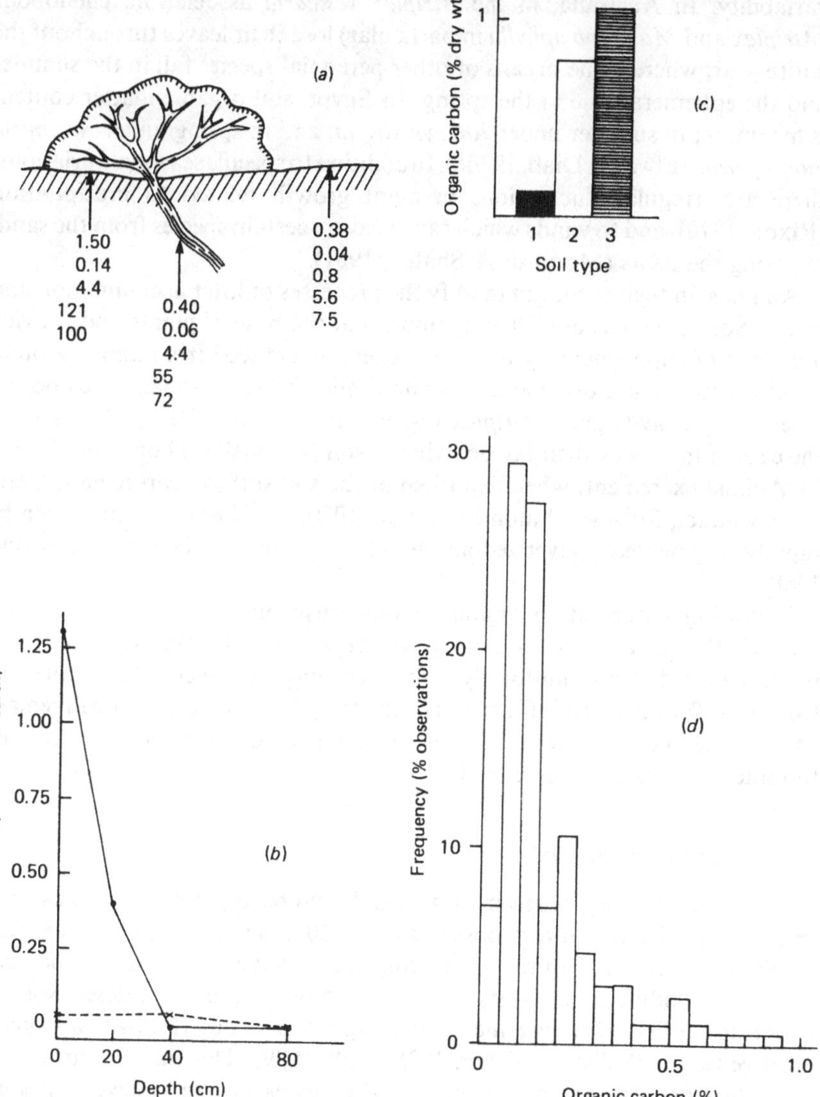

Fig. 13.1. Organic matter and carbon content of desert soils. (*a*) Characteristics of soil at different distances from the base of *Anabasis aretioides*. The values indicate, going from top to bottom: amount of carbon (%); amount of nitrogen (%); amount of phosphorus (mg 100 g^{-1} soil); amount of potassium (mg 100 g^{-1} soil); intensity of the nitrogen fixation, the value of 100 being attributed to the point situated at the soil surface, under the plant canopy. (After Vargues, 1953 & Ozenda, 1958.) (*b*) Amount of organic matter of soil at different depths under *Tamarix* (solid line) and from a non-vegetated area (broken line) from the area north of the Negev. (After Litav, 1957.) (*c*) Mean amount of organic carbon of soils of Tassili n'Ajjer in the central Sahara. 1, soil with sparse plant cover; 2, soil under *Cupressus dupreziana*; 3, soil of a palm grove with primitive management. (*d*) Frequency of the amount of organic carbon of the soils of the Tassili n'Ajjer, analyzed by Leredde (1957).

variability. In Australia, in the *Atriplex vesicaria* association, chenopods (*Atriplex* and *Maireana aphylla* in particular) lose their leaves throughout the entire year, whereas the organs of other perennial species fall in the summer and the ephemerals die in the spring. In Egypt, soil organic matter content is maximum in summer under *Rhazya stricta*, and in spring under *Artemisia monosperma* (Elwan & Diab, 1970*b*). In addition to specific seasonal variations, there are irregular fluctuations in plant growth related to precipitation (Rixon, 1970), and to winds which can excavate certain species from the sand, exposing the roots (Montasir & Shafey, 1951).

Animals, in their turn, can modify the processes of litter accumulation and decay. Some, such as ants, live in mounds at the base of plants and provide a deposit of supplementary organic matter; others feed from above-ground plant organs or soil organic debris and diminish litter formation, as occurs when sheep heavily graze *Atriplex vesicaria* in Australia (Rixon, 1971), or in the case of insects or their larvae living in soil (Cloudsley-Thompson, 1964*a*, *b*). Animal excrement, when deposited at the soil surface, can remain there, loosely intact, for several months (Rixon, 1970) or, after desiccation, can be rapidly fragmented, pulverized and finally dispersed by the wind (Ranzoni, 1968).

Following its deposition, organic matter is distributed diversely throughout the soil. If a product of roots or of soil organisms, the organic matter will be distributed quite uniformly until reaching relatively deep horizons (Montasir & Foda, 1956). If derived from litter, the organic matter will remain concentrated close to the soil surface since rains are either too infrequent or too intense to carry it far into the soil.

Microorganisms of arid soil

A census of microorganisms in arid soils should be approached with caution (Fig. 13.2). Usually, direct observation is 50 times more important than results obtained by culture, considering the fact that one usually does not know the conditions necessary to obtain the development of desert species (Thornton, 1953). However, measuring respiration of desert soils always gives positive results (Killian & Feher, 1935, 1938, 1939). This fact, confirmed by the dehydrogenase activity (Skujins, 1973), leads one to believe that soils completely lacking active microorganisms do not exist. Rougieux (1966) found only 90 microorganisms in 1 g of barren reg soil, containing 0.4% water,

Fig. 13.2. Abundance of microorganisms in arid soils. (*a*) Seasonal variation of the number of microbial organisms contained in 1 gram of moist, Saharan soil (after Killian & Feher, 1935, 1938). 1, soil of the oasis of Aoulef el Arah; 2, a reg near Adrar; 3, sandy desert near In Salah; 4, Tanezrouft. (*b*) Seasonal variation in the number of bacterial organisms and the rhizosphere effect in Arabia. (After Elwan & Diab, 1970*d*). 2, sample between 20–30 cm in depth; 3, sample between 30–40 cm in depth. (*c*) Relation between the number of microorganisms (bacteria and fungi) contained in 1 gram of moist, Saharan soil as a function of the amount of water. (After Killian & Feher, 1935, 1938.)

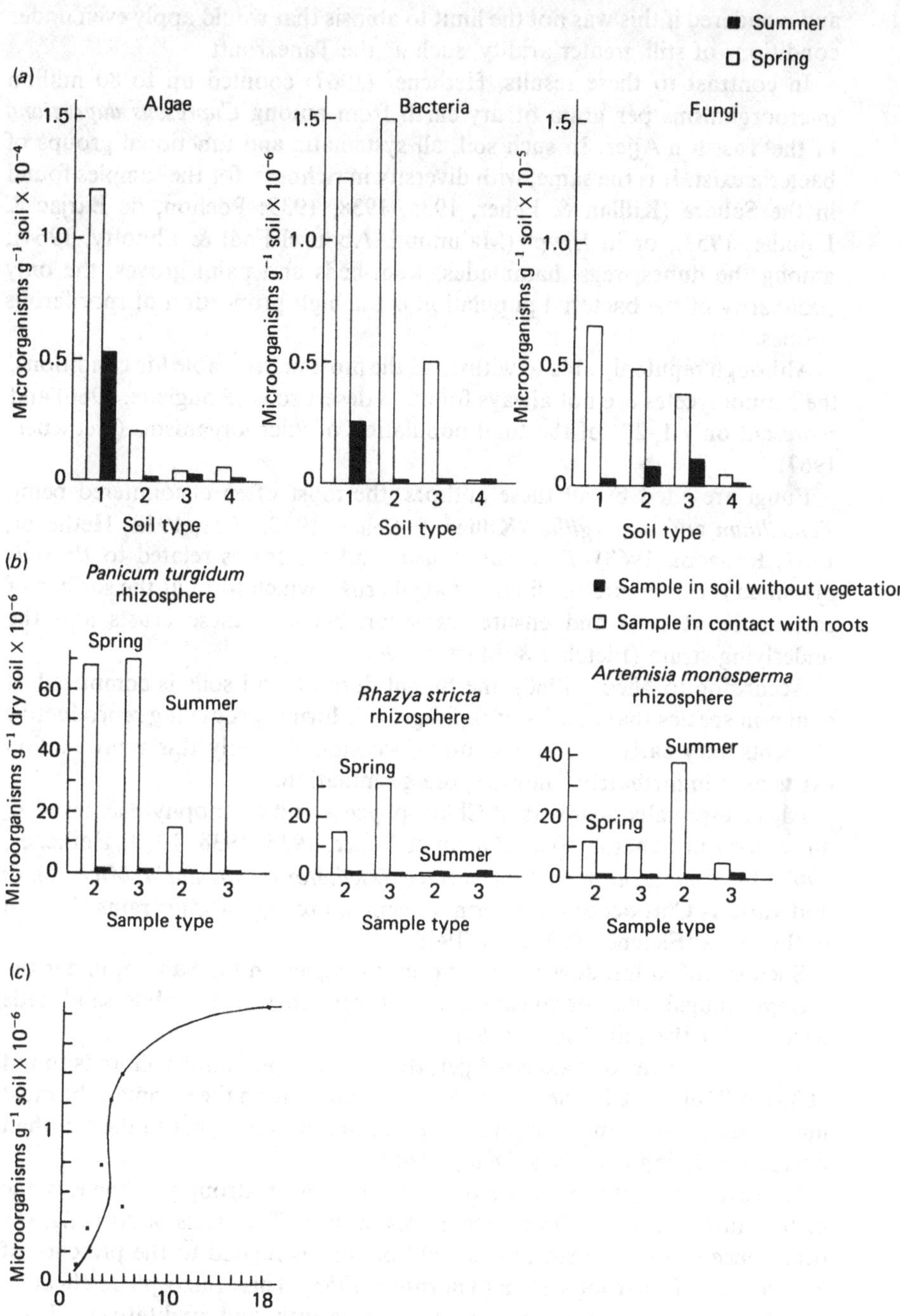
Summer
Spring
(a)
Algae
Bacteria
Fungi
Microorganisms g⁻¹ soil × 10⁻⁴
Microorganisms g⁻¹ soil × 10⁻⁶
Microorganisms g⁻¹ soil × 10⁻⁵
1.5
1.0
0.5
0
1 2 3 4
Soil type
(b)
Panicum turgidum rhizosphere
Sample in soil without vegetation
Sample in contact with roots
Microorganisms g⁻¹ dry soil × 10⁻⁶
Spring
Summer
60
40
20
0
2 3 2 3
Sample type
Rhazya stricta rhizosphere
Artemisia monosperma rhizosphere
(c)
Microorganisms g⁻¹ soil × 10⁻⁶
1
0
0 10 18
Water (%)

and wondered if this was not the limit to abiosis that would apply even under conditions of still greater aridity, such as the Tanezrouft.

In contrast to these results, Hethener (1967) counted up to 80 million microorganisms per gram of dry earth from among *Cupressus dupreziana* of the Tassili n'Ajjer. In such soil, all systematic and functional groups of bacteria exist. It is the same, with diversity in richness, for the samples found in the Sahara (Killian & Feher, 1935, 1938, 1939; Pochon, de Barjac & Lajudie, 1957), or in Egypt (Mahmoud, Abou el-Fadl & Elmofty, 1964); among the dunes, regs, hammadas, wadi-beds and palm groves, the only peculiarity of the bacterial population is the high proportion of sporiferous species.

Although reputedly able to withstand the most unfavorable life conditions, the actinomycetes are not always found in desert soils (Rougieux, 1966) and represent only 1–2% of the total population of microorganisms (Hethener, 1967).

Fungi are cited by all these authors, the most often encountered being *Penicillium* and *Aspergillus* (Killian & Feher, 1935, 1938, 1939; Hethener, 1967; Ranzoni, 1968). *Rhizopus*, *Mucor* and the forms related to *Botrytis pyramidalis* can assure the filling of algal crusts which form at the surface of certain desert soils and ensure transition between these crusts and the underlying strata (Fletcher & Martin, 1948).

According to Nicot (1960), the fungal flora of arid soils is comprised of common species that consist of rapid-growth forms, producing reproductive elements very early and abundantly. Occasionally, only this reproduction exists as it immediately follows spore germination.

Algae, especially members of Chlorophyceae and Cyanophyceae, occur in all seasons in Saharan soils (Killian & Feher, 1935, 1938, 1939; Hethener, 1967). In Arizona, at the red soil surface, *Oscillatoria, Nodularia, Microcoleus* and various Chroococcaceae form a deep-colored crust after rains, 2.4 mm in thickness (Fletcher & Martin, 1948).

Such stratification does not exist only for algae. In the Sahara, under the canopy, fungal microorganisms are concentrated in the mobile sand beds which cover the soil (Nicot, 1960).

In the oasis soils of Tassili n'Ajjer, the most abundant microflora is found at 30 cm (Dubost & Hethener, 1966). In Arabia, during the summer, the most numerous bacterial microorganisms are found buried deeper in the soil than during the spring (Elwan & Diab, 1970*d*).

However, it is the presence of roots that most strongly influences the distribution of bacterial microorganisms in the soil. In soils of Arizona, the occurrence of microorganisms in soil horizons is related to the presence of a large part of the root system (Thornton, 1953). This 'rhizosphere effect' is readily observed in desert soils, both quantitatively and qualitatively. Thus, in summer, the bacteria with complex sustenance requirements occur only in the presence of *Rhazya stricta*, whereas in spring they are found throughout the soil (Elwan & Diab, 1970*b*). The percentage of sporiferous species in contact with the roots of *Moltakea callosa* is 270 times less than the number found in bare soil, and the cellulolytics, *Azotobacter* and *Clostridium*, exist only in the rhizosphere. Fungi of the rhizosphere generally include *Alternaria*, *Fusarium* and *Aspergillus*, while *Penicillium* dominates in bare soil (Mahmoud *et al.* 1964).

The importance of the rhizosphere effect is a function of season. The effect is maximum in the winter in the case of *Rhazya stricta*, whose roots produce bacterial inhibitors which can be eliminated only by winter rains (Elwan & Diab, 1970*b*; see Fig. 13.2). Around the roots of *Artemisia monosperma*, microbial activity decreases in summer after the sagebrush has set fruit, which coincides with a considerable slowing down of metabolic activity (Elwan & Diab, 1970*a*). On the other hand, the rhizosphere effect of *Panicum turgidum* is maximum in summer (Elwan & Diab, 1970*c*).

Also, without necessarily agreeing with the conclusions of Sabinin & Minina (1932), according to whom desert soils would be sterile without root contact, one can consider that those roots create around themselves a favorable environment for intense microbial activity, and in that region, chemical transformations of various elements may take place during any season.

By modifying the distribution of organic matter in the soil, animals also influence the kinds and numbers of microflora present (termites for example; Moureaux, 1965). Feces become a preferential place for microorganism development, particularly fungi (Ranzoni, 1968). In the oasis, the cultural practices of man aid in the development of the microflora (Dubost & Hethener, 1966).

In many habitats characterized by aridity, high light intensity and extreme temperatures, many researchers adopt the approach of relating the sensitivity of microorganisms in these habitats to the factors of water, light and temperature. Killian & Feher (1935, 1938, 1939), Dommergues (1962) and Dubost & Hethener (1966) believe that the actinomycetes and fungi require, in general, less water than bacteria. But this rule has, without doubt, exceptions, such as those shown in the analysis of a number of Saharan soils by Rougieux (1966). At any rate, it is clear that in the course of soil desiccation, microbial activity and hence, transfer of chemical elements is greatly modified. Dommergues (1962) distinguishes: (*a*) the hyperxerophiles, including the majority of species responsible for the ammonification of tyrosine, glycolysis, sulfhydrogenization, mineralization of glycerophosphate and the degradation of plant material such as leaves of the peanut and teak; (*b*) the xerophiles, especially responsible for cellulolysis and sulfur oxidation; (*c*) the hygrophiles, degrading carboxymethyl-cellulose or fixing atmospheric nitrogen.

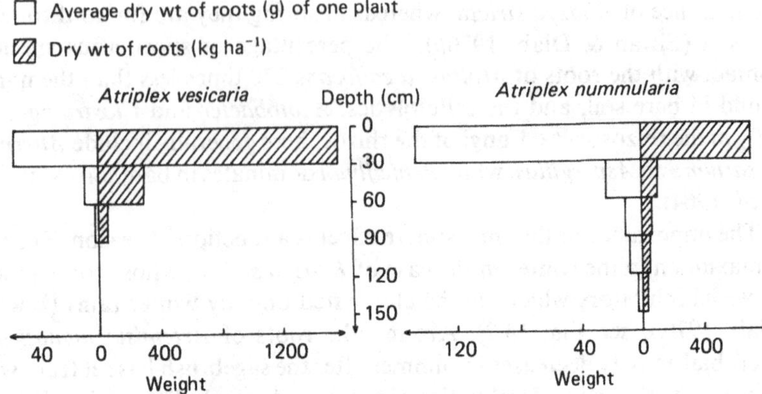

Fig. 13.3. Distribution of average dry weight of roots of a plant and of the dry weight of the collection of roots observed in a hectare as a function of depth under *Atriplex vesicaria* and *Atriplex nummularia* in Australia. (After Jones & Hodgkinson 1970.)

Among fungi, Dubost & Hethener (1966) state that moisture has little effect on *Aspergillus*, whereas *Penicillium* is not found in dry soils. The Mucorales cannot tolerate harsh desiccation and, as a result, do not live in upper soil horizons. At those levels, microorganisms could benefit from water condensation in the capillary spaces due to quick temperature changes (Killian & Feher, 1935, 1938, 1939) or from dew on the soil surface several hours before sunrise (Elwan & Diab, 1970*a*).

It is necessary to note that, according to Voinova-Raikova (as cited by Quastel, 1965), the effect of water content on bacteria in the process of ammonification and on the actinomycetes, could be a function of the carbon: nitrogen ratio existing in the soil.

The light sensitivity of arid soil microorganisms is much less well known or understood. Durrell & Shields (1960) in Nevada, and Nicot (1960) in the Sahara, observed the high frequency of black pigments among desert and semidesert fungi of these regions. These pigments could be a protective screen against solar radiation, ultraviolet rays in particular. The black spores of *Stemphylium ilicis* are still living after 1 h of exposure to radiation of 257 nm. The pigments that can be extracted from these spores impede all radiation with wavelengths between 200 and 2000 nm.

Contrary to expectations, there are not a great number of thermophiles in Saharan soils (Pochon *et al.*, 1957) but, as in the case of moisture, the different microorganisms are rapidly affected by a rise in temperature; these are the most sensitive of the algae (Killian & Feher, 1935, 1938). In sandy areas of Egypt populated by *Panicum turgidum*, the number of bacteria with simple nutritional needs increases with temperature, whereas bacteria requiring a complex environment are at a disadvantage during hot weather (Elwan & Diab, 1970*c*).

In an arid habitat, seasonal soil changes in moisture, light and temperature are probably responsible for quantitative and qualitative modification of the microflora, and accordingly, for the release of soil minerals (Figs. 13.3 and 13.4).

The importance of the root system in mineral dynamics

The importance of the root system in mineral dynamics is shown by the dominant role played by roots in distributing microorganisms in arid soils. In arid habitats, the morphological features of roots can imply certain characteristics of mineral dynamics. Leredde (1957) observes that if the assimilative mineral reserves from the soil itself are low per unit volume, they do not constitute a limiting growth factor for perennial xerophytes, given their extensive root system. These root systems can concentrate minerals from a large volume of soil into the above-ground structure. The roots grow slowly, as the soil is often difficult to penetrate. The absorption of mineral matter is equally slow, although continuous (Krause, 1958). The difficulties of penetration and the mode of mineral circulation are particularly important in the case of easily cracked rocks relative to rock for which the penetration of shrub roots plays an important role in local soil formation (Kassas & Imam, 1954). It is not the same situation for the ephemerals, whose root systems must find an adequate supply of minerals in a short time and within a relatively restricted volume of soil. The roots of ephemerals rapidly disappear, ultimately facilitating soil aeration and circulation of soil solution (Kachkarov & Korovine, 1942).

Different species absorb minerals from different soil horizons. Shreve (1942) states that for *Artemisia tridentata* and *Larrea tridentata* communities in North America, the dominant and subdominant species complement one another in the distribution of root systems and in their periods of activity. Certain species are particularly adept at taking up available minerals after a light rain. This is the case with *Atriplex nummularia* and *A. vesicaria* in Australia, both of which exhibit fine ephemeral roots that branch within the first 10 cm of soil and join with adventive roots which originate from stems bedded in the soil (Osborn *et al.*, 1932; Jones and Hodgkinson, 1970; Fig. 13.3). In Egypt, *Fagonia arabica* is capable of absorbing soil minerals from the surface and at depth; in fact, it has a principal root greater than 3 m long which produces branches and numerous radicles (immature roots) throughout its length.

These root systems are also more easily modeled with regard to edaphic conditions. With *Zygophyllum album*, the drier the soil the deeper the roots penetrate, and it is in sandy soil that the root biomass: leafed stems ratio is the smallest (Montasir & Foda, 1955, see Table 13.1).

The reserve 'root buds' (radicles) depicted by certain xerophytes, particularly those under saline conditions (Killian, 1951), are responsible for certain

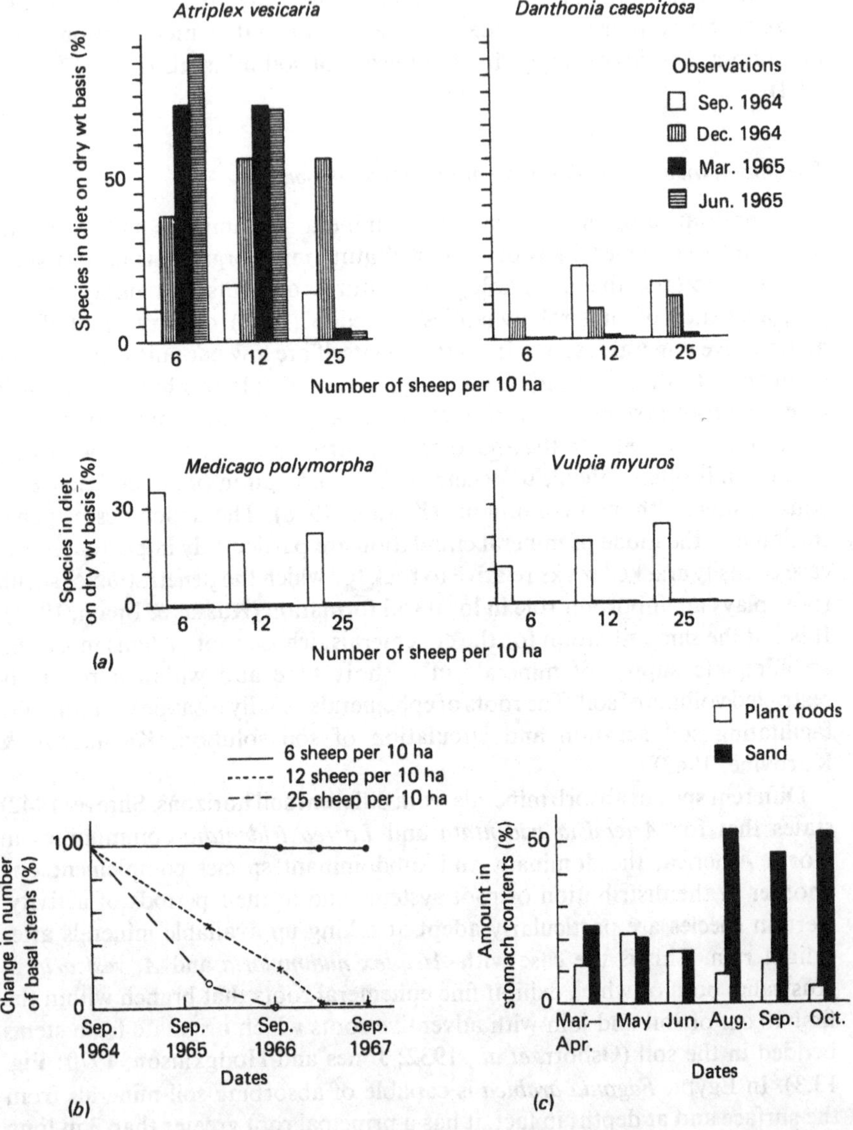

Fig. 13.4. Animal–plant relations in desert habitats. (*a*) Seasonal dietary pattern of sheep in relation to intensity of grazing. (After Wilson, Leigh & Mulham, 1969.) (*b*) Change in number of basal stems of *Atriplex vesicaria* from 1964–67, in relation to grazing pressure. (After Wilson *et al.*, 1969.) (*c*) Annual change in diets of *Scincus scincus* in the Grand Erg Occidental in the Algerian Sahara. Shows percentages of plant foods and of sand found in stomach contents. (After Vernet & Grenot, 1972.)

Table 13.1. *Modifications of the root system of* Zygophyllum album *under the influence of soil conditions.* (After Montasir & Foda, 1955)

Soil type	Amount in water (% saturation)	Plant age (days)	Length (cm) of the principal root	Dry wt (mg) of plant roots	Root:shoot ratio (on dry wt basis × 100)
Clay	60	60	22·0	29·5	15
Silt	60	60	18·0	11·6	8
Limestone	60	60	21·8	10·9	9
Sand	60	60	23·5	9·6	7
Sand and clay	80	60	13·0	8·4	14
	60	60	21·0	19·6	10
	30	60	21·5	12·7	11
	15	60	23·0	8·0	40
	Level of cover in the soil, depth (cm) below the surface				
Sand and clay	25	50	19·5	18·0	12
	15	50	12·5	7·0	9
	5	50	6·0	2·0	26

rhythmic aspects of mineral absorption from the soil. The radicles, formed deep in the cortical parenchyma of the parent roots, remain enclosed there all the time under dry or saline conditions, but grow quickly as external conditions permit. In contrast, the roots of *Haloxylon scoparium* maintain numerous hairs in good functional condition all year long. Therefore, absorption of mineral material seems to persist during the dry season (Killian, 1941).

Finally, one wonders to what extent the presence of 'root sheaths' of certain xerophytes (e.g., *Aristida* in a sandy soil and *Zygophyllum simplex* in a saline habitat) modifies the degree and location of mineral absorption in dry soils (Killian, 1947*a*).

The importance of animals in mineral dynamics

So far in this chapter, it has been noted that animal activity can modify the distribution of organic matter and microorganisms in arid soils, but this is only one aspect of the importance of animals in the mineral dynamics of arid habitats.

Certain animals mediate in the food chain by consuming vegetation. Certain xerophytes provide excellent nourishment because of high protein content (above-ground organs of *Atriplex* and *Maireana* contain more than 20% protein; Wilson, 1966), even during periods of dryness (as in the case of *Acacia aneura* of Australia; Kozlowski, 1968). Certain herbivorous animals make no dietary distinction between desert plants. Thus, one can find 50 different species in the stomach contents of *Uromastix acanthinurus* which, in a period of scarcity, can satisfy itself with halophytes or harsh vegetation such as *Zilla spinosa* or *Launaea arborescens* (Dubuis, Faurel, Grenot & Vernet, 1971; Grenot & Vernet, 1972, 1973). On the other hand, the existence of rodent herbivores (gerbils and jumping mice) is associated with succulent chenopods (Grenot & Niaussat, 1967), and the ant *Acantholepis frauenfeldi* exploits flowers and cochineals of *Tamarix* (Delye, 1968).

This plant nourishment of certain animals undergoes both quantitative and qualitative variation throughout the course of a year. Therefore, directly after a rain, plant recovery is beneficial to phytophagous insects (caterpillars, heteropterous insects) or flower-dwelling insects (butterflies, certain hymenopterans; Grenot & Vernet, 1972, 1973). During winter and spring in Australia, sheep mostly graze the herbaceous, perennial species (*Danthonia caespitosa*) or annuals (*Medicago polymorpha*, *Vulpia myuros*) with *Atriplex vesicaria* then constituting only 10% of their diet. In summer it is just the opposite; *Atriplex* is consumed, especially if the summer is dry and the pasture is stressed (Leigh & Wilson , 1970; Wilson *et al.*, 1969; Fig. 13.4). As a result, animals are often involved with modifying the nature of equilibria between plane species and, therefore, the composition of the community. In Utah, grazing is beneficial to the rarely grazed *Artemisia tridentata*, compared to species which normally compete with it for water. It is the same in *Maireana vestita* communities, where the *Maireana* is left intact by the sheep but *Poa sandbergii* is uprooted or consumed to the point that it cannot flower and therefore almost disappears (Kearney *et al.*, 1914). In the arid zone of Rajasthan, the importance of grazing can be recognized by observing the vegetation. When grazing is severe, desirable species are rare (*Dichanthium annulatum*, *Lasiurus sindicus*) and the unpalatable species dominate (*Tephrosia purpurea*, *Crotalaria burhia*; Gupta & Saxena, 1971). In Australia, intense grazing by sheep in *Atriplex vesicaria* communities causes the death of a number of *Atriplex vesicaria* stems because this species is incapable of forming new branches other than on palatable young stems (Leigh & Wilson, 1970; Fig. 13.4). Under light grazing, more *Atriplex* stems will remain alive and leafy during the dry season. If the grazing is very light, plant density will increase and many plants will remain quite small as the competition is severe (Osborn *et al.*, 1932). From measures of soil respiration one finds that soil microbiological activity is diminished following severe grazing, no doubt because the soil organic matter deposits are less under a thinner stand of *Atriplex*. Moreover, the suppression of bushes reduces the number of favorable habitats for soil fauna and diminishes the activity of decomposers. A modification of the animal–plant equilibrium is involved with the dynamics of chemical elements (Rixon, 1970). Sometimes (e.g., in the desert of Betpak-Dala) the vegetation is so scarce that it cannot efficiently sustain granivorous birds. Herbivorous insects are sometimes also limited in number and serve as prey for scarce insectivorous birds, several species of lizards and snakes (Kachkarov & Korovine, 1942).

Certain animals eat dead plant material. In the Sahara, when aridity has killed numerous plants, these dead plants become the prey of indiscriminate termites who even consume those species recognized as particularly poisonous, such as the Asclepiadaceae and *Calotropis procera.* In 1944, in the south of the Fezzan, and in 1949 in the center of the Tassili n'Ajjer, *Psammotermes* caused a very rapid disappearance of pasture lands of *Calligonum comosum.* In the Fezzan, in a reg deprived of vegetation, one finds *Psammotermes* down several centimeters in the soil, where they gnaw on subfossil *Tamarix* trunks covered up by alluvium after a humid Pleistocene (Bernard, 1954). Thus, in an arid habitat, even where there are no living plants, withered plant material and a subterranean fauna can be found (Cloudsley-Thompson, 1964*b*) which participate in litter decomposition (Bullock, 1967). In dune hollows or any reg hollow, there is an accumulation of plant debris which constitutes an important alluvial deposit, constantly renewed for diverse plant-eating species, in particular the thysanurans, who make up one of the first food chain links (Grenot & Niaussat, 1967; Vernet & Grenot, 1972). In the stomach contents of certain species (e.g., *Scincus scincus*, whose diet is normally composed of grains and other animals), there are significant quantities of sand which can comprise a source of mineral elements for the animal (Vernet & Grenot, 1972; Fig. 13.4), and the digestive system of dry-region ungulates harbors bacteria and protozoa which participate in cellulose digestion (Macfarlane, 1964).

In arid regions animals participate in the dynamics of chemical elements by virtue of their excrement. They can modify the location of this impact since certain animals, *Uromastix acanthinurus* for example, have special places for defecation where excrement accumulates (Grenot, 1969; Grenot & Vernet, 1972). The example of termites is even more suggestive of this, whether it be *Tamarix* mounds (Killian, 1947*b*) in the dark, subarid soils in West Africa (Moureaux, 1965), or in Australia (Lee & Wood, 1971), where the incorporation of insect fecal material into the soil contributes to carbon and nitrogen deposits and stimulates the cycling of elements that termites digest very efficiently – the compound plant polyosidics. When fecal matter is deposited at the soil surface it decomposes very slowly and certain elements are locked up for quite a long time. For example, in the arid regions of Australia 70% of sheep excrement is still on the soil surface after seven months. During this time it adds 3–4 mg kg^{-1} of surface soil, whereas the amount of excrement remaining on the surface ranges from 40–100 mg kg^{-1}. A great part of mineralized nitrogen in the feces remains in these residues a long time and is inaccessible to vegetation while the excrement is not incorporated into the soil. Rixon (1970) calculated that with a livestock density of 0.6 sheep, ha^{-1} excrement production in one year is 44 kg ha^{-1}, containing around 1.3 kg of nitrogen, of which only 1% goes into the soil.

Particular case studies

Nitrogen dynamics

Amount of nitrogen in desert soils

All arid soils analysed are poor in nitrogen: 0.008 to 0.064% in Saharan soils (Killian & Feher, 1935); 0.005 to 0.040%, with a median of 0.020% in soils of the Tassili n'Ajjer of central Sahara, the greatest values being found where fine particles (silt and clay) are abundant (Leredde, 1957; Fig. 13.5*a*); 0.005 to 0.050% in cultivated soils or in oasis fallows of the Tassili n'Ajjer (Dubost & Hethener, 1966); 0.14 to 1.42% in the soils of Hoggar (Rivking, cited by Rougieux, 1966); a median of 0.60% in arid Australian soils (Charley & Cowling, 1968); less than 0.05% in dry regions of California (Russell, 1968).

When the soil is analyzed in different layers, several authors (Lipman, 1915 and Wells, 1967 in California, cited by Garcia-Moya & McKell 1970; Charley & Cowling, 1968 in Australia) report a rapid decrease of these amounts as one goes deeper into the soils (Fig. 13.5b, c).

The presence of a relatively larger amount of nitrogen in the upper soil horizon is most often directly linked to the nature and occurrence of the vegetation. In the Tassili n'Ajjer, the greater the amount of soil nitrogen, the denser the population of *Cupressus dupreziana* (Hethener, 1967). In Australia, Rixon (1970) found more nitrogen in the soil under *Atriplex vesicaria* than outside the canopy of this species. It is necessary to consider not only the deposition of litter generated by *Atriplex*, but also that produced by other species, in particular *Medicago*, which is four times more abundant under *Atriplex* than elsewhere.

In certain cases, however, this maxim can be misleading. In Arizona, for example, the amount of soil nitrogen is maximum under *Larrea tridentata*, medium under *Acacia greggi* and minimum under *Cassia armata*; the quantities of nitrogen contained in these three species bringing 6.47, 5.06 and 11.39 kg ha^{-1}, respectively. The variation of soil nitrogen concentration as a function of soil depth and position in relation to the plant axis, is much less pronounced for *Cassia armata* than for the other two species.

Garcia-Moya & McKell (1970) believe that in order to explain this phenomenon, it is necessary to consider herbaceous species which grow in the favorable habitat created by the shrub canopy and fallen litter, and which

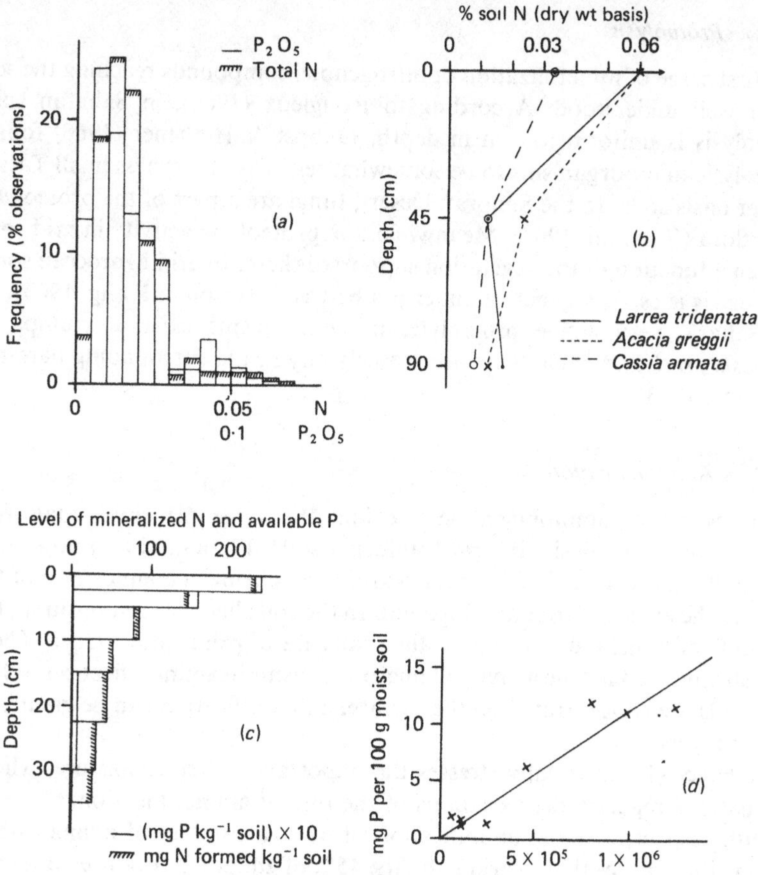

Fig. 13.5. Nitrogen and phosphorus in desert soils. (*a*) Frequency of the amount of total nitrogen (%) and assimilative P_2O_5 (%) in soils of Tassili n'Ajjer analyzed by Leredde (1957). (*b*) Variation in the amount of nitrogen (% dry soil weight) in relation to depth (cm) in the California desert under *Larrea tridentata*, *Acacia greggii* and *Cassia armata*. (After Garcia-Moya & McKell, 1970).) (*c*) Variation in the intensity of mineralization of nitrogen compounds (after an incubation period of nine weeks) and of the quantity of assimilative phosphorus in relation to depth in the arid Australian soil. (After Charley & Cowling, 1968.) (*d*) Relation between the amount of assimilative phosphorus in Saharan soils and the number of microorganisms (bacteria and fungi) found in 1 g of moist soil. (After Killian & Feher, 1935.)

produce root systems that are capable of rapidly exploiting a large volume of soil. In addition, Charley & Cowling (1968) point out in regard to *Atriplex vesicaria*, that decaying organs convey very little nitrogen away from the plant due to the fact that before abscission the greater part of the nitrogen has returned to live plant organs.

Proteolysis

This first stage of mineralization of nitrogenous compounds reaching the soil is not well understood. According to Rougieux (1966), in Saharan soils proteolysis is uniform to 1 m in depth. Dubost & Hethener (1966) found proteolytic microorganisms to be somewhat sensitive to dryness in all Tassili n'Ajjer oasis soils. In the Sonoran Desert, fungi are a part of the proteolytic microflora (Ranzoni, 1968). Meanwhile, the proteolytic activity in arid soils is often a function of the vegetation supported there. In arid Moroccan soils, proteolysis is especially active under jujube-trees (Sasson & Kung, 1963). In the Sahara, very active proteolytic microorganisms exist at clumps of *Anabasis aretioides* while there are scarcely any in the surrounding bare reg (Vargues, 1953).

Ammonification

With respect to ammonification, we find the same principles that were outlined for proteolysis. In arid California soils, Lipman (1912) notes that ammonification is a function of the texture and chemical composition of the soil and the extent of root development. In the soil chemical composition, the amounts of humus, water and alkaline salts are of prime importance. These salts strongly reduce the activity of microorganisms in ammonification. When such soils are concentrated on the surface, ammonification can be greater at deeper layers.

Hethener (1967) equally stresses the importance of soil moisture, whose fluctuations regulate the variations of the rate of ammonification.

With respect to the ammonification of urea, a product of animal urine, Rixon (1970) notes that in arid habitats, 45% of ammonia produced does not remain in the soil. There is volatilization of the ammonia which is more significant as it becomes hotter because higher temperatures inhibit the nitrification process (Watson & Lapins, 1968).

Nitrification

In certain desert soils, it is impossible to detect the presence of nitrates (under *Fagonia arabica* in Egypt, for example; Montasir & Shafey, 1951). On the other hand, important quantities of nitrates are found in some desert soils, such as arid soil of Utah or Colorado. However, this fact does not necessarily mean that these soils exhibit intense nitrification, because often it is a question of secondary nitrate accumulations with an ancient or marine origin, as demonstrated by the nature of salts (sodium chloride in particular) which accompany the nitrates into the soil (Stewart, 1913).

In assessing nitrification, it is much better to rely on the activity of microorganisms rather than the amount of soil nitrates. The soils of *Peganum harmala*, a species reputed to be a nitrophile, are not always particularly rich in nitrates (Ozenda, 1954), although the nitrification rate is high (Killian, 1941). In fact, nitrifying bacteria are found in almost all Saharan soils analyzed by Killian & Feher (1935, 1939; Table 13.2). The bacteria are not present in the sand dune area of the Grand Erg Occidental, a reg without vegetation in Tanezrouft, soil with fissures between the steep rocks of the Tamanrasset region (Killian & Feher, 1939), and soils of the Béni-Ounif region (Pochon *et al.*, 1957). According to Lipman (1915), the low level or absence of nitrification found in certain desert soils is correlated with the amount of organic matter which serves as a source of energy for the nitrifiers or the source of NH_4^+ that can be oxidized.

According to Lipman (1912) and Rougieux (1966), another factor likely to limit the activity of nitrifying microorganisms would be the small amount of water in these soils. The effect of high temperatures often found in arid habitats is more complex without doubt. In the arid soils of Israel, the transformation of ammonium sulphate to nitrates is maximum at 28 °C. This transformation is greatly inhibited at 37 °C since the development of *Nitrosomonas* and *Nitrobacter* is reduced at that temperature. But if ammonium sulfate is not added to these soils, nitrification is greater at 37 °C than at 28 °C. This last process is insensitive to chloromycetin (25 mg 100 g^{-1} soil) and potassium chlorate at 10^{-3} M, whereas nitrification from the NH_4^+ is suppressed by those two substances. In arid soils, the formation of nitrates could then be due to phenomena other than oxidation of NH_4^+. Perhaps, intervention by *Achromobacter* or *Aspergillus* could take place (Etinger-Tulczynska, 1969; Table 13.2).

The study of nitrification in relation to soil depth also suggests the intervention of microorganisms particular to arid soils during the nitrification process and is not to be overlooked. Although Dubost & Hethener (1966) believe that one can interpret the nitrification distribution regarding depth by taking into consideration the sensitivity of nitrifying microorganisms to humidity and their strict aerobic character, and although Rougieux (1966) suggests that the maximum nitrifying organisms are found between 25 and 50 cm in these Saharan soils because at this level aeration is still good and variations in temperature are reduced, Lipman (1912) has observed that nitrification can continue up to 3 m in depth, and Thornton (1953) emphasizes the infrequent presence of *Mycobacterium* among the nitrifying organisms found in the upper soil layers.

Vegetation is equally involved in modification of nitrification. In Australia, the amount of soil nitrates is the same under *Atriplex vesicaria* as in the interspaces, but nitrification is two times greater under *Atriplex* than elsewhere. It is under these bushes that nitrogen cycling is most rapid (Rixon, 1970, 1971; Table 13.2).

Table 13.2. *Nitrification and fixation of nitrogen in desert soils.* In each series of experiments, the value of 100 was designated to be the greatest intensity

Influence	Relative values	Source
Vegetation	*Nitrification intensity*	
Under *Atriplex vesicaria*		
September 1968	93	Rixon, 1970
February 1969	100	
June 1969	70	
Bare soil		
September 1968	32	
February 1969	27	
June 1969	48	
Temperature		
Israeli desert soil		
Soil alone		Etinger-Tulczynska, 1969
28 °C	5·9	
37 °C	8·0	
Soil + $(NH_4)_2SO_4$		
28 °C	100	
37 °C	4·1	
Alkaline salts		
California semiarid soils		
with $(NH_4)_2SO_4$	100	Kelly, 1916
+0.05% Na_2CO_3	62	
+0.1% Na_2CO_3	37	
+0.5% Na_2CO_3	1	
+0.1% Na_2SO_4	86	
+0.5% Na_2SO_4	46	
	N_2 fixation intensity	
Water		
Algal crusts in Arizona		Mayland & McIntosh 1966; Mayland, *et al.*, 1966
Moist crusts	100	
Dry crusts	55	
	Number of bacteria per gram of moist soil	
Rhizosphere effects (Egyptian desert)		
Rhizosphere of *Moltakea callosa*		
Nitrifiers	150000	Mahmoud, *et al.*, 1964
Azotobacter	350000	
Clostridium	52000	
Bare soil		
Nitrifiers	100	
Azotobacter	0	
Clostridium	0	

Finally, it is necessary to note the great sensitivity of nitrifying microorganisms to soil alkalinity (Lipman, 1912; Thornton, 1953). Not only is there volatilization of ammonia in alkaline habitats and subsequent inefficiency of nitrification due to lack of NH_4^+ (Sindhu & Cornfield, 1968), but when the pH exceeds 7.7 the activity of nitrifying microorganisms ceases (Martin, Buehrer & Caster, 1942). The nature of the alkaline salts is involved, sodium carbonate being more toxic than sodium sulfate (Kelley, 1916).

Atmospheric nitrogen fixation

The fixation of atmospheric nitrogen by simple physical phenomena (photochemical transformation, adsorption by soil colloids) does not seem to have any importance in desert regions (Garcia-Moya & McKell, 1970), in contrast to biological fixation.

Fixation by free bacteria

The presence of nitrogen-fixing bacteria in desert soils seems very variable with two exceptions, a sandy reg without vegetation in the Adrar region and gypseous sand without vegetation close to In Salah. All Saharan soils analyzed by Killian & Feher (1938, 1939) contain nitrogen-fixing bacteria, whereas Pochon *et al.* (1957) conclude that there was an absence of nitrogen-fixing bacteria in Saharan soils. In 11 arid California soils tested by Lipman (1915), only three contained no *Azotobacter*. In the semiarid soils of Washington State, however, *Azotobacter* density seemed too low for these bacteria to contribute efficiently to nitrogen fixation (Vandecaveye & Moodie, 1942).

In fact, in desert habitats, the development of nitrogen-fixing bacteria is often limited either by aridity or by lack of organic matter, or yet by poor aeration of such arid soils. In the oasis of Tassili n'Ajjer, Dubost & Hethener (1966) note that the aerobic nitrogen fixers are affected by water stress and insufficient aeration, whereas the anaerobic bacteria are less sensitive to aridity. For mounds at the base of *Tamarix*, Killian (1947*b*) states that nitrogen fixation ceases when soils contain less than 10% water. Although sensitive to soil moisture, the frequency of *Azotobacter* and *Clostridium* in Saharan soils corresponds to the amount of soil organic matter. According to Hethener (1967), these microorganisms would assure an enrichment in nitrogen of this organic matter, which would benefit cellulolytic and lignolytic organisms. The organic matter is then broken down, more or less directly dependent on the activity of nitrogen-fixing bacteria, and is itself the limiting factor in the development of these bacteria (Thornton, 1953). Rougieux (1966) also notes the presence of *Azotobacter chroococcum* in soils relatively rich in organic matter where an important cellulolytic activity is revealed. This dependence of nitrogen-fixing bacteria on soil organic matter explains, without doubt, their frequency under the jujube-trees of Morocco (Sasson & Kung, 1963), in sand retained by the lower branches of *Anabasis aretioides* in the Sahara (Vargues, 1953), and at root contact with *Moltakea callosa* in Egypt (Mahmoud *et al.*, 1964).

Finally, according to Anderson's (1958) study of arid soils in Idaho, *Azotobacter* would scarcely develop in soils in which the pH is less than 6.2.

Fixation by the symbiosis of phanerogam bacteria

Bacterial nodules exist on Zygophyllaceae (Sabet, 1946). They could be responsible for the nitrogen-richness of soils populated by *Zygophyllum album* (Montasir & Foda, 1956). On *Zygophyllum coccineum*, nodulation is specific (Mostafa & Mahmoud, 1951), and is favored by dryness and high temperatures (Montasir & Sidrak, 1952).

Little is known of the efficiency of these nodules (Thornton, 1953). Garcia-Moya & McKell (1970) raise this problem in stating that in Arizona the leguminous shrubs do not have tissues particularly high in nitrogen content, and do not contribute more than other species to soil nitrogen enrichment.

Fixation by algae and lichens

The algal crusts of desert soils in Arizona contain up to eight times more nitrogen than the soil underneath. When they become moist they are capable of fixing 2 g of nitrogen per 100 square meters per day, because of the presence of *Nostoc*, *Scytonema* and *Anabaena* (Fletcher & Martin, 1948; Cameron & Fuller, 1960). This fixation is less if these crusts are alternately moistened and dried. If they are completely dry, fixation ceases. Of the nitrogen thereby absorbed 40–45% is again found in the crust in the form of amines, 10% in NH_4^+, and 5% in the form of hexosamines. Water can transport NH_4^+ and glutamic and aspartic acids (Mayland, 1965) away from the crust where they can be utilized by higher plants (Shields & Durrell, 1964; Mayland & McIntosh, 1966; Mayland, McIntosh & Fuller, 1966).

In arid regions of Australia, algal and lichen crusts sometimes cover 30% of the soil surface. Since they contain nigrogen-fixing species (*Collema coccophorus*, for example) their importance in nitrogen cycling is, without doubt, very great (Rogers & Lange, 1966). It is further necessary to note that crusts of algae improve the underlying soil structure by protecting it from water and wind erosion (Alexander, 1969).

This fixation of nitrogen by algal and lichen crusts occurs at all latitudes. In Antarctica, Fogg & Stewart (1968) showed, with the help of $^{15}N_2$, that fixation is effective as soon as the temperature is greater than 0 °C.

Denitrification

In general, denitrifying bacteria exist in desert soils. A single exception is cited by Killian & Feher (1938, 1939) in reg sand close to Béni-Abbès in the Sahara. Rougieux (1966) notes that it is a question especially of *Bacillus* and *Pseudomonas*. In the oasis of Tassili n'Ajjer, Dubost & Hethener (1966) place these organisms mostly in the middle soil horizon; they become more numerous where the vegetation becomes more abundant (Hethener, 1967).

Phosphorus dynamics

According to Charley & Cowling (1968), the total phosphorus content of arid soils is comparable to soils in humid regions (around 600–700 mg P kg^{-1}). The Australian arid soils are an exception to this rule with an average of 240 mg P kg^{-1}. In any case, these quantities represent a large reserve compared to the amounts of phosphorus immobilized in desert vegetation. In soil under *Atriplex vesicaria* in Australia, there are 1082 kg P ha^{-1} to a depth of 45 cm, and only 2.3 kg P ha^{-1} in standing vegetation and litter deposited at the soil surface.

In fact, it seems that exceptions are much more numerous and variability much greater than Charley & Cowling (1968) supposed. Killian & Feher (1935, 1938) found only 17–30 mg of total P kg^{-1} in Saharan soils. These authors were precise in describing chemical forms of soil phosphorus and found that there are only 6.33–6.45 mg P soluble in citric acid per kilogram of soil. In the Tassili n'Ajjer, however, Leredde's (1957) values for this same form of phosphorus go from 30–1660 mg kg^{-1} soil.

Although the results of quantitative analysis are very diverse, considering the methods of analysis utilized, all the authors agree that desert soils are, in general, deficient in assimilative forms of phosphorus. Thus, in microbiological tests conducted on arid Moroccan soils (Sasson & Kung, 1963), and on the oasis soils of the Tassili n'Ajjer (Dubost & Hethener, 1966), the addition of phosphates is always beneficial to microbiotic activity. According to Charley & Cowling (1968), low levels of assimilative phosphorus compounds in certain desert soils could be responsible for the deficiency of these same soils in assimilative nitrogen compounds, the phosphorus being indispensable to the activity of the microorganisms involved in the nitrogen cycle. Rixon (1970) is therefore led to believe that phosphorus and nitrogen are the elements most limiting for productivity in the majority of arid and semiarid regions.

Higher plants and animals contribute to the phosphorus cycle in the soil. The animals leave excrement, often rich in organic phosphorus compounds (as in the case of sheep; Rixon 1970). Perennial vegetation can create local accumulations of phosphorus either at the bases of plants or around their roots, as in the case of *Anabasis aretioides* (Vargues, 1953; Fig. 13.1), but the litter is not always rich in phosphorus as a great proportion of the element leaves senescent organs and returns to the living part of the plant. The phosphorus then remains directly available for future organs, unless soil uptake is essential (Charley & Cowling, 1968).

Killian & Feher (1935, 1938) state that in the summer, Saharan soil is richer in phosphorus compounds soluble in citric acid than in winter. These authors believe that this could be due to weak aestival activity of root absorption or to a seasonal fluctuation of microorganisms. One observes a clear relationship between soil richness in these phosphorus compounds and microbiological activity, without identifying whether it is these compounds that regulate the rapid reproduction of soil microorganisms or whether it is their activity which is responsible for the movement of phosphorus in soluble form (Fig. 13.5*d*). In desert soils there are microorganisms capable of dissolving phosphates. In Egyptian soils, it is mostly a question of sporulant microorganisms and *Streptomyces*; that is to say, microorganisms reputed to be resistant to high temperatures and dryness (Taha Mahmoud, el Damaty & Abd el-Hafez, 1969). Often these microorganisms are concentrated around roots and are apparent only in summer, as in the case of the rhizosphere of *Rhazya stricta* (Elwan & Diab, 1970*b*), *Zilla spinosa* and *Pulicaria crispa* (Elwan & Khodair, 1967, cited by Elwan & Diab, 1970*c*, *d*).

Potassium dynamics

Saharan soils contain only 0.0010–0.0057% of potassium soluble in citric acid, the total potassium being 11–35 times more abundant. Despite a convenient potassium reserve, these soils are deficient in available potassium if compared to moderately fertile soils. It therefore seems that in the desert, the transformation of insoluble soil potassium into soluble compounds under the action of meteorological or biological agents is very slow (Killian & Feher, 1935, 1938).

Some desert vegetation can accumulate significant quantities of potassium in above-ground organs. In *Grayia spinosa* leaves of arid regions in the western United States, K^+ is the most abundant cation. Before abscission, potassium remains entirely in the leaves, which consequently carry appreciable quantities of it to the soil (around 1 g K m^{-2} soil yr^{-1}). The plants that develop at the base of *Grayia* (*Bromus tectorum*, for example) are abnormally rich in potassium because the soil where they root themselves has a high amount of exchangeable K^+ (Rickard, 1965). At the base of *Anabasis aretioides* also, and around its roots, there is enrichment of the soil in potassium (Vargues, 1953; Fig. 13.1).

The K^+ and, in a general way, the cationic state of the soil, can influence the vegetation and vice versa. In Australia, the density of young *Atriplex vesicaria* increases at the same time as the (Na + K):Ca ratio in the soil. For older plants, the rule is inversed (Anderson, 1970). Potassium dynamics are therefore a function of plant age. For *Zilla macroptera*, potassium appears as a very mobile ion; it can return to the soil via the roots during winter (Binet, 1955).

Animals, by their excrement, can equally modify soil cations including those which enter the exchangeable phase in the soil. In Australia, soil in termite mounds is particularly rich in K^+ and Ca^{2+} and other exchangeable cations, because fecal material is incorporated into the soil by termites for their next construction (Lee & Wood, 1971).

Dynamics of sodium chloride and other soluble salts

According to Aubert (1960), sandy soil must be considered saline when the amount of soluble salts reaches 0.2%. The threshold for salinity is 0.4% for a clay-type soil or for a soil in which conductivity of the saturated soil extract is greater than 4 micromhos cm^{-1}. Such soils are often found in arid habitats where soluble salts can be carried away by water, whether horizontally or vertically. Due to intense evaporation, saline deposits occur, beginning with slightly soluble salts (calcium carbonate and calcium sulfate), then the very soluble salts, in particular chloride and carbonates of sodium. Concentration of soil solution can occur very rapidly if the weather is very hot and dry, and if within several meters there is a gradient from a nonsaline soil to a soil very rich in soluble salts (as in the case of the Judean Desert; Evenari & Guggenheim, 1938).

Evidently, high levels of soluble salts in the soil will permit the development of a very specialized flora which can avoid salty soil layers or live in the presence of high salt concentrations. Annual species have a cycle of development which is contained entirely within humid periods, i.e., when there is dilution of the soil solution. *Suaeda asphaltica* requires that only the first 5 cm of soil be weakly saline because it is in the upper horizons that the roots develop (Evenari & Guggenheim, 1938). *Artemisia tridentata* requires that 1 m of its surrounding soil contain few alkaline salts. On the other hand, *Sarcobatus vermiculatus* (Kearney *et al.*, 1914), *Zygophyllum dumosum* (Evenari & Guggenheim, 1938), *Atriplex halimus* (Killian, 1941), *Nitraria tridentata* (Kassas & Imam, 1954) and *Zygophyllum album* (Montasir & Foda, 1956) can tolerate strong soil salinity.

The kind of salts in solution in the soil is equally important. *Zygophyllum album* resists sulfates well, but does not tolerate carbonates (Montasir & Foda, 1956). *Zygophyllum waterlotii* and *Randonia africana* are species that are gypsophiles (Quezel & Simmoneau, 1963). *Atriplex halimus* tolerates a high concentration of sodium chloride (Killian, 1941).

Once installed in these saline soils, the flora contributes to the distribution and movement of the constituent elements of the salts in solution. Merely the presence of vegetation can disturb these movements. In the Egyptian desert, *Pennisetum dichotomum* protects the soil against all wind erosion and thereby produces a progressive augmentation of the salinity of the surface soil (Kassas & Imam, 1954). In other cases, Ozenda (1954) notes that by reducing winter floods and summer evaporation, the vegetation diminishes the vertical migration of salts.

It is the same at the base of *Halocnemum*, where sand carried by wind accumulates on the soil, upon which *Suaeda fruticosa* can develop as it is isolated from the saline clay directly underneath.

However, each species contains, and so absorbs, proportions of mineral elements which are characteristic of it, and therefore participates uniquely in the cycling of diverse elements (Killian, 1947*a*). In the Negev, *Tamarix* plants in deep, salty soil take up significant quantities of sodium chloride which accumulate in the leafed branches. Once dead, these leaves gather at the base of the tree and mix into the first centimeter of soil, whose enrichment in sodium chloride manifests itself in the development of a halophilous and ruderal flora (Litav, 1957; Fig. 13.6*a*). A comparable phenomenon takes place with various species of *Atriplex* (Jones & Hodgkinson, 1970), and with *Sarcobatus vermiculatus*, whose leaves are particularly rich in sodium and under which 1 m^2 of soil can receive 3.75 g of sodium each year from dead plant organs (Rickard, 1965). In this last case, the soil accumulates sodium and is alkalized, because in the leaves of *Sarcobatus*, Na^+ is accompanied by organic anions which are transformed in the soil into carbonate and bicarbonate ions which precipitate Ca^{2+} and Mg^{2+}. The sodium then displaces the calcium and magnesium adsorbed onto soil colloids from a pH augmentation (Fireman & Hayward, 1952; Fig. 13.6*b*).

This raising of deep salts by such a process is, however, not general. It does not occur with *Acacia cyanophylla*, whose decaying plant organs do not accumulate at the base of the tree but are immediately carried away by the wind (Litav, 1957). In this case, one can say that the salts from deeper horizons are dispersed over the surface of the landscape by wind.

When the above-ground plant organs, in particular the leaves, receive large quantities of soluble salts, there are often salt secretions at specialized glands, as in the case of *Tamarix* (Zohary & Orshansky, 1949; Campbell & Strong, 1964; Thompson, Berry & Liu, 1969), *Fagonia* and *Zygophyllum simplex* (Killian, 1974*a*, *b*), or in superficial hairs, as in *Atriplex* (Jones & Hodgkinson, 1970; Osmond *et al.*, 1969; Mozafar & Goodin, 1970). These structures create saline crusts at the leaf surfaces, which in desert habitats are rarely washed off, and fall concurrently with the encrusted organ. In certain cases the secretion of salt can take place from the roots as in the Fezzan around the roots of *Zygophyllum simplex*, in which the sand is concentrated because of rejection by plant organs of anions (Cl^- and SO_4^{2-}), cations (Ca^{2+} and Mg^{2+}) and pectic compounds (Killian, 1947*a*, *b*).

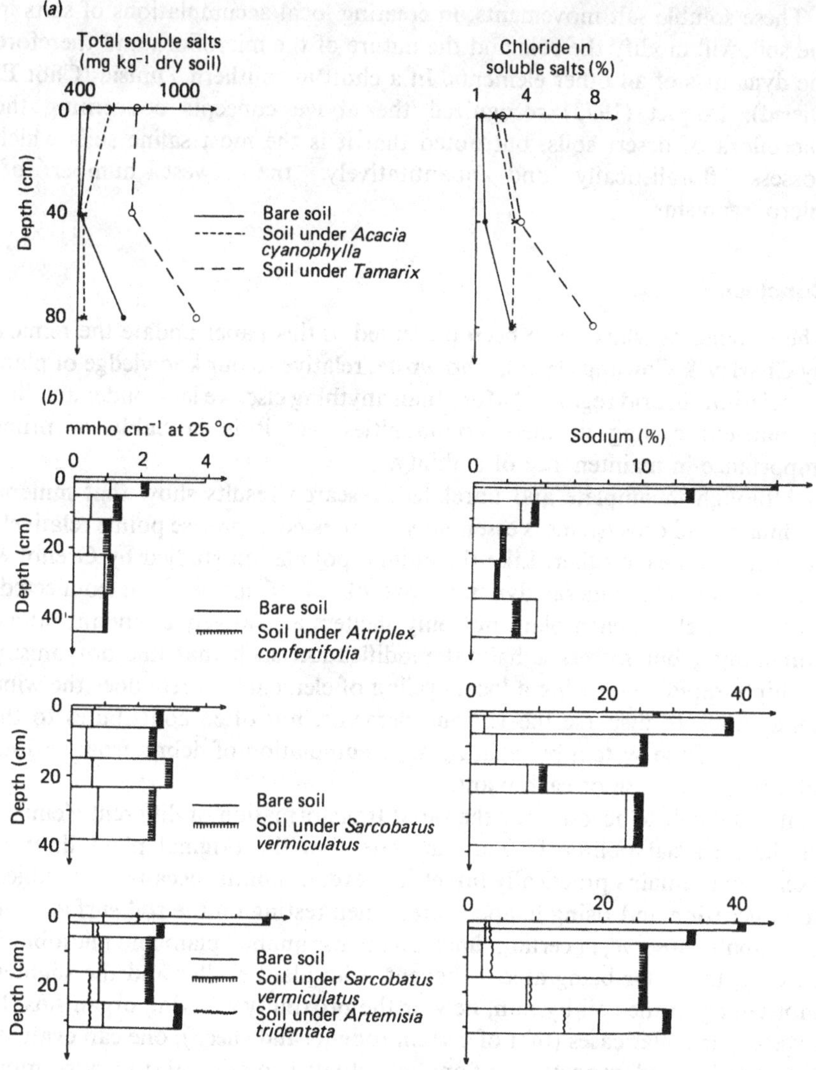

Fig. 13.6. Dynamics of sodium chloride and soluble salts in desert soils. (*a*) Variation in the quantity of soluble salts and of the percentage of Cl^- in soluble salts in relation to depth in the northern Negev. (After Litav, 1957.) (*b*) Variation in the specific electrical conductivity from an extract of saturated soil (in water), and the percentage of sodium in the cationic phase of exchange in relation to depth in the desert soils of Utah. (After Fireman & Hayward, 1952.)

These soluble salt movements, in creating local accumulations of salts in the soil, will modify the role and the nature of the microflora and therefore the dynamics of all other elements. In a chott in southern Tunisia (Chot El Djerid), Loquet (1972) recognized the above concepts concerning the microflora of desert soils, but noted that it is the most saline soils which possess, floralistically and quantitatively, the fewest number of-microorganisms.

Conclusions

The statements which have been presented in this paper update the remark by Charley & Cowling (1968), who wrote, relative to our knowledge of plant associations in arid regions: 'More than anything else, we lack understanding of nutrient cycling in these communities, yet it is probably of prime importance in maintenance of stability.'

Although incomplete and unrelated, research results show that mineral cycling in arid ecosystems is essentially manifested at precise points relatively isolated from each other. Like the animal population studied by Grenot & Niaussat (1967), mineral dynamics are of the 'dispersed and connected' pattern. In effect, each plant not only shelters a relatively abundant faunal community, but fosters a habitat modification such that microorganisms multiply rapidly, allowing a local cycling of elements. Rarely does the wind come along to disperse the various elements, but often contributes to the coalition of the system by creating an accumulation of debris from various sources at the base of each plant.

It is difficult to account for the rapid transformation of different elements at these special points. If one can state that the original plant litter or excrement remains practically intact for several months because it is subject to desiccation and rising temperatures when resting on the soil surface, one can equally observe, at certain times, an intense animal, plant and microbiotic activity, the latter being able to benefit from plant shelter and the slightest moistening of the soil by rain, dew or the proximity of living organisms. In several particular cases (that of certain rodents and sheep), one can evaluate the duration and importance of animal activities in mineral transport; more often it is difficult, however – indeed impossible – to pinpoint the role of most animals in mineral dynamics. As to higher vegetation, if we are to stress its importance it is necessary to realize that information is lacking in the following areas: seasonal and daily variations of mineral absorption and their relation to water movement; the rhythm of the shedding of deciduous organs; the importance of leaching of above-ground organs; root competition in relation to soil nutritive elements, or competition between roots and micro-organisms. Each focal point in mineral dynamics should be an object of integrated study in which the utilization of radioisotopes could, without doubt, be of great service.

References

Alexander, M. (1969). Microbiologial problems of the arid zone. In: *International conference on global impacts of applied microbiology, Addis Abeba, Ethiopia, 1967*, pp. 285–91. Wiley, New York.

Anderson, D. J. (1969). Analysis of patterns in *Atriplex vesicaria* communities from the Riverine plain of New South Wales. In: *The biology of* Atriplex (ed. R. Jones), pp. 63–8. CSIRO Division of Plant Industry, Canberra.

Anderson, G. R. (1958). Ecology of *Azotobacter* in soils of the Palouse region. I. Occurrence. *Soil Science*, **86**(2), 57–61.

Aubert, G. (1960). *Les sols de la zone aride. UNESCO, Colloque de Paris, Communication No. 5*. UNESCO, Paris.

Bernard, F. (1954). Rôle des insectes sociaux dans les terrains du Sahara. *Travaux de l'Institut Recherche Sahariennes, Alger*, **12** (2), 29–39.

Binet, P. (1955). Action du climat désertique sur le développement, la forme, la structure et le métabolisme de *Zilla macroptera* Coss. Thesis, University of Paris.

Bullock, J. A. (1967). The insect factor in plant ecology. *Journal of Indian Botanical Society*, **46** (4), 323–30.

Cameron, R. E. & Fuller, W. H. (1960). Nitrogen fixation by some Algae in Arizona soils. *Proceedings of the Soil Science Society of America*, **24**, 353–6.

Campbell, C. J. & Strong, J. E. (1964). Salt gland anatomy in *Tamarix pentandra* (Tamaricaceae). *South Western Naturalist*, **9** (4), 232–8.

Charley, J. L. & Cowling, S. W. (1968). Changes in soil nutrient status resulting from overgrazing and their consequences in plant communities of semi-arid areas. *Proceedings of the Ecological Society of Australia*, **3**, 23–38.

Cloudsley-Thompson, J. L. (1964*a*). Terrestrial animals in dry heat: introduction. In: *Handbook of Physiology* (ed. D. B. Hill), Section 4, pp. 447–9. American Physical Society, Washington.

Cloudsley-Thompson, J. L. (1964*b*). Terrestrial animals in dry heat. In: *Handbook of Physiology* (ed. D. B. Hill), Section 4, pp. 451–65. American Physical Society, Washington.

Delye, G. (1967). Recherches sur l'ecologie, la physiologie, et l'ethologie des Fourmis du Sahara. Thesis, Université d'Aix-Marseille.

Dommergues, Y. (1962). Contribution à l'étude de la dynamique microbienne des sols en zone semi-aride et en zone tropicale sèche. Thesis, University of Paris.

Dubost, D. & Hethener, P. (1966). Aperçu microbiologique des sols de deux oasis du Tassili N'Ajjer: Djanet et Iheria. *Travaux de l'Institut Recherche Sahariennes*, **25** (1–2), 7–27.

Dubuis, A., Faurel, L., Grenot, C. & Vernet, R. (1971). Sur le régime alimentaire du Lézard saharien *Uromastix acanthinurus* Bell. *Comptes Rendus hebdomadaire des séances de l'Académie des Sciences, Paris, Série D*, **273**, 500–3.

Durrel, L. W. & Shields, L. M. (1960). Fungi isolated in culture from soils of the Nevada Test site. *Mycologia*, **52** (4), 636–41.

Elwan, S. H. & Diab, A. (1970*a*). Studies in desert microbiology. II. Development of bacteria in the rhizosphere and soil of *Artemisia monosperma*. Del. in relation to environment. *United Arab Republic Journal of Botany*, **13** (1), 97–108.

Elwan, S. H. & Diab, A. (1970*b*). Studies in desert microbiology. III. Certain aspects of the rhizosphere effect of *Rhazya stricta* Decn. in relation to environment. *United Arab Republic Journal of Botany*, **13** (1), 109–19.

Elwan, S. H. & Diab, A. (1970*c*). Studies in desert microbiology. IV. Bacteriology of the root region of a fodder xerophyte in relation to environment. *United Arab Republic Journal of Botany*, **13** (2), 159–69.

Elwan, S. H. & Diab, A. (1970*d*). Studies in desert microbiology. V. Certain patterns of bacterial development in relation to depth and environment. *United Arab Republic Journal of Botany*, **13** (2), 171–9.

Etinger-Tulczynska, R. (1969). A comparative study of nitrification in soils from arid and semi-arid areas of Israel. *Journal of Soil Science*, **20** (2), 307–17.

Evenari, M. & Guggenheim, K. (1938). Etude de la distribution de quelques plantes du désert de Judée en fonction des conditions du sol. *Bulletin des Societies Botanique, Genève, 2ème série*, **29**, 43–71.

Fireman, M. & Hayward, H. E. (1952). Indicator significance of some shrubs in the Escalante Desert, Utah. *Botanical Gazette*, **114** (2), 143–55.

Fletcher, J. E. & Martin, W. P. Some effects of Algae and Molds in the rain-crust of desert soils. *Ecology*, **29**, 95–100.

Fogg, G. E. & Stewart, W. D. P. (1968). In situ determinations of biological nitrogen fixation in Antarctica. *British Antarctic Survey Bulletin*, **15**, 39–46.

Garcia-Moya, E. & McKell, C. M. (1970). Contribution of shrubs to the nitrogen economy of a desert-wash plant community. *Ecology*, **51** (1), 81–8.

Grenot, Cl. (1969). Sur le régime alimentaire de la vipère à cornes. *Science et Nature*, **96**, 21–4.

Grenot, Cl. & Niaussat, P. (1967). Aperçu écologique sur une région hyperdésertique du Sahara central (Reggan). *Science et Nature*, **81**, 2–16.

Grenot, Cl. & Vernet, R. (1972). Les reptiles dans l'écosystème au Sahara Occidental. *Comptes Rendus des séances de la Société Biogéographie*. **4**, 96–112.

Grenot, Cl. & Vernet, R. (1973). Sur une population *d'Uromastix acanthinurus* Bell. isolée au milieu du Grand Erg occidental (Sahara algérien). *Comptes Rendus hebdomadaire des séances de l'Académie des Sciences, Paris, Série D*, **276**, 1349–52.

Gupta, R. K. & Saxena, S. K. (1971). Ecological studies on the protected and overgrazed rangelands in the arid zone of W. Rajasthan. *Journal of the Indian Botanical Society*, **50** (4), 289–300.

Hethener, P. (1967). Activité microbiologique des sols à *Cupressus dupreziana* A. Camus au Tassili N'Ajjer (Sahara central). *Bulletin de la Société d'Histoire Naturelle de l'Afrique du Nord, Algérie*, **58** (1–2), 39–100.

Jones, R. & Hodgkinson, K. C. (1970). Root growth of rangeland chenopods: morphology and production of *Atriplex nummularia* and *Atriplex vesicaria*. In: *The biology of* Atriplex (ed. R. Jones), pp. 77–85. CSIRO Division of Plant Industry, Canberra.

Kachkarov, D. N. & Korovine, E. P. (1942). *La vie dans les déserts*. Payot, Paris.

Kassas, M. & Girgis, W. A. (1970). Habitat and plant communities in the Egyptian Desert. VII. Geographical facies of plant communities. *Journal of Ecology*, **58** (2), 335–50.

Kassas, M. & Imam, M. (1954). Habitat and plant communities in the Egyptian Desert. III. The wadi bed ecosystem. *Journal of Ecology*, **42** (2), 424–41.

Kearney, T. H., Briggs, L. J., Shantz, H. L., McLane, J. W. & Piemeisel, R. L. (1914). Indicator significance of vegetation in Tooele Valley, Utah. *Journal of Agricultural Research*, **1** (5), 365–417.

Kelley, W. J. (1916). Nitrification in semiarid soils. *Israeli Journal of Agricultural Research*, **7** (10), 417–37.

Kellog, C. E. (1953). Potentialities and problems of arid soils. In: *Desert Research. Proceedings of an International Symposium, Jerusalem. May 1952*, pp. 19–40. Wiley, New York.

Killian, Ch. (1941). Sols et plantes indicatrices dans les parties non irriguées de oasis de Figuig et de Beni-Ounif. *Bulletin de la Société d'Histoire Naturelle de l'Afrique du Nord*, **32** (8), 301–14.

Killian, Ch. (1947*a*). Biologie végétale au Fezzan. I. Observations sur la biologie de quelques plantes fezzanaises. *Travaux de l'Institut Recherche Sahariennes, Alger*, **4**, 3–63.

Killian, Ch. 1947*b*). Biologie végétale au Fezzan. II. Sur la nebka à *Tamarix aphylla* au Centre Fezzanais et son humidification en particulier. *Travaux de l'Institut Recherche Sahariennes, Alger*, **4**, 67–107.

Killian, Ch. (1951). Observation sur la biologie d'un halophyte saharien, *Frankenia pulverulenta. Travaux de l'Institut Recherche Sahariennes, Alger*, **7**, 87–109.

Killian, Ch. & Feher, D. (1935). Recherches sur les phénomènes microbiologiques des sols sahariens. *Annales de l'Institut Pasteur*, **55**, 573–623.

Killian, Ch. & Feher, D. (1938). Le rôle et l'importance de l'exploration microbiologique des sols sahariens. In: *La vie dans la région désertique nord-tropicale de l'ancien monde*, pp. 81–106. Société de Bio-géographie, Paris.

Killian, Ch. & Feher, D. (1939). Recherches sur la microbiologie des sols désertiques. *Encyclopédie Biologique, Paris*, **21**, 1–123.

Kozlowski, T. T. (1968). *Water deficits and plant growth*. Academic Press, New York.

Krause, W. (1958). Boden und Pflanzengesellschaften. *Handbuch der Pflanzenphysiologie*, **4**, 807–50.

Lee, K. E. & Wood, T. G. (1971). Physical and chemical effects on soils of some Australian termites and their pedological significance. *Pedobiologia*, **11** (5), 376–409.

Leigh, J. H. & Wilson, A. D. (1970). Utilization of *Atriplex* species by sheep. In: *The biology of* Atriplex (ed. R. Jones), pp. 97–104. CSIRO Division of Plant Industry, Canberra.

Leredde, C. (1957). *Etude écologique et phytogéographique du Tassili n'Ajjer*, vol. 2. Institut des Recherches Sahariennes, Alger.

Lipman, C. B. (1912). The distribution and activities of bacteria in soils of the arid regions. *University of California Publications in Agricultural Science*, **1** (1), 1–20.

Lipman, C. B. (1915). The nitrogen problem in arid soils. *Proceedings of the National Academy of Sciences, USA*, **1** (9), 477–80.

Litav, M. (1957). The influence of *Tamarix aphylla* on soil composition in the northern Negev of Israel. *Bulletin of the Research Council of Israel*, **6** D (1), 38–45.

Loquet, M. (1972). Etude pédologique de deux sols halomorphes. (Estuaire de la Somme et Chott El Djerid: Sud Tunisien.) III. Ecologie des Champignons du sol (étude de trois stations). *Annales de l'Institut Pasteur*, **122** (6), 1151–70.

Macfarlane, W. V. (1964). Terrestrial animals in dry heat: ungulates. In: *Handbook of Physiology* (ed. D. B. Dill), Section 4, pp. 509–39. American Physical Society, Washington.

Mahmoud, S. A. Z., Abou el-Fadl, M. & Elmofty, M. Kh. (1964). Studies on the rhizosphere microflora of a desert plant. *Folia Mikrobiologica*, **9**, 1–8.

Martin, W. P., Buehrer, T. F. & Caster, A. B. (1942). Threshold pH value for the nitrification of ammonia in desert soils. *Proceedings of Soil Science Society of America*, **7**, 223–8.

Mayland, H. F. (1965). Isotopic nitrogen fixation by desert algal crust organisms. *Dissertation Abstracts* (*Agriculture*), **26** (3), 1268.

Mayland, H. F. & MacIntosh, T. H. (1966). Availability of biologically fixed atmospheric nitrogen-15 to higher plants. *Nature, London*, **209**, 421–2.

Mayland, H. F., MacIntosh, T. H. & Fuller, W, H. (1966). Fixation of isotopic nitrogen on a semiarid soil by algal crust organisms. *Proceedings of the Soil Science Society of America*, **30** (1), 56–60.

Migahid, A. M., Abd el Rahman, A. A., el Shafei Ali, M. & Hammouda, M. A. (1955). Types of habitat and vegetation at Ras El Hikma. *Bulletin de l'Institut du Desert d'Egypte*, **5** (2), 107–90.

Montasir, A. H. & Foda, H. A. (1955). Effect of soil moisture, water table and soil structure on the development of *Zygophyllum album* L., *Bulletin de l'Institut du Désert d'Egypte*, **5** (1), 16–34.

Montasir, A. H. & Foda, H. A. (1956). Edaphic habitat of *Zygophyllum album* L., *Bulletin de l'Institut du Désert d'Egypte*, **6** (1), 74–97.

Montasir, A. H. & Shafey, M. (1951). Studies on the autecology of *Fagonia arabica*. *Bulletin de l'Institut de Désert d'Egypte*, **1** (1), 55–73.

Montasir, A. H. & Sidrak, G. H. (1952). Root nodulation in *Zygophyllum coccineum* L. *Bulletin de l'Institut Fouad I du Désert*, **2** (1), 68–70.

Mostafa, M. A. & Mahmoud, M. Z. (1951). Bacterial isolates from root nodules of Zygophyllaceae. *Nature, London*, **167**, 446–7.

Moureaux, Cl. (1965). Glycolyse et activité microbiologique globale en divers sols ouest-africains. *Pédologie*, **3** (1), 43–77.

Mozafar, A. & Goodin, J. R. (1970). Vesiculated hairs: a mechanism for salt tolerance in *Atriplex halimus* L. *Plant Physiology*, **45** (1), 62–5.

Muller, C. H. (1953). The association of desert annuals with shrubs. *American Journal of Botany*, **40** (2), 53–60.

Nicot, J. (1960). Some characteristics of the microflora in desert sands. In: *The ecology of soil fungi* (ed. D. Parkinson & J. S. Waid), pp. 94–7. Liverpool University Press.

Osborn, T. G. B., Wood, J. G. & Paltridge, T. B. (1932). On the growth and reaction to grazing of the perennial saltbush, *Atriplex vesicaria*. An ecological study of the biotic factor. *Proceedings of the Linnean Society of New South Wales*, **57** (5–6), 243–4, 377–402.

Osmond, C. B., Lüttge, U., West, K. R., Pallaghy, C. K. & Shacher-Hill, B. (1969). Ion absorption in *Atriplex* leaf tissue. II. Secretion of ions to epidermal bladders. *Australian Journal of Biological Science*, **22**, 797–814.

Ozenda, P. (1954). Observations sur la végétation d'une région semi-aride: les Hauts-Plateaux du Sud-Algérois. *Bulletin de la Société d'Histoire Naturelle de l'Afrique du Nord*, **45**, 189–225.

Ozenda, P. (1958). *Flore du Sahara septentrional et central*. Centre National Recherches Scientifique, Paris.

Pochon, J., de Barjac, H. & Lajudie, J. (1957). Recherches sur la microflore des sols sahariens. *Annales de l'Institut Pasteur*, **92** (6), 833–6.

Quastel, J. H. (1965). Soil metabolism. *Annual Review of Plant Physiology*, **16**, 217–40.

Quezel, P. & Simmoneau, P. (1963). Les peuplements d'Acacia du Sahara nord-occidental. Etude Phytosociologique. *Travaux de l'Institut Recherches Sahariennes*, **20**, 79–121.

Ranzoni, F. V. (1968). Fungi isolated in culture from soils of the Sonoran desert. *Mycologia*, **60** (2), 356–71.

Rickard, W. H. (1965). Sodium and potassium accumulation by greasewood and hopsage leaves. *Botanical Gazette*, **126** (2), 116–19.

Rixon, A. J. (1970). Cycling of nutrients in a grazed *Atriplex vesicaria* community. In: *The biology of* Atriplex (ed. R. Jones), pp. 87–95. CSIRO Division of Plant Industry, Canberra.

Rixon, A. J. (1971). Oxygen uptake and nitrification by soil within a grazed *Atriplex vesicaria* community in semiarid rangeland. *Journal of Range Management*, **24** (6), 435–9.

Rogers, R. W. & Lange, R. T. (1966). Nitrogen fixation by lichens of arid soil crusts. *Nature, London*, **209**, 96–7.

Rougieux, R. (1966). Contribution à l'étude de l'activité microbienne en sol désertique (Sahara). Thesis, University of Bordeaux.

Russel, J. S. (1968). Implications of the effect of nitrogen on daily pasture growth rates in the semi-arid sub-tropics. *Journal of the Australian Institute of Agricultural Science*, **34** (3), 169–70.

Sabet, Y. S. (1946). Bacterial root nodules in the Zygophyllaceae. *Nature, London*, **157**, 656–7.

Sabinin, D. A. & Minina, E. G. (1932). Das mikrobiologische Bodenprofil als zonales Kennzeichen. In: *Proceedings of the 2nd International Congress of Soil Science, Leningrad–Moscow*, pp. 224–35.

Sasson, A. & Kung, F. (1963). Sur l'activité biologique des sols arides du Maroc. *Bulletin des Sociétés Sciences Naturelle Physique, Maroc*, **43** (1–2), 55–77.

Shields, L. M. & Durrel, L. W. (1964). Algae in relation to soil fertility. *Botanical Review*, **30** (1), 92–128.

Shreve, F. (1942). The desert vegetation of North America. *Botanical Review*, **8** (4), 195–246.

Sindhu, M. A. & Cornfield, A. H. (1968). Effect of simulated reclamation on nitrification of ammonium sulfate applied to saline and alkali soils from West Pakistan. *Journal of Science of Food and Agriculture*, **19** (11), 648–50.

Skujins, J. (1973). Dehydrogenase: an indicator of biological activities in arid soils. *Bulletin of the Ecological Research Committee, Stockholm*, **17**, 235–41.

Stewart, R. (1913). The intensity of nitrification in arid soils. *Zentralblatt für Bakteriologie und Parasitenkunde*, **2/36**, 477–90.

Taha, S. M., Mahmoud, A. Z., el-Damaty, H. A. & Abd el-Hafez, A. M. (1969). Activity of phosphate dissolving bacteria in Egyptian soils. *Plant and Soil*, **31** (1), 149–60.

Thompson, W. W., Berry, W. L. & Liu, L. L. (1969). Localization and secretion of salt by the salt glands of *Tamarix aphylla*. *Proceedings of the National Academy of Sciences, USA*, **63** (2), 310–17.

Thornton, H. G. (1953). Some problems presented by the microbiology of arid soils. In: *Desert research, proceedings of the International Symposium, Jerusalem, May 1952*, pp. 295–300. Research Council of Israel, Special Publication No. 2.

Vandecaveye, S. C. & Moodie, C. D. (1942). Occurrence and activity of *Azotobacter* in semiarid soils in Washington. *Proceedings of the Soil Science Society of America*, **7**, 229–36.

Vargues, H. (1953). Etude microbiologique de quelques sols sahariens en relation avec la présence d'*Anabasis aretioides* Coss, et Moq. In: *Desert research, Proceedings of the International Symposium, Jerusalem, May 1952*, pp. 318–24. Research Council of Israel, Special Pubication No. 2.

Vernet, R. & Grenot, C. (1972). Etude du milieu et structure trophique du peuplement reptilien dans le Grand Erg Occidental. (Sahara algérien). *Comptes Rendus de la Société Biogéographie*, **4**, 112–23.

Watson, E. R. & Lapins, P. (1968). Losses of nitrogen from urine on soils from south-western Australia. *Australian Journal of Experimental Agriculture and Animal Husbandry*, **9**, 85–91.

Wells, W. F. (1967). Aspects of shrub–herb productivity in an arid environment. MS thesis, University of California, Berkeley.

Wilson, A. D. (1966). The value of *Atriplex* (Saltbush) and *Kochia* (bluebush) species as food for sheep. *Australian Journal of Agricultural Research*, **17**, 147–53.

Wilson, A. D., Leigh, J. H. & Mulham, W. E. (1969). A study of Merino sheep grazing a bladder saltbush (*Atriplex vesicaria*) – cotton-bush (*Kochia aphylla*) community on the Riverine Plain. *Australian Journal of Agricultural Research*, **20**, 1123–36.
Zohary, M. & Orshansky, G. (1949). Structure and ecology of the vegetation in the Dead Sea region of Palestine. *Palestine Journal of Botany* **4** (4), 177–206.

Manuscript received by the editors September 1975

14. Long-term dynamics in arid-land vegetation and ecosystems of North Africa

H. N. LE HOUÉROU

Introduction

Arid lands are here considered in a broad sense, that is to say those lying below the 400 mm isohyet. Long-term dynamics are understood to be those that occur over periods of more than one year. The time scale and units may be defined as in Table 14.1.

The time scale in Table 14.1 is not arbitrary. There is little information on earlier human cultures and the environments and ecosystems in which they lived. This is the case, for instance, for Chellean and Acheulean cultures dating back from 1000000 to 100000 B.P. and contemporary to the European glaciations of Mindel and Gunz or to the North African rainy periods or 'pluvial' Saletian and Villafranchian–Moulouyan. Information is becoming much more precise with the Aterian cultures dating back from 50000 to 20000 B.P. and with the Ibero–Maurusian and Capsian cultures dating back from 17000 to 5000 B.P., times for which there is good evidence from pollen analysis. This is true again for the neolithic period, 5000 to 2500 B.P., where rock paintings and engravings give supplementary information mostly on the fauna and human populations.

The historic period starts about 2500 B.P. with documents written by Herodotus. This period corresponds to Phoenician and Roman history which in North Africa ends by 1300 B.P. when the whole country was taken over by the Arab conquerors. Modern history starts about 100 years ago when the country was open to European influence during the colonial period. During Pleistocene and prehistoric times, vegetation and environmental changes were mainly due to important climatic changes, whereas there is no evidence of climatic changes during the historic period. Changes in vegetation and environment during the historic period are mainly due to human actions (Monod, 1958; Le Houérou, 1968).

Methodology

The methodology followed is complex and makes use of results from various sciences and techniques. These, in turn, depend upon the periods concerned as shown in Table 14.2. This is obviously not the place to review these various

Table 14.1. *Time scale and units for long-term dynamics*

Periods	Time spans (years B.P.)	Units of time scale (yr)
Early and middle Pleistocene	3000000 to 50000	10000 to 100000
Prehistoric and Protohistoric	50000 to 2500	100 to 1000
Early historic	2500 to 1300	10 to 100
Late historic	1300 to 100	1 to 10
Present time	100 to now	1

Table 14.2. *Methodology used to study long-term dynamics*

Period	Sciences and techniques used
Pleistocene	Geology, geomorphology, paleontology, palynology, prehistory, radiochronology
Prehistoric and Protohistoric	Palaeontology, palynology, prehistory, archaeology, radiochronology
Historic (early)	Archaeology, history, epigraphy, palynology, dendrochronology
Historic (late)	History, ethnology, documents written by explorers and early scientists
Present time	History, inventories, surveys, maps, remote sensing, research, experimentation, monitoring

sciences and techniques nor their detailed contribution to the knowledge of vegetation dynamics. However, techniques used in surveys or present day research warrant some comments. More detailed and reliable data are obtained from climatic studies, ecological surveys, palynology research, phytosociological research, vegetation and soil mapping, monitoring and experimental field work. Botanical exploration started during the eighteenth century, whereas vegetation studies were initiated only in this century.

In Western Europe, cadastral surveys and detailed cadastral maps (with scales of 1:5000 to 1:20000) date back to the eighteenth century, and sometimes earlier, giving invaluable and precise information on long term dynamics (Kuhnholtz-Lordat, 1945; Barry, 1961). However, in North Africa, vegetation mapping was not undertaken before the present century – the earlier small scale maps were drawn between 1920 and 1930. Medium to large scale vegetation maps were drawn later, after the Second World War and particularly since the mid-fifties.

Detailed vegetation maps of various scales (1:10000 to 1:500000) cover the areas shown in Table 14.3. A small country like Tunisia (160000 km^2) has a total of more than 1000 terrestrial plant communities which have all

Table 14.3. *Countries for which detailed vegetation maps are available*

Country	Area (ha × 10^{-3})
Morocco	7000
Algeria	10000
Tunisia	16000
Libya	3000
Total	36000

been identified, described, classified and mapped at large to intermediate scale (1:50000 to 1:500000) over the whole country.

In the course of vegetation studies, numerous plant communities have been defined and their dynamic relationships established through the concept of 'vegetation series'. This concept has mainly been developed by Gaussen (1952, 1954) and by Kuhnholtz-Lordat (1945).

A vegetation series is a sequence of plant communities which may replace each other in a definite order from the stable primeval type little influenced by man to communities strongly affected such as cultivated fields. Each series corresponds to specific ecological conditions of climate and soil. Each term of a given series is thus a result of a certain type and intensity of human action and land use on a specific physical environment which is common to all the terms of the series.

Present-day vegetation

The main vegetation types encountered in North African arid lands are described in Volume 1, Chapter 3.

Main plant communities

Aleppo pine forests

This type of vegetation covers an area of about 100000 ha in the arid zone where it occupies hills and low mountains in the 200–400 mm rainfall belt, mainly in the 300–400 mm belt. Aleppo pine forests are mainly developed in the semi-arid and sub-humid climates (400–800 mm rainfall) where they occupy about 1 × 10^6 ha (Morocco, 60000 ha; Algeria, 700000 ha; Tunisia, 240000 ha). The plant communities vary inside the Aleppo pine forests according to climatic and edaphic conditions.

Biomass and productivity

The standing above-ground phytomass of these forests is about 20000–40000 kg ha^{-1} dry matter. The above-ground productivity is 800–1600 kg $ha^{-1} yr^{-1}$ dry matter. Timber production is 0.5 $m^3 ha^{-1} yr^{-1}$ and the total wood production (timber+fuel) 300–600 kg ha^{-1}. Browsed and grazed vegetation amounts to 300 to 600 kg $ha^{-1} yr^{-1}$ and the litter to 200–600 kg $ha^{-1} yr^{-1}$ dry matter.

Degraded forests and garrigues of Juniperus phoenicea *and* Rosmarinus officinalis

This type of shrubland (matorral, chapparal) occupies about 1×10^6 ha within the arid zone. It is derived from the Aleppo pine forest (or from the *Tetraclinis* forest, along the coast) where the trees have been removed either by cutting or burning (or both) together with browsing. Floristic differences from the original forest are of two types.

(1) Qualitative: (*a*) disappearance of *Pinus halepensis* and of a few other species closely linked to shade (Sciaphytes) and to high content of organic matter in the top soil (humicolous species); (*b*) appearance of some steppic heliophilous species such as *Artemisia herba alba*, *Atractylis serratuloides*, *Helianthemum kahiricum* and *Herniaria fontanesii*.

(2) Quantitative: Shrub species, which play a minor role in vegetation structure of the forest, become dominant, such as; *Juniperus phoenicea*, *Rosmarinus officinalis*, *Globularia alypum*, *Cistus libanotis*, *Fumana thymifolia* and *Thymus hirtus*. Also *Stipa tenacissima* (alfa grass) often becomes dominant.

Soil types are degraded forms of those of the Aleppo pine forest with lower organic matter content (2.5 to 5%) and a somewhat degraded and less stable structure. Geological substrata are identical.

Biomass and productivity

Above-ground phytomass varies considerably according to the degradation status – between 5000 and 15000 kg ha^{-1} dry matter. Above-ground dry matter production ranges from 400–1200 kg $ha^{-1} yr^{-1}$.

Alfa grass steppes

These occur between the 100–400 mm isohyets and today cover about 5.5×10^6 ha (Libya 0.4, Tunisia 0.6, Algeria 3.0, Morocco 1.5). However, alfa grass steppes covered more than 8×10^6 ha at the beginning of this century. This reduction in surface is due to over-exploitation (alfa grass is used for making luxury paper; from *c.* 1860 until recently it was exported to Europe; it is now transformed into paper paste in Algeria and Tunisia), by clearing for cultivation and by burning for grazing. *Stipa tenacissima* is dominant, covering 10–60% of the ground (according to the depletion status of the steppe). Soil is often shallow and barren between the tussocks of alfa grass. A few perennial small shrubs are found, however, such as: *Artemisia herba alba, Thymus hirtus, Helianthemum ruficomum, Hippocrepis scabra* and *Hammada scoparia.*

Annuals develop, in springtime mainly, inside the tussocks where soil is deeper. In many places, sparse individuals of forest remnants may still be found, such as *Juniperus phoenicaea, Rosmarinus officinalis, Genista microcephala,* and *Cistus libanotis.*

When the alfa steppe is degraded (which is the usual case) steppic shrubs tend to replace the alfa grass occupying the ground left barren by the disappearance of the grass tussocks. These small shrubs are mainly: *Artemisia herba alba, Helianthemum ruficomum, Atractylis serratuloides, Thymelaea tartonraira, Atractylis humilis, Thymelaea nitida, Helianthemum kahiricum, Thymelaea hirsuta, Hammada scoparia, Thymelaea microphylla* and *Noaea mucronata.* This invasion of thinned and degraded alfa steppe by under-shrubs gives way progressively to chamaephytic steppes.

When the alfa steppe is cleared for cultivation the land is invaded by annual weeds and forbs. When cultivation either ceases or is practised at intervals of several years, small shrubs tend to dominate again through several dynamic stages. These stages depend on frequency of cultivation. The dominant annual weeds with various periods of cultivation are as follows:

(*a*) Cultivation every year or second year: *Chrysanthemum coronarium; Anacyclus clavatus.*

(*b*) Cultivation every 3–5 yr: *Artemisia campestris; Cynodon dactylon.*

(*c*) Cultivation every 5–20 yr: *Artemisia herba alba, Hammada scoparia; Plantago albicans.*

(*d*) Cultivation every 20–50 yr: *Lygeum spartum.*

In fact, succession may be stopped at any stage and then restart from stage (*a*). however, succession from stage (*c*) or (*d*) towards the alfa grass steppe has never been reported.

Once alfa grass has been removed it never comes back. Regeneration can only occur under the shade of open forest or inside long-protected enclosures (10–20 yr). This is why chamaephytic steppes cover such huge surfaces in the North American arid zone today.

Alfa grass steppes occur on many types of substrata, for example, limestone, sandstone, marl, gypsum, calcareous crusts (calcrete), and sandy alluvia. The only common features of all sites occupied are very good drainage, alkaline pH, lack of hydromorphy and, usually, lack of salinity.

Within given climatic conditions, a specific floristic composition of the steppe is linked to each type of substratum. There are thus many related but different plant communities which have a common physiognomic feature – the dominance of *Stipa tenacissima.* All these various plant communities, floristically and ecologically characterized, present several dynamic stages according to their degradation status. Each stage in each plant community has a characteristic range of biomass and productivity.

Standing above-ground biomass may vary from less than 1000 kg ha^{-1} dry matter to over 10000 kg ha^{-1} dry matter according to climate, soil, depletion status and floristic composition (qualitative and quantitative). In the same way, and linked to the same factors, production may vary from 100 to 1000 kg ha^{-1} yr^{-1} dry matter.

Alfa production for fibre (paper paste) averages about 200 kg ha^{-1} dry matter. But it used to be three times more at the beginning of this century. This reduction in productivity is a result of over-exploitation and other misuses such as burning to develop early grazing. Alfa grass has low palatability and only for a short period in winter time, destruction by fire encourages other more palatable species; in addition the young sprouts appearing after burning are highly nutritious and palatable. However, the plant dies if burning occurs twice within 3–5 yr.

Chamaephytic steppes

As suggested above, chamaephytic steppes result from the degradation of gramineous steppes. The main types of chamaephytic steppes are those dominated by *Artemisia herba alba* and *A. campestris.*

Artemisia herba alba

They cover immense areas from Spain (Zaragoza, Murcia) through all North Africa and the Near East to Iran, Afghanistan and southern Russia. In North Africa alone they cover 8–10 × 10^6 ha. All these steppes are characterized by medium to heavy textured top soil (often shallow) associated with average rainfall below 400 mm.

Associated species are numerous and depend on climatic conditions, soil conditions, floristic zone or region, human and grazing pressure. In many cases *A. herba alba* itself is removed by overgrazing or fuel gathering and replaced by other still more hardy species such as *Salsola vermiculata* and *Anabasis oropediorum* or by ephemeroids: *Poa bulbosa* and *Carex pachystylis* (Near East only for the last species).

The above-ground standing biomass varies from over 3000 kg ha^{-1} dry matter to less than 300 kg ha^{-1} dry matter according to climate, soil, and depletion status; a good stand on a deep soil between the 350–400 mm isohyets would have a phytomass of 3000 to 4000 kg ha^{-1} dry matter whereas a poor stand on shallow soil on the 100 mm isohyet would have a phytomass of 200–300 kg ha^{-1} dry matter.

Production varies in the same proportions and according to the same factors from 800–1000 kg ha^{-1} yr^{-1} dry matter to 50–100 kg ha^{-1} yr^{-1} dry matter.

Psammophilous steppes of Artemisia campestris

These occupy 3–5 × 10^6 ha in North Africa. Great areas of this type of steppe have been converted to orchards especially in Tunisia, Libya and Morocco. These dry land orchards (olives, almonds, figs, apricots) cover close to 1.5 × 10^6 ha of which 60% is in Tunisia. The dynamic stages observed in these areas are as follows.

(*a*) Culture and fallow characterized by annual weeds: *Cutandia divaricata, Lotus pusillus, Cutandia dichotoma, Koeleria salzmanni, Schismus calycinus, Launaea resedifolia, Senecio gallicus, Ifloga spicata, Silene colorata, Brassica tournefortii, Hippocrepis bicontorta, Anthemis pedunculata* and *Anacyclus cyrtolepidioides.*

(*b*) Post-cultural steppe aged 3–5 yr characterized by: *Cynodon dactylon, Onopordon arenarium, Atractylis candida, Cleome arabica, Carduus gaetulus, Centaurea dimorpha, Euphorbia terracina, Diplotaxis simplex* and *Pituranthos tortuosus.*

(*c*) Steppes over 5 yr old characterized by: *Helianthemum lippii, Echiochilon fruticosum, Elizaldia violacea, Hammada schmittiana, Thymelaea microphylla, Rhantherium suaveolens, Aristida obtusa, Eragrostis papposa, Aristida plumosa, Plantago albicans, Polygonum equisetiforme, Nolletia chrysocomoides* and *Argyrolobium uniflorum.*

Soils are always sandy (medium textured to coarse) calcareous and water storage capability is always fair to good.

Above-ground phytomass may vary from 5000 to 500 kg ha^{-1} dry matter depending on climate (average rainfall) and depletion stage. Production is 1500–500 kg ha^{-1} yr^{-1} dry matter according to the same criteria. Owing to water storage capability production is higher and more regular than in the *A. herba alba* steppes.

Halophilous crassulescent steppes

These types of steppes are numerous and complex. Their common feature is the dominance of halophilous chenopod shrubs such as: *Salsola vermiculata, Atriplex halimus, Salsola zygophylla, Atriplex glauca, Salsola tentranda, Atriplex malvana, Salsola tetragona, Atriplex mollis, Suaeda fructicosa, Arthrocnemum glaucum, Suaeda pruinosa, Halocnemum strobilaceum,* and

Suaeda mollis. These generally occur in depressions with some periodic flooding and more or less deep saline water table.

Above-ground phytomass may vary from 1000 to 15000 kg ha^{-1} dry matter. Production may be from 5000 to 300 kg ha^{-1} yr^{-1} dry matter according to ecological conditions.

Most of these crassulescent steppes, except in the more halophilous environments, seem to be derived from woodland dominated by tree and tall shrub species such as: *Tamarix gallica*, *Tamarix aphylla*, *Tamarix africana*, *Tamarix boveana*, *Nitraria retusa*, *Phoenix dactylifera*, *Phragmites communis*, *Nerium deander*, *Atriplex halimus*, and *Arundo plinii*.

Pseudo steppes

These comprise tall shrubs or small trees such as: *Ziziphus lotus*, *Retama raetam*, *Nitraria retusa*, *Calligonum comosum*, *Rhus tripartitum*, *Calligonum azel*, *Rhus pentaphyllum*, *Calligonum arich*, *Acacia raddiana*, *Genista saharae*, *Acacia gummifera*, *Periploca loevigata*, *Ephedra slata*, and *Argania spinosa*. The shrubs are 0.5–5 m high and occupy terraces and depressions or dunes. Tall shrubs are sparse and interspread with chamaephytic and other life forms.

Soils are alway deep and have a favourable water balance due to topography, permeability, run-on, periodical floodings, or water tables.

In depressions and terraces, this vegetation is derived from a riverine woodland; *Acacia raddiana* and/or *Pistacia atlantica* were the dominant species in the primeval woodland. Wood cutting and fuel gathering have been responsible for this degradation over many centuries. These plant communities are also often cleared for cultivation and this results in cultigene plant communities dominated by weeds. However, *Ziziphus lotus* hummocks remain in the middle of cultivated fields as testimonies of the previous natural vegetation.

Summary of present-day dynamics

The scheme shown in Fig. 14.1 shows that dynamics from forest to degraded forest works both ways. That is,any protected degraded forest would evolve towards a forest without problem whenever protection is long enough.

Evolution from cultivated field towards chamaephytic steppes is also quite usual. But no evolution from chamaephytic steppe to alfa steppe has been reported. Similarly, evolution from alfa steppe towards forest or garrique has not been proved (to my knowledge) once the forest remnants have disappeared.

Summarizing, one could say that degradation is reversible at the top and at the bottom, but not in the intermediate part of Fig. 14.1.

This can be explained by the biology of *Stipa tenacissima* and of the forest

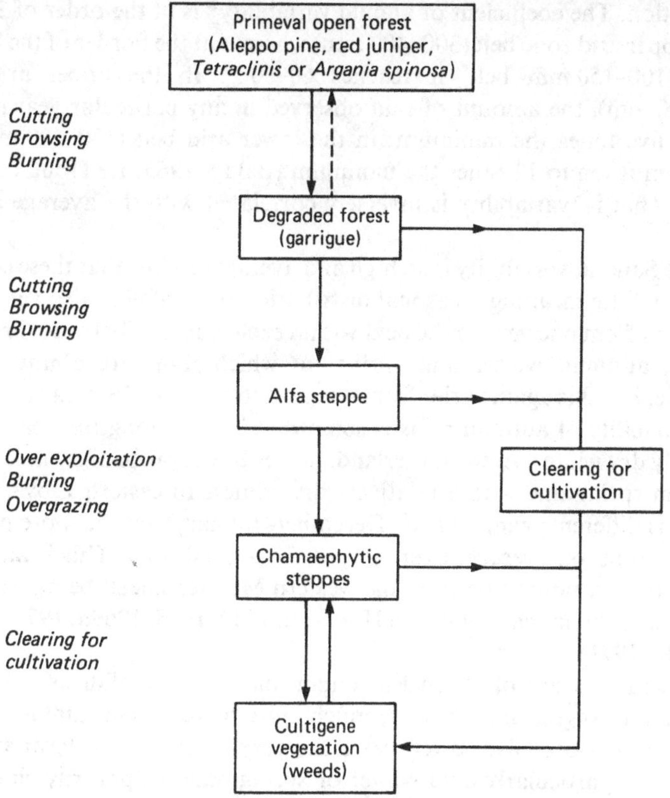

Fig. 14.1. Scheme summarizing present-day dynamics.

species. Both need shade and long protection, otherwise their seedlings can neither withstand browsing, grazing, trampling, nor high temperature and evapotranspiration demand. These regeneration conditions are met in the forest but not in the sun burnt, scorching steppes.

This shows the fragility of arid environments. Once certain thresholds are passed, the degradation is irreversible. This is what we have called 'steppization' and 'desertization'. (Le Houérou, 1969*a*).

Vegetation and ecosystem changes due to climatic fluctuations over periods of a few years

General: variability of climatic factors

These changes are mainly due to the variability either in rainfall amount and distribution from one year to the next, or from sequences of rainy and dry years. The variability is not only in total annual amount but also in seasonal distribution. The coefficient of annual variability* is of the order of 30–40% in the upper arid zone belt (300–400 mm) whereas at the border of the Sahara, in the 100–150 mm belt, it reaches 60–80%. In the upper arid zone (300–400 mm), the amount of rain observed in any particular year is about four to five times the minimum; in the lower arid belt (100–200 mm) this maximum is ten to 12 times the minimum (Baldy, 1965; Le Houérou, 1959, 1969*a*). That is, variability is inversely correlated with the average amount of rain.

In the Sahara, variability is so high and averages so low that these concepts have very little meaning. Seasonal distribution of rainfall is also very erratic and varies from one year to the next within each season. There are three rainy seasons, autumn, winter and spring, in which rains are almost equally probable. But this again varies from one area to the next. Generally speaking, the probability of autumn rains is somewhat higher along the coast, with a relatively dry winter. In the hinterland, however, the probability is somewhat higher in springtime with a relatively dry winter. In eastern Libya, rainfall pattern is different: winter rains (December–January) are the more probable and represent, on average, over 50% of the annual total. This is an eastern Mediterranean rainfall pattern, the western Mediterranean being characterized by an autumn maximum (Le Houérou, 1959, 1965, 1969*a*, 1971; Akman & Daget, 1971).

Temperatures are of much less importance than rainfall as a factor in vegetation changes, since they are much more regular than rainfall: temperature averages over five to ten years are very close to long term averages. However, a particularly cold winter or spring may temporarily eliminate a few species or favour others.

Direct effect of rainfall fluctuations

This may be considered in at least three ways: (*a*) the effect of rainfall on the botanical composition of plant communities; (*b*) the effect of rainfall on the number of individuals of each species and on plant cover; and (*c*) the effect of rainfall on species development: biomass and production.

The effect of rainfall on botanical composition (understood as presence–absence) of plant communities is very strong, especially on annual species. Sequences of rainy years will favour the expansion of species that are usually found under less arid conditions. On the contrary, series of drought years will induce a northward expansion of desert species; however, this concerns mainly annual or biannual species, as one may expect. But this also depends on soil types. In drought years, there are few or no annuals on silty to clayey soils and on shallow soils, whereas their number is reduced on sandy soils.

* $V = \sigma/P \times 100$, where P is average annual precipitation in mm and σ is the standard deviation of annual precipitations.

On the contrary, during a rainy year, numerous annuals will develop especially on silty to heavy textured soils. The effect of rainfall on perennials is much less important, although during a sequence of several dry years, many individuals from some species may die in some sites.

The effect of rainfall on number of individuals in each species is very strong both on annuals and perennials. The latter will mainly reseed during the rainy years and therefore their density will increase during those years. Plant cover also varies in the same way from almost nil to 100%.

Biomass and production vary in the same proportions as plant cover: from practically nil to several thousand kg ha^{-1} yr^{-1} dry matter, especially on medium to heavy textured soils. This again depends on soil types: the variability is much lower on sandy soils. Plant size varies in the same proportions; for instance, *Erucaria vesicaria* may either not germinate or reach a size of 15 mm at maturity or up to 1500 mm in a particularly favourable year.

Indirect effects of rainfall fluctuations

The indirect effects of climate fluctuations are probably more important and certainly much better known that its direct effects. These indirect effects are mainly concerned with: (*a*) clearing the steppe for cereal cultivation; (*b*) grazing conditions; (*c*) effects on the fauna; and (*d*) effects on man.

When the autumn and early winter rains are favourable, almost all the arable lands are tilled and sown with cereals (mostly barley). The surfaces tilled depend on the intensity of the rains and their distribution before and during tilling time (October–January). The vegetation is also affected by the tilling technique. The surface cultivated may vary between years from one to 10 million hectares.

Tilling by means of the traditional swing plough will kill some perennials but many of those will survive and re-seed if tilling does not take place during consecutive years. The natural vegetation will then recover rather quickly. The regeneration will in turn depend on several factors: on climatic conditions during subsequent years, on whether tillage will take place or not during consecutive years, on animal pressure (stocking rate) and on the entropy* of the systems present (steppe versus cultigene vegetation). However, tilling will kill most individuals of some species, for instance *Stipa tenacissima*. As this

* Entropy is understood here (Le Houérou, 1969*a*, p. 241) as the relative depletion status of an ecosystem in respect to its ability to come back to its initial stage. For instance, when in a particular site most of the *Artemisia herba alba* steppe has been cleared and is actually cultivated every year, the potential of the cultigene vegetation to evolve back to the *A. herba alpa* steppe is low due to permanent cultivation and low seed availability, hence the entropy is high. When, on the contrary the proportion of cultivated steppe is small and the cultivation takes place once in several years, the entropy is low since time is long enough between two cultivations to come back to the *A. herba alba* steppe, and seeds are plentiful.

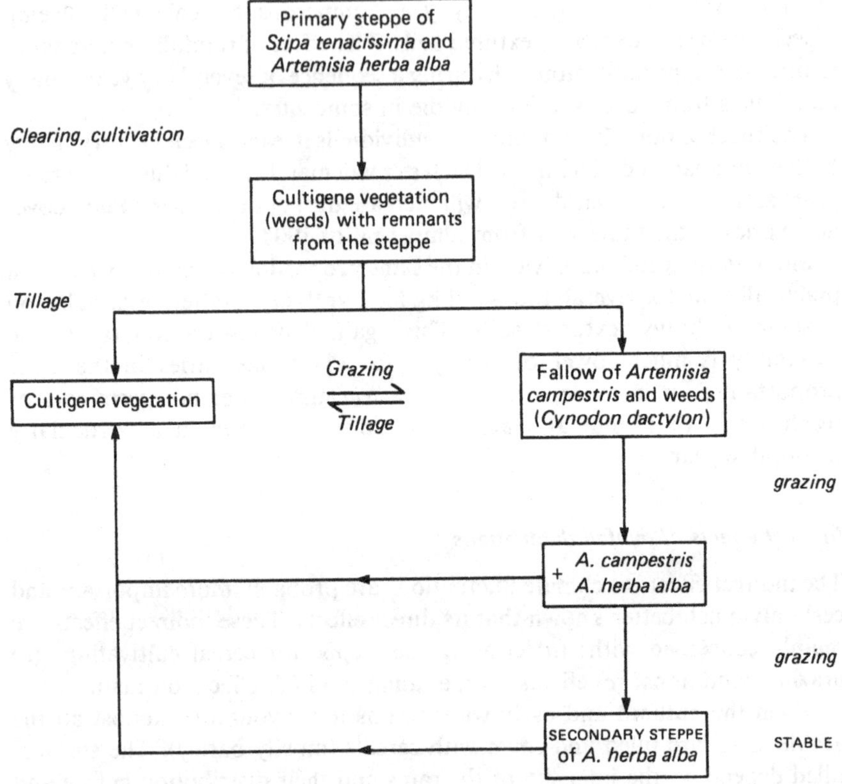

Fig. 14. 2. Dynamic succession on silty and deep soils.

species does not re-seed in the steppe (rhyzomatous expansion is the only method of propagation for *S. tenacissima* in the steppe), tillage will eliminate it for ever. This is why *S. tenacissima* steppes can only be found on shallow rocky soils, today, since these are non-arable land. On silty, more or less deep soils, the dynamic succession is as in Fig. 14.2. This, again, is the reason why the border between the *Stipa tenacissima* steppes and the *Artemisia herba alba* steppes is very often a sharp line which corresponds to the past maximum expansion of cultivation (Le Houérou & Froment, 1966, p. 85).

How often a particular piece of land is cultivated and hence its dynamic stage) depends also on soil-topographic conditions. In the major depressions, with regular run-on waters, cultivation occurs almost every year, whereas on the pediments it occurs irregularly, depending on climatic and soil-topographic conditions.

The dynamics between steppe and cultigene vegetation may be summarized with the help of a few concepts: stability, permanence, resilience and entropy

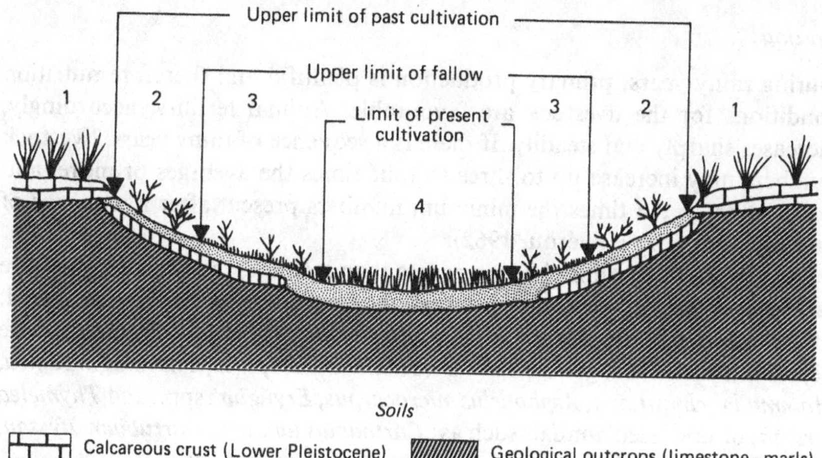

Fig. 14.3. The steppic catena. *Vegetation*. 1. Primary steppe of *Stipa tenacissima* on shallow soil (usually calcareous crust). 2. Secondary steppe of *Artemisia herba alba* on silty soil with or without a deep calcrete layer. 3. Steppe/fallow of *Artemisia campestris* and *Cynodon dactylon* on deep silty soils. 4. Cultigene vegetation (weeds): cereals and recent fallows. Deep silty soils with some run in.

(Fig. 14.3). These have been discussed by many authors, for instance, Noy-Meir (1974) and Le Houérou (1969*a*).

The *Stipa tenacissima* steppe is permanent but not resilient, since once it has been destroyed, it does not re-establish; therefore, it is not stable in the wider sense. On the contrary, the *Artemisia herba alba* steppes are stable since they are both permanent and resilient. They may be considered as a neoclimax (Le Houérou, 1969*a*), the time of regeneration being of three to ten years with an intermediary stage of *Artemisia campestris*.

Forests and degraded forests are also stable up to a certain point. However, these concepts should be completed with the notion of 'entropy' (Le Houérou, 1969*a*) which is inversely related to resilience. If, for instance, in a given and large enough area, *Artemisia* steppes cover 90% of the land and cultivated fields 10% (low entropy), the resilience will be high. In the reverse situation, it will be low when the entropy is high. If the steppe has totally disappeared over large areas, resilience will also tend to disappear. It is a matter of seed availability and timing between successive tillages. The tilling technique is also of primary importance. Generalized mechanical tillage will very soon lead to a homogenous cultigene vegetation and total disappearance of steppe remnants, hence resilience tends to zero. Entropy becomes maximum if tillage occurs regularly every one to three years.

Grazing

During rainy years, primary production is plentiful and therefore nutrition conditions for the livestock are favourable. Animal fertility, accordingly, increases sharply and steadily. If there is a sequence of rainy years, livestock numbers may increase up to three to four times the averages or more, and up to seven to ten times the minimum numbers present after a sequence of drought years (Le Houérou, 1962).

Therefore, when rainfall becomes less than average, the stocking rates are far beyond the carrying capacity and the vegetation is badly overgrazed. Many perennials are destroyed and progressively replaced by unpalatable perennials such as: *Peganum harmala, Hammada scoparia, Thapsia garganica, Hammada schmittiana, Asphodelus microcarpus, Eryngium* spp., and *Thymelea hirsuta*; or ungrazed annuals such as: *Carthamus lanatus, Marrubium alysson, Malva parviflora, Carlina involucrata*, and *Cleome arabica*.

If the drought lasts several years, many animals will die (70% in 1946–48) and a new cycle takes place.

To what extent this vegetation of unpalatable species is stable is not known. However, one may suspect that it was not stable in the past; otherwise, it would now cover most of the arid zone as such cycles have been operating for centuries. But it seems also that this type of vegetation is becoming permanent under the present day conditions, due to the demographic explosion which characterized the second half of this century and which results in a high and constant animal and human pressure on the ecosystems. The average stocking rate in the arid zone of North Africa (between the 100 and 400 mm isohyets) is 0.5 sheep equivalents per hectare. This is about twice what it should be for long-term sustained productivity as shown by many experiments over the last 40 years in the area and in comparison with more or less similar conditions of northern America, southern Asia, southern Africa, and southern and Western Australia.

Productivity from overgrazed ecosystems is currently 20–33% of that when the same ecosystem is rationally grazed (Rodin & Vinogradov, 1970; Delhaye, Le Houérou & Sarson, 1974; Floret & Pontanier, 1974). In some circumstances, it is only 10%.

Overgrazing acts in the following ways:

(*a*) by reducing plant cover and biomass, mainly of perennials;

(*b*) by reducing the number of annuals in the long term through lack of seed production;

(*c*) by reducing the number of palatable species and individuals;

(*d*) by replacing palatable species by unpalatable ones;

(*e*) by strongly reducing primary production and still more drastically secondary production;

(*f*) by increasing erosion as a result of a reduced plant cover.

All experiments show that after a few years of total protection, the chamaephytic steppes are dominated by palatable perennial grasses such as: *Stipa lagascae*, *Stipa parviflora*, *Stipagrostis ciliata*, *S. plumosa*, *S. acutiflora*, *Dactylis hispanica*, *Hyparrhenia hirta*, *Cenchrus ciliaris* and *Digitaria nodosa*, which together may cover up to 80% of soil surface.

Evidence of this fact is given in long-term exclosures (5–15 yr) in Libya, Tunisia, Algeria or Morocco (Le Houérou, 1962, 1969*a*, *b*; 1970).

Fauna

Populations of wild animals follow a similar pattern to domestic livestock. Populations increase steadily during rainy years. This is particularly the case for rodents; for instance, in good years, densities of over 100 individuals ha^{-1} have been reported for *Psammomys obesus* (Bernard, 1969; Franclet & Le Houérou, 1971). Then they suddenly almost disappear, but drought is probably not the only factor involved. The same applies to other mammals as well as to insects, particularly grasshoppers and locusts.

Even sea ecocystems are strongly affected by rainfall fluctuations. During the rainy years, important amounts of run-off (billions of cubic metres) reach the sea, carrying considerable quantities of dung which favour the plankton; professional fishermen know that good fishing seasons will ensue.

Man

Even man is affected by climatic fluctuations. The nomads and shepherds whose nutrition depends largely on livestock production, show a much higher fertility (higher birth rates and lower infant mortality) during sequences of good years when milk is plentiful. There have been, to my knowledge, no studies on the subject in North Africa, although this is common knowledge.

Allogenic succession

This is ecological succession due to geomorphic processes of active glyptogenesis; either by erosion (hydric, aeolian) or by sedimentation (floodings, dune formation).

Erosion

Generally speaking, erosion is due to, or linked with, anthropozoogenic actions. However, after a certain stage has been reached, the phenomenon sometimes becomes irreversible, even if the initial causative factor is discontinued. This is the case, for instance, on steep sloping gypsiferous marls eroded in badlands. Even with a total protection, erosion will often continue, as perennial vegetation is not able to establish itself, owing to the speed of the erosion process.

In the erosion processes, the dynamics of plant successions show a progressive change from communities dominated by trees, then by shrubs, then by perennial herbs and finally by annuals. This is a classical sequence common to many areas besides North Africa and the Mediterranean basin and does not therefore call for more comment.

However, there is a very little known process (I observed it only twice; in the Messad region, in the steppic high plains of Algeria and in the Bergent region of eastern Morocco); that is the 'steppe tigrée', analogous with Clos-Arceduc's (1956) 'brousse tigrée' of Niger and Mali and similar to the 'vegetation arcs' described in Somalia (Boaler & Hodge, 1962, 1964; Hemming, 1965) and in Jordan (White, 1969). The process is similar to the one described by Boudet (1971) in the Sahelian 'brousse tigrée' although climatic vegetation and soils conditions are quite different here from those of the African dry tropics.

The 'steppe tigrée' consists of strips a few metres wide of alternatively dense and sparse plant cover which develops on silty soils over calcareous crusts on pediments. The slope is 3–7%. The strips develop in contour as a result of sheet erosion on the almost barren strips alternating with sedimentation on the intermediary strips. It is a sheet-flood or sheet-wash process over distances of 2–10 m with alternate sheet erosion and silting. This has been observed in *Artemisia herba alba* steppes. The eroded strips have a cover of 5–10% and are dominated by *Hammada scoparia* (Algeria) or *Anabasis aphylla* (Morocco) whereas the sedimentation strips have a plant cover of 30–60% and are dominated by *Artemisia herba alba*.

Sheet erosion affects a layer of 1–5 cm thick in the top silty soil which becomes hard, sealed on the surface and unfavourable for plant development. Sedimentation affects a layer of about the same thickness, but as water intake and plant cover are much higher, conditions are much more favourable to plant life. The eroded strips, besides having a much sparser plant cover, are dominated by species considered as more xerophytic than those of the flooding strips (*Hammada scoparia* versus *Artemesia herba alba* for instance).

The strips move slowly from downslope to upslope by a process of regressive erosion. The speed of the phenomenon is not known: it is probably a few centimetres per year on average. The strips are conspicuous on aerial photographs at the 1:25000 scale on which they induce images which resemble a tiger skin, hence the name.

It seems quite possible that, considered in a geological perspective of tens or hundreds of thousand years, this mechanism might have contributed to the making of the pleistocene pediments which cover huge areas in the North African arid zone.

Wind erosion

Excluding the true desert, wind erosion in the arid zone is mostly due to a combination of climatic and human factors. It is mainly active in the 50–250 mm belt. It will be examined further on p. 377 under desertization.

Sedimentation

Very large changes in vegetation and ecosystems may be induced by sedimentation occurring with catastrophic floodings. This was the case in central Tunisia in September–October, 1969 where close to 50 000 ha were covered with a layer of 20–200 cm of sediments (from coarse sand to heavy clay).

The natural vegetation is then usually killed except for a few tall shrubs or small trees. The new barren sediments are progressively colonized by pioneer species which vary greatly with: climatic conditions, the nature of the sediment (clay, sand) and its thickness. The pioneer species are often unpalatable weeds including naturalized species such as *Xanthium spinosum*.

The nature and composition of the plant communities which ensue depend on many factors, including anthropogeny. The result after stabilization, a few years later, may be totally different from the natural vegetation that existed before the sedimentation.

Autogenic succession

This concerns changes in vegetation due to the micro-environmental effects of the vegetation itself or to the effect of fauna.

(*a*) Effect of plant cover and shade on the structure and botanical composition of the community.

(*b*) Effect of hummock building by some species.

(*c*) Effect of dune fixation by natural or artificial vegetation.

(*d*) Ant-hill effect.

Effect of plant cover

The size, density and ground cover of the taller perennials affect strongly the composition of plant communities either *directly* – many species are linked to shade (sciaphytes) or to the open sun (heliophytes); *indirectly* – as soil properties under shade are different from those under open sun. They have more balanced temperatures, higher organic matter content, higher structure stability, higher permeability and hence water intake and balance, and so on.

This is why 'desertization', and the reverse process of 'biological recovery' are self-catalysing mechanisms. This will be dealt with below under the headings 'Desertization' and 'Steppization'.

The 'nebka effect'

'Nebka' is an Arabic word for a hillock made by accumulation of sand around a shrub or a tree, which grows up as the sand accumulates. This accumulation of sand is slow enough to permit deposition of alternating layers of sand and organic matter from the leaves shed from the shrub or tree. The nebka is therefore not drifting sand but is almost stable. Both run-off and sand storms play a role in this nebka building since they occur in depressions and along the stream network. The mechanism of nebka formation and the special biotope it constitutes has been studied by Killian (1945) and by Long (1954). A nebka may be 1–5 m in height and from 2–10 m in diameter; it usually shows an ovoid or ellipsoid shape when seen from above. The plants which create nebka are mainly *Ziziphus lotus* or *Tamarix* spp. and sometimes *Nitraria retusa*, *Atriplex halimus*, or *Limoniastrum guyonianum*.

The nebka is a special biotope where the fauna is very rich: insects, amphibians, rodents, lizards and snakes find shelter and food. The soil of the nebka is very rich in organic matter, particularly nitrogen-rich substances, which results in specific plant communities dominated by nitratophytes such as: *Sisymbrium irio*, *Mesembrianthemum cristallinum*, *Emex spinosus*, *Pegamum harmala*, *Diplotaxis muralis*, and *Malva parviflora*.

Whenever areas with nebkas are cleared for cultivation, crops are always of better quality in the place formerly occupied by nebkas. This is known to ecologists and agronomists as the '*Ziziphus* effect' (since the nebkas are usually built around *Ziziphus lotus* shrubs).

Dune stabilization

Dune stabilization is achieved in two ways: natural and artificial. Natural stabilization occurs under total protection and may take a very long time. I have never seen natural stabilization in areas with mean rainfall below 100 mm. Artificial stabilization is induced by mechanical interventions to reduce sand mobility and the introduction of fast growing shrubs and trees adapted to sandy habitats.

In both cases, the vegetation changes include colonization by pioneer species such as *Silene arenaroides*, *Scorophularia saharae*, *Astragalus gombo*, *Onopordon arenarium*, *Astragalus gombiformis*, *Rumex tingitanus*, *Nolletia chrysocomoides* and *Aristida pungens*. At a later stage of stabilization, other species will develop such as: *Retama raetam*, *Euphorbia guyoniana*, *Artemisia campestris*, *Helianthemum lippii*, and *Echiochilon fruticosum*, whereas artificial dune fixation will use: *Acacia cyanophylla, Calligonum azel, Acacia ligulata, Calligonum arich, Saccharum spontaneum, Ricinus communis,* and *Eucalyptus* spp.

Anthill effect

In many plant communities, particularly degraded communities, one may notice, mainly in spring time, dark green spots in the middle of lighter vegetation. This is often the effect of ant-hills where botanical composition, as well as plant development and density, are different from surrounding communities: production is also usually much higher. In very degraded conditions as, for instance, in the *Anabasis oropediorum–Salsola vermiculata* steppe of the Hodna region of Algeria, ant-hill vegetation (*Erucaria vesicaria–Medicago truncatula*) represents 1–2% of the total area, but with a production 10–20 times higher than that of the surrounding community. This seems due to several factors linked to ants' activities: (*a*) increased permeability and water intake due to improved soil structure; (*b*) increased fertility from ants catabolism; and (*c*) increased amount of seeds gathered in the surroundings.

Steppization

This term is used to designate a radical change in vegetation and soil from forest to steppe (Le Houérou, 1969*a*, pp. 332–58). This change may be slow and progressive or fast and drastic. It is slow under the usual conditions of forest degradation due to over-cutting, over-browsing, distillation (rosemary), burning and other mismanagements. It is sudden and violent when clearing for cultivation occurs, followed, as often happens, by abandonment and replacement of cultigene by steppic communities.

What happens to the vegetation?

The structure of the vegetation becomes simpler: from a multi-layer structure (trees, shrubs, undershrubs, herbs, mosses) it tends to a two-layer structure: undershrubs and herbs. In the case of clearing for cultivation, the original complex structure is suddenly replaced by an even simpler one characterized by weeds (cultigene). When cultivation ceases, and when entropy is not too high, the cultigene vegetation is replaced by steppic vegetation characterized by undershrubs, perennial grasses, herbs and ephemeroids.

The floristic composition changes drastically together with structure (progressively or suddenly) from forest remnants and companions (sciaphytes and heliophytes) such as: *Rosmarinus officinalis, Cistus villosus, Globularia alypum, Phillyrea media, Juniperus phoenicea, Fumana ericoides, Thymus hirtus, Fumana thymifolia, Thymus capitatus* and *Rhamnus lycioides*; to steppic species such as: *Artemisia herba alba, Atractylis serratuloides, Hammada scoparia, Atractylis humilis, Gymnocarpos decander, Helianthemum kahiricum*; or weeds such as *Erucaria vesicaria, Chrysanthemum coronarium, Anacyclus clavatus, Launaea nudicaulis, Coronilla scorpioides, Centaurea nicaeensis*. This succession shows increasing xerophily.

What happens to the soils?

Changes in soils are parallel to those of vegetation. The amount of organic matter decreases from 5–15% in forest to 2–3% in steppe and 0.5–2% in cultigene vegetation (basic soil and climatic conditions remaining identical). The nature of organic matter changes; the carbon:nitrogen ratio decreases from 15–20% to 8–10% and the relation of humic acids to total organic matter increases from less than 50% to over 70%. The distribution of the organic matter in the soil also changes. In forest and degraded forest soils, the organic matter is concentrated in the upper layers (rendzina, brown calcareous soils) whereas in steppic soils, organic matter is more evenly distributed and with a progressive decrease in content from upper to deep layers (isohumic soils).

Structure stability decreases sharply from forest to steppic and to cultigene soils, especially in the upper layers. Linked to structure, content and nature of organic matter, permeability and water intake decreases from forest to steppe soils. Inversely, run-off increases and water balance deteriorates. This is very clearly reflected in the vegetation which becomes more and more xerophilous as steppization progresses.

Causes of steppization

The causes of steppization are well known and linked to human actions. They are:

(*a*) clearing forest for cultivation and abandonment after a few years;
(*b*) forest fires;
(*c*) wood cutting (firewood mainly);
(*d*) exploitation for distillation (rosemary, junipers);
(*e*) overbrowsing; and
(*f*) over-exploitation of alfa grass (papermills).

All conditions being equal, the process is more rapid in the more arid areas. Below the 300 mm isohyet, and even below the 400 mm isohyet, the process often seems irreversible. This is why the 400 mm isohyet is the upper limit of steppic vegetation today, although there are still forest remnants in a few sites, bearing witness of the past situation, down to the 200 mm isohyet.

Although there is no evidence of any trend of rainfall reduction over the historic period, there is an obvious worsening of the bio-climatical conditions during the last few decades, owing to increased human pressure on the land.

Desertization

Desertization* has been defined as an irreversible reduction of plant cover in arid areas, leading to the creation or the extension of desert land forms and landscapes such as regs, hammadas and dunes (Le Houérou, 1962).

In North Africa, the phenomenon has been outlined by Quezel (1958), Le Houérou (1962, 1968, 1969*a*, 1973, 1974), Flohn & Kettata (1971), Kassas (1968, 1970), Grove (1970), Floret & Le Floch (1972), Ionesco (1972), Rapp (1974).

Desertization is not due to an increased climatic aridity. There is no evidence of any systematic trend towards a reduction in rainfall during the period of instrumented recording (Dubief, 1953; Le Houérou, 1968; Flohn & Kettata, 1971), that is, over the last 130 years.

There is no evidence either of any systematic trend of rainfall reduction since the beginning of historic times, that is, over the last 2500 years (Monod, 1958; Dubief, 1956, Le Houérou, 1968; Quezel, 1960; Grove, 1970). Evidence of this is given by archaeologic, historic and palynologic findings and also by written documents. All these are in agreement with the similar studies conducted in the Near East (Monod, 1958; Evenari, Shanan & Tadmor, 1971).

Desertization is a result of human actions on fragile environment, particularly during sequences of dry years. The main human actions concerned are steppe clearing for cultivation, overgrazing, and fuel gathering. The intensities of these actions have increased parallel to human density. Population has increased six-fold since the beginning of this century. The area cultivated and the number of animals raised have grown in about the same proportions. Average population density in the arid zone (100–400 mm) (including urban) is 25 people km^{-2}. Animal density is 0.5 sheep equivalent per hectare (50 sheep km^{-2}).

Desertization mainly occurs in the 50–250 mm belt on the northern border of the desert (100 mm). Evaluations based on remote sensing studies of sampling areas give an average figure of several tens of thousands of hectares turned sterile every year at the northern edge of the Sahara (Le Houérou, 1968, 1974). More accurate figures will be available soon from studies which are being undertaken now.†

In summary, desertization is a recent phenomenon due to the population explosion (started about 1930) which results in the rupture of the centuries-old equilibrium between man and natural resources. Long drought periods are obviously a worsening factor but not the major cause.

* The word 'desertization' is preferred to 'desertification' since it has been clearly defined whereas the latter term has never been clearly defined and is currently used with many different meanings including the regression of vegetation in sub-humid and humid zones of Africa.

† Note added in proof: about 1% of the land surface per annum in the 100–200 mm isohyet zone, according to Floret & Le Floch, 1980.

Ecological changes over long periods

Changes due to climate

During the Pleistocene period there was a series of climatic changes consisting of alternating humid and dry periods: the 'pluvials' and 'interpluvials' contemporary with the European glaciations of Gunz, Mindel, Riss, Würm and Bühl. Some of these pluvials were warm and others cool. The ecosystems of the early Pleistocene (Villafranchian) appear to have constituted a transition between the tropical ecosystems of the Pliocene and the semi-arid to humid mediterranean ecosystems that characterize the middle Pleistocene.

The Villafranchian flora in northern Tunisia for instance, comprised 52% Mediterranean species, 26% tropical species and 21% boreal species. Of those, 47% belong to the present flora of this region, such as: *Quercus faginea, Q. suber, Q. ilex, Olea europea, Certonia siliqua* and *Laurus nobilis*. They were mixed with tropical relicts from the Pliocene, such as: *Sapindus* sp., *Cassia* sp., *Pittosporum* sp., *Salix canariensis* and some boreal species that today do not exist in North Africa: *Fagus silvatica* and *Ulmus scabra*.

The Villafranchian fauna includes also many tropical remnants from the late Tertiary: *Rhinoceros simus, Hyaena striata, Alcephalus bubalus, Equus mauritanicus, Elephas atlanticus, Gorgon taurinus prognu, Taurotragus derbyanus*, together with more boreal elements such as *Ursus arctos, Sus crofa, Rhinoceros merki*, and *Bos primigenius*. Similar to the flora, the tropical fauna progressively disappeared in middle and late Pleistocene.

During the pluvial periods of middle to late Pleistocene, the flora became typically Mediterranean: *Cedrus* sp., *Pinus halepensis, Tilia* sp., *Quercus ilex, Q. suber, Q. coccifera, Cupressus* sp., *Juniperus* sp., *Abies* sp., *Pistacia lentiscus, Olea europea, Ulmus* sp., *Chamerops humilis, Erica scoparia, Fraxinus* sp., *Smilax aspera, Rubus ulmifolius, Populus alba, Phillyrea media*, and *Spartium junceum*. The corresponding fauna is of a savanna/forest type: *Camelus thomasi, C. dromedarius, Gazella cuvieri, G. dorcas, Bos ibericus, Ovis tragelaphus, Ovis* sp., *Panthera leo, Panthera pardus, Felis lynx, Canis anthus, Struthio camelus*, and *Acynonix jubatus*.

The interpluvial periods are characterized by steppic or desertic families and genera such as *Artemisia, Nitraria, Thymelea*, Chenopodiaceae, Compositae and numerous grasses, together with arid fauna such as: *Hystrix cristata, Ctenodactylus gundi, Meriones shawi* and *Jaculus jaculus*. These arid periods numbered at least three and corresponded to the establishment of dune systems.

For the last 10000 years, the following changes have been described by Quezel (1960).

12000 B.P. Upper Würm. Dry period, steppic vegetation, probably of Mediterranean type.

10000 – 8000 B.P. Ibero-Maurusian palethnies. Forests of cedrus and deciduous Mediterranean forest in higher elevation and Mediterranean forest at lower elevation. Probable rainfall 500–2000 mm.

8000–4700 B.P. Saharan neolithic, Bovidian period (wall printings), Mediterranean forest; aleppo pine, junipers, holm oak, lentisk, olive; comparable to the vegetation found in the 300–600 mm belt today. Savannah fauna: giraffe, elephants, lions, gazelles, etc.

4700–2500 B.P. Savannah with spiny species of *Acacia* and *Ziziphus* similar to the present vegetation of the Sahel.
Savannah fauna; general reduction of rainfall.

2500 B.P. to present. Increased aridity. Vegetation similar to that now encountered in remote places.

Since the beginning of the historic times (2500 B.P.) the climate has probably not changed. However, changes in vegetation have occurred due to human actions.

Changes due to land use

At the beginning of the Phoenician and Roman occupations (2500 and 2200 B.P.), the arid zone of North Africa was probably covered by forest vegetation down to the present edge of the desert (100 mm) (Le Houérou, 1969*a*). Indirect evidence of this is given by historic texts ('Elephantos fert Africa utra Syrticas Solitudines' – Plinius) from Carthaginian, Greek and Roman writers (Le Houérou, 1959) and also by the knowledge of the dynamics of the present day vegetation as well as by numerous experiments of reafforestation (Le Houérou, 1968, 1969*a*, *b*).

During the Carthaginian, Roman and Byzantine periods, which lasted 1150 years (2500 to 1330 B.P.), the arid zone was extensively cultivated. The density of human settlements measured on topographic maps reaches 0.25 km^{-2} in western central Tunisia (Le Houérou, 1969*a*). As early as 2200 to 2170 B.P., Massinissa, King of Numidia, sold the Romans 20000 t of cereals. He sent the Roman army in Macedonia a gift of 7000 t of wheat and felt offended when his Roman friends proposed to reimburse him (Camps, 1960).

This period has thus witnessed a general and intense degradation of the natural vegetation, and it is therefore, very likely that it was a country already ruined by erosion and a collapsing society that the Arab conquerors took over in 1330 B.P.

A pastoral society then developed for 1230 years (1330–100 B.P.) characterized by a very low population density kept stable by periodic or permanent warfare between tribes and periodic starvation when drought periods occurred. This was certainly a period very beneficial to natural vegetation and natural ecosystem stability or progress.

About 100 years ago, forests covered more than twice the area they now do (six to eight million hectare instead of three million hectare). The productivity of these ecosystems was also about twice what it is now. For instance, statistics on alfa grass show that production at the end of the last century was 400–500 kg ha^{-1} yr^{-1} whereas it hardly reaches 200 kg ha^{-1} yr^{-1} today (Le Houérou, 1969*a*).

The same occurs in grazing ecosystems. Experiments show that protected plant communities have a primary production three to five times higher than under current management practices (Rodin & Vinograd, 1970; Le Houérou, 1971; Floret & Pontanier, 1974).

Similarly, the fauna has receded very rapidly during this century. Ostriches were common on the northern edges of the Sahara one hundred years ago: lion, cheetah, leopard, addax, oryx and Mohor's gazelle became extinct 40–80 yr ago. Hyaena, Cuvier's gazelle, ibex and barbary sheep have become extremely rare. Dorcas gazelle which were plentiful and could be seen in herds of dozens in the arid steppes 30 years ago, are now very rare and can be seen only in the remotest areas of the Sahara. All this is due to the same reason: exponential growth of human population and consecutive high pressure on the land and ecosystems.

Changes after release of human and stock pressure

Such changes depend upon the type of ecosystems considered. In the lower arid bio-climatic substage (100–200 mm rainfall) degradation on shallow soils is irreversible (desertization). For instance, the tracks left by the tanks of the German, English and French armies in Libya and Tunisia in 1942–43 are still conspicuous in the landscape (Le Houérou, 1959, 1968, 1969*a*, 1974; Floret & Le Floch, 1972) on the shallow soils, even in areas where human and stock pressures are locally very low. The perennial species have not re-established themselves on those soils and under these rainfall conditions.

On deep soils, the conditions are quite different. The perennials re-establish rather quickly and the final stage after a few years is either a steppe dominated by perennial grasses or a tall shrubs pseudo-steppe or even an open woodland according to local conditions and entropy of the systems present (Long, 1954; Le Houérou, 1969*a*).

Succession from cultigene vegetation to chamaephytic steppe is common as well as from chamaephytic steppe to gramineous steppe. The same applies to succession from garrique to forest. However, succession from steppe to garrique and forest has never been reported; that is, steppization is irreversible.

Is the soil–plant–animal system stable, apart from rainfall fluctuations?

Without human influence

This is a very difficult question to answer with certainty because we have no areas which have been totally protected from man and animals for very long periods. However, according to observations made for relatively short periods (10–20 yr), it does seem that such an equilibrium exists. The flora and the vegetation in the arid zone are probably very well adapted and able to recover after long drought periods. They do have a high resilience (Noy-Meir, 1974) under natural conditions with a very low entropy, except perhaps on very shallow soils in areas receiving less than 100 mm average annual precipitation.

With man and animals

Here the answer depends upon the intensity of human and animal actions. In a pastoral or agricultural/pastoral society, I would think that a stocking rate of 20–25 sheep equivalents km^{-2} would be an average compatible with sustained productivity of the ecosystems in the 100–400 mm rainfall belt. The carrying capacity for human population in a pastoral/agricultural society would be about 0.2 person km^{-2}. That is to say that the rural population the area could support under long term sustained production and without deterioration of the ecological equilibrium would be about one million people for the entire North African arid zone. This is about what it was 100 years ago and what, very likely, it used to be during the past centuries (1200–100 B.P.).

In the case of a technologically well-developed society such as existed towards the end of the Roman times when water and run-off were almost totally controlled and used, the carrying capacity could probably be much higher, about 1 person km^{-2}, if we accept the example of population density attained on European farms which developed in the area for about 50 years. This, in turn, depends upon the economic conditions and the standard of living of the population concerned.

Conclusions

Vegetation changes in North Africa today are mainly due to the exponential population growth and the pressure on the land which results from this population explosion. Climatic factors and especially drought periods are a worsening factor, but not the real primary cause.

The vegetation changes witnessed result in a general degradation of natural ecosystems, their production and their productivity, in the rupture of the centuries-old ecological equilibrium. This rupture provokes an accelerated erosion and the extension of the desert over several tens of thousands of hectares every year.

However, regeneration is in many cases possible whenever the pressure on the land is decreased. This regeneration can be speeded-up artificially by sound land use practices and proper use of the natural resources so as to provoke a 'biological recovery'. This 'biological recovery' is obtained through higher plant cover, higher organic matter content in the soil, and better water intake which, in turn, result in a high production and productivity of the arid-land ecosystems.

References

Akman, Y. & Daget, Ph. (1971). Quelques aspects synoptiques des climats de la Turquie. *Bulletin de la Société Languedocienne de Geographie*, **5** (3), 269–300.

Baldy, Ch. (1965). *Climatologie de la Tunisie centrale*. Food and Agriculture Organization, Rome.

Barry, J. P. (1961). Contribution à l'étude de la végétation de la région de Nimes. *Annales Biologiques*, **1961**, 309–550.

Bernard, J. (1969). *Les mammiféres de Tunisie et des régions voisines. I. Les rongeurs. Bulletin de la Faculté Agronomie, Tunis*, pp. 24–5, 37–172.

Boaler, S. B. & Hodge, C. A. H. (1962). Vegetation strips in Somaliland. *Journal of Ecology*, **50**, 465–74.

Boaler, S. B. & Hodge, C. A. H. (1964). Observations on vegetation arcs in the northern region, Somali Republic. *Journal of Ecology*, **52**, 511–44.

Boudet, G. (1972). Desertification de l'Afrique tropicale sèche. *Adansonia*, **2**, 505–34.

Camps, G. (1960). Massinissa, ou les débuts de l'histoire. *Libyca*, **8**, 1–320.

Clos-Arceduc, M. (1956). Étude sur les photographies aériennes d'une formation végétale africaine, la brousse tigrée. *Bulletin de l'Institut française d'Afrique Noire*, **18** (3), 677–84.

Delhaye, R., Le Houérou, H. N. & Sarson, M. (1974). Amélioration des pâturages et de l'elevage. *Etude des ressources naturelles et experimentation et demonstrations agricoles dans la region du Hodna, Algèrié*. Food and Agriculture organization, Rome.

Dubief, J. (1953). *Rapport sur l'évolution des régions arides dans le passé et à l'époque actuelle*. UNESCO/NS/AZ/112.

Dubief, J. (1956). Note sur l'evolution du climat Saharien au cours des derniers millénaires. *Proceedings of the 4th International Congress on Quaternary Research*, **2**, 848–51. Rome-Pisa.

Evenari, M., Shanan, L. & Tadmor, N. H. (1971). *The Negev: the challenge of a desert*. Harvard University Press, Cambridge, Massachusetts.

Flohn, H. & Kettata, M. (1971). *Étude des conditions climatiques de l'avance du Sahara tunisien*. World Meteorological Organization, Technical Note No. 16., Geneva.

Floret, C. & Le Floch, E. (1972). *Désertisation et ressources pastorales dans la Tunisie présaharienne*. Ministry of Agriculture, Tunis.

Floret, C. H. & Pontanier, R. (1974). *Etude de trois formations végétales naturelles du Sud Tunisien. Production bilan hydrique des sols*. Ministry of Agriculture, Tunis.

Franclet, A. & Le Houérou, H. N. (1971). *Les Atriplex en Tunisie et en Afrique du Nord*. Food and Agriculture Organization, Rome.

Gaussen, H. (1952). Le dynamisme des biocenoses végétales. *Colloques Internationaeux Ecology*, 9–22, CNRS, Paris.

Gaussen, H. (1954). *Geographie des plantes*, 2nd edn., A. Colin, Paris.

Grove, A. T. (1970). Climatic change in Africa in the last 20000 years. In *Colloque IGU, Ouargla*, vol. 2, Institut de Geographie, Alger.

Hemming, C. F. (1965). Vegetation arcs in Somaliland. *Journal of Ecology*, **53**, 57 – 67.

Ionesco, R. (1972). *Pastoralisme et desertisation.* Ministry of Agriculture, Tunis.

Kassas, M. (1968). Dynamics of desert vegetation. *IBP/CT Colloque Hammamet (Mimeo).*

Kassas, M. (1970). Desertification versus potential recovery in circum-Saharan territories. In: *Arid lands in transition* (ed. H. E. Dregne), pp. 123–39. American Association for the Advancement of Science, Washington, D.C.

Killian, Ch. (1945). Un cas très particulier d'humification au désert due à l'activité des microorganismes dans le sol des Nebkas. *Revue Canadienne de Biologie*, **4** (1), 3–36.

Kuhnholtz-Lordat, G. (1945). La silva, le saltus et l'ager. *Annales de l'École nationale d'Agriculture de Montpellier*, **26** (4), 1–84.

Le Houérou, H. N. (1959). *Recherches écologiques et floristiques sur la végétation de la Tunisie mérédionale.* Institut Recherches Sahariennes, Alger.

Le Houérou, H. N. (1962). *Les paturages naturels de la Tunisie aride et désertique.* Paris Institute de Science Economique Appliquée.

Le Houérou, H. N. (1965). *Improvement of natural pastures and fodder resources. Report to the Government of Libya. Expanded Technical Assistance Programme, Report No. 1979.* Food and Agriculture Organization, Rome.

Le Houérou, H. N. (1968). La désertisation du Sahara septentrional et des steppes limitrophes (Libye, Tunisie, Algérie). *Colloque Hammamet et Annales Alger de Geographique*, **3** (6), 2–27.

Le Houérou, H. N. (1969*a*). La végétation de la Tunisie steppique. (Avec référence aux végétations analogues d'Algérie, de Libye et du Maroc). *Annales de l'Institut National de la Recherche Agronomique de Tunisie.* No. 42.

Le Houérou, H. N. (1969*b*). *Principes, méthodes et techniques d'amélioration fourragère et pastorale en Tunisie.* Food and Agriculture Organization, Rome.

Le Houérou, H. N. (1970). *North Africa: past, present, future.* In *Arid lands in transition* (ed. H. E. Dregne), pp. 227–78. American Association for the Advancement of Science, Washington, D.C.

Le Houérou, H. N. (1971). An assessment of the primary and secondary production of the arid grazing lands ecosystems in North Africa. *Ecophysiological foundation of ecosystems productivity in arid zone* (ed. L. E. Rodin), pp. 168–72. Nauka, Leningrad.

Le Houérou, H. N. (1973). *Ecological foundations of agricultural and range development in Western Libya.* Division of Plant Protein Production, Food and Agriculture Organization, Rome.

Le Houérou, H. N. Deterioration of the ecological equilibrium in the arid zones of North Africa. *Special Public No. 39*, pp. 45–57. Agricultural Research Organization, Volcani centre, Bel-Dagan, Israel.

Le Houérou, H. N. & Froment, D. (1966). Définition d'une doctrine pastorale pour la Tunisie steppique. *Bulletin de l'Ecole Nationale Supérieure d'Agronomie, Tunis*, **10–11**, 72–152.

Long, G. A. (1954). Contribution à l'étude de la végétation de la Tunisie centrale. *Annales du Service botanique et Agronomique*, Tunis, No. 27.

Monod, Ch. (1958). Les parts respectives de l'homme et des phenomènes naturels dans la dégradation des paysages et le déclin des civilisations à travers le monde méditerranéen L.S. avec les deserts et semi-déserts adjacents au cours des derniers millénaires. *Union National pour la Conservation de la Nature (Comptes Rendu), 7è Réunion Technologique, Athènes.*

Noy-Meir, I. (1974). Stability in arid ecosystems and the effects of man on it. *Proceedings of the 1st International Congress on Ecology.* The Hague.

Quezel, P. (1958). Quelques aspects de la dégradation du paysage végétal au Sahara et en Afrique du Nord. *Union National pour la Conservation de la Nature (Comptes Rendu), 7è Réunion Technique, Athènes.*

Quezel, P. (1960). Flore et palynologie saharienne. *Bulletin de l'Institut francais d'Afrique Noire*, **22** (2), 353–9.

Rapp, A. (1974). *A review of desertisation in Africa – water, vegetation, man.* Secretariat for International Ecology, Sweden, Stockholm.

Rodin, L. & Vinogradov, B. (eds.) (1970). *Etude geobotanique des paturages du secteur ouest du département de Medea (Algèrie).* Nauka, Leningrad.

White, L. P. (1969). Vegetation arcs in Jordan. *Journal of Ecology*, **57** (2), 461–4.

Manuscript received by the editors December, 1974

15. The modelling of arid ecosystem dynamics

D. W. GOODALL

Introduction

Ecosystem modelling has been prominent in the International Biological Programme (IBP). Before the Programme was initiated, experience of ecosystem modelling was rather limited, but since 1967 it has become a subject of widespread interest, largely as a result of the impetus given by the IBP. Arid-land ecosystems have had a share of this interest. Indeed, it may be said that there is much common ground in the modelling of all terrestrial ecosystems, and some is shared even with the modelling of aquatic systems.

A model is a representation of a system in simpler terms – a mapping of the system, in the mathematical sense. In a model, certain elements of the system and certain relationships between them are selected, and are represented by elements of different types which stand for them, and which have relationships analogous to those between the real-life elements. We are here specifically concerned with dynamic models that attempt to represent changes in the ecosystem through time.

The simplest type of dynamic model is perhaps the word model, in which the elements of the system are represented by nouns or phrases, and their relations by verbs and syntactical constructions. In this sense all good 'natural-history' descriptions of processes and changes in ecosystems are word models. They may be supplemented or replaced by diagrammatic models, where the elements may be represented (for instance) by rectangular boxes, and the processes by lines connecting them.

Diagrammatic models have usually been excessively simple, the elements represented separately often being limited to trophic levels. The interrelations between elements, and the dynamics of the system they make up, depend on more than feeding; to treat species as combined in trophic levels may omit important features of the system. This has been well emphasised by Georgievsky & Kuznetsov (1976), who developed a considerably more complex word and diagram model of arid-land ecosystems. They divided herbivores, for instance, according to which organs of the plants they eat, and whether their net effect on the plant is favourable or unfavourable. In all, they recognize functional categories of organisms in the system in place of the conventional four or five trophic levels. Others have gone further and recognized individual species or small groups of species as making distinctive contributions to system dynamics.

Verbal or diagrammatic models can represent only very crudely, if at all, quantitative aspects of system dynamics. For this purpose one needs to progress to mathematical models, where elements and processes are represented by mathematical symbols. A great deal of biological knowledge may be incorporated in such a model – though it is easy for the sophistication of the mathematics to out-step the firm biological data. From the mathematical model it is a brief step to the analogue or digital computer. And, since mathematical models of ecosystem processes are often analytically intractable, the computer may be a necessary tool for their implementation. It is in these developments that most progress has been made over the past decade, thanks largely to the widespread availability of the digital computer and interest in its potentialities.

In the International Biological Programme, verbal and diagrammatic models have usually been regarded as preliminary steps towards mathematical models which can be implemented on the computer, and it is with such models that this chapter will mainly be concerned. We are not concerned here with models of particular processes (photosynthesis, predation, and competition are examples of processes where the modelling approach has proved valuable), but rather with models of the system as a whole, where a variety of processes, and interactions among diverse elements, are involved.

One of the first such models specifically concerned with arid-land ecosystems was that developed by Goodall (1965, 1967, 1969, 1970) for vegetation in Australia in the 250 mm rainfall belt, subjected to extensive grazing by domestic livestock. The model considered an area grazed as a management unit (a fenced paddock) which could be divided into sections, differing in topography, soils and vegetation, or in distance from a water supply or fenceline. These were considered to be factors, apart from vegetation, that determine grazing behaviour of livestock. Within each section changes in the soil water content were modelled as a function of rainfall, topography (determining run-on or run-off) and temperature which, with the plant cover and soil moisture, determined losses by evapotranspiration. Photosynthesis was calculated separately for each plant species or species group. It was assumed that the photosynthetic portion of the plant was either identical with, or proportional to, the portion available as forage, and that other portions of the plant did not need to be explicitly included in the model. Seed production and germination were not included, so the species considered were in effect all perennials. The photosynthesis was calculated as a function of air temperature and soil moisture (expressed as a percentage of the 'available' range), the constants in the calculation being species-specific. Light was regarded as non-limiting.

Grazing pressure was distributed among the sections into which the paddock was divided in proportion to a function expressing their relative attractiveness to the livestock. This function depended on the one hand on the weight per unit area of forage in different palatability categories, on the other hand on the distance from water and the current temperature (expressing the water stress on the livestock), and on the distance from the paddock boundary.

Within each section of the paddock, the amount of forage removed per animal-day of grazing depended on the amount of forage of different palatability available, and this consumption was apportioned among the species in proportion to its palatability and the quantity present.

While no attempt was made to trace the fate of the material consumed in excreta and respiration, the gain or loss in weight of the livestock was calculated as a function of their forage consumption and their current average weight. Mortality was also a function of the current average weight.

The exogenous variables used in this model were the mean monthly temperature, and rainfall which was treated stochastically. For each month, two frequency distributions of daily rainfall were used, one for days following days without rain, and another for days following days with rain. Random samples from the appropriate distribution for each daily time step enabled storm sequences to be simulated with some realism.

In its original form this model specifically included sheep as the herbivore, and in one version included a function for wool growth. Moreover, the sheep were regarded as mature, and non-reproducing – a flock of wethers, for instance. In subsequent development the model was made more general. Reproduction was introduced, the animal population being divided into monthly age-groups or cohorts. Each age group had its own constants for food consumption, growth and mortality, though all had the same food preferences. The mature animals also had a function for reproductive rate, month-specific, which led to a proportion of biomass being transferred to the first cohort. At the beginning of each month, the individuals in each cohort were transferred into the next older cohort.

Still further development enabled the model to simulate grazing by herbivores of more than one species, each species having its own feeding pattern, spatial distribution of feeding activity, and functions for growth, mortality and reproduction. In this way, management with mixed domestic livestock, or competition between domestic livestock and wild herbivores such as sheep and rabbits, could be modelled.

Data for implementation of this model were scanty, and its development was undertaken more to show the potentialities of the technique than for its direct value as a management tool. In exercising it for this purpose, it was used to show the effects of different stocking rates, alternate versus continuous grazing, different rainfall patterns, changes in mean temperature, and capital improvements. Some sensitivity tests were also performed to examine the response of some output values to changes in the value of certain parameters.

IBP modelling work in the United States

Ad hoc models

Introduction

Turning now to the period of the International Biological Programme, the Desert Biome project within the United States (like other Biome programmes in that country) emphasized ecosystem modelling from the start. For the first two years, attention was concentrated (under the direction of Dr Kent W. Bridges) on simulation models designed to answer specific questions raised by scientists working in the project (Bridges, 1971*a*), such as:

(1) what would be the effects of grasshopper control on the quantitive and qualitative aspects of forage availability to domestic livestock?
(2) what are the effects of variations in the intensity and vertical patterning of graze removal on the pattern of root growth in a desert community?
(3) what combination of domestic ungulates (cattle), native ungulates (pronghorn), and exotic ungulates (camel) will maximize secondary production in a given desert area?

The simulation programmes were written in the general programming language PL1, and were designed to be easily understandable by the field biologist. Mathematical sophistication was kept to a minimum, and functional expressions were often replaced by graphical representation of the dependence of changes on the factors influencing them.

To illustrate the type of model, a brief description will be given of that designed to answer the last question above (Bridges, 1971*b*). A number of plant species or types are included, as well as the three types of herbivore. Changes in forage quantity are calculated monthly. The relative growth rate for the *i*th forage type in the *j*th month is given by

$$\exp\{a_i(b_j - \Sigma x_i)/b_j\} \qquad [15.1]$$

where x_i is the quantity of forage of the *i*th type, b_j is the maximum total forage possible on that site in that month, and a_i is a constant for that forage type.

For the *k*th herbivore type, the food consumption per individual is:

$$r_k = F_{1_k}(Y_k) \qquad [15.2]$$

where Y_k is the mean herbivore body weight and F_{1_k} is an empirical function, the value of which is derived by interpolation in a graph. This consumption is apportioned among the plant types in proportion to

$$d_{ik}\, x_i\{P_{ij} + (1 - P_{ij})\, c_i\} \qquad [15.3]$$

where d_{ik} expresses the preference of that herbivore for the *i*th plant type, P_{ij} is the proportion of the forage from the *i*th plant type which is in the green state during the jth month, and c_i is a constant expressing the relative availability as forage of the non-green material. In order to determine consumption over the whole area, these figures are multiplied by the herbivore density.

The change in weight of each herbivore during the month is calculated by

$$\Delta Y_k = F_{2_k}(r_k)\,F_{3_k}(S_k)\,F_{4_k}\left(\frac{g_k+1}{2}\right) - F_{5_k}(Y_k) \qquad [15.4)]$$

where S_k is the mean age of the kth herbivore, g_k is the proportion of green material in its diet, and F_{2_k}, F_{3_k}, and F_{4_k} are empirical functions, used in the form of graphs, expressing the dependence of weight gain on forage consumption, age, and diet quality respectively, while F_5 is a similar function expressing respiration rate as a function of body weight.

Within each herbivore type, all individuals are taken to be of the same age, and no reproduction occurs. In respect of the plants, possible participation of non-forage organs in re-growth is not provided for. Possible weather effects are not taken into account. All these deficiencies in the model were recognized; it was frankly intended as a preliminary exercise, for further subsequent development if it was thought desirable.

Strategies of annual plants

One of these models that was in fact developed further was concerned with the evolutionary 'strategies' of annual plants in arid regions.

Wilcott (1973) developed three models addressing this question, each operating with an annual time step. The yearly life cycle of an annual plant was considered from the initial supply of seeds in the soil to its replenishment from a successful seed crop. A proportion G of the seeds germinate, a proportion P are eaten by animals. Of those which germinate, a proportion K succeed in growing to mature plants, of which the average seed crop is S. In the first model, it was assumed that these four parameters were constant from year to year, and that seed cohorts were of uniform behaviour, or in any case indistinguishable. At equilibrium

$$GKS = P + G - PG \qquad [15.5]$$

Examining this equilibrium condition showed that, provided the environment permitted 30% or more of the seed reserves to germinate ($G \geqslant 0.3$), the seeds produced per seed germinating (KS) need to be little more than 1.0 for stability, and that the population is then insensitive to seed losses (P). If the proportion of seed germinating per annum is considerably less, the population becomes sensitive to seed losses; but even with $G = 0.05$ and $P = 0.95$ (i.e. 95% of the seeds lost annually, and only 5% of the seeds germinating), a replacement rate of $KS = 20$ is sufficient to maintain stability, and such a value is well within the normal range for desert annuals.

The second of Wilcott's models also treated the environment as constant from year to year, but distinguished seed cohorts of different ages which enabled dormancy, and seed unavailability to animals through burial, to be taken into account. Each of the four parameters of the preceeding paragraph was allowed to take values specific to the different cohorts. Using subscripts to indicate these different values, with subscript 0 representing the seed crop of the current year, 1 for the previous year and so forth, the equilibrium condition is now:

$$G_0 K_0 S_0 = \sum_{i=1}^{\infty} G_i K_i S_i \sum_{j=0}^{i-1} (1-G_i)(1-P_j) \qquad [15.6]$$

When the behaviour of this function with different age curves for G_j and P_j was explored, it was found that drastic differences in the age structure of the seed population, and in the population dynamics of the species, would result. If the model was run with given values for G_j and P_j, a stable age distribution was reached within five to ten generations.

Unlike these two models already described, Wilcott's third model took account of the dependence on environmental factors of the parameters used: specifically, on rainfall. P and G remained functions of seed age; K (survival of seedlings to maturity) was a function of total rainfall during the growing season, and of seedling density; S (seed production per plant) was also a function of total rainfall and of density of mature plants; G was modified by the rainfall during that part of the year suitable for germination; and rainfall values were generated stochastically.

The dependence of germination on rainfall in the third model was linear between two threshold values. The dependence of seedling survival and seed production on rainfall and on density was included in the model in the form of two empirical sets of points between which interpolation took place. For seed losses by animals it was assumed that the major loss occurred in the first year, and that thereafter a constant low rate continued. Repeated runs of the model showed large changes in age distribution of seeds, largely reflecting the rainfall history of recent years.

In order to explore the strategies available to desert annuals in terms of this model, the different parameter sets in the model were varied. With values for G representing very prolonged dormancy, for instance, the species became extinct in a century or so. On the other hand, different patterns of dormancy had surprisingly little effect on the total seed population. Varying the germination response to rainfall indicated an advantage (in terms of total seed population) in responding to low rainfall, despite the risk of subsequent losses if growing rains did not follow. Varying the curve of seed losses with age showed little disadvantage from a heavier loss in the first year, provided the rate of loss diminished to acceptable proportions in subsequent years.

Varying the survival from seedling to maturity (K) had relatively little effect on the total seed density (because of the adverse effect of crowding on seed production) as compared with varying seed production per plant. This suggests as advantageous strategies either a re-allocation of biomass to inflorescences in preference to roots, or a decrease in mean seed weight.

Combining these variations in different ways showed the complementary character of the dormancy curve, and seed losses. The more prolonged the dormancy, the less could seed losses be tolerated; thus, the plant could compensate for increasing seed losses by a reduction in dormancy. The optimum value of G for any particular value of P_j is in fact numerically similar to it; and it is little affected by the prevailing rainfall. The importance of low rates of loss for older seeds suggests that adaptations to increase the chance of burial (such as small seed size) may be of considerable value to the species.

A grazing model

Also among the *ad-hoc* models produced under the US/IBP Desert Biome programmes is one for which Wilkin & Norton (1974) have been responsible. This is a management-oriented model of the relationship between range vegetation and the livestock subsisting on it. Its purpose is to provide fairly long-term predictions of forage productivity, with a view to facilitating decisions on livestock management. It is a simple model, in the sense that the only state variables are the biomasses of forage of different species in an area considered as homogeneous. But the year is divided into a small number of periods which constitute the time steps, and the need to account for the effects of utilization on forage growth and the balance between forage species over time steps of several weeks or months introduces complexities into the algorithms which describe changes in the system.

The model takes no account of changes in the livestock themselves. They are treated as an input quantity, consuming or destroying a fixed amount of forage per animal per day, and distributing this consumption among the various forage species in a way which depends only on their productivity.

Underlying the model is the assumption that, without domestic livestock, the productivity of each species would be at a certain equilibrium value, collectively constituting the 'climax' vegetation for that site. The effect of any particular grazing practices is to shift the equilibrium towards a new set of values, determined for that site only by those grazing practices.

The model calculates changes from year to year as if the precipitation for each year were the average for the locality. The productivity of each forage species for the initial year, which is provided as input data, is adjusted to the average rainfall pattern by dividing by a factor

$$F_i = \sum_j \alpha_{ij} p_j \qquad [15.7]$$

where p_j is the precipitation over some period during or preceding that year, and α_{ij} is a constant for forage plant and period,* by a factor

$$T = \beta_1 + \beta_2 t \qquad [15.8]$$

where t is the mean temperature of the growing season; and by a factor expressing the effect of grazing utilization:

$$V_i = 1 + \gamma_{1_i} \cdot L_i + \gamma_{2_i} \cdot L_i^2 \qquad [15.9]$$

where L_i is the proportion of forage of the ith species destroyed by grazing during that season, and is an input value. Thus, the values for current productivity, normalized for average conditions, with which the simulation starts, are

$$y_i = \frac{Y_i}{F_i \cdot T \cdot V_i} \qquad [15.10]$$

where Y_i are the input productivity values.

At the end of the simulation process, the normalized values are adjusted for current conditions in the final year of simulation by reversing the process:

$$Y_i = y_i \cdot F_i \cdot T \cdot V_i \qquad [15.11]$$

The core of the programme is a calculation for each period of the total amount of vegetation utilized by the herbivores, and its distribution among the various forage types. The total forage consumed or destroyed by the herbivores is calculated on the basis of herbivore population present, at a rate independent of season, so that the amount removed during the period is proportional to the length of the period. Calculation of the distribution of this utilization among the different forage types then proceeds. Where the amount of the ith type of forage available during the period is x_i, and the relative amount is r_i, that is

$$r_i = x_i / \sum_i x_i \qquad [15.12]$$

then the proportion of this forage utilized is given by

$$v_i = A^{r_i(\delta_{1_i} + \delta_{2_i} A + \delta_{3_i} r_i + z_j)} \qquad [15.13]$$

where *A is the overall proportion utilized,* δ_{1_i} is a constant for the locality based on the rainfall pattern there; δ_{2_i}, δ_{3_j} are constants expressing the relationship to utilization and composition; and z_j is an arbitrary value fitted for the current (jth) period by an iterative process.

The total amount of the ith forage type utilized is

$$u_i = v_i \cdot x_i \qquad [15.14]$$

* Throughout this section, Greek letters are used to symbolize constant parameters of the model.

and A, the overall proportion utilized, is

$$A = \sum_i u_i / \sum_i x_i \qquad [15.15]$$

Thus the calculations of A and of v_i are interdependent, and are performed iteratively.

The amount of forage available, x_i, is calculated by adding to the amount remaining at the end of the previous period the expected growth during the current period. The latter is affected by the expected degree of utilization since the beginning of the growing season. Let

$$U_i = \sum_j u_{ij} \qquad [15.16]$$

where the summation (subscript j) extends over all periods distinguished from the beginning of the year to the present, and let the cumulative growth over the same period be

$$G_i = \sum_j g_{ij} \qquad [15.17]$$

where g_{ij} is the growth of species i during period j. Then, if Z_i is the expected growth for the whole year in the absence of grazing, the modification in this quantity through grazing can be expressed as

$$W_i = Z_i\{1 + \epsilon_{1_i} \cdot U_i/G_i + \epsilon_{2_i}(U_i/G_i)^2\} \qquad [15.18]$$

where ϵ_{1_i}, ϵ_{2_i} are constants to specific species i. The expected growth during the jth period is

$$\eta_{ij} W_i \qquad [15.19]$$

where η_{ij} is a constant proportion of the total annual growth for that species and period, given standard weather conditions. Since calculation of x_i thus requires knowledge of u_i for each species, a further iterative process is required to complete the utilization calculations for the period.

It will be noted that preferences by the herbivores among the different forage types enter into the calculations in the form of the constants δ_{1_i}, δ_{2_i} and δ_{3_i} in expression [15.13]. If more than one type of herbivore is grazing the pasture, these constants are averaged over the different herbivores with weights proportional to their total forage use.

The forage suffers losses from native herbivores and natural attrition, in addition to grazing by livestock. These losses are taken to follow a constant seasonal pattern, so that the loss from a particular forage species during a particular period is a constant proportion of the expected productivity for that species over the year as a whole.

At the end of each year of simulation, the basic productivity of the pasture as a whole is re-calculated. A figure for the average annual productivity of that type of pasture in the absence of grazing has been read in (P). The basic productivity for the past year ($D = \Sigma y_i$) is then compared with a quantity

$$C = P\left\{1-\left(\frac{m}{n}\right)^4\right\} \quad [15.20]$$

where m is the total utilization over the past year and n the total production. If they differ appreciably, the basic productivity for the coming year, D', is set at

$$\zeta^s D \quad [15.21]$$

where ζ is a constant greater than unity and

$$s = \frac{C-D}{\sqrt{[\text{abs}\,(C-D)\cdot(C+D)]}} \quad [15.22]$$

This basic productivity D' is then apportioned among the forage species, to give new values for y_i. For this purpose, the model uses figures which have been supplied for the relative abundance of species in a pasture of that type in an ungrazed condition ($w_i, \sum_i w_i = 1$). The expected relative productivity for each species is calculated as

$$R_i = w_i \prod_j (1 + \theta_{1_i} \cdot v_{ij}^{\frac{1}{2}} + \theta_{2_i} \cdot v_{ij} + \theta_{3_i} \cdot v_{ij}) \quad [15.23]$$

Then, if $R_i D'$ differs appreciably from the existing value for y_i, the new value for y_i is calculated as

$$\zeta^{q_i} \cdot y_i \quad [15.24]$$

where

$$q_i = \frac{R_i D' - y_i}{\sqrt{[\text{abs}\,(R_i D' - y_i)(R_i D' + y_i)]}} \quad [15.25]$$

Wilkin & Norton (1974) describe an application of this model to salt-desert shrub vegetation in south-western Utah grazed by sheep, in an area under the control of the US Forest Service for which long-term data were available (Hutchings & Stewart, 1953; Holmgren & Hutchings, 1971). Since matched grazed and ungrazed areas were available, the ungrazed potential (P and W_i) could be estimated directly.

The parameters of the model were estimated from data obtained in the same experimental area over a period of 30 years. Simulation of the herbage yield over this same period gave excellent agreement with observed results. The 'acid test', of course, will be to apply the model, using these same parameter estimates as far as appropriate, to similar conditions elsewhere.

General-purpose models

Introduction

The US/IBP Desert Biome Programme has also developed models of complete desert ecosystems which are not designed for a particular purpose, but rather are intended to be applied to a variety of purposes as occasion demands, and to be available for refinement for these particular purposes as needed.

The principles underlying this modelling activity have been described by Goodall (1973*a*, 1975*a*). These models envisage a system which can be defined afresh on each occasion in terms of the sets of species and groups and organ types to be distinguished, the stages of development to be modelled, the chemical constituents to be tracked through the system, the time step, and parameters expressing the rates and factor-dependence of the various processes. Furthermore, the main programmes (Goodall, 1973*b*; Goodall & Gist, 1973) constitute a general framework for ecosystem modelling, not unlike the ELM model for grasslands developed by Innis and his collaborators in Colorado (Anway *et al.*,1972), or the FLEX1 programmes of Overton, Colby, Gourley & White (1973). These main programmes set up the structure of the system, and provide for input and output; but the ecosystem processes themselves are handled by sub-routines which may be combined in various ways, modified or replaced as occasion demands. The 'package' of programmes also includes sub-routines for testing the sensitivity of specified model outputs to variations in particular initial conditions and parameter values, separately and in combination (Noy-Meir & Goodall, 1973).

The rates of processes are expressed by difference equations over a time step which is commonly one day, but may be varied at will, and may differ for different processes. Some process rates are calculated by separate sub-routines and, when combined, they may be inconsistent with overall constraints (such as non-negative values of most state variables) particularly since the sub-routines may have been developed independently; provision is made for a time step to be divided where this happens.

The main sub-routines into which the programme is divided deal with processes in plants, in animals, and in the soil. Each of these has been developed at different levels of detail and sophistication. At the time of writing, there are four alternative sub-routines for plant processes (Valentine, 1973, 1974) one of which uses constant coefficients common to all plant groups; a second differentiates them by plant groups; in a third the coefficients are time-varying – that is, they have fixed values over periods of the year determined by date or by phenological stage; in a fourth alternative version, the rates of processes depend on current environmental conditions, either provided as input or computed elsewhere in the model. This fourth version of the sub-routine for plant processes will be described in some detail as an example.

Plant processes

In the fourth version of the plant sub-model, the various groups of processes – phenological changes, photosynthesis, respiration, translocation, mineral uptake, and death – are dealt with in different sub-routines. In each of these, each plant species group is considered separately, and may have different parameter values or even (in some cases) use different function types to determine the rates of processes.

To illustrate the type of function used in the models, one may take the daily rate of photosynthesis, which is treated as a function of the total daily irradiation (L), the mean leaf temperature (T), and the mean leaf water stress (W). It is calculated as:

$$A_i = \frac{L}{\alpha_{1_i}+L}\cdot \max\left\{0, \min\left(1, \frac{\alpha_{2_i}-W}{\alpha_{2_i}-\alpha_{3_i}}\right)\right\}\cdot F(T)\cdot \alpha_{4_i} \qquad [15.26]$$

where

$$F(T) = \left[\frac{\alpha_{5_i}-T}{\alpha_{5_i}-\alpha_{6_i}}\right]^{\alpha_{7_i}}\cdot \exp\left[\frac{\alpha_{7_i}}{\alpha_{8_i}}\left\{1-\left(\frac{\alpha_{5_i}-T}{\alpha_{5_i}-\alpha_{6_i}}\right)^{\alpha_{8_i}}\right\}\right] \quad T \leqslant \alpha_{5_i} \qquad [15.27]$$

$$F(T) = 0, \qquad T > \alpha_{5_i}$$

the Greek symbols being constants, and the subscript i referring to the species and age group.

Respiration rate for the various organs, per unit carbon content, is calculated by

$$R_i = \{\beta_{1_{ij}} - \beta_{2_{ij}}\cdot \exp(\beta_{3_{ij}}{}^{T})\}\cdot \min\left(1, \frac{\beta_{4_{ij}}-W}{\beta_{4_{ij}}-\beta_{5_{ij}}}\right) \qquad [15.28]$$

where W and T are the water stress and temperature within the organ in question, and $\beta 1_{ij}$ etc. are constants for the jth organ of the ith plant group.

Translocation from the leaves is dependent on the concentration of labile carbon in the leaf tissues as well as on leaf temperature and water status:

$$M_i = \min\left[1, \max\left(0, \frac{\gamma_{1_i}-W}{\gamma_{1_i}-\gamma_{2_i}}\right)\right]\cdot\left(\frac{\gamma_{3_i}-T}{\gamma_{3_i}-\gamma_{4_i}}\right)^{\gamma 4_i}\cdot$$

$$\exp\left[\frac{\gamma_{5_i}}{\gamma_{6_i}}\cdot\left\{1-\left(\frac{\gamma_{3_i}-T}{\gamma_{3_i}-\gamma_{4_i}}\right)^{\gamma 6_i}\right\}\right]\cdot\{\gamma_{7_i}+\gamma_{8_i}\exp\gamma_{9_i}S\} \qquad [15.29]$$

where S is the ratio of labile to total carbon in the leaves, γ_{1_i} etc. are constants, and T and W have the same meanings as before.

During certain phenological stages, translocation products may be derived from organs other than leaves. If the species is germinating, the translocation products are derived from seed reserves; if the plant is a perennial, at the phenological stage of leafing-out, they are derived from roots and woody stems – again at a rate depending on the ratio of labile to total carbon in the donor organs.

The organs constituting 'sinks' for translocation are also determined by the phenological stage. The ratio of reserve carbon to total carbon in the entire plant is used as the argument in a series of demand functions of the form

$$\max[0, \min\{1, \delta_{1_i} + \delta_{2_i} \exp(\delta_{3_i} D)\}] \qquad [15.30]$$

to determine the proportion of translocation product allotted to these various sinks. These functions are calculated organ by organ in a sequence determined by the phenological stage, so that, if the supply of translocation product is limited, only the 'sinks' higher in the priority list receive it. To determine the allocation of carbon received by the 'sink' between new protoplasmic growth, cell-wall material, and labile reserves similar functions are calculated depending on the existing proportions of these components in the organ, with parameters specific to the organ, species and age-group in question. Where:

$$p = \epsilon_{1_i} + \epsilon_{2_i} \exp(\epsilon_{3_i}{}^{E}) \qquad [15.31]$$

$$q = \epsilon_{4_i} + \epsilon_{5_i} \exp(\epsilon_{6_i}{}^{F}) \qquad [15.32]$$

E being the ratio of labile to total carbon in the organ in question, and F that of protoplasmic to total carbon, then the proportions of translocate added to these categories are

$$\min\{p, p/(p+q)\} \qquad [15.33]$$

$$\min\{q, q/(p+q)\} \qquad [15.34]$$

respectively, while the residue (if any) constitutes growth in the structural component.

Changes in phenological stage are activated by a special sub-routine. Germination may, for instance, depend on soil temperature, water potential, or photoperiod exceeding specified threshold values, and in some cases may occur only if rain has fallen during the previous three days. Changes to a reproductive state may be actuated by water tension or temperature in specific soil horizons, by photoperiods within specified limits, or by a ratio of labile to total carbon in the plant above a given threshold. Dormancy is initiated when the proportion of labile to total carbon in the plant, or the water content or temperature of specified soil horizons, fall below specified thresholds.

For each plant organ, the mineral nutrients required to maintain constant the ratio of nutrient to total carbon (or, in the case of nitrogen, protein carbon) are calculated; the total quantity of nutrients required is then withdrawn from the various soil horizons in proportion to the root biomass they contain.

The death rate of different organs (proportional mortality per time unit) is calculated as

$$\eta_{1_i} + \eta_{2_i} + \exp(\eta_{3_i} t) \qquad [15.35]$$

where t is the time since the start of the current phenological stage. Additional mortality may result if soil temperature or soil water potential fall below specified critical values.

Animal processes

The animals (Payne, 1973) are divided into a number of species groups, as far as possible combining species of similar feeding and reproductive habits. Each of these may be divided into several stages of development, a provision of great importance for many of the insect groups, which may change their habits entirely at a metamorphosis. Food consumption is based primarily on the carbon content of the animal itself, and of its food sources. The total consumption of carbon has a maximum depending on the weight of carbon in the animal, but is reduced below this maximum as the amounts of different foods available decrease. These amounts are weighted by their acceptability to the animal in question. The alternative food sources may include the different plant species groups, all animals (including the possibility of cannibalism), and all types of litter. This organization of the model makes it unnecessary to distinguish different trophic levels among the animals – distinctions which indeed are often biologically unrealistic.

The same table of acceptability is used to determine how the total food consumption is distributed among the various alternative food sources. Once the amount consumed from each source is calculated, the appropriate amount of each chemical constituent is transferred from these sources. Not all, however, becomes part of the consumer biomass. Each animal group is respiring, egesting and excreting – the first at a rate dependent on temperature, and on the mean weight per individual. The proportion of the intake of each constituent which is assimilated is assumed constant, and the balance is lost as egesta which enter the detritus.

Development from one stage to another may be determined by various factors, but it is usually a function of temperature. In one of the alternatives provided in the sub-routine a temperature accumulator is started at a fixed point in time. When this accumulator exceeds a certain threshold, transfer from one stage to the next begins, the proportion transferred increasing with the excess of the accumulator above the threshold until transfer is complete. Since a record is being kept of the animal population numbers concurrently with their biomass, transfer of population along with biomass is required. For this purpose, it is assumed that the biomass per individual within each stage of development has a known distribution and that the individuals transferred are always the largest ones. In arthropods, a transfer from one stage to another is often associated with a moult, or other loss of material, which is then transferred to detritus.

Reproduction is essentially a transfer of material from one stage of development to another (from adult to egg or new-born young), and is accordingly treated in the same way as other such transfers, except that there is a creation rather than a transfer of population. Other function types are available too, to determine the proportion of biomass transferred, as alternatives to the temperature accumulation mentioned.

Soil processes

The soil sub-model (Radford, 1973; Lommen, 1974) covers the activity of micro-organisms in the soil, as well as physical and chemical changes.

Changes in soil water, temperature, and ionic concentrations are modelled by partial differential equations, with the vertical dimension and time as the two variables of differentiation. At present, these models of physical processes assume that the soil is homogeneous with depth in relevant respects. Adequate treatment of sites where this is not true will call for further development of the model.

Temperature at different depths is calculated by a sub-model described by Hanks, Austin & Ondrechen (1971). It is based on the partial differential equation

$$\frac{\partial T}{\partial t} = \frac{\partial}{\partial z}\left[\sigma \frac{\partial T}{\partial z}\right] \qquad [15.36]$$

where σ is the thermal diffusivity (which in general may be a function of time and depth). The thermal diffusivity is equal to the ratio of thermal conductivity to heat capacity.

Potential evapotranspiration, required for the soil–water sub-model, is calculated by the Blaney & Criddle (1950) technique. The changes in soil water are then calculated by a sub-model developed by Griffin, Hanks & Childs (1973), in which the changes depend on the relations between soil water content and diffusion pressure on the one hand, and conductivity on the other, together with a term accounting for the extraction of water by roots.

The theoretical aspects of this sub-model can be described by the following relationships. It involves the numerical solution of the one-dimensional general flow equation with a plant-root extraction term, $A(z)$, as given by Nimah & Hanks (1973):

$$\frac{\partial W}{\partial t} = \frac{\partial}{\partial z}\left\{K(W)\frac{\partial H}{\partial z}\right\} + A(z) \qquad [15.37]$$

$A(z)$ is defined as:

$$A(z) = \frac{\{P + \alpha \cdot z - S(z) - O(z)\} \cdot E(z) \cdot K(W)}{\Delta z} \qquad [15.38]$$

where W is the volumetric water content, t is time, z is depth, K is the hydraulic conductivity, H is the hydraulic head, P is the effective water potential in the root at the soil surface where z is considered zero, α is a constant depending on the flow characteristics of the soil, $S(z)$ and $O(z)$ are the soil pressure head and osmotic potential (in head units) at depth z, and $E(z)$ is the fraction of total active roots in the depth increment Δz. Any excess of water at the soil surface over the infiltration capacity is lost as run-off.

Solute movement is modelled partly as a diffusion process, partly as dependent on the volumetric flux of solution; interactions with the solid phase of the soil (adsorption, etc.) are not taken into account at this stage of model development. The equation for the one-dimensional transient salt concentration was derived by Bresler (1973):

$$\frac{\partial}{\partial t}[Wc] = \frac{\partial}{\partial z}\left[D(V, W)\frac{\partial c}{\partial z}\right] - \frac{\partial(qc)}{\partial z} \qquad [15.39]$$

where c is the concentration of the solute, $D(V, W)$ is the combined diffusion–dispersion coefficient, q is the volumetric flux of solution, V is the average interstitial flow velocity, and other symbols have the same meanings as before.

The sub-models of physical processes were tested separately by a run of 28 days, using field data in 1971 from Curlew Valley, Utah, and showed excellent agreement between measured and predicted soil parameters.

The nitrogen sub-model (Parnas & Radford, 1974*a*) describes nitrogen transformations in the soil horizons by determining the growth and death rates of the micro-organisms responsible for the transformations. The processes included are: (*a*) symbiotic fixation of nitrogen; (*b*) heterotrophic fixation of nitrogen; (*c*) autotrophic fixation of nitrogen; (*d*) oxidation of ammonium to nitrite; (*e*) oxidation of nitrite to nitrate; (*f*) denitrification; and (*g*) volatilization of ammonia.

As an example, consider the heterotrophic fixation of nitrogen, First g, the maximum growth rate of these organisms under the present environmental conditions, is calculated:

$$g = G \cdot F_1(T) \cdot F_2(pH) \cdot F_3(S) \cdot F_4(W) \qquad [15.40]$$

where G is the maximum growth rate under optimum conditions, and $F_1 \ldots F_4$ are simple trapezoidal functions of temperature, pH, salinity and soil water potential respectively, varying between 0 and 1. The actual growth rate R can then be calculated:

$$R = \frac{gS}{S+\kappa} \qquad [15.41]$$

where S is substrate concentration (total soil organic carbon for heterotrophs) and κ is the Michaelis–Menten constant for the reaction. The death rate D for these fixers has one of two values: (*a*) a high value if $S = 0$, and (*b*) a low value if $S > 0$. The change in biomass, B, of heterotrophic fixers between time

t and time $(t+\Delta t)$ is

$$B(t+\Delta t) = B(t)\cdot \exp(R-D) \qquad [15.42]$$

The sub-model of the decomposition processes (Parnas & Radford, 1974*b*) follows the same conception as in the nitrogen sub-model: namely, the rate of decomposition of a substance is proportional to the rate of growth of its decomposers. Growth rates are calculated in a manner identical to that used in the nitrogen sub-model. Decomposition is calculated by 'environmental zones', i.e. soil horizons, soil surface (litter and animal residues) and above the surface (standing dead). The decomposition rate for each 'environmental zone' depends on a specific population of decomposing organisms, and is affected by temperature, soil moisture, salinity, and the carbon: nitrogen ratio of the substrate.

All these microbial processes are accompanied by increase in microbial biomass. Carbon dioxide, ammonia and other products may be released, but the material incorporated in the microbial biomass becomes available for recycling only when the micro-organisms die which, as mentioned above, is a function of substrate availability.

Various other processes affecting the soil are also included in this sub-model. The surface of a desert soil is often partially covered by a crust formed of lichens or algae (predominantly Cyanophyceae), and the growth of these crustal organisms is modelled. If the surface soil is not too wet or too dry, the crustal organisms grow at a rate proportional to their existing biomass (expressed as carbon), relative growth rate varying with season.

In the Great Basin Desert, the soil commonly freezes during the winter. In the model, the entire soil is considered frozen if the average air temperature is less than a minimum and snow is absent, or if the soil is already frozen and snow depth is greater than a minimum, or if snow depth is less than the minimum and average temperature is less than a second minimum different from (usually less than) the one already mentioned.

Not all precipitation reaches the soil surface; a proportion is lost by interception. In the sub-model under discussion, below a minimum daily amount of rainfall (depending on average vegetation height) all rainfall on vegetation is intercepted. Above this minimum a constant fraction is intercepted.

Precipitation falling as snow, and not intercepted by the vegetation, rests on the soil surface; the depth of snow is tracked in the model as a state variable. Snow-melt is proportional to the temperature difference between average daily temperature and a seasonally determined minimum. It is also proportional to the quantity of liquid water in contact with unmelted snow. If snow melt is impossible, snow blowing can occur. The amount lost from a snowfall at the time of fall is a seasonal fraction of the snow falling.

Wind and water erosion processes are handled similarly. For water erosion, the amount eroded is assumed to increase exponentially with run-off. Wind erosion increases exponentially with wind speed. Erosion depends, in addition, on wind gusts, average vegetation height, type of soil, amount of ground cover and whether or not the soil is frozen.

Implementation of models

The prime need in implementing these general-purpose models has, of course, been to develop simulations for each of the sites where the US/IBP Desert Biome programme has been taking complete inventories of the ecosystem and monitoring the changes it has undergone, i.e. the *validation sites*. There are five of these, and for three of them effective simulations have been performed at the time of writing, using the third version of each of the process sub-models. These three are: Rock Valley in the Mohave Desert; Curlew Valley in the Great Basin; and the Jornada site in the Chihuahuan Desert. In the last two cases, separate simulations were performed for more than one distinct ecosystem within the same site. These simulations have been reported in detail by Gist & Goodall (1974). A number of sensitivity tests have also been performed on some of these implemented models, by varying particular parameters of the model and studying the effects of these changes on the subsequent behaviour of the model; for this purpose, use was made of the special sub-routines developed by Noy-Meir & Goodall (1973).

During a 'Workshop' meeting in 1972, the opportunity was taken to assemble limited sets of data from deserts in other continents – central Australia; the Rajasthan Desert in India; the Repetek Reservation in Turkmenistan; the coastal desert of Egypt; the northern Negev of Israel; and southern Tunisia – and implement the general-system model using these data. The first version of each process sub-routine was used, and the results have been published in part (Anon. 1972).

The models were also implemented in respect of two alpine-tundra ecosystems in Norway, using data collected under the Norwegian IBP programme, and kindly provided by Dr F. Wielgolaski during a stay in Logan. Very satisfactory simulations were performed, and this success with a very different type of ecosystem is convincing evidence of the generalizability of the models.

Modelling work in Israel

During the period under review, there have been several attempts in Israel to model ecosystems in the arid and semi-arid zones, mainly as an aid in grazing and water management.

The model NEGEV (Seligman, Tadmor, Noy-Meir & Dovrat, 1971) was initiated in 1971. In many ways it resembles the PASTOR model of Goodall mentioned above (p. 386–7). Like that, it divides the landscape into topographically distinct land units, and is primarily concerned with the interaction between grazing animal and vegetation, largely under the control of soil moisture. The NEGEV model makes use of data for rainfall, temperature and pan evaporation. The soil moisture balance involves first a calculation of run-off R from up-slope units as:

$$R = \alpha_1(P-\alpha_2) \qquad [15.43]$$

where P is the daily precipitation, α_1 is the run-off coefficient and α_2 the depression storage. Run-off from the slopes becomes run-on into wadi land units. The soil is divided into four horizons, and it is assumed that the infiltration water $(P-R)$ first increases the water content of the uppermost horizon to field capacity, and that any excess penetrates immediately to deeper layers. Water loss from each horizon by evapotranspiration is calculated over periods of ten days from pan evaporation, soil moisture and vegetation cover, and the resultant change in soil moisture is the balance between this figure and gain by infiltration.

The vegetation biomass is divided into annuals and perennials, and dead material from each is kept separate. The relative growth rate for each plant type is calculated as

$$\beta_{1_i}[1-\exp\{\beta_{2_i}(T-\beta_{3_i})\}]\cdot\{1-\exp(\beta_{4_i}W)\} \qquad [15.44]$$

where T is the average daily temperature, W is the proportion of available soil moisture, averaged over the rooting zone, and β_{1_i} etc. are constants, where the subscript i indicates the plant type. The relative death rate is

$$\exp[-\gamma_{1_i}(W+\gamma_{2_i})] \qquad [15.45]$$

and specifies the rate of transfer from living biomass to the corresponding category of dead material. A minimal value is imposed on each type of living plant material, representing a reserve of seeds or buds from which regeneration can take place.

Dead material is lost by weathering at a constant relative rate.

The sheep population is divided into age groups. The potential consumption of the ith plant type by the jth animal group is

$$P_{ij} = \frac{\delta_{1_{ij}}}{1+\delta_{2_{ij}}\exp(\delta_{3_{ij}}V_i)} \qquad [15.46]$$

where V_i is the biomass of the plant type and $\delta_{1_{ij}}$ etc. are constants. Each animal type has an upper limit ϵ_j for food consumption per unit live weight. The food types are then taken in order of preference: live annuals, dead annuals, live shrubs, dead shrubs. If P_{ij} for the first type exceeds ϵ_j, only this type is eaten, to the amount of ϵ_j; if not, the next type is taken as well, and so on until a total of ϵ_j has been eaten.

The weight increase of sheep in each group is a linear function of food intake in excess of maintenance energy requirements, taken as proportional to $W_j^{.75}$, where W_j is the average weight of that group. If intake falls below that figure, weight loss is another linear function of the discrepancy. Mortality in all age groups, and fecundity and rates of miscarriage in ewes, are calculated as functions of their body weight.

The programmes provide for alternative management practices to be applied to the grazing systems simulated, including movement of animals between land units, reduction or increase in numbers of animals, and supplementary feeding.

Tests of the model against actual sequences of data in the absence of grazing showed good general agreement. Some sensitivity tests were also performed. It was found, for instance, that the response of vegetational biomass in the model to variation in β_{4_i} and γ_{1_i} was greater than to similar variation in γ_{2_i}.

When sheep were introduced into the model, and kept on a single land type, they eventually were exterminated as a result of an adverse year despite the fact that the parameters were varied between reasonable limits in a fruitless attempt to avoid this result of the simulation. As the authors indicate, restriction of the animals to a single habitat type and lack of regulation of sheep numbers are unrealistic management constraints; one still wonders, however, whether this may not point to a fault in the model.

A quite distinct model, known as ARIDCROP (Tadmor & Noy-Meir, 1973; van Keulen, 1975), was also developed for Israeli ecosystems, based on experience with modelling of agricultural crops in the Netherlands (de Wit, Brouwer & Penning de Vries, 1970; de Wit & Goudriaan, 1974). It is designed for annual vegetation only, such as is found in much of the grazing land of the northern Negev, and is mainly concerned with primary production as a function of climate variables. Models of nitrogen cycling in this type of vegetation have also been developed.

Modelling of arid ecosystems in Australia

During the period under review, a considerable amount of attention has been given in Australia to simulation modelling. Much effort, for instance, has gone into hydrological models, and some of these are relevant to ecosystem modelling. One of these hydrological models in particular, WATBAL (McAlpine, 1970; Winkworth, 1970), has had considerable influence on ecosystem models developed in Australia. Most of the ecosystem models in question have, however, been directed towards pastures in higher-rainfall country (see, for instance, Paltridge, 1970; Freer, Davidson, Armstrong & Donnelly, 1970; Rose *et al.*, 1972; Australian Society of Animal Production, 1972), and, though some of the models may be applicable to arid situations, no reports of such applications have come to hand in most cases.

Several simulation studies have been concerned with the arid shrub-dominated grazing lands of western New South Wales and northern South Australia. Chudleigh (1971) used a simple model of an ecosystem subject to grazing for a study of pastoral management in the Broken Hill region, but based the model on the manager's estimates of the condition of, and changes in the forage rather than on direct measurement.

Noble (1970) used a model rather similar to Goodall's (pp. 386–7) to simulate an ecosystem subject to grazing at Yudnapinna, South Australia. Unlike Goodall, he included ephemeral plants without distinction of species; their growth was treated over a number of discrete stages of development, taking temperature and soil moisture into account. Dependence of growth on temperature was sinusoidal. In respect of consumption, Noble assumed that the animals exercised no preference among the various forage types available.

As part of a project to apply dynamic programming to the management of arid grazing systems in western New South Wales, Fisher (1974) developed an ecosystem model based broadly on that of Goodall (see p. 387). Rainfall input was again generated stochastically, but using a log-normal distribution. A more sophisticated treatment of run-off was adopted, based on four separate moisture storages, and run-off to the lower-lying parts of the paddock depended on run-off from the other parts, whereas in Goodall's model they were independent.

In the vegetation sub-model, provision is made for mortality and other losses from causes additional to consumption by sheep. These losses, relative to the biomass of the ith species, are given by

$$\alpha_{1_i} T\{1-\exp(\alpha_{2_i} L)\} \quad [15.47]$$

and

$$\alpha_{3_i} \exp(\alpha_{4_i} N) \quad [15.48]$$

respectively, where T is the mean weekly temperature, L the number of consecutive days during which water stress has prevented growth, and N the number of mature grazing stock, while α_{1_i} etc. are constants. It will be noted that losses from causes other than consumption by sheep (for instance, by native herbivores) are inversely related to the grazing pressure by the sheep themselves, α_{4_i} being negative.

Germination of ephemerals occurs when the soil water potential has exceeded the wilting point for two weeks; their growth is proportional to the difference between the current biomass and a maximum. They mature in four weeks, but in each week are subject to mortality if the soil water potential is below wilting point.

Fisher's treatment of forage consumption was virtually the same as Goodall's, but his treatment of animal productivity was more sophisticated and more realistic. If the gross intake per animal is I and the intake of digestible dry matter is H then the metabolizable energy available for maintenance is

$$M = \beta_1 H + \beta_2 \quad [15.49]$$

and the gross metabolizability is

$$G = \gamma HI \quad [15.50]$$

The efficiency of energy conversion for maintenance is a linear function of gross metabolizability:

$$E = \delta_1 G + \delta_2 \quad [15.41]$$

Energy requirements arise partly from the fasting metabolic rate, partly from exercise. Hence the metabolizable energy required to maintain the sheep in energy balance is

$$R = (\epsilon_1 W + \epsilon_2 + \epsilon_3 Wt)/E \quad [15.52]$$

where W is the body weight and t the time spent in grazing. If M is greater than R, the efficiency of energy conversion is somewhat different from E:

$$E' = \eta_1 G + \eta_2 \quad [15.53]$$

and the surplus energy available is

$$S = (M-R)/E' \quad [15.54]$$

Then the change in body weight is

$$\Delta W = S(\zeta_1 W^2 + \zeta_2 W + \zeta_3)/[1 - \zeta_4 S \cdot \max(0, \zeta_5 + \zeta_6 W)] \quad [15.55)]$$

Wool growth is proportional to the intake of digestible dry matter:

$$\xi_1 H \quad [15.56]$$

As in Goodall's model, Fisher made mortality a function of body weight:

$$(1-\mu_1)\exp\{\mu_2(W-\mu_3)+\mu_4\} \quad [15.57]$$

The model was made stochastic by adding a random variable to the net change calculated for each three months in certain of the state variables. By running the model repeatedly with different initial sets of state variables, Fisher obtained sets of initial and final values over the three-month time step, which could then provide statistical distributions serving as input to the stochastic dynamic-programming routine. In.this way an optimum management procedure could be devised, in the sense of an algorithm determining stocking rates in the light of vegetation conditions.

The ecosystem work in Kunoth Paddock, in the Alice Springs area, is built around the concept of ecosystem modelling. Up to the present, a number of partial models have been built, though none in which the behaviour of the ecosystem as a whole is simulated. Winkworth & Fleming (personal communication) are modelling water redistribution in the paddock. Ross (personal communication) developed a model of growth of *Eragrotis eriopoda*, including the extraction of water from the soil.

In Ross's model, which he called GAME, the water in the soil is lost by evaporation and drainage at a rate which is a linear function of the square root of the time since it was last recharged. Losses by transpiration (L) are proportional to the product of available moisture, plant biomass, and pan evaporation rate, so

$$\frac{dW}{dt} = \alpha_1 + \alpha_2 t^{\frac{1}{2}} + L \qquad [15.58]$$

where

$$L = \alpha_3 WPE \qquad [15.59]$$

and W is the available soil moisture, P the plant biomass per unit area, E the pan evaporation rate, t is the time since recharge, and α_1 etc. are constants.

Photosynthesis is assumed proportional to transpiration, and respiration to be a linear function of temperature, so that the change in vegetation biomass is

$$\frac{dP}{dt} = \beta_1 L - \beta_2 PT \qquad [15.60]$$

where T is the mean daily temperature, and β_1, β_2 are constants.

Conclusions

It is clear that considerable activity in the modelling of arid-land ecosystems has developed during the International Biological Programme, and that a considerable variety of approaches have been used. So far, this has taken the form of a proliferation of ideas rather than firm achievement. It is to be expected that, during the coming decade, these ideas will crystallize around a few types of models that seem particularly promising, that these models will then be tested exhaustively against field data, and that they will then be applied successfully to the problems of managing arid-land ecosystems to the long-term benefit of mankind.

Acknowledgements

I am indebted to a number of people who have made unpublished work available for this review. I may mention particularly the personnel of the US/IBP Desert Biome programme; Dr I. Noy-Meir in Israel; and the late Dr M. Ross and Mr R. E. Winkworth in Australia.

References

Anon. (1972). *Report of the International Arid Lands Workshop, September 9–19, 1972.* US/IBP Desert Biome, Logan, Utah.

Anway, J. C., Brittain, E. G., Hunt, H. W., Innis, G. S., Parton, W. J., Rodell, C. F. & Sauer, R. H. (1972). *ELM: Version 1.0.* US/IBP Grassland Biome Technical Report No. 156.

Australian Society of Animal Production. (1972). Symposium – systems modelling. *Proceedings of the Australian Society of Animal Production*, **9**, 1–138.

Blaney, H. G. & Criddle, W. D. (1950). *Determining watering requirements in irrigated areas from climatological and irrigation data.* USDA Soil Conservation Service. TP-96.

Bresler, E. (1973). Simultaneous transport of solute and water under transient unsaturated flow conditions. *Water Resources Research*, **9**, 975–86.

Bridges, K. (1971*a*). *CAMEL, Version 1.* US/IBP Desert Biome Modelling Report Series, **3.1.10**, 1–27.

Bridges, K. (1971*b*). *QUESTIONS, Version 1.* US/IBP Desert Biome Modelling Report Series, **3.1.11**, 1–3.

Chudleigh, P. D. (1971). Pastoral management in the West Darling region of New South Wales. PhD thesis, University of New South Wales.

de Wit, C. T. & Gourdriaan, J. (1974). *Simulation of ecological processes.* Pudoc, Wageningen.

de Wit, C. T., Brouwer, R. & Penning de Vries, F. W. T. (1970). Prediction and measurement of photosynthetic productivity. In: *Proceedings of IBP/PP Technical meeting, Trebon, 1969*, pp. 47–50. Pudoc, Wageningen.

Fisher, I. H. (1974). Resource optimization in arid grazing systems. PhD thesis, University of New South Wales.

Freer, M., Davidson, J. L., Armstrong, J. S. & Donnelly, J. R. (1970). Simulation of summer grazing. *Proceedings of the XI International Grassland Congress*, pp. 914–17. University of Queensland Press, St Lucia.

Georgievsky, A. B. & Kuznetsov, V. (1976). *Trophofunctional approach to an ecosystem model.* US/IBP Desert Biome Translation No. 1. Utah State University, Logan.

Gist, C. S. & Goodall, D. W. (1974). *Simulation of Desert Biome sites using the general purpose model.* US/IBP Desert Biome Research Memorandum RM 74-53.

Goodall, D. W. (1965). Computer methods in range management. Paper read to Australian Arid Zone Research Conference, Alice Springs, September 1965.

Goodall, D. W. (1967). Computer simulation of changes in vegetation subject to grazing. *Journal of Indian Botanical Society*, **46**, 356–62.

Goodall, D. W. (1969). Simulating the grazing situation. In: *Concepts and models of biomathematics; simulation techniques and methods* (ed. F. Heinmets), pp. 211–36. Marcel Dekker, New York.

Goodall, D. W. (1970). Use of computer in the grazing management of semi-arid lands. *Proceedings of the XI International Grassland Congress*, pp. 917–22. University of Queensland Press, St Lucia.

Goodall, D. W. (1973*a*). Ecosystem simulation in the US/IBP Desert Biome. *1973 Summer Computer Simulation Conference*, pp. 777–80.

Goodall, D. W. (1973*b*). *Terrestrial models – Introduction.* US/IBP Desert Biome Research Memorandum RM 75-53. Utah State University, Logan.

Goodall, D. W. (1975). Ecosystem modelling in the Desert Biome. In: *Systems analysis and simulation in ecology*, vol. 3 (ed. Bernard C. Patten), pp. 73–94. Academic Press, New York & London.

Goodall, D. W. & Gist, C. (1973). *Terrestrial models: Introduction and MAIN programs.* US/IBP Desert Biome Research Memorandum RM 73-53. Utah State University, Logan.

Griffin, R. A., Hanks, R. J. & Childs, S. (1973). *Model for estimating water, salt, and temperature distribution in soil profile.* US/IBP Desert Biome and Soil Science and Biometeorology Department. Utah State University, Logan.

Hanks, R. J., Austin, D. D. & Ondrechen, W. T. (1971). *Soil model – heat, water and salt flow.* US/IBP Desert Biome Research Memorandum RM 71-18. Utah State University, Logan.

Holmgren, R. C. & Hutchings, S. (1971). Salt desert shrub response to grazing use. In: *Wildland shrubs, their biology and utilization* (ed. C. M. McKell, J. P. Blaisdell & J. R. Goodin), pp. 153–64. USDA Forest Service, General Technical Report INT-1.

Hutchings, S. S. & Stewart, G. (1953). Increasing forage yields and sheep production on intermountain winter ranges. *USDA Circular 925*, pp. 1–63.

Lommen, P. (1974). *Terrestrial model: soil processes.* US/IBP Desert Biome Research Memorandum RM 74-51. Utah State University, Logan.

McAlpine, J. R. (1970). Estimating pasture growth periods and droughts from simple water balance models. *Proceedings of the XI International Grassland Congress*, pp. 484–7. University of Queensland Press, St Lucia.

Nimah, M. N. & Hanks, R. J. (1973). Model for estimating soil water, plant, and atmospheric interrelations. I. Description and sensitivity. II. Field test of model. *Proceedings of the Soil Science Society of America*, **37**, 522–32.

Noble, I. R. (1970). A computer simulation of a model of the sheep-vegetation system in the arid zone. BSc (Hons.) thesis, University of Adelaide.

Noy-Meir, I. & Goodall, D. W. (1973). *Sensitivity analysis.* US/IBP Desert Biome Research Memorandum RM 73-58. Utah State University, Logan.

Overton, W. S., Colby, J. A., Gourley, J. & White, C. (1973). *FLEX1 user's manual.* US/IBP Coniferous Forest Biome Internal Report No. 126.

Paltridge, G. W. (1970). A model of a growing pasture. *Agricultural Meteorology*, **7**, 93–130.

Parnas, H. & Radford, J. (1974*a*). *Nitrogen submodel.* US/IBP Desert Biome Research Memorandum RM 74-65. Utah State University, Logan.

Parnas, H, & Radford, J. (1974*b*). *Decomposition submodel.* US/IBP Desert Biome Research Memorandum RM 74-64. Utah State University, Logan.

Payne, S. (1973). *Terrestrial models: animal sub-models.* US/IBP Desert Biome Research Memorandum RM 73-55. Utah State University, Logan.

Radford, J. (1973) *Terrestrial models: soil processes (Versions I, II, III).* US/IBP Desert Biome Research Memorandum RM 73-56. Utah State University, Logan.

Rose, C. W., Begg, J. E., Byrne, G. F., Torssell, B. W. R. & Goncz, J. H. (1972). A simulation model of growth–field relationships for Townsville stylo (*Stylosanthes humilis* H.B.K.) pasture. *Agricultural Meteorology*, **10**, 161–83.

Seligman, N. G., Tadmor, N. H., Noy-Meir, I. & Dovrat, A. (1971). An exercise in simulation of a semi-arid Mediterranean grassland. *Bulletin de Recherche Agronomique Gembloux, Semaine d'étude des problèmes méditerraneens*, 138–43.

Tadmor, N. H. & Noy-Meir, I. (1973). Methodology for the study of productivity in arid ecosystems. *Mexican–Israeli Symposium on Arid Zones, Saltillo, May, 1973.*

Valentine, W. (1973). *Terrestrial models: plant processes (Versions I, II, III).* US/IBP Desert Biome Research Memorandum RM 73-54. Utah State University, Logan.

Valentine, W. (1974). *Terrestrial models: plant sub-models – the fourth version.* US/IBP Desert Biome Research Memorandum RM 74-56. Utah State University, Logan.

van Keulen, H. (1975). *Simulation of water use and herbage growth in arid regions.* Simulation Monographs, Pudoc, Wageningen.

Wilcott, J. C. (1973). A seed demography model for finding optimal strategies for desert annuals. PhD dissertation, Utah State University, Logan.

Wilkin, D. C. & Norton, B. E. (1974). *Resource management.* US/IBP Desert Biome Research Memorandum RM 74-67. Utah State University, Logan.

Winkworth, R. E. (1970). The soil water regime of an arid grassland (*Eragrostis eriopoda* Benth.) community in central Australia. *Agricultural Meteorology* **7**, 387–99.

16. Spatial effects in modelling of arid ecosystems

I. NOY-MEIR

Spatial heterogeneity and ecosystem modelling

The problem

The specification and implementation of a mathematical model of an ecological system involves the assignment of numerical values to all variables and parameters defined in the model (biomasses, concentrations, etc.). Usually the model is implemented at the site level, and a single value is given to each variable and parameter. This single value is implicitly or explicitly defined and measured as the 'typical' value for the site. Since the site is usually not a point, but an area of finite size ($10^2 - 10^5\ m^2$ in most ecosystem studies), this practice implicitly assumes that the values of all variables and parameters do not vary significantly within the area of the site, or that such spatial variations, if they do exist, do not affect the behaviour of the mathematical model. How valid is either of these assumptions in real ecosystems?

In most natural ecosystems (and in many pastoral and agricultural ones) there is considerable horizontal spatial heterogeneity in topography, soil, plant and animal populations at all scales. Thus within a 'site' there will, in general, be considerable local variation in the values of most variables and parameters. In some cases this variation may not be critical for the properties of the site-level model, and the use of typical site values in the model may not affect the results very much. But in other cases neglecting within-site variation may distort the model considerably and undermine its validity. The problem is that in any specific situation it is difficult to say *a priori* which is the case. Thus the degree of spatial heterogeneity within any specific site, and the importance of its possible effects on any specific model of that site, need to be carefully examined before a decision is made to neglect them.

There are three main approaches to this problem: homogeneous sites, spatial compartmentalization and mean values.

Selection of homogeneous sites

The problem will be minimized if only sites that are internally as homogeneous as possible are selected for validation and implementation of ecosystem models. However, in many ecosystems strong small-scale spatial heterogeneity is the norm and a very homogeneous site may either be impossible to find, or it will be very atypical of the system. It may be possible to reduce heterogeneity by reducing the area of the site; but this is limited by some minimum area of the site dictated by practical considerations, and by the need to include in it large enough populations of all species of interest (particularly if large animals are to be included). Thus, site selection can reduce spatial heterogeneity problems in some cases, but it cannot eliminate them in all cases.

Spatial compartmentalization

Alternatively, one may decide to do justice to the spatial variation within the ecosystem by representing it within the model. Spatial variation can either be represented by defining continuous spatial gradients or by dividing the area into discrete spatial compartments (or 'strata'). Models with continuous spatial gradients become very complex mathematically if more than very few interdependent variables have to be considered, of if the pattern of variation is anything but very simple. Therefore, the more feasible solution is to divide the ecosystem into spatial compartments which are relatively homogeneous internally and to model the processes in each compartment separately (e.g. Goodall, 1967, 1971; Seligman, Tadmor, Noy-Meir & Dovrat, 1971). Instead of one value for each variable and parameter in the model, there will be a set of values corresponding to the various compartments. If the processes in the different compartments can be assumed to be functionally identical, and independent of the processes in other compartments, the transition from the single-compartment to the multi-compartment model is easy (e.g. replacement of each variable by a vector in a computer simulation program, or use of a frequency distribution instead of one mean value). It becomes more complicated, if different processes operate in different compartments, or more importantly, if there are dynamic interactions and feedbacks between compartments. In many spatially heterogeneous ecosystems, such interactions by flows of populations, energy and materials between compartments are important. For some (or all) compartments, these spatial transport processes may be at least as important as the internal processes within the compartment. In those cases a spatially compartmentalized model, if it is to be realistic, must be rather more complex than just a multiple of the single-compartment model.

In principle, the realism of an ecosystem model can be increased by dividing it into any indefinite number of spatial compartments. But there are severe limitations. The number of necessary measurements of variables and parameters increases proportionally to the number of compartments recognized in the model. The model itself becomes larger, more complex and more cumbersome to experiment with and therefore less efficient in serving its original purpose of providing a manageable simplification of the complex real system. Beyond a certain limit, computer space and time limitations may appear even with today's largest and fastest computers; in any case, computing costs increase drastically with the size of the model. Therefore, it is necessary to reduce the number of spatial compartments in a model as much as is possible without introducing 'serious' distortions of reality. How many compartments, and which type of compartments, does this mean in any specific case? The question may be easier to approach if we ask for each of the possible spatial stratifications in the system of interest: how different will the results of this model be if these compartments are combined into one rather than treated separately? The answer will obviously depend on the degree of spatial variability, but also on the nature of the processes involved.

Mean values

It is thus clear that even if we try to restrict ourselves to homogeneous sites, or if we recognize in the model a feasible number of spatial compartments, there will still be residual spatial variability within the site, or within each compartment. This variability will have to be covered by the single 'typical' value of each variable and parameter for the site or the compartment. The most logical choice is to define this 'typical' value as the mean of the values of the variable over the area of the site (compartment) and to estimate it in the field (when necessary) as the mean of a statistically representative sample from this area. Indeed, such mean values are used extensively either as inputs to ecosystem models or as validation data for them, often without asking the question: how is the behaviour of the model affected by using the mean of a spatially varying parameter or variable?

Consider a very simple case: only two spatial compartments, each internally homogeneous, between which there is no feedback; and a mathematical model of the static relation between two variables, an independent (input) variable x and a dependent (output) variable y. A unique functional relation between x and y, at a given point in time in any internally uniform area i may be derived from theory or from a controlled experiment:

$$y_i = F(x_i) \qquad [16.1]$$

thus for the two compartments,

$$y_1 = F(x_1) \qquad [16.2]$$

$$y_2 = F(x_2) \qquad [16.3]$$

Now suppose we combine both compartments into one and take the weighted mean of the input variable for the whole site

$$\bar{x} = a_1 x_1 + a_2 x_2 \qquad [16.4]$$

where a_1, $a_2 = 1 - a_1$, are the proportions of the area in compartments 1, 2. If

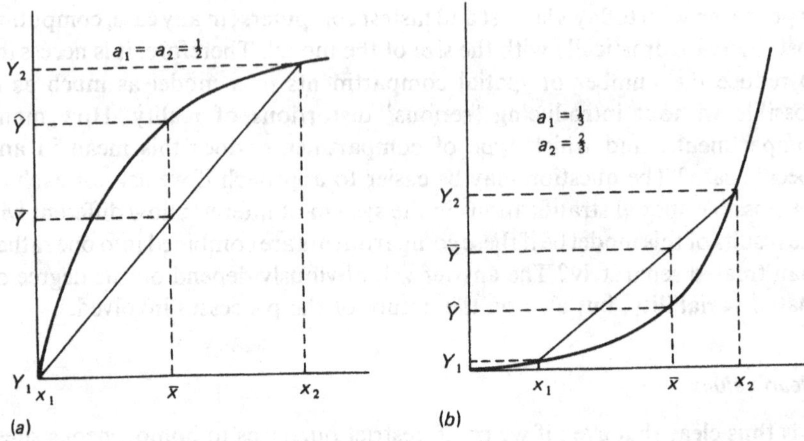

Fig. 16.1. The effect of two-compartment spatial heterogeneity in an input x on the average response $\bar{y} = a_1 y_1 + a_2 y_2$ and its relation to $\hat{y}$, the response estimated, from the average input $\bar{x} = a_1 x_1 + a_2 x_2$, when the response function $y(x)$ is: (*a*) convex; (*b*) concave.

we apply the established functional relation F to predict the output variable for the whole site, the result is:

$$\hat{y} = F(\bar{x}) = F(a_1 x_1 + a_2 x_2) \qquad [16.5]$$

If we measure the output variable as the mean of a representative sample of the whole area the result is:

$$\bar{y} = a_1 y_1 + a_2 y_2 = a_1 F(x_1) + a_2 F(x_2) \qquad [16.6]$$

In general, $\bar{y}$ needs not to be equal to $\hat{y}$, that is, the actual mean output may be either larger or smaller than the output calculated from the mean input. The two will be equal only in some special cases, particularly if the functional relation F is linear in the range x_1 to x_2:

if $$y_i = F(x_i) = c + dx_i \qquad [16.7]$$

then $$\bar{y} = a_1(c + dx_1) + a_2(c + dx_2) = c + d(a_1 x_1 + a_2 x_2) \qquad [16.8]$$

and $$\hat{y} = c + d\bar{x} = c + d(a_1 x_1 + a_2 x_2) = \bar{y} \qquad [16.9]$$

It can be shown by simple graphical constructions (Fig. 16.1), that if F is generally concave between x_1 and x_2 then the actual mean output $\bar{y}$ will be greater than the computed $\hat{y}_i$. F is generally convex, $\bar{y}$ is smaller than $\hat{y}$. If F is sigmoid then $\bar{y}$ may be greater than, equal to, or smaller than $\hat{y}$, depending on the areal proportions of the compartments.

The effect will be more complex if there are more than two compartments, more than two variables, or if dynamic relations (differential equations) are considered. But, in general, 'lumping' together in the model spatially differentiated areas and their representation by mean values will cause distortion and produce results deviating from reality. This deviation may be erroneously interpreted as evidence for rejection of the basic (site-level) model, even when the basic model is really correct. The distortion due to heterogeneity may be small or large, depending on the magnitude of differences between the ignored spatial strata and on the degree of non-linearity of the basic functional relations.

In modelling any specific process in any specific situation it is therefore necessary to decide which spatial compartments it is absolutely necessary to keep separate, and which can be safely coalesced and represented by means with little effect on the results. This decision may be based on *a priori* qualitative considerations along the lines discussed (degree of heterogeneity, degree of non-linearity and additivity), or on the examination of the magnitude of the spatial effect in some pilot model of the specific situation.

In the following section some of the spatial effects which are specifically important in arid ecosystems will be described and discussed. For some of them the implications for modelling will also be discussed in some detail.

Some spatial effects in arid ecosystems

Redistribution of water input by run-off

The primary limiting factor for most biological processes in arid ecosystems is water. Characteristically, in arid zones the water input (rain) occurs in discrete events which are both infrequent and subject to much random variation (Noy-Meir, 1973). Each substantial rain event, or cluster of a few rain events, produces a rapid 'pulse' of activation, growth and reproduction in plant and animal populations. These processes decline equally rapidly as the water reserves are exhausted, and in the dry period until the next rain most organisms remain in a state of very low (or zero) activity and production. The magnitude of the production and population puse produced during the wet period is mainly proportional to the total quantity of the water input; it also depends on its distribution in time and on the levels of other factors (temperature, nutrients) during the wet season. Relations between total production (or reproduction) and total rainfall in a given rain season have been empirically established in some arid ecosystems, particularly for primary production (e.g. Walter, 1962, p. 275; Beatley, 1969; Lieth, 1972). Animal populations directly dependent on green vegetation or high soil moisture probably show similar responses. In general, this relation is very close to linear over a wide range of water input, between a minimum threshold input (below which production is zero) and some maximum input where the response to additional water levels off, either gradually or abruptly. Actually, in many typical arid and semi-arid sites this upper (saturation) limit may never be reached (or reached only in exceptionally wet seasons) so that the relation appears to be simply linear above a threshold.

In some situations, for example a sandy plain, all the rain infiltrates into the soil on the spot, so that the actual water input remains uniformly distributed over the whole area. In many arid ecosystems, topographic and soil conditions are such that only part of each rainfall infiltrates *in situ*. A substantial part is redistributed by run-off (and to some extent by sub-surface flow) and accumulates and infiltrates only in sink (run-on) areas. Thus the actual water input (after any rain event) may vary considerably in space between different areas. Some examples of arid ecosystems in which water redistribution by run-off at small distances is important are:

(*a*) plains and alluvial fans with a dense network of small runnels and channels;
(*b*) plains and gentle slopes with a regular pattern of slight depressions, e.g. gilgais (Williams, 1955) or the mulga grove–intergrove system (Slatyer, 1961);
(*c*) rocky slopes interspersed with soil pockets (Danin, Orshan & Zohary, 1975); and
(*d*) areas with strong microvariation in surface soil characteristics, in particular infiltrability (Shanan, 1974).

What are the implications of the unequal redistribution of water input for plant and animal production, and for models of these processes?

First, internal water redistribution within a study site will increase the variance of production measurements between sampling units (e.g. quadrats), thus making exact model validation more difficult (van Keulen, 1975). However, it may also have effects on the mean production from the site.

A simple model: assumptions

The effects of water redistribution by run-off on mean plant production may be demonstrated with the aid of a model which is very simple indeed, but is a good prototype of more complex and realistic models of this situation. The following assumptions are made.

(1) The production Y (yield, reproduction, etc.) resulting from a rain event (or a rain period of several successive events) in a given homogeneous site is a unique function of the total water input R to the site at the event (or over the period).
(2) The function relating Y to R is linear between a threshold R_t and a saturation limit R_s (Fig. 16.2).

$$Y = \begin{cases} 0 \quad \text{if} \quad R < R_t \\ b(R - R_t) \quad \text{if} \quad R_t < R < R_s \\ Y_s = b(R_s - R_t) \quad \text{if} \quad R_s < R \end{cases} \qquad [16.10]$$

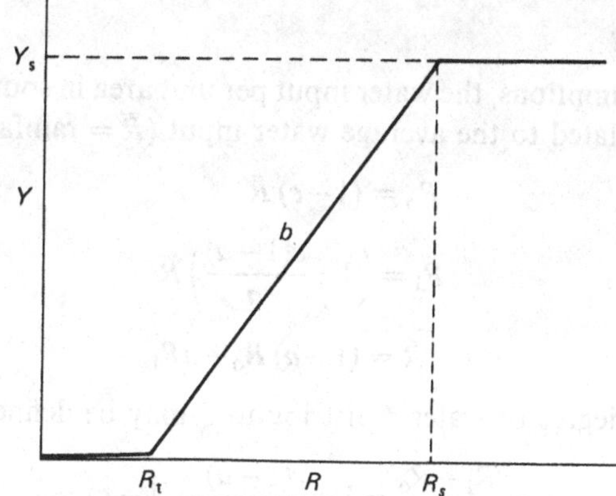

Fig. 16.2. The function assumed to relate the yield Y in a homogeneous site to the water input ('rain'). R. R_t, threshold input; R_s, saturation input; Y_s, saturation yield; b, production/water use efficiency.

The threshold R_t may be interpreted as the part of the water input which is ineffective in producing growth. It depends mainly on the total amount of water lost by direct evaporation from the surface, and possibly on a minimum amount of water required for compensation of plant maintenance and a minimum water input required to initiate growth (germination).

The slope b expresses the efficiency of plant production per unit water actually used for growth. It depends on intrinsic properties of the plant, as well as on climatic (evaporative demand) and edaphic (nutrient supply) conditions during the growing period (de Wit, 1958).

The upper limit of the linear relation is determined either by the maximum effective water capacity of the soil, R_s (depth of soil or of rooting zone $\times$ volumetric moisture capacity), of by the maximum production capacity of the vegetation, Y_s, as a resource other than water (e.g. nutrients) becomes limiting.

(3) The landscape may be divided into only two phases: source areas (o) which always produce run-off, and sink areas (i) which always receive and absorb all the run-off from the source areas. The relative areal proportions of source and sink are constant, a_o $(= 1 - a)$ and a_i $(= a)$.

(4) Run-off is a constant proportion, c, of the rainfall in the source areas.

(5) The production/water function (parameters R_t, b, R_s) is identical for source and sink areas.

Results

With these assumptions, the water input per unit area in source (R_o) and sink (R_i) areas is related to the average water input ($\bar{R}$ = rainfall) by

$$R_o = (1-c)\,\bar{R} \quad [16.11]$$

$$R_i = \left(1+\frac{c(1-a)}{a}\right)\bar{R} \quad [16.12]$$

so that:

$$\bar{R} = (1-a)\,R_o + aR_i \quad [16.13]$$

A relative 'degree of water redistribution' may be defined as

$$q = \frac{R_i - R_o}{\bar{R}} = 1+\frac{c(1-a)}{a} - (1-c) = \frac{c}{a} \quad [16.14]$$

The production $Y(R)$ per unit area in the source $Y_o = Y(R_o)$ and the sink $Y_i = Y(R_i)$ may be combined to the average production per unit area

$$\bar{Y} = (1-a)\,Y_o + aY_i \quad [16.15]$$

$$\bar{Y} = (1-a)\,Y[\bar{R}(1-c)] + aY\left[\bar{R}\left(1+\frac{c(1-a)}{a}\right)\right] \quad [16.16]$$

This average production will obviously not always be equal to the production corresponding to the average water input

$$\hat{Y} = Y(\bar{R}) = Y[(1-a)\,R_o + aR_i] \quad [16.17]$$

Simple graphical (Fig. 16.3) or mathematical considerations lead to a set of conclusions about the relative values of $\bar{Y}$ and $\hat{Y}$.

(*a*) There are three possible cases in which $\bar{Y} = \hat{Y}$, that is, there is no effect of run-off redistribution on production. (i) Rainfall is low and the 'degree of redistribution', q, is also low (low c, high a), so that even in the sink areas the water input is below the threshold, $R_i < R_t$; then $\bar{Y} = \hat{Y} = 0$. (ii) Rainfall is high and the degree of redistribution is low, so that even in the source areas input is above saturation, $R_o > R_s$; then $\bar{Y} = \hat{Y} = Y_s$. (iii) Rainfall is in the range of linear production response (R_t to R_s), and the degree of redistribution is low, so that the input to both source and sink areas is also within that range. Then

$$Y_o = b(R_o - R_t) \quad [16.18]$$

$$Y_i = b(R_i - R_t) \quad [16.19]$$

$$\begin{aligned}\bar{Y} &= (1-a)\,bR_o - (1-a)\,bR_t + abR_i - abR_t \\ &= b[(1-a)\,R_o + aR_i - R_t] = b[\bar{R} - R_t] = \hat{Y}\end{aligned} \quad [16.20]$$

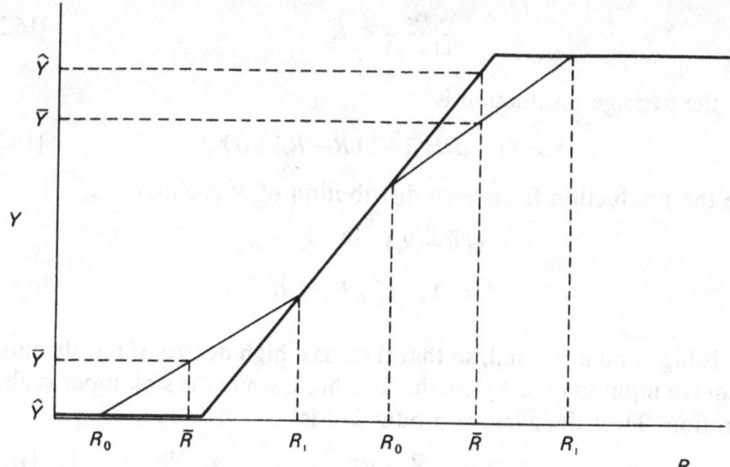

Fig. 16.3. Two examples of the graphical construction which is equivalent to the calculation of the average yield $\bar{Y}$ from the basic $Y(R)$ function, the actual water inputs to the two compartments, R_o and R_i, and the areal ratio between them $(1-a)/a$.

(*b*) Redistribution by run-off always increases average production, $\hat{Y} < \bar{Y}$, if the input to the source area is below the threshold and the sink input is in the linear range; that is, if

$$\frac{R_t}{1+\dfrac{c(1-a)}{a}} < \bar{R} < \frac{R_t}{(1-c)} \qquad [16.21]$$

and

$$\bar{R} < \frac{R_g}{1+\dfrac{c(1-a)}{a}} \qquad [16.22]$$

The average production is then

$$\bar{Y} = aY_i = ab(R_i - R_t) = b[a(\bar{R} - R_t) + c(1-a)\bar{R}] \qquad [16.23]$$

while the production corresponding to an even distribution of rainfall is either

$\hat{Y} = 0$, if $\bar{R} < R_t$

or $\hat{Y} = b(\bar{R} - R_t)$, if $R_t < \bar{R}$

(*c*) Redistribution by run-off always decreases average production, $\bar{Y} < \hat{Y}$, if the source input is in the linear range but the sink input is above saturation; that is, if

$$\frac{R_g}{1+\dfrac{c(1-a)}{a}} < \bar{R} < \frac{R_g}{(1-c)} \qquad [16.24]$$

and
$$\frac{R_t}{(1-c)} < \bar{R} \qquad [16.25]$$

Then the average production is

$$\bar{Y} = (1-a)\,b[(1-c)\,\bar{R} - R_t] + aY_s \qquad [16.26]$$

while the production from even distribution of $\bar{R}$ is either

$$\hat{Y} = b(\bar{R} - R_t) \quad \text{if} \quad \bar{R} < R_s$$

or
$$\hat{Y} = Y_s \quad \text{if} \quad R_s < \bar{R}$$

(*d*) If c is high and a is small, so that there is a high degree of redistribution, the source input may be below the threshold, while the sink input is above saturation. Then the average production is

$$\bar{Y} = aY_s \qquad [16.27]$$

which may be either higher or lower than the production expected from an evenly distributed rainfall, $\hat{Y}$. If $\bar{R} < R_t$, always $\hat{Y} < \bar{Y}$; if $R_s < \bar{R}$, always $\bar{Y} < \hat{Y}$; or if $R_t < \bar{R} < R_s$ is in the linear range either $\hat{Y}$ or $\bar{Y}$ may be greater, depending on whether $\bar{R}$ is greater or less than $(a/b \cdot Y_s + R_t)$.

(*e*) The response function to rainfall $\bar{R}$ of the average production $\bar{Y}$ of a site with internal run-off redistribution may be of either of two forms (both of which are different from the basic response function at the homogeneous site level). (i) If the degree of redistribution is high enough for the sink to reach saturation at a lower rainfall than the one at which the source reaches the threshold, that is, if

$$\frac{R_s}{1+\dfrac{c(1-a)}{a}} < \frac{R_t}{(1-c)} \qquad [16.28]$$

or
$$\frac{R_s}{R_t} < \frac{1+\dfrac{c(1-a)}{a}}{1-c} \qquad [16.29]$$

$$\frac{R_s - R_t}{R_t} < \frac{c}{a(1-c)} \qquad [16.30]$$

In this case the average production/rainfall function will be a two-step saturation function (Fig. 16.4*a*). (ii) If the degree of redistribution is not very high, so that the linear response ranges of the source and the sink overlap in part,

$$\frac{c}{a(1-c)} < \frac{R_s - R_t}{R_t} \qquad [16.31]$$

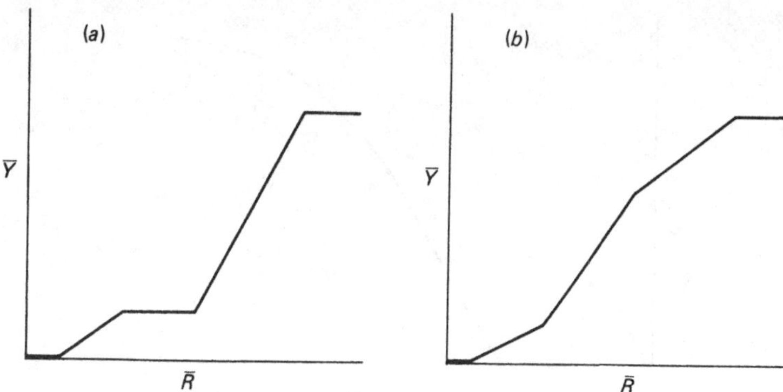

Fig. 16.4. The shape of the function relating the average yield $\bar{Y}$, in an heterogeneous area consisting of one source and one sink compartment, to the average water input (the rainfall) $\bar{R}$: (*a*), when the sink reaches saturation before the source reaches the threshold; (*b*), otherwise.

Then the middle part ofthe average production response function will consist of three linear segments with positive slope:

$$\bar{Y} = \begin{cases} b[(a+c-ac)\bar{R} - aR_t] \\ b[\bar{R} - R_t] \\ b(1-a)[(1-c)\bar{R} - R_t] \end{cases} \qquad [16.32]$$

The slopes of the first and third segments are always less than b (in fact the sum of the two equals b), so that the function has a sigmoid-like shape (Fig. 16.4*b*).

Extension and conclusions

This very simplified model may be made more realistic by several modifications of the original assumptions, for instance: (i) the function relating yield to water input above a threshold is a continuous saturation function rather than a discontinuous ramp function; (ii) the landscape is divided into $k > 2$ compartments, between which water redistribution occurs, so that the actual input to compartment j in a given rain, or rain period is $R_j = c_j \bar{R}$. The values of relative input coefficients c_j may have any distribution (provided that the weighted mean is 1).

These two modifications will affect the previous results only in that the response function of average production $\bar{Y}$ to rainfall $\bar{R}$ will be a smooth continuous function (Fig. 16.5) rather than a piecemeal linear function. The general conclusions will still be valid:

(*a*) The production/rainfall function of a spatially heterogeneous site will generally be a sigmoid curve (or a double saturation curve, if there is very

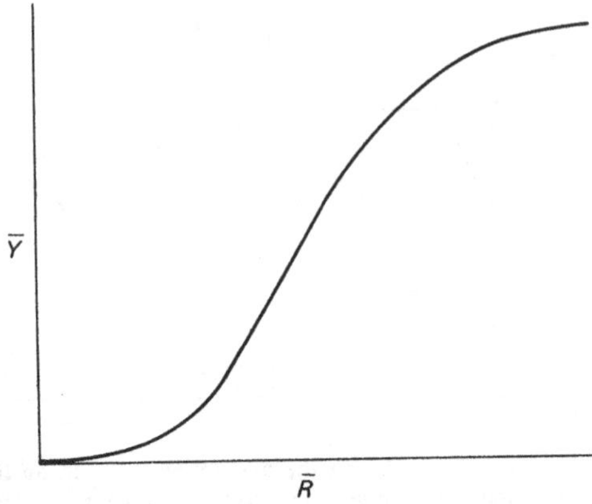

Fig. 16.5. The shape of the $\bar{Y}(\bar{R})$ function in a heterogeneous area consisting of many compartments with a continuous distribution of relative water input values.

strong source/sink differentiation), rather than the simple saturation curve in a uniform site:

(*b*) The average production in a spatially heterogeneous site will be above that expected in a uniform site (at the same rainfall) in a range of moderately low rainfalls, and below it in a range of moderately high rainfalls; the size of the discrepancy is related to the variability of water inputs within the site; and

(*c*) The threshold rainfall for plant production in a spatially heterogeneous site will be below that of a uniform site, and will be determined by the compartment which receives the highest relative water input.

These facts must be considered in the interpretation of any results of a model which is based on uniform-site functions and applied to an area that may be heterogeneous by internal water redistribution; neglecting them could lead to serious errors of interpretation.

The modelling of water redistribution

The important effects that horizontal water redistribution in arid ecosystems can have on biological processes, and on their modelling, point to the necessity for a good understanding of the processes governing the degree and pattern of redistribution, and for an ability to express these in realistic mathematical models. In the previous section a given redistribution pattern has been assumed, which was expressed by parameters a and c (or a_j, c_j in a multi-compartment model). However, the relative water input ($c_j = R_j/\bar{R}$) that any point (or compartment) receives will in general not be a constant; rather it will relate to properties of the rain event or the rain period, in particular amount and intensity distribution.

Two main types of mathematical run-off/rainfall models have been developed, and neither of them are easily applicable to the problem of within-site redistribution in most ecosystem studies (Noy-Meir, 1973). On the one hand, soil physicists have developed basic models of infiltration rates into the soil (e.g. Hanks & Bowers, 1962), which are applicable to a soil surface which is horizontally homogeneous in its physical properties and moisture content; in nature that usually means areas of the order of 1 m^2. In principle, if the relevant physical parameters and the direction of run-off were known for each point or small homogeneous area, this model could be extended to predict water redistribution within a larger heterogeneous area in any rain event, from data on rainfall intensities at a scale of minutes. However, such an extension requires data in quantity and detail which often are not available for ecosystems.

At the other extreme, hydrologists have been interested in predicting the total amount and time course of run-off flow in major channels from watershed areas of the order of square kilometres (e.g. Chapman, 1970) as a function of rainfall and general properties of the watershed. However, much of the water redistribution which is of ecological importance at the 'site' level occurs at the intermediate areal scales (10–10000 m^2) and does not necessarily involve channel flow (Yair, 1973). Models of run-off and water redistribution specifically directed to this intermediate level would seem to be most useful to ecologists. Shanan (1974) has developed a model for predicting channel run-off from small (100–100000 m^2) watersheds in arid zones, which takes into account factors such as areal variability of infiltration rates and length of overland flow on slope.

In so far as ecologists have attempted to model run-off they have usually assumed daily run-off from (or run-on to) each spatial compartment to be a unique function of daily rainfall (as a measure also of intensity) (Goodall, 1967, 1971; Seligman *et al.*, 1971). Such empirical run-off/rainfall functions, either curvilinear, or linear above a threshold, have been derived from experimental data from watersheds of various sizes (e.g. Tadmor & Shanan, 1969). Their use as functions in a model of water redistribution may be justifiable, provided that the parameter values used have been taken from areas of the same size as well as of the same soil and surface-cover properties. A much more detailed and realistic model for run-off has been used by Radford (1973) and Goodall (this volume) in an arid ecosystem model. This model calculates run-off rates over short periods by comparing rainfall intensity, infiltration rate and surface storage.

But even having properly modelled the amount of run-off *from* each compartment there still remains the problem of the further fate of this water, either as run-on to other compartments or as outflow from the system. In model PASTOR (Goodall, 1967, 1971) the possibility of compartments which receive run-on rather than lose run-off was allowed by run-off/rainfall functions of negative slope; but the direct interaction between source-run-off and sink-run-on was not represented. In model NEGEV (Seligman *et al.*, 1971) the run-off from two source compartments was added to the rainfall to give the 'effective rainfall' input to the sink compartment, before the run-off from the latter was computed.

In real ecosystems there may be some simple situations in which some areas are always run-off sources to the same sink areas. However, often there are more than two links (compartments) in each run-off–run-on chain; for example, down a slope. Then each of the intermediate compartments may act as both a source and a sink, and also as a channel for transmitting run-off from upper compartments to lower ones. The relative importance of these roles and the net effect on the final water distribution depend on rainfall amount and intensity at each event, on the length, topography and surface properties of each compartment and on the relative spatial arrangement of the compartments. Such fairly complex redistribution patterns seem to have considerable ecological importance in many arid ecosystems; their modelling is a challenge which seems to be unanswered as yet.

Redistribution of soil, organic matter and nutrients by wind and water

The potential for wind erosion is highest in arid regions, and that for water erosion in semi-arid regions (Marshall, 1973), mainly due to the sparse and low vegetative cover. Not all soils of arid ecosystems are easily eroded; but in those which are (sand dunes, 'duplex' soils), processes of wind and water erosion may be quite important in removing the upper layers of soil from some areas and depositing the material in other areas. Since in arid soils a large proportion of nutrients and organic matter are concentrated in the upper five to ten centimetres (Charley & Cowling, 1968) this erosion will involve a substantial horizontal redistribution of nutrients. Furthermore, even if the soil surface itself is resistant to erosion, wind and water forces will often be sufficient to move and redeposit much of the organic material on top of the soil (plant litter and standing-dead material, animal excretions) and thus redistribute the nutrients in it.

Observations in the Negev and Judaean Deserts in Israel indicate that the first heavy rain in each season removes most of the litter and much of the finer herbaceous standing-dry material (e.g. annual grasses) from areas with smooth (not stony) surfaces on steep and even on moderate slopes. Concentrations of this material can be found in those parts of runnels and channels where the water flow is slowed down by the topographical structure or by obstacles (shrubs, terraces). After a heavy flood in a major watercourse, there may be continuous bands of organic remnants (up to 1 m wide and up to 20 cm deep, including much coarse plant remnants and animal faeces) along the banks of the high water level, as well as large heaps of these remnants behind each shrub, terrace and fence (see also Ludwig & Whitford, this volume).

During dry periods, much of the finer plant litter (and seeds) is picked up and redeposited by wind, particularly by the occasional strong 'dust-storms' and by the frequent local turbulent 'whirl-winds'. Though much of this material may be redeposited at short distances and essentially circulate within the same site, there seems to be some net effect of trapping of wind-borne material (sand and dust as well as litter and seeds) within the canopy of low-branching shrubs and in small steep-sided depressions (e.g. channels). Around shrubs this tends to create hummocks of coarser texture and higher content of organic matter (Weinstein, 1975, Binet, this volume). The sorting by wind of fine organic material from sands and its deposition in concentrated 'strings' in places of lower wind speed has been described for the Kalahari Desert (Brinck, 1956).

Thus in many arid ecosystems there is significant transport and horizontal redistribution of organic matter and nutrients, often at scales smaller than 'the site'. This uneven distribution may effect many biological processes, such as plant growth, activity of microbial decomposers (Binet, this volume) and of detritivorous animals (Brinck, 1956) which may be much more intensive in the sinks, hummocks or bands where organic matter is concentrated. If the response of these processes to the concentration of organic matter or nutrients is non-linear, the use of mean concentrations in a model of the whole site may cause the same type of distortion as the neglect of internal water redistribution.

While it is true that water is the primary limiting factor for plant growth in arid zones, nutrients (in particular nitrogen and phosphorus) may become locally and temporarily limiting where and when water is abundant (West, this volume). In terms of the simple yield/water model described above (Equation 16.10), the saturation yield Y_s may often be determined by nutrient availability. Sometimes the water and nutrient factors may have an interactive effect; for example, the yield/water efficiency (the slope b) may be increased by nitrogen availability (Kafkafi, 1973).

There will most probably be an interaction between the effects of spatial heterogeneity on water redistribution by run-off and on soil and nutrients distribution by erosion. The average production will be larger if the run-on 'sink' areas also have a deeper soil and higher nutrient content (higher Y_s), than if the spatial distributions of water and nutrients (and soil depth) are inversely or randomly related. In general, redistribution of soil and nutrients by water erosion will be more or less positively correlated with the redistribution of water itself; but for wind redistribution this may not be the case. In any case the results of the site-level model will be affected also by the joint distribution of water, soil and nutrients between compartments.

Another factor which may limit plant production in arid zones in soil salinity. In so far as the quantity of water input is positively correlated with the depth of leaching of salts, the joint distribution of water input and depth of the low-salinity soil will be near-optimal for maximum average production. However, in conditions of impeded drainage an inverse correlation may exist (the wettest sites will also be the most saline).

Redistribution of organic matter and nutrients by plants and animals

The local concentration of organic matter and nutrients by litterfall under trees and shrubs, and by patches of animal activity (excretion, nests and burrows) is not specific to arid ecosystems. However, it may be ecologically more important there than in some other ecosystems (e.g. forests or dense grasslands), because populations of plants and animals are much sparser and therefore the relative concentration effect is stronger. In most arid and semi-arid ecosystems, trees and shrubs cover less than 5–10% of the area. In these small patches the concentration of organic matter, nitrogen and phosphorus may be two to ten times higher than in the rest of the area (Weinstein, 1975; Binet, this volume; West, this volume) and often there is a concentration gradient from the canopy outwards. Similar local concentrations also occur around ant nests in arid zones. The possible effects of such uneven distribution of nutrients on biological processes, and on the use of mean concentrations in mathematical models of these processes, have already been discussed above.

Plant survival, dispersal and recolonization

A characteristic feature of arid climates is the large variability between years, particularly in the amount and distribution of rainfall. In some years, rainfall is well above average and seed production by all plants is prolific. But there is a high frequency of drought years. In those years seed production is near zero and mortality is very high in most plant populations. Apparent complete extinction can be observed in the populations of many annuals and shrubs in many, or most, desert habitats (particularly if two or three droughts occur successively). Nevertheless, most of the extinguished populations have been observed to recover rapidly within one or a few years of better rainfall (Hall, Specht & Eardley, 1964; Luria, 1975). In some populations, particularly annuals, the recovery is mostly from a reserve of seeds in the soil, which retain viability and dormancy for many years. But many desert shrubs and some annuals have only short-lived seeds. The survival of the species depends on the spatial heterogeneity of the environment. In a set of drought years it may be extinguished locally in some or most habitats, but it survives in some more favourable habitats or micro-habitats (usually run-on areas, or steep, shaded slopes). In subsequent good years the seeds produced by the refuge population disperse and recolonize the less favourable habitats.

In many desert plant populations, particularly those near the drier limit of the tolerance of the species, most of the demographical changes seem to be associated with such spatial contraction and expansion at scales from 1 m to 1 km. Thus a model of the population in one of the less favourable and only intermittently occupied 'sites' would be totally inadequate without at least including an 'immigration' term. Preferably the model should explicitly represent population changes in each of the relevant spatial compartments, within a larger area, as well as dispersal between them. A simpler alternative might be to model the entire population in the larger area (including all types of sites) and to incorporate the spatial variation in probability distributions of survival and reproduction. This might be adequate as long as density-dependent (non-linear) effects on reproduction are not very important.

The problem was discussed here with reference to plant populations, but similar considerations apply to populations of animals in which the mobility range of the individual is small compared with the 'grain' of the spatial variation in the attributes of the environment that determines drought survival.

Nomadic migrations

Populations of larger and more mobile desert animals can avoid extinction in drought-affected sites by moving out in search of areas that are more favourable in terms of the food and/or water supply. The favourability of these areas may either be permanent, and determined by the topographical hydrological or long-term climatic features, or temporary and caused by random spatial variation in rainfall (Sharon, 1972). Accordingly, the routes of migration may either be recurrent or opportunistic. Populations of most large desert mammals migrate over distances from a few kilometres to hundreds of kilometres in response to changing conditions (e.g. Newsome, 1965). Migrations of similar extent are (or were) characteristic of human populations in arid zones, particularly of hunter-gatherers in Australia, and pastoralists with their domestic animals in Africa and Asia (Noy-Meir, 1974).

In populations of some arid zone birds (Keast, 1959) and of the desert locust (Bodenheimer, 1958) cyclic or opportunistic nomadic migrations may extend even over thousands of kilometres; that is, over whole continents.

The inclusion of such nomadic populations in mathematical models of arid ecosystems involves some problems. In site-level models they can at best be represented as external 'input' variables which appear and disappear at pre-determined or random times. Any model of changes in the nomadic populations themselves must operate at a regional level; that is, they must include an area with dimensions of 10–1000 km differentiated into several spatial compartments.

Utilization of several habitats by animals

This problem is not specific to arid ecosystems but rather to any ecosystems with strong fine-grained spatial variation. But there are many arid ecosystems in which the phenomenon is important: the same animal may obtain food, water and refuge from heat and drought (or different components of its food) in different habitats. Therefore, its success depends on conditions in all of them, and its activity in turn affects each of them. In modelling this situation, the conflict again arises between the requirement to distinguish the spatial compartments relevant to each animal species and the need to keep the number of compartments to a minimum. An optimal solution must be sought in every specific case by *a priori* considerations of the type discussed above and maybe by some experiments or models with different types of compartmentalization.

Animal pressure around water sources: the piosphere effect

This is a special case of the previous problem, which may be the most important one in arid zones. Drinking water normally occurs there only in widely scattered sources which are usually point sources (springs, wells, cisterns, seasonal 'waterholes'); sometimes linear sources (streams). Most small desert animals and maybe some large ones (e.g. gazelles, Carlisle & Ghorbial, 1968) are not dependent on free drinking water and can obtain all the water they need from their food or from soil and atmosphere moisture (including dew), even in dry periods. However, most large animals, in particular the domestic ruminants which are the dominant herbivores in many arid ecosystems, need to drink water (at least when eating dry or saline food) at intervals of: 12–24 h for sheep (Squires, 1970) and cattle, 3–5 days for goats (Shkolnik, Borut & Choshniak, 1972) and to 5–10 days for camels (Gauthier-Pilters, 1972), depending possibly also on climatic conditions. When eating green food with a high moisture content, the drinking interval can be much longer, or the animals may not need to drink at all. On the other hand, drinking requirements increase with the salt content of the food.

Thus around each water point there is a 'sphere' of animal activity, the outer radius of which depends on the maximum distance the animals can move from the water (which depends on the maximum drinking interval and the daily walking distance). Along any radius within this sphere, there is a gradient of decreasing animal activity from the water point outward. Lange (1969) has coined the term *piosphere* (from the Greek *pios*, to drink) and examined the centrifugal gradients of some of the effects of animal activity within it. The gradients are evident in the density of animal tracks and of faeces (Lange, 1969), in the effects of trampling on the soil surface and its lichen cover (Rogers & Lange, 1971). But most important is the effect of the gradient in grazing pressure on the quantity, composition and productivity of the vegetation (Osborn, Wood & Paltridge, 1932; Barker & Lange, 1969).

A central core near the water may be completely bare of any vegetation, the next annulus had only wholly unpalatable (and nitrophilous) species. Going outwards there is usually a steady increase in total plant cover and biomass, together with a shift of composition from species that are relatively unpalatable, or resistant to grazing, to a higher proportion of species that are both palatable and susceptible to grazing damage. The effect of herbivores on the vegetation may diminish to near zero in the areas sufficiently distant from any water source. This observed spatial pattern is only an expression of the dynamic plant–herbivore interaction in the piosphere, which changes as plant growth rate, animal density and maximum grazing radius vary between seasons and between years.

Range managers in arid zones have recognized for a long time that the density and spatial pattern of water-points, through the piosphere effect, can have significant effects on vegetation condition and animal production from an area. But the question of what is the optimal spacing and pattern of waters has not been an easy one to answer, and the solutions recommended on the basis of experience and intuition, by different people, are sometimes conflicting. In principle, this is the kind of question which mathematical models may help to solve, and indeed some attempts in this direction have already been made. On the other hand, to neglect the piosphere effect in models of arid ecosystems (e.g. by taking average values) could in some situations cause serious errors in the prediction of plant–herbivore (and plant–nutrient) interactions. This is because the effect of herbivore grazing pressure on vegetation biomass and production is almost certainly always non-linear; often it may even be discontinuous (Morley, 1966; Noy-Meir, 1975).

A water-point and a piosphere located in an otherwise homogeneous environment, might be one of the few spatial phenomena in ecosystems that could be feasibly described by continuous gradient models. Lange (1969) and Barker & Lange (1969) found that piosphere gradients in the density of sheep tracks, faecal pellets and of several plant species could be described by significant linear regression equations based on the distance from water. However, they also found that there were marked irregularities in these gradients, resulting from obstacles to sheep movement (dense shrubs, fences) and from variations in vegetation due to topographic or edaphic factors.

An alternative approach is to sub-divide the piosphere into several discrete spatial compartments or strata, each with a different range of animal activity levels and with different vegetation Then any irregularities caused by physical obstacles and by spatial variation in vegetation due to factors other than grazing pressure can be accounted for in the delimitation of strata. This solution was used by Goodall in his first model of a semi-arid grazing ecosystem, PASTOR (Goodall, 1967, 1971), where the area of a grazing paddock was sub-divided into three 'grazing pressure' strata depending on distance from water and from fences. Goodall (1971) also did simulation experiments with this model to test the effect of number and position of water-points in a paddock on total plant and animal production. The results showed an increase in production when one water-point in one corner was supplemented with a second in an opposite corner, or when the single water-point was moved to the centre of the paddock.

Conclusions

This preliminary survey indicates that problems of spatial heterogeneity are likely to be of considerable importance in attempts to develop models of arid ecosystems, at almost any scale. The effects of horizontal water redistribution on production, and of the piosphere and of nomadic migrations on the interactions between plants and large herbivores must certainly play a central role in such attempts in most arid ecosystems. The effects of nutrient redistribution may be less frequently critical, but should still be carefully considered.

Some attempts at developing models of the spatial phenomena characteristic of arid ecosystems have already been made. But much further work is needed, particularly if ecosystem models are to be applied to practical problems, which usually appear at spatial scales larger than that of a small 'site'.

Acknowledgements

I am grateful to Professor C. T. de Wit for a discussion some years ago which stimulated my interest in the effect of water redistribution on plant production; and to the Ford Foundation for financial support (Grant 7/E-3 through Israel Foundation Trustees).

References

Barker, S. & Lange, R. T. (1969). Effects of moderate sheep stocking on plant populations of a black oak-bluebush association. *Australian Journal of Botany*, **17**, 527–37.

Beatley, J. C. (1969). Biomass of desert winter annual populations in southern Nevada. *Oikos*, **20**, 261–73.

Bodenheimer, F. S. (1958). Animal ecology to-day. *Monographiae Biologicae*, **6**, 1–276.

Brinck, P. (1956). The food factor in animal desert life. In: *Bertil Hanström; zoological papers in honour of his 65th birthday* (ed. K. G. Winstrand), pp. 120–37. Zoological Institute, Lund, Sweden.

Carlisle, D. B. & Ghorbial, L. I. (1968). Food and water requirements of Dorcas gazelle in the Sudan. *Mammalia*, **32**, 570–6.

Chapman, I. G. (1970). Optimization of a rainfall–run-off model for an arid zone catchment. *Bulletin of the International Association of Scientific Hydrology*, **96**, 127–44.

Charley, J. L. & Cowling, S. L. (1968). Changes in soil nutrient status resulting from overgrazing and their consequences in plant communities of semi-arid areas. *Proceedings of the Ecological Society of Australia*, **3**, 28–38.

Danin, A., Orshan, G. & Zohary, M. (1975). The vegetation of the northern Negev. *Israel Journal of Botany*, **24**, 118–72.

Gauthier-Pilters, H. (1972). Observations sur la consommation d'eau du Dromadaire en été dans la region de Deni-Abbés (Sahara nord-occidentale). *Bulletin de l'Institut Fondamental d'Afrique Noire, Dakar*, **34A**, 219–59.

Goodall, D. W. (1967). Computer simulation of changes in vegetation subject to grazing. *Journal of the Indian Botanical Society*, **46**, 356–62.

Goodall, D. W. (1971). Extensive grazing systems. In: *Systems analysis in agricultural management* (ed. J. B. Dent & J. R. Anderson), pp. 173–87. Wiley, Sydney.

Hall, E. A., Specht, R. L. & Eardley, C. M. (1964). Regeneration of the vegetation on Koonamore vegetation reserve, 1926–1962. *Australian Journal of Botany*, **12**, 205–64.

Hanks, R. J. & Bowers, S. A. (1962). Numerical solution of the moisture flow equation for infiltration into layered soil. *Proceedings of the Soil Science Society of America*, **26**, 530–4.

Kafkafi, U. (1973). Nutrient supply to irrigated crops. In: *Arid zone irrigation. Ecological Studies, vol. 5* (ed. B. Yaron, E. Danfors & Y. Vaadia), pp. 177–88. Springer-Verlag, Berlin.

Keast, A. (1959). Australian birds: their zoogeography and adaptations to an arid continent. *Monographiae Biologicae*, **8**, 89–114.

Keulen, H. van (1975). *Simulation of water use and herbage growth in arid regions.* Simulation Monographs. Pudoc, Wageningen.

Lange, R. T. (1969). The piosphere: sheep track and dung patterns. *Journal of Range Movement*, **22**, 396–400.

Lieth, H. (1972). Uber die Primärproduktion der Pflanzendecke der Erde. *Angewandte Botanik*, **46**, 1–37.

Luria, M. (1975). A study of annual plant populations and communities in a loessial plain in the Negev. MSc thesis, Hebrew University, Jerusalem (Hebrew with English summary).

Marshall, J. K. (1973). Drought, land use and soil erosion. In: *Drought* (ed. J. V. Lovett), pp. 55–77. Angus & Robertson, Sydney.

Morley, F. H. W. (1966). Stability and productivity of pastures. *Proceedings of the New Zealand Society of Animal Production*, **26**, 8–21.

Newsome, A. E. (1965). The abundance of red kangaroos, *Megaleia rufa* in central Australia. *Australian Journal of Zoology*, **13**, 269–87.

Noy-Meir, I. (1973). Desert ecosystems: environment and producers. *Annual Review of Ecology and Systematics*, **4**, 25–52.

Noy-Meir, I. (1974). Desert ecosystems: higher trophic levels. *Annual Review of Ecology and Systematics*, **5**, 195–214.

Noy-Meir, I. (1975). Stability of grazing systems: an application of predator–grey graphs. *Journal of Ecology*, **63**, 459–81.

Osborn, T. G. B., Wood, J. G. & Paltridge, T. B. (1932). On the growth and reaction to grazing of the perennial saltbush, *Atriplex vesicaria*: an ecological study of the biotic factor. *Proceedings of the Linnean Society of New South Wales*, **57**, 377–402.

Radford, J. (1973). *Terrestrial process models: soil submodels (versions I, II, III).* US/IBP Desert Biome Research Memorandum RM 73-56. Utah State University, Logan.

Rogers, R. W. & Lange, R. T. (1971). Lichen populations on arid soil crusts around sheep watering places. *Oikos*, **22**, 93–100.

Seligman, N. G., Tadmor, N. H., Noy-Meir, I. & Dovrat, A. (1971). An exercise in simulation of a semi-arid Mediterranean grassland. *Bulletin Recherche Agronomique Gembloux. Semaine d'étude des problémes méditerranéens*, 1971, 138–43.

Shanan, L. (1974). Rainfall and run-off relationships in small watersheds in the Avdat region of the Negev Desert highlands. PhD thesis, Hebrew University, Jerusalem.

Sharon, D. (1972). The spottiness of rainfall in a desert area. *Journal of hydrology*, **17**, 161–75.

Shkolnik, A., Borut, A. & Choshniak, J. (1972). Water economy of the bedouin goat. *Symposia of the Zoological Society of London*, **31**, 229–42.

Slatyer, R. O. (1961). Methodology of a water balance study conducted on a desert woodland (*Acacia aneura*) community. *Arid Zone Research*, **16**, 15–26.

Squires, V. R. (1970). Grazing behaviour of sheep in relation to watering points in semi-arid rangelands. In: *Proceedings of the XI International Grassland Congress*, pp. 880–4. Queensland University Press, St Lucia.

Tadmor, N. H. & Shanan, L. (1969). Run-off inducement in an arid region by removal of vegetation. *Proceedings of the Soil Science Society of America*, **33**, 790–4.

Walter, H. (1962). *Die Vegetation der Erde*, vol. I. G. Fischer, Jena.

Weinstein, N. (1975). The micro-habitats created by the desert shrub *Hammada scoparia* and their effect on herbaceous plants. MSc thesis, Hebrew University, Jerusalem (Hebrew with English summary).

Williams, O. B. (1955). Studies in the ecology of the Riverine plain. I. The gilgai microrelief and associated flora. *Australian Journal of Botany*, **3**, 99–112.

Wit, C. T. de (1958). *Transpiration and crop yields.* Institut voor Biologischen Scheikundig Onderzoek van Landbouwgewassen, Wageningen, Netherlands, Mededeling 59.

Yair, A. (1973). Sources of runoff and sediment supplied by the slopes of a first-order drainage basin in an arid environment. *International Symposium of Recent Geomorphic Processes*, IGY, Commission of Present-Day Geomorphological Processes, Göttingen.

Manuscript received by the editors December 1974

17. Simulation of plant production in arid regions: a hierarchical approach

H. VAN KEULEN & C. T. DE WIT

Introduction

The increasing interest in the agricultural production potential of the world, reflected among other things by the enormous efforts put into the IBP programmes in recent years, has diverted much research into the development of tools suitable for the calculation of potential productivity. This has coincided with growing accessibility to high speed computers which enable relatively easy handling of rather complicated systems, resulting in the development of a variety of computer models of crop, growth, plant production and agricultural potential (e.g. Seligman, 1976). In many cases, these models are physiologically based and deal with situations where growth conditions, particularly water and nutrient availability are optimal, so that the productivity is mainly determined by climatological conditions. It appears that this approach has not been widely applied under more marginal conditions, for example in arid regions with low and erratic rainfall, where extensive grazing is the dominant way of exploitation. It seems, however, that the application of systems analysis followed by model building is also a most promising way to indicate means of improving the situation under these conditions either by changing the existing system or by hinting at better ways of exploiting the present system. In order to arrive at such recommendations the entire field of primary production, grazing, animal production and grazing management must be taken into account. The basis for such a systems approach is a thorough understanding of the principles that govern primary production. Therefore, a joint Dutch–Israeli research project was initiated in 1970 to assess, as a first step, how herbage production in arid regions is dependent on weather, soil conditions and type of plant cover (Van Keulen & De Wit, 1975).

One of the results of this project is described in this paper: a simulation model,* which deals with the calculation of plant production in situations where water is the main limiting factor. Obviously, it should be possible for these conditions to simplify the model with respect to the detail of physiological

* All models described in this chapter are written in CSMP (Continuous System Modelling Program) developed by IBM for its 360 and 370 series of machines. These or similar languages (according to the CSSL definition) are now also available for other machines.

processes, which may be of importance for the understanding of plant behaviour but have relatively little influence on production. However, before simplifications can be applied information should be available on the relative importance of the various processes. Otherwise, simplification leads to restricted applicability of the model for predictive and extrapolation purposes. A possible way to achieve that aim is the hierarchical approach, that is, a method where at each level of interest, models are developed, linking an explanatory level to an explainable level (De Wit, 1970). The results of such a model may then, after thorough validation, be incorporated into lower resolution models by means of functions or analytical expressions. An example of this aproach is the work by Penning de Vries (1974), who calculated respiratory losses on the basis of biochemical transformations necessary to produce and maintain plant tissue from primary photosynthetic products. The results of that model are entered in models for crop growth in the form of 'production value' parameters (that is weight efficiency) for the main plant constituents (De Wit *et al.*, 1978). Another example is a model for the calculation of daily gross photosynthesis from the radiation climate, the geometrical and optical properties of the leaves, and the photosynthesis–light response curve of individual leaves (De Wit, 1965), the results being used in a tabulated form in models for plant production (Van Keulen, 1976*a*). The basic principle behind this approach is that it is in general undesirable to incorporate several explanatory and explainable levels in the same model because this renders it complicated and unwieldy.

In the model described in this chapter the transpiration coefficient, that is the ratio between dry matter production and transpiration, is used as the parameter to link it to a detailed model of canopy photosynthesis, respiration and transpiration (De Wit *et al.*, 1978). In the first section of this chapter attention is paid, therefore, to measurement of the transpiration coefficient and to its calculation by means of the detailed model, of which a brief description is also given. In the last section the application of the crop growth model to arid conditions is described.

Experimentation and simulation of the transpiration coefficient

The relation between dry matter production and water use by plants has been the subject of experimental work ever since the classical experiments by Briggs & Shantz (1913). In the framework of the joint research project on 'Actual and potential herbage production in semi-arid regions', such experiments were carried out at the Avdat Desert Research Station of the Hebrew University, in the central Negev desert of Israel. Plants from both the natural vegetation and cultivated species were investigated. The aim of the experiments was to study possible differences in water use efficiency between different species under varying climatological conditions.

Material and methods

The plants were grown in plastic buckets filled with local loess, which had been mixed with sufficient fertilizer to ensure optimum nutrient conditions. The surface was covered with 5 cm of coarse gravel to prevent direct soil evaporation and to avoid the deleterious effects of repeated watering on the surface layer of the soil which is susceptible to crust formation. The plants were watered regularly to avoid water stress, the exact timing being determined by the transpirational losses. Approximately once a week, all buckets were weighed and water was added to the original weight, corrected for the amount of biomass present. The plants were harvested after various growing periods, preferably in the vegetative stage. However, this was not possible with some of the species from the natural vegetation, which flowered very soon after sprouting. The aerial parts were dried for some days at 70 °C and the dry weight determined. The roots were washed free of soil over a narrow sieve, dried at 70 °C and, after careful removal of adhering sand and stones, weighed.

Meteorological data were collected concurrently in a standard weather station of the Israeli Meteorological Service, located near the experimental site.

Results and discussion

In Tables 17.1(*a*) and 17.1(*b*) the results of two series of the experiments are given, for the seasons 1971/72 and 1972/73, respectively. They include both cultivated species, for which local varieties were used, and species from the natural vegetation. The transpiration coefficients (TRC) were obtained by dividing the total amount of water transpired by the total dry matter harvested.

In the first series (Table 17.1*a*), the values for the cultivated species and the species from the natural vegetation are all within 20% of the average, except for oats (*Avena sativa*). Most of the differences between the other species can be attributed to differences in the growth pattern of the various plants, which resulted in a different distribution of the production in time. The weather conditions varied considerably during the experimental period: high radiation, high temperatures and low humidity in autumn and low radiation, low temperatures and high humidity in winter. Hence, the efficiency of water use varies in the different periods, as it is correlated with the evaporative demand of the atmosphere (De Wit, 1958).

There is no reason to attribute the very low value for *Avena sativa* measured in winter 1971/72, which is reminiscent of the values measured for C_4-species (De Wit, 1958), to experimental errors. A possible explanation is, that in this case the stomatal conductivity of the plants was governed by net assimilation

Table 17.1(*a*). *Measured transpiration coefficients in Avdat from 8 October 1971 to 25 January 1972*

Species	No. of pots	Dry shoot wt per pot (g)	Dry root wt per pot (g)	Total dry wt per pot (g)	Transpiration coefficient
Avena sativa	8	175	79	254	79
Hordeum sativum	8	98	71	169	104
Triticum sativum	6	110	47	157	129
Reboudia pinnata	6	35	7	42	146
Medicago hispida	10	47	17	64	157
Hordeum murinum,	10	47	31	78	129
Avena sterilis	10	75	65	140	124

Table 17.1(*b*). *Measured transpiration coefficients in Avdat in 1972/73*

Species	Growing period	No. of pots	Dry shoot wt per pot (g)	Dry root wt per pot (g)	Total dry wt per pot (g)	Transpiration coefficient
Avena sativa	4 Oct. 72–26 Jan. 73	2	341	73	414	221
Hordeum sativum	5 Oct. 72–23 Feb. 73	3	373	136	509	224
Triticum sativum	17 Nov. 72–16 Feb. 73	3	161	51	212	131
Reboudia pinnata	10 Dec. 72–16 Feb. 73	3	37	4	41	168
Medicago hispida	18 Oct. 72–2 Mar. 73	3	72	11	83	192
Hordeum murinum	27 Oct. 72–16 Feb. 73	2	220	68	288	178
Avena sterilis	1 Nov. 72–26 Jan. 73	3	96	45	141	129
Phalaris minor	1 Nov. 72–16 Feb. 73	2	178	105	283	162

in such a way, that the internal CO_2-concentration is maintained at a constant value. Such behaviour, which allows plants to combine a high growth rate with a relatively low transpiration rate, and is therefore of special importance for production in (semi-)arid regions, has been reported before (De Wit *et al.*, 1978; Goudriaan & Van Laar, 1978). It is not clear, however, how widespread this phenomenon is, nor is it understood, whether varietal or environmental factors are responsible for its occurrence. Further research on the subject is of prime importance.

In the second series (Table 17.1*b*) the differences between cultivated and natural species are within 10% of the average. It is somewhat more difficult to compare these values, as the growing periods do not match exactly, due to a slightly different experimental set-up. These experiments (see also Van Keulen, 1975) indicate that, under normal circumstances, no basic difference in water use efficiency exists between the present-day cereals and their wild relatives.

Simulation of the transpiration coefficient

The simulation model

To test whether it is possible to calculate the transpiration coefficients, a number of experimental periods were simulated with the basic crop growth simulator (BACROS), developed at the Department of Theoretical Production Ecology, Wageningen. A detailed report on the model and its validation has been published (De Wit *et al.*, 1978). It is sufficient here to give a brief outline of the model.

The crop considered is in the vegetative phase and is well supplied with water and nutrients. Growth of the crop is defined as increase in dry weight of the structural plant material, that is, exclusive of the organic components that are classified as reserves. The model is based on physical, chemical and physiological processes so that it can be used without geographical limitation.

Micro-weather is calculated from daily weather data measured at screen height, the extinction of radiant energy from sun and sky within the crop being taken into account. The infrared radiation in the canopy is also computed. The distribution of radiation over the leaves is required in order to calculate assimilation and transpiration. This distribution is determined by the architecture of the crop, which has to be defined. The extinction of turbulence in the canopy is also considered, so that transfer of heat, vapour and carbon dioxide can be computed. The ratio of latent and sensible heat exchange regulates the micro-weather to a large extent, and this ratio is mainly determined by stomatal behaviour. Soil temperature is not simulated here, it is assumed that it follows the air temperature with a delay of 4 h and the associated decrease in amplitude. More detail on the simulation of micro-weather is given by Goudriaan (1977).

The assimilation of carbon dioxide by the canopy is calculated by adding the assimilation rates of the variously exposed leaves in successive leaf layers. The latter rates are dependent on light intensity, carbon dioxide concentration in the ambient air and the resistance for carbon dioxide diffusion from the atmosphere towards the active sites. This resistance is the sum of the resistance for turbulent transport in and above the canopy, the resistance of the laminar layer around the leaves, the stomatal resistance and the mesophyl resistance.

Respiration is the sum of maintenance and growth respiration. The latter is the result of the conversion of primary photosynthates into structural plant material and is therefore proportional to the growth rate. The intensity of growth respiration is affected by the chemical composition of the newly formed material. This intensity is independent of the temperature, but growth respiration is indirectly influenced by temperature through its effect on the growth rate. Usually, carbon dioxide evolution originating from transport processes is included in the growth respiration. The rate of maintenance respiration depends on the turnover rates of proteins, the re-synthesis of other degraded compounds and the maintenance of ionic gradients. This respiration process is therefore dependent on the chemical composition of the plants and is, moreover, affected by temperature.

The growth rate of shoot and root is dependent on the amount of reserves present and the temperature. Under internal water stress, growth of above-ground organs is retarded by making a larger proportion of the reserves available for growth of roots. In this way a functional balance is maintained between shoot and root (Brouwer, 1963).

Up to now, no satisfactory solution has been found for the simulation of the growth of leaf surfaces in relation to the growth of shoot weight, so that these aspects are mimicked rather than simulated: their description is based on information obtained from field trials and not from knowledge of the underlying processes.

The water status of the crop is determined by the balance between transpiration and water uptake from the soil. The transpiration rate is found by adding the transpiration rates of the variously exposed leaves in successive leaf layers of the crop. The latter rates are found from the absorbed radiation, the resistance of the laminar layer, the humidity and temperature of the ambient air and the stomatal resistance. Contrary to the version published by De Wit *et al.* (1978) stomatal resistance in this case was governed either by light intensity or by the water status of the crop. The calculations also provide leaf temperatures, which are used in the photosynthesis section and averaged to give canopy temperature, applied in growth and respiration calculations. Water uptake is determined by the conductivity of the root system, the water status of the plant and that in the soil. The soil is assumed to be in optimum moisture conditions ('field capacity') so that transport in the soil can be neglected. The conductivity of the root system is determined by its weight, the degree of suberization of the roots and the soil temperature,

This model describes quite accurately the rates of photosynthesis, transpiration and respiration, when executed with time-steps in the order of minutes (enclosure experiments: Van Keulen & Louwerse, 1975; Van Keulen, Louwerse, Sibma & Alberda, 1975; De Wit *et al.*, 1978) and seasonal, above-ground dry matter production when the time-steps are 1 h (Goudriaan, 1973; Dayan & Dovrat, 1977; De Wit *et al.*, 1978).

Calculated and measured transpiration coefficients

The model described in the previous section had to be adapted to calculate transpiration coefficients of single plants, growing in containers. The leaf area index, which is normally entered as a forcing function loses most of its meaning in the pots as the radiation climate and the wind speed are distinctly different from that in a closed green crop surface. The simulation experiments

Table 17.2. *Comparison between measured and simulated transpiration coefficients*

Species	Growing period	Measured	Simulated	Ratio of simulated:measured
Oats (*Avena sativa*)	8 Oct. 71–22 Nov. 71	170	165	0.97
	8 Oct. 71–25 Jan. 72	79	149	1.86[a]
	4 Oct. 72–17 Nov. 72	235	257	1.10
	4 Oct. 72–5 Jan. 73	194	206	1.07
Wheat (*Triticum sativum*)	8 Oct. 71–25 Jan. 72	129	147	1.14
	17 Oct. 72–27 Nov. 72	217	228	1.05
	17 Nov. 72–16 Feb. 73	131	218	1.67[a]
	5 Jan. 73–16 Mar. 73	166	202	1.22
Barley (*Hordeum sativum*)	8 Oct. 71–22 Nov. 71	173	125	1.07
	8 Oct. 71–25 Jan. 72	104	149	1.44[a]
	5 Oct. 72–17 Nov. 72	217	246	1.14
	5 Oct. 72–23 Feb. 73	224	209	0.94

[a] Variation greater than 25%.

were therefore done with a constant leaf area index of two for the whole period. Physiological data used were those of Van Keulen & Louwerse (1975) for wheat. For more details on the calculation procedure see Van Keulen (1975).

In Table 17.2 a comparison between measured and calculated transpiration coefficients is given. Only experiments with cultivated species are shown here, as the differences with the natural species are very small (Table 17.1). In general the measured and calculated values agree within 25%, except where indicated. The latter values could again be explained by the existence of stomatal regulation through internal carbon dioxide concentration. The deviations in the other cases do not seem to be systematic and several factors could have caused the observed discrepancies: the problems, mentioned already concerning the radiation climate, or the influence of respiration (especially maintenance respiration, which is proportional to the amount of biomass present, a value unknown during the growing period).

From the overall agreement it seems acceptable to use the calculated values of the transpiration coefficient when the production potential has to be calculated. In the crop growth model described in the next section, a procedure is incorporated to calculate the daily value of the transpiration coefficient from daily weather observations. This procedure is based on the results of BACROS (Van Keulen, 1975).

The simulation model 'Arid Crop'

A detailed description of the model, a complete listing and an evaluation of its behaviour are given by Van Keulen (1975).

Outline of the model

The model simulates the growth curve of a canopy and the water balance in the soil below it.

At the beginning of the growing season germination starts after initial wetting of the soil and continues until the temperature sum required for establishment is reached, provided that the upper soil layers do not dry out before then. In the latter case the seedlings die and a new wave of germination starts after re-wetting. After establishment, at which moment a constant initial biomass is assumed, the daily growth rate of the canopy is calculated from daily transpiration and the value of the transpiration coefficient.

Daily transpiration is a function of the evaporative demand of the atmosphere, determined by radiation and the combined effect of wind speed and humidity, the leaf area of the canopy and the interactive effect of the distribution of roots and moisture in the soil profile. For the latter relation an approach similar to that of Viehmeyer & Hendrickson (1955) is used, that is, soil water is almost freely available above 'wilting point' after which a sudden drop in availability occurs. When part of the root system is in dry soil layers, its reduced activity may be compensated by enhanced uptake from wetter layers. The transpiration coefficient is the ratio between potential growth rate, determined by the balance between gross photosynthesis and respiration and potential transpiration dependent on the same factors as the actual rate except for the soil component. The value of the transpiration coefficient is assumed to be independent of moisture conditions in the soil. This is an oversimplification (Lof, 1975) but the value of the transpiration coefficient has little effect on dry matter production because of the low transpiration rate under dry soil conditions.

The growth rate of leaf area is obtained from the increase in dry weight, assuming a temperature-dependent leaf area ratio. This simplification is mainly applied because many of the basic processes governing plant morphogenesis are insufficiently understood. The physiological age of the crop is defined in its development stage, being 0 at establishment and 1 at maturity. The rate of development is a linear function of temperature, the relation being constructed from field experience. When maturity is reached, the canopy ceases to grow, even when water is still available in the soil.

Root growth may refer to both vertical extension of the root system and to increase in dry weight. Due to lack of reliable data, there is no feedback of the one process to the other. For the vertical extension, it is assumed that a homogeneous root front moves downward with a rate dependent on soil temperature, till a dry layer is reached, after which growth ceases. The increase in dry weight is obtained from the growth rate of the crop by allocation of the new material between shoot and root: the ratio is dependent on the development stage, from 0.5 at the beginning of the season to 0.025 at the end.

To simulate the water balance in the soil, the potential rooting zone is divided into a number of homogeneous compartments (De Wit & Van Keulen, 1972) increasing in thickness from the top downwards. The total amount of water infiltrating into the soil, rain corrected for run-off or run-on, is distributed in such a way, that the compartments are consecutively filled to field capacity. Redistribution of soil moisture under the influence of potential gradients originating from uneven drying or wetting of the profile, is not taken into account as that has little influence on the availability of water for plant roots.

Direct evaporation from the soil surface, which may be a substantial source of moisture loss (especially with unfavourable rainfall distribution) is obtained from the potential rate, calculated according to Penman (1948), corrected for the dryness of the upper soil layer. The reduction factor, as a function of the moisture content, is again calculated with a detailed model: this is another example of the hierarchical approach (Van Keulen, 1975). Total evaporative losses are withdrawn from the various compartments, 'mimicking' redistribution by means of an extinction coefficient, depending on soil physical properties and moisture distribution, which is also obtained from the model just mentioned.

Water uptake by the roots depends on root distribution and moisture distribution in the soil, and is furthermore influenced by soil temperature, in its combined effect on root conductivity and viscosity of the water.

The temperature of the soil is calculated as a ten-day running average of the air temperature.

Time-steps of one day are employed, which is one-tenth of the minimum value of the time constant for the growth of vegetation (the maximum relative growth rate is ± 0.1 day^{-1}, cf. De Wit & Goudriaan, 1974), and the rectilinear integration method is used.

Results and discussion

Data to validate the model, were collected during three years of experimentation at the Tadmor Experimental Farm in the northern Negev desert of Israel. The site has a Mediterranean-type climate with an average annual rainfall of 250 mm, concentrated in winter (November–March). The vegetation under study is an abandoned cropland vegetation, consisting of herbaceous annuals (Tadmor, Eyal & Benjamin, 1974). Meteorological data were collected at a weather station of the Meteorological Service, located at about 8 km from the experimental site, except for rainfall, which was recorded at the site. Standing crop was determined approximately every fortnight, using a double sampling technique, based on visual estimates (Tadmor, Brieghet, Noy-Meir & Benjamin, 1975). Soil moisture was recorded every fortnight, and after heavy rains, by the neutron moderation technique complemented by gravimetric sampling of the upper 30 cm.

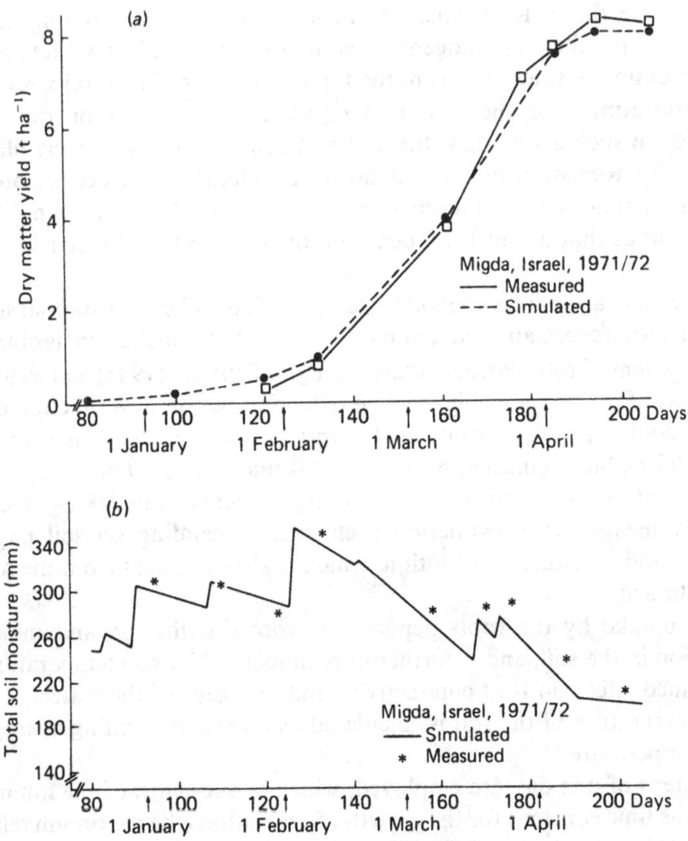

Fig. 17.1. Comparison between measured and simulated values in a site under natural vegetation in Migda, 1971/72. (*a*) Dry matter production. (*b*) Total soil moisture.

In Figs. 17.1 and 17.2 the measured and calculated values of above-ground dry matter production and total soil moisture content are given for the seasons 1971/72 and 1973/74. Comparison of the two figures shows some interesting features: it is striking, that the agreement for 1971/72 is much better than for 1973/74 (and likewise 1972/73, which is not shown here). The main reason for this difference must be that the 1971/72 data were continuously used to test the performance of the model during its development. This illustrates clearly the importance of using strictly independent data during model development and model validation, even in situations where a minimum of empirical relations is included (Van Keulen, 1976*b*). The deviations at the end of the 1973/74 season must be mainly attributed to the nitrogen situation in the field, as it is likely that at that time shortage did occur. Under natural conditions at the Migda site, nitrogen appears to be equally

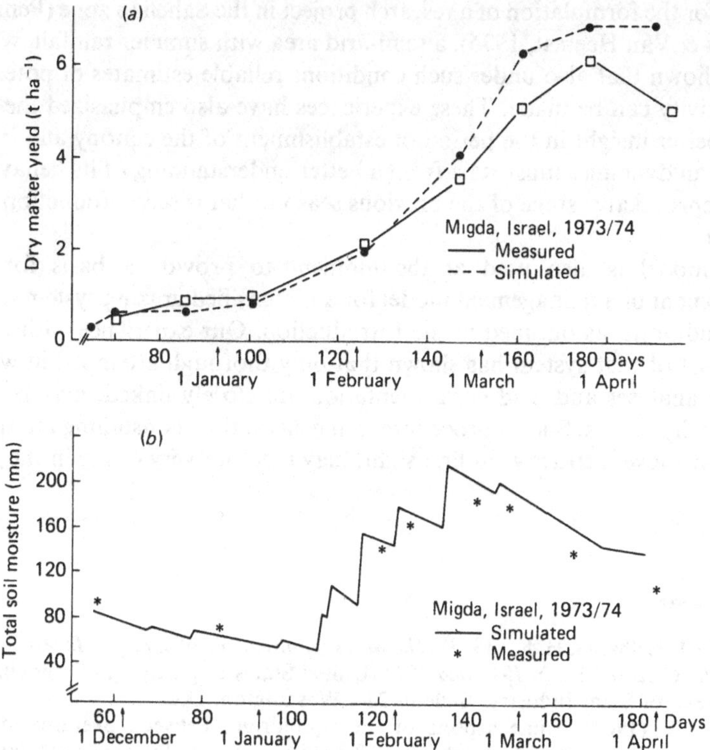

Fig. 17.2. Comparison between measured and simulated values in a site under natural vegetation in Migda, 1973/74. (*a*) Dry matter production. (*b*) Total soil moisture.

important as a growth limiting factor as water (Harpaz, 1975). Total soil moisture compares fairly well in both seasons, after taking into account that due to heterogeneity of the soil, the accuracy of measurement seldom exceeds 10% of the field mean.

Conclusions

Comparison of the results of the simulation model with the measurements, shows that the behaviour of herbaceous vegetation growing undisturbed under optimum nutrient conditions can be predicted reasonably accurately.

It is also clear, that several processes need further examination and better understanding (early development, morphogenesis) before the model may be used in connection with other models to predict the behaviour of a semi-arid ecosystem under its normal grazing management scheme. Formulation of this model, however, has increased our insight into the relevance of various factors that play a role in production under semi-arid conditions. It has served as a basis for the formulation of a research project in the Sahelian zone (Penning de Vries & Van Heemst, 1975), a semi-arid area with summer rainfall, where it was shown that also under such conditions reliable estimates of potential productivity can be made. These experiences have also emphasized the fact that a better insight in the period of establishment of the canopy and in the vegetation dynamics must start from a better understanding of its behaviour in the reproductive stage of the previous season, that is seed production and survival.

The model is also used at the moment to provide a basis for the development of a management model for an intensified grazing system under arid conditions, as outlined in the Introduction. Our experience with but a small part of that system has shown that only thorough attempts, in which systems analyses and field experimentation are closely linked, may lead to trustworthy results. Such a procedure is in general time-consuming but quick results, however attractive at first sight, may turn out very costly in the long run.

References

Briggs, L. J. & Shantz, H. K. (1913). *The water requirements of plants. I. Investigations in the Great Plains in 1910 and 1911.* United States Department of Agriculture, Bureau of Plant Industry, Bulletin 284. Washington, D.C.

Brouwer, R. (1963). Some aspects of the equilibrium between overground and underground plant parts. Jaarboek 1962, Instituut voor Biologisch en Scheikundig Onderzoek van Landbouwgewassen, pp. 31–9.

Dayan, E. & Dovrat, A. (1977). *Measured and simulated herbage production of Rhodes grass.* Report of the Hebrew University of Jerusalem, Faculty of Agriculture, Rehovot.

de Wit, C. T. (1958). *Transpiration and crop yields.* Verslagen van landbouwkundige Onderzoekingen 64.6, Pudoc, Wageningen.

de Wit, C. T. (1965). *Photosynthesis of lead canopies.* Verslagen van landbouwkundige Onderzoekingen 663, Pudoc, Wageningen.

de Wit, C. T. (1970). Dynamic concepts in biology. In: *Prediction and measurement of photosynthetic productivity. Proceedings IBP/PP Technical Meeting, Trebon, 1969*, ed. I. Setlik, pp. 17–23. Pudoc, Wageningen.

de Wit, C. T. & van Keulen, H. (1972). *Simulation of transport processes in soils.* Simulation Monographs, Pudoc, Wageningen.

de Wit, C. T. & Goudriaan, J. (1974). *Simulation of ecological processes.* Simulation Monographs, Pudoc, Wageningen.

de Wit, C. T. *et al.* (1978). *Simulation of assimilation, respiration and transpiration of crops.* Simulation Monographs, Pudoc, Wageningen.

Goudriaan, J. (1973). Crop simulation and experimental evaluation. In: *Proceedings Summer Computer Simulation Conference, Montreal*, vol. 2, pp. 827–9. Simulation Councils, La Jolla, California.

Goudriaan, J. (1977). *Crop micrometeorology: a simulation study.* Simulation Monographs, Pudoc, Wageningen.

Goudriaan, J. & van Laar, H. H. (1978). Relations between leaf resistance, CO_2 concentration and CO_2 assimilation in maize, beans, lalang grass and sunflower. *Photosynthetica*, **12**, 241–9.

Harpaz, Y. (1975). Simulation of the nitrogen balance in semi-arid regions. PhD thesis, Hebrew University, Jerusalem.

Lof, H. (1975). Water use efficiency and competition between arid zone annuals, especially the grasses Phalaris minor and Hordeum murinum. Agricultural Research Report (Verslagen van landbouwkundige Onderzoekingen) 853, Pudoc, Wageningen.

Penman, H. L. (1948). Natural evaporation from open water, bare soil and grass. *Proceedings of the Royal Society of London, Series A*, **193**, 120–46.

Penning de Vries, F. W. T. (1974) Substrate utilization and respiration in relation to growth and maintenance in higher plants. *Netherlands Journal of Agricultural Science*, **22**, 40–4.

Penning de Vries, F. W. T. & van Heemst, H. D. J. (1975). Potential primary production of unirrigated land in the Sahel. *Proceedings of a seminar on Evaluation and mapping of tropical African Rangelands*, pp. 323–27. International Livestock Centre for Africa, Addis Ababa.

Seligman, N. G. (1976). A critical appraisal of some grassland models. In: *Critical evaluation of systems analysis in ecosystems research and management* (ed. C. T. de Wit & G. W. Arnold), Simulation Monographs, pp. 60–97, Pudoc, Wageningen.

Tadmor, N. H., Eyal, E. & Benjamin, R. W. (1974). Plant and sheep production on semi-arid annual grassland in Israel. *Journal of Range Management*, **27**, 427–32.

Tadmor, N. H., Brieghet, A., Noy-Meir, I. & Benjamin, R. W. (1975). An evaluation of the calibrated weight-estimate method for measuring production in annual vegetation. *Journal of Range Management*, **28**, 65–9.

van Keulen, H. (1975). *Simulation of water use and herbage growth in arid regions.* Simulation Monographs, Pudoc, Wageningen.

van Keulen, H. (1976*a*). *A calculation method for potential rice production.* Contributions Central Research Institute for Agriculture, Bogor, No. 21.

van Keulen, H. (1976*b*). Evaluation of models. In: *Critical evaluation of systems analysis in ecosystems research and management* (ed. C. T. de Wit & G. W. Arnold), pp. 22–9. Simulation Monographs, Pudoc, Wageningen.

van Keulen, H. & Louwerse, W. (1974). Simulation models for plant production. Proceedings Symposium 'On agrometeorology of the wheat crop', pp. 196–209. World Meteorological Organization, Bulletin no. 396.

van Keulen, H., Louwerse, W., Sibma, L. & Alberda, Th. (1975). Crop simulation and experimental evaluation – a case study. In: *Photosynthesis and productivity in different environments.* International Biological Programme, vol. 3. Cambridge University Press.

van Keulen, H. & de Wit, C. T. (1975). Actual and potential herbage production in arid regions: an annotated bibliography of a joint Dutch–Israeli research project (1972–1975). Report of the Department of Theoretical Production Ecology, Institute of Biological and Chemical Research on Field Crops and Herbage (IBS), Wageningen.

Viehmeyer, F. J. & Hendrickson, A. H. (1955). Does transpiration decrease as the soil moisture decreases? Transactions of the American Geophysical Union, 36, 525–8.

Manuscript received by the editors December 1974

18. Understanding arid ecosystems: the challenge

I. NOY-MEIR

Arid ecosystems have been, and probably will always be, of considerable economic importance to man. Their greatest challenge is a practical one: how to use and manage their resources for the long-term benefit of man. The management problems are discussed elsewhere in this volume. However, the end of this part on ecosystem dynamics and modelling seems to be an appropriate point to evaluate the purely scientific aspects of our efforts to 'understand the structure and function' of arid ecosystems.

The attitudes of scientists towards the study of arid ecosystems range from their facile dismissal as 'uninteresting' to almost mystical preoccupation. A somewhat less extreme view, which may be used as a reason either for or against their study, is that arid ecosystems are relatively simple in structure and function, compared with other terrestrial ecosystems. The arguments that could be brought forward in favour of this hypothesis are: (*a*) a single factor, water, is limiting for most biological processes; (*b*) the vertical stratification of vegetation is poorly developed; (*c*) due to its low cover and height, vegetation has little feedback effects on microenvironmental conditions, and (*d*) diversity tends to be low.

If indeed deserts are 'simple systems' we should now, at the end of the IBP effort, be able to understand their workings quite well and to apply this understanding in models of predictive value. What is the situation in reality, as reflected in the preceding papers?

To 'understand' an ecosystem one must first observe the dynamic behaviour and interaction of its components over at least a few years. Such observations were carried out within the IBP, as a quantitative and integrated programme in the main Desert Biome sites in the United States, and to a lesser extent in arid-land sites in other countries. Chapter 11 by Ludwig & Whitford on short-term dynamics of water and energy flows in deserts is based mainly on the results and experience obtained in one of these intensive sites in three years. What emerges from this study is that the Chihuahuan arid ecosystem is far from simple. Moisture is indeed the primary limiting resource and most biological processes occur in 'pulses' (Westoby, 1972; Bridges *et al.*, 1972) induced by rainfall events, as may be regarded 'typical' for deserts. But there is nevertheless tremendous diversification among both plants and animals in the patterns of response, and of resource use. The Chihuahuan ecosystem also demonstrates the overriding effects of spatial heterogeneity in water distribution on plant production and through it on animal activity. A preliminary comparison with results from other sites indicates that differences in rainfall seasonality and land use history may decisively modify structure and function of arid ecosystems. In all, the chapter by Ludwig & Whitford shows that, when studied carefully, the water and energy dynamics in every arid ecosystem reveal remarkable complexity, and that their integration and understanding already present a considerable challenge. Many more studies of this type are needed for world-wide comparisons and integration but the authors already attempt some interesting tentative conclusions, in particular regarding the seed-eating communities in arid ecosystems.

Ludwig & Whitford also stress the importance of termites and ants for localized decomposition and nutrient cycling in arid ecosystems. This theme is expanded in the chapters by West and Binet, both of which show that though decomposition and nutrient dynamics may have some rather special features in arid ecosystems, they are by no means unimportant in determining primary energy flows. In fact, most of the processes involved in cycling in more humid ecosystems (except deep leaching) seem to be significant and relevant also in arid ecosystems, even though some of them are more highly concentrated. In particular, both West and Binet bring evidence that the shrubs in arid ecosystems fulfil essentially all the same functions 'in miniature' as the trees in forest and woodland systems. Thus nutrient dynamics in arid ecosystems seem to be no less complex than elsewhere; if anything they are more complex, due to the importance of spatial distribution and transport processes.

Le Houérou's chapter on long-term dynamics of the arid ecosystem of North Africa draws on information from many sources and disciplines and exposes the problems of understanding arid ecosystem dynamics on a historical and geographical scale. How can the gap be bridged between this scale and intensive, short-term site-level studies? A hint may be contained in Ludwig & Whitford's chapter, who observe that significant 'carry-over' effects may occur between seasons and 'pulses', mediated for instance by the amount of seeds on the surface, or in ant granaries. The 'reserve' components in arid ecosystems (Westoby, 1972) are responsible for both the memory of the system between pulses and for its long-term resilience.

Mathematical models are expected, among other things, to complement both short-term intensive and long-term extensive observations and to help integrate them into a coherent theoretical framework for understanding arid ecosystems functioning. Goodall's review chapter and the additional chapters on modelling show the considerable diversity of approaches that have been taken to model arid ecosystems or their sub-systems. They also indicate that our knowledge of a fair number of basic processes in these systems is apparently sufficient to allow them to be formulated in fairly detailed and sophisticated mathematical expressions. On the other hand, some of the modelling problems that are especially important in arid ecosystems, such as unpredictable temporal variation and spatial heterogeneity, have only begun to be treated. Most of the effort so far has been in constructing, developing and testing models, relatively less in experimenting with models and deriving new conclusions from them. The only exceptions are the earliest arid ecosystem model, PASTOR (Goodall, 1967), and some of the '*ad hoc* models' of limited subsystems (e.g. Wilcott's (1973) model of annual desert plants reproduction) which have indeed been intensively utilized. Much of the benefit from mathematical models, in deepening the 'understanding' of the systems they represent, is to be expected at the utilization and experimentation stage. Thus most of their potential in relation to arid ecosystems still remains to be realized. There is reason to hope that this will be done in the near future, though the problems involved should not be underestimated.

The present 'state of the art' as regards the integrated study of arid ecosystems as whole systems is well reflected in this part: initial and promising synthesis of intensive site observations, a quite distinct synthesis of long-term extensive observations, a good start on modelling in several directions. A true synthesis of these three lines of research cannot yet be reported, though arid-land scientists in several groups are probably making progress towards it. Achievement of this synthesis remains at least as great an intellectual challenge in arid ecosystems as elsewhere, since the results so far indicate some special sources of complexity in their structure and dynamics.

References

Bridges, K. W., Wilcott, C., Westoby, M., Kickert, R. & Wilkin, D. (1972). *Nature: a guide to ecosystem modelling*. Presented at the Ecosystem Modelling Symposium, American Institute of Biological Sciences Meeting, Minneapolis.

Goodall, D. W. (1967). Computer simulation of changes in vegetation subject to grazing. *Journal of the Indian Botanical Society*, **46**, 356–62.

Westoby, M. (1972). *Problem-oriented modelling: a conceptual framework*. Presented at the Desert Biome Information Meeting, Tempe.

Wilcott, J. C. (1973). A seed demography model for finding optimal strategies for desert annuals. PhD thesis, Utah State University, Logan.

PART III

Management of arid lands

19. Introduction

R. A. PERRY

In using arid lands, man must recognize the limitations imposed by a harsh environment and manage accordingly. Rainfall is not only low but falls in short periods interspersed with long dry periods. Plant growth, dependent on available soil moisture, is limited to short bursts separated by long periods of quiescence. Herein lies the most basic aspect of arid land management; plant material produced in short bursts must be conserved and its use spread out over long non-productive periods.

Vegetation has a key role in arid lands. It provides forage for animals and a standing vegetative cover to protect the landscape surface from wind and water erosion. Overuse of vegetation exposes the surface to erosion, reducing future productivity and future landscape protection. Regeneration takes a long time. Treatments to accelerate regeneration normally are not economic because of low productivity per unit area.

Traditional systems of grazing such as nomadism and transhumance have been developed from experience to cope with low productivity and spatial and temporal variability. Historically such systems have worked well with the level of human and animal population controlled at low levels by periodic droughts. However, in the twentieth century, traditional systems in many countries have been disrupted either by aid programs which reduce population control by droughts or by agricultural developments restricting the use of part of the transhumance transect. In many countries human and animal populations have increased several fold and are now well above the capability of arid ecosystems to support them.

Increasing population has increased pressure to use arid land for growing short season cereal and pulse crops and for agriculture using water harvesting or run-off diversions. The areas cropped during the short growing season are bare and vulnerable to wind erosion during the long dry period.

In the last few decades tourism and recreation have become an important use for arid lands. Tourists are attracted by the relatively low population density, clear clean air, and plenty of sunshine. They tend to concentrate in areas of spectacular scenery or where the remains of past civilizations occur. Protecting these areas from overuse is a problem as are the use of off-road vehicles and disruption to the hydrologic regime caused by tourist infrastructure.

Increasing populations have brought a need for greater development of sparse water resources both surface and subsurface. Water resource development needs to be associated with comprehensive total land use and social planning. Without adequate controls overdevelopment of water resources leads ultimately to underdevelopment of the region.

In many arid lesser developed countries there are social and economic conflicts between rural and urban populations. The sparse rural populations of the arid areas are different culturally from those in urban areas and are isolated from the power structure which governs them. Their needs are poorly understood and some policies which advantage the urban populations, e.g. controlled low food prices, are detrimental to the rural population. Generally, experience in lesser developed countries shows a tendency for administration to be overoptimistic in planning and to develop strategies beyond the potential of the resources.

20. Management of arid-land resources for domestic livestock forage

C. M. McKELL & B. E. NORTON

Perspective and introduction

Vegetation and soil are the most conspicuous resources of arid lands. Both are vulnerable to management practices imposed by man through the grazing of domestic livestock. Overgrazing of plants by animals and the subsequent effects on soil by the loss of a vegetative cover and trampling are some of the major causes of deterioration of these fragile lands.

As man has learned through experience over the millenia and more recently by scientific experimentation, his use of arid lands must recognize the limitations imposed by a harsh environment. This chapter therefore has a two-fold purpose: first, to describe the forage-producing ability of arid lands for domestic animals, outline values and constraints of arid land plants as forage, and discuss ecological effects of animal use of individual plants and communities; second, to examine the adaptation of grazing management to arid land constraints and evaluate some alternative methods for improving productivity.

There are two ways to approach the subject of desert forage production: (*a*) to survey the forage resources of the world's arid lands and list amounts and quality of the vegetation types and their distribution, and simultaneously catalogue various management objectives, strategies and procedures; (*b*) to examine the biological and ecological properties of arid ecosystems in relation to forage production and determine appropriate principles for managing livestock on the resource. The former approach would probably be preferred by the range manager (for example, Williams, Allred, Denio & Paulson, 1968, Lewis 1969, Stoddart, Smith & Box, 1975) who begins from an *a priori* standpoint of land use and attempts to achieve management goals through utilization and perhaps manipulation of the available resources. The latter approach is closer to the theme of the International Biological Programme, determining the biological basis of productivity with an emphasis on understanding the processes rather than the consequences of production (IBP Secretariat, 1965). For this discussion we have adopted the perspective of the IBP and will consider resource management from the point of view of the ecology of desert ecosystems. Our attention will focus on 'arid' ecosystems, from 60–100 to 150–250 mm mean annual precipitation which, as defined by

Noy-Meir (1973) following McGinnies, Goldman & Paylore (1968), is too arid for dryland farming. In 'semiarid' ecosystems the differences from grasslands and savannahs is often obscure, while 'extreme arid' ecosystems have insufficient productivity for domestic livestock interests.

Ecology of forage production

An axiom for the desert environment is that as mean annual precipitation decreases, the variability of that precipitation rises (Reitan & Green, 1968; Raikes, 1969; Evenari, Shanan & Tadmor, 1971; Noy-Meir 1973; Wallen & Gwynne, 1978). Analyzing data from many stations in the Mediterranean Basin with over 50 years of record, Le Houérou & Hoste (1977) reported a clear negative correlation, $r = 0.73$, between average rainfall and standard deviation of the precipitation. In addition to its high year-to-year variability, desert precipitation shows considerable spatial variation (Sharon, 1972), especially at sub-tropical latitudes (Wallen & Gwynne, 1978). In the Negev, Evenari *et al.* (1971) found that within a 10 ha area the highest rain gauge measurement was three to five times the lowest on the same day. Further heterogeneity of water distribution is generated by topographic features such as wadi depressions (Kassas & Imam, 1954; Obeid & Mahmoud, 1971) and the slope–shelf landscape gradient of mulga ecosystems in central Australia (Slatyer, 1973). Differences in edaphic parameters influencing soil water relationships, such as ion concentration and particle size, are often correlated with topographic patterns (Ayyad & El-Ghareeb 1972; Dawson & Ahern, 1973) and may accentuate the patchiness of the water resource. It is not surprising that water availability under high evaporative stress is the principal constraint on plant growth in desert ecosystems (Martin, 1975; McGinnies & Arnold, 1939).

The extensive literature on floristics, structure and physiology of desert vegetation includes the relevant chapters of this volume and Volume 1, specific monographs arising from IBP research (e.g. Simpson, 1977; Mabry, Hunziker & DiFeo, 1977), the recent work by Petrov (1976) and the comprehensive survey by McGinnies (1968) on desert vegetation, Levitt's (1972) tome on physiological resistance to stress, and Hadley's (1975) collection of papers on surviving in a desert environment. Our particular concern is with the characteristics of primary production that reflect the constraints to management of livestock in the arid environment.

Plant production that reflects the constraints of the arid environment

Desert plants must combine opportunism under suitable growing conditions with survival tactics during adversity. Some perennials remain evergreen and conserve water by restricting their stomatal conductance, but these plants may sacrifice the opportunity for high carbon fixation rates in the growing season. Others exhibit the drought-deciduous habit, losing high-transpiring leaves, which permit greater gas exchange rates in the spring, when they become a liability in the summer. The ecophysiology of shrubs with various survival strategies has been studied at the hot desert sites in the US/IBP (e.g., Bamberg, Kleinkopf, Wallace & Vollmer, 1975; Szarek & Woodhouse, 1976, 1977, 1978). Similar research in the cold-winter Great Basin desert (Caldwell, White, Moore & Camp, 1977) identified below-ground turnover of the root system as a dominant aspect of the carbon balance of shrubs which must withstand severe stresses in both winter and summer (Caldwell & Camp, 1974). Other adaptations to aridity, such as sclerophylly, microphylly or cladophylly and the special morphological and physiological features of cacti, are well known. Even a brief reflection on the survival mechanisms of desert perennials will suggest that the primary products of photosynthesis during the growing season may be shortly caducous, or indurated, or concentrated in non-foliage parts of the plant.

Adaptation to desert life is perhaps most evident among the desert annuals. Their rapid growth in response to water adequate to supply their needs over a contracted life cycle, makes them a sensitive barometer of seasonal favor and a fascinating study for desert ecologists (Went, 1949; Beatley, 1967; Mott, 1972; Mulroy & Rundel, 1977). Ephemerals have very high photosynthetic rates under suitable conditions (Halvorsen & Patten, 1975; Mooney, Ehleringer & Berry, 1976; Patten, 1978) and are capable of enormous fluctuations in productivity from year to year (Beatley, 1969; Norton, 1974). This dynamic feature of desert vegetation is particularly important to forage management and will be referred to again. But first, let us separate desert plant production into two principal components: biota and habitat. The flexibility of energy fixation in desert ecosystems is a product of species opportunism in response to water availability and the spatial arrangement of biotic and edaphic components.

Perennial plants in the desert are sparsely distributed in a regular or, more usually, clumped pattern (Barbour, MacMahon, Bamberg & Ludwig, 1977). Total ground-cover amounts to about 20% or less, depending on the vegetation type. At the Mohave Desert study site of the US/IBP, where precipitation averages around 100 mm, the vegetation is dominated by creosote bush (*Larrea tridentata*) and the cover is 22% (Turner, 1975), while 200 km north a salt desert shrub community has perennial cover of 10% with a mean annual precipitation of 156 mm (Norton, 1978). The pattern of scattered plants generates a two-part mosaic of shrubs (or grasses if the bunches are long lived) and interspaces. Plant detritus and aeolian materials deposit around the shrubs and a mound develops relative to the interspace depressions (Eckert *et al.*, 1978). Soils beneath the shrubs have higher nutrient ratings than do the interspace soils (Garcia-Moya & McKell, 1970; Charley,

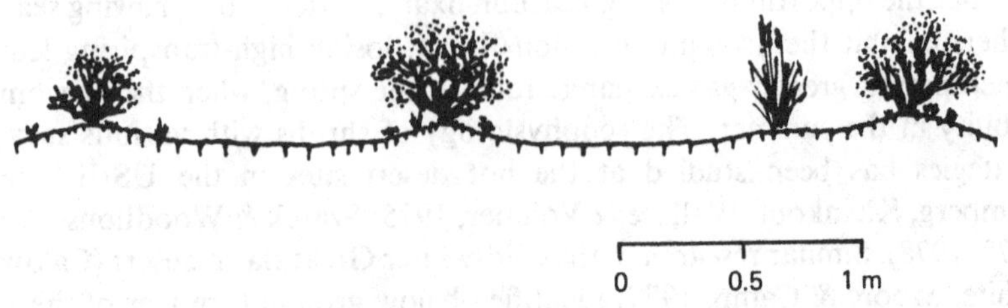

Fig. 20.1. The shrub–interspace mosaic in many desert ecosystems is defined by accumulation of materials and soil improvement beneath shrubs, and deflation from interspace depressions. The frequency and depth of soil surface cracking across a profile in the Great Basin Desert is shown. (Adapted from Eckert *et al.*, 1978.)

1972; Tiedemann & Klemmedson, 1973; Charley & West, 1975), eliciting the term 'islands of fertility' (Garcia-Moya & McKell, 1970; West & Skujins, 1978). The surface soil morphology of polygonal peds separated by cracks, the vesicular development and platy structure in the crust (Hugie & Passey, 1964; Eckert *et al.*, 1978), varies along a shrub-interspace gradient of soil physicol-chemical properties and water relations. In the soil crust interspace of the central Nevada desert, the vesicular horizon is unstable; infiltration is reduced and sediment production rises (Blackburn, 1975). Under shrubs in the Sonoran Desert, the infiltration rate was three times that in the open spaces, the organic matter content was higher and the bulk density was lower (Lyford & Qashu, 1969). Sandy soils may not exhibit such spatial contrasts in infiltration. An illustration of the two-phase, shrub-interspace mosaic is shown in Fig. 20.1.

The microtopographic relief and soil texture pattern causes an uneven distribution of water input. In North American deserts, the shrub mound habitat is more favorable for annual plant growth than are the interspaces (Went, 1942; Muller, 1953; Halvorsen & Patten, 1975). In Australia, however, ephemeral productivity may be greater in the depressions of 'gilgai' microtopography, where ponding occurs (Wilson & Leigh, 1964). The gilgai formation is often aligned to the contour in the Australian aridzone, producing a run-off/run-on redistribution of rainfall matching the intergrove/grove pattern of vegetation (Perry, 1970). Although fresh seed is blown or washed onto the bare intergrove patches every year and germinates successfully, rapid desiccation of the intergrove (run-off) soil severely restricts productivity and inhibits maturation (Mott, 1973). Plant growth is confined to locations of better water infiltration.

Although the interspace areas appear bare for most of the year, the soil is thoroughly explored by the root systems of perennial plants, except for the surface layers which rapidly dry out. The perennials initially draw water from the soil beneath their canopies and progressively extract water at increasing distances from the plant, so that in the middle of the dry season the soil may

Table 20.1. *Annual and perennial plant production*[a] *(kg ha^{-1} dry wt) and corresponding precipitation for five consecutive years at an IBP study site in the Mohave Desert on the Nevada Test Site of the US Energy Research and Development Administration*

Production	1972	1973	1974	1975	1976
Annuals	3	674	17	150	146
Perennials	183	573	181	144	331
Total	186	1247	198	294	477
Precipitation (mm) (Oct.–Sep.)	89	249	100	118	192

[a] Data summarized from annual progress reports of research at the Mohave Desert intensive study site (Site Coordinator: F. B. Turner) of the US/IBP Desert Biome program: Research Memoranda 73-2, 74-2, 75-2, 76-2, and 77-2. Utah State University, Logan.

be drier underneath a shrub than between shrubs, despite the higher infiltration rate of shrub-mound soil (Cable, 1977). In terms of water use, the soil volume of the bare interspaces represents a reservoir for the dry season. The interconnections among distribution of the soil resources, microtopography and perennial plant location, optimizes survival for perennial species. When water is scarce, the ephemerals fail to germinate or die; in a good season they utilize surplus water, generate substantial biomass, boost mineral cycling rates and flush the animal populations, but their rapid maturation minimizes competition with perennials. With such sparse above-ground perennial plant cover, space and light are not likely to be limiting factors to ephemeral productivity.

The desert ecosystem is pre-eminently geared to respond to precipitation. Since the water input that drives the system often comes in 'spurts' of a wet week or a downpour, the response behavior may be characterized as a pulse triggered by an environment cue. Instead of ecosystem processes being regulated by the amounts of materials in ecosystems compartments, in the 'pulse-and-reserve' paradigm developed in the US/IBP Desert Biome and discussed by Noy-Meir (1973), a trigger event or accumulation of stimuli initiates a burst of biological activity that 'automatically' continues through sequential phenophases. This concept is more applicable to hot deserts than cool deserts such as the Great Basin of the western United States, where temperature plays a more significant role in growth and development, and phenology is protracted (West & Gasto, 1978).

Strong correlations between precipitation and primary production in deserts emphasizes the overriding role of water in the system (Hutchings & Stewart, 1953; Wagner, 1976; Le Houérou & Hoste, 1977; Webb *et al.*, 1978). In a desert system one would expect great annual variations in rainfall, and therefore in plant growth; in fact, the variability in plant production exceeds

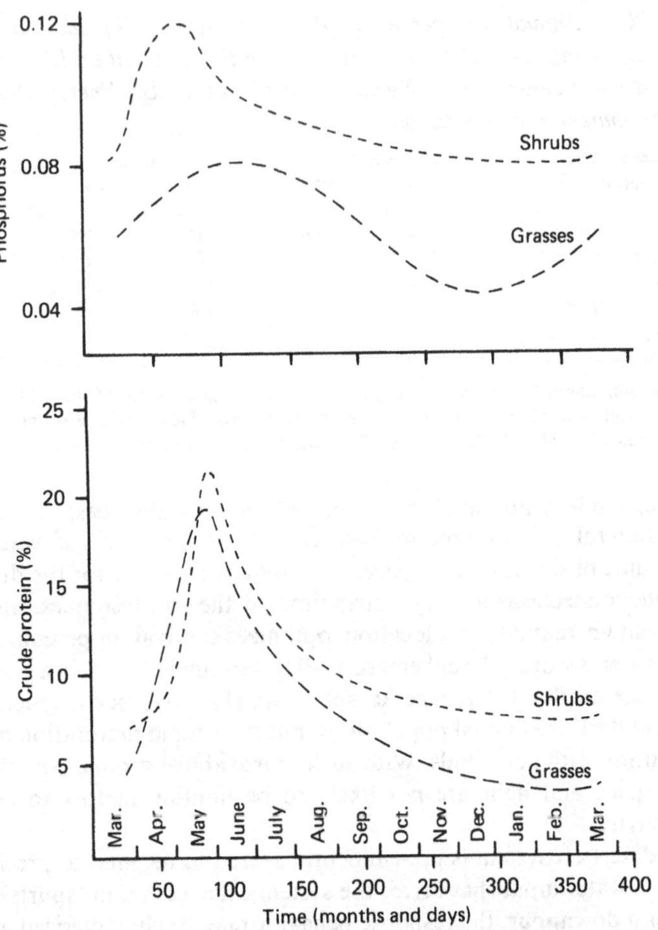

Fig. 20.2. The seasonal variation in phosphorus and nitrogen contents for shrub shoots and perennial grass foliage at the Desert Experimental Range in southwestern Utah. Vegetation is a sparse salt desert shrub type with mean annual precipitation of 157 mm. (Curves drawn from data in Gutierrez-Garza, 1978 and Cook, Stoddart & Harris, 1959.)

that of precipitation (Cook & Sims, 1975; Wallen & Gwynne, 1978). Production and precipitation data from the driest site of the US/IBP Desert Biome studies show that it is the opportunistic and at times abundant growth of ephemerals that accounts for such high variability (Table 20.1).

So far we have not distinguished between primary production and forage, which has a certain feed value. Let us assume, for the moment, that all foliage growth is palatable and consider it as a nutritional resource. The emphemeral biomass is green and nutritious for only a short period until maturity, whereupon the hayed remnants may provide a source of energy for an extended period. Shrub growth, on the other hand, maintains reasonable nutrient levels thoughout the year and acts as a nutritional reserve during the dry season (Goodin & McKell, 1971). Perennial grasses in the desert rapidly lose nutritional value in the dry season (especially in the tropics), and may constitute an energy feed complementary to shrub foliage. The annual cycle of nutritional values of perennial grasses and shrubs is plotted in Fig. 20.2 for salt desert shrub range in southwestern Utah, where vegetation is only grazed during the winter months, mostly by sheep. When seasonal trends in palatability of herbaceous species are superimposed on the cycle of biomass and nutrient availability, the temporal variation in forage is accentuated.

Compared with grasslands, plant production in a desert is low (Rodin & Bazilevich, 1967; Caldwell, 1975), while production variability is high. Compared with perennial grasslands, ephemerals may comprise a large share of the production. Compared with the relative homogeneity of grasslands, desert soils tend to develop structured patterns in microrelief and horizontal mosaics of nutrient distribution. Compared with the higher maintenance levels of grasslands, plant growth in deserts occurs in short-duration pulses separated by long periods of quiescence.

Using arid-land information for improving management

Generation of information

Over the years, arid lands of low productivity have received considerably less study than agricultural lands whose higher productivity earned them major programs of scientific and applied reseaıch. Even so, there is a considerable body of knowledge about food and fiber production of arid lands [McGinnies, Goldman & Paylore, 1971). Information on desert ecology has been greatly expanded by the results from the International Biological Programme, Desert Biome research.

Information use in decision making

Rational decisions are presumably more easily made when adequate information is available. The problem facing those making decisions about arid-land forage management is the lack of integration that often exists in the information base. Decisions on one aspect of management have immediate impacts on other areas of arid-land resources and unless these interactions are well known and allowances made to deal with them the results can be serious. An example is the decision to increase livestock forage by reducing sagebrush plant density by herbicide application. In dense sagebrush stands, the shrub cover could be removed with resulting benefits for both forage production and sage grouse. However, where sagebrush constituted less than

15% of the overstory, any increased productivity of forage would not justify the loss of sage grouse habitat (Phillips, 1971).

Because arid lands are so vulnerable to abuse, it is important to avoid taking unnecessary chances in resource management decisions. Recovery is slow and mistakes may not be rectified for many years. Because of the high variability in weather factors, especially periodic oscillations in precipitation, drought is an ever-present specter. Wise managers use the low part of precipitation cycles as their norm for making decisions. Overpopulation, either of man or animals or both, creates pressures for over-use of arid lands that can only result in disaster as occurred in the African Sahel in 1974–76 (Novikoff, 1976).

Ecosystem management approach

Future use of arid lands must increasingly follow an ecosystem management approach that includes all aspects of the biotic and abiotic systems. Integration of research and management will bring benefits to the land resource not realized in the past when plans for a single resource use were developed without considering impacts on other uses or long term effects. Interdisciplinary research such as developed in the International Biological Programme offers many opportunities for a broader understanding of the natural interrelationships among various ecosystem components and the implications for management.

Resource-compatible management strategies

Historical perspective

The kinds and numbers of plants growing in arid regions are surprisingly varied for so harsh an environment. Yet, few of these species are suitable for direct human consumption. Early man undoubtedly recognized that some animals were adapted to arid lands and were able to consume the rough forage from grasses, forbs and shrubs and thus maintain themselves. Man in turn, found he could maintain himself in the inhospitable land by using various products such as milk, meat, hides, fur and even motive power from the adapted animals (Lundholm, 1976). For primitive cultures in arid lands, livestock products still provide the basis for existence. Over-use of desert rangelands in many places has thus resulted in serious consequences to the human population. To prevent abuse and live within the limits of productivity imposed by the harsh environment man developed 'rules' or practices that helped him use the land within the constraints of desert ecosystems.

Over many seasons of watching animals graze on plants and walk over the land, herders learned much about plant palatability, parts of plants preferred, which plants regrow when used and how much use particular plants and communities could withstand in a given year and still produce forage in subsequent years. These observations were the basis for the traditional grazing practices that were basically tied to the seasonal and cyclical patterns of plant growth and water availability. Livestock numbers as well as human numbers were limited by the cylical and meager productivity of the arid lands they occupied.

In most desert areas, forage should not be grazed until the plants are physiologically capable of sustaining such use and there is sufficient feed available for animals. This condition is referred to as *range readiness* (Stoddart *et al.*, 1975). The inherent limit to the numbers of animals that can safely be grazed in a given area or range unit without causing a deterioration in the plant cover or soil is called the *carrying capacity*. The duration or amount of grazing use should not exceed the tolerance of plants to sustain defoliation and should be designed to allow key plant species to maintain vigor and reproduce. Grazing should be minimal or avoided during critical periods such as early growth, flowering or seedling establishment. Uniform *distribution* of animals over an area is another management strategy that can provide greater forage utilization with minimal impact on a range ecosystem. Periods of non-use or rest for an entire growing season (Shiflet & Heady, 1971) may be essential to help heavily grazed species of high palatability recover their vigor.

Evolution of grazing systems

Faced with the need to utilize arid lands at the most opportune time, man gradually developed systems for management of his animals on these lands. Where dependable sources of water were near such as a river, lake, or oasis, permanent settlements developed and livestock were herded out into the arid hinterland (Johnson, 1969). Where forage and water could be found over a large area but in limited and unpredictable amounts, a pattern of cyclic migration developed whereby herders and their families moved with the animals to areas of favorable production. Grazing livestock on arid lands required patterns of use that were generally opportunistic and easily adjusted when precipitation was greater in one desert valley than another.

Because of their relatively low productivity, seasonal differences and fragility, arid lands must be used in a manner compatible with their limitations. Primitive grazing systems were simplistic but generally did not reduce the productivity of desert ecosystems except where animal populations exceeded the natural capacity of the land to produce. In recent times, however, livestock numbers have regularly and almost universally been in excess of that capacity. Periods of cyclic droughts are characteristic of arid and semiarid lands (Nicholson, 1978; Schulman, 1956; Meigs, 1953), and must be considered in setting the upper limit of livestock numbers to be grazed (Stoddart *et al.*, 1975).

Numerous systems of grazing use evolved in response to the variable nature of arid lands. Nomadism and transhumance were among the first.

Nomadism

Nomadism occurs when man and his animals wander in search of pasture, but the wandering is usually in a pattern set according to areas of feed availability and tradition. Johnson (1969) described nomadism as a livelihood form that utilizes marginal resources that would otherwise be lost because the areas are too dry or steep for crop agriculture. In favorable years, pastoralism and agricultural activity occur in combination, but in dry years only the herding of flocks remains. In his report to the government of the Sudan, Sheppard (1968) argued for nomadism as an effective adaptation of livestock management to recurring erratic variations in forage production due to climatic factors and seasonal shortages of drinking water. Avoidance of seasonal insect, disease or soil problems may also be a basis for nomadism. Thus, nomadism is an opportunistic but logical system of use in arid regions where continued grazing of any one location is not practical. Efforts to achieve greater stability by developing water sources (wells, reservoirs etc.) have not been completely successful because the additional water has encouraged over-stocking and a concentration of animals around the water sources (Novikoff, 1976). Nomadism is still important in North Africa and the Near East, often in the transhumant form of wet season/dry season grazing.

A modern adaptation of shifting livestock use in an arid region is the 'best pasture' system advocated for the southwestern United States by Herbel & Nelson (1969) in which animals are moved to the grazing units where rainfall has stimulated the forage plants to grow better than in other locations.

Transhumance

This is a form of nomadism that involves trailing of livestock between two locations where forage is available in relation to elevation or latitude in a set pattern according to seasonal changes (Johnson, 1969). The avoidance of insect problems such as the tsetse fly in Central Nigeria (Davies, 1962) or a seeking of livestock watering places are examples of factors that encourage transhumance.

Transhumance is commonly practiced in many desert areas that have adjacent mountains such as the Great Basin of the western United States and the Karadj Basin of Iran. Such places have topographic gradients that allow graziers to spend the winter months in low desert elevations and move into mountains for summer grazing. These movements involve a fixed base of operations in an area of dependable productivity, usually centered around irrigated agriculture.

A critical problem of transhumance, aside from the social disruption and instability that may occur if family units are moved, is the herding of livestock onto plant communities that are not ready for grazing or staying so long in specific places that plant vigor and reproduction are reduced. Livestock numbers can easily exceed the carrying capacity of areas traditionally used for grazing. Under such conditions more structured grazing controls are needed. Deterioration of mountain and desert rangelands in Utah early in this century prompted an eminent ecologist to ask 'Is Utah Sahara bound?' (Cottam, 1947). Overgrazing of winter desert ranges and migration to mountain ranges before they were ready were the major causes of the drastic decline in range and watershed condition.

Regulated grazing systems

A common management strategy is to set the starting date for grazing in relation to the time when desirable forage species are vigorous enough to sustain use. To provide some early grazing at the beginning of the grazing season, use can be rotated among pastures or grazing areas (Anderson, 1967). Where several graziers use one rangeland area, some form of use control must be exercised through tribal, user organizations or a government agency.

Grazing or fodder reserves have been established to provide a buffer against periods of low production (Le Houérou, 1973; Lundholm, 1976), but maintaining such reserves by relying upon voluntary restraint is difficult. Protecting areas from grazing to create a feed reserve such as the traditional 'Hema' of the Arabian peninsula (Draz, 1969) requires special social or political constraints.

When traditional prohibitions break down and common use becomes 'use for all' with no individual restraints, areas and time of use must be assigned. To further implement strict controls, fences may be necessary but they can only be established where there is a recognition of property rights and boundaries. Ultimately, some unit of society such as a tribal leader, a users association or the civil government may have to introduce a system of traditional or written permits to control the various detrimental aspects of grazing such as excess numbers of animals, and inappropriate periods of use (Hickey, 1966).

Given the ability to control grazing use, a number of more intensive methods for grazing management have been developed (Heady, 1975). These methods usually attempt to manage grazing in relation to: capacity of a range area to provide feed for, or carry, a number of animals; amount of forage removed from plants; season of grazing; animal distribution; and physiological requirements of plants. Each method seeks to regulate one or more aspects of range use. Much of the research dealing with these methods has been developed at research stations in relatively productive range areas rather than arid-desert areas.

Continuous or year-long grazing

In areas having a relatively uniform seasonal climate, where various plant species can be grazed in sequence, rangelands can be utilized at a minimal level for the entire year. This is not an intensive system of management because the animals are responding to seasonal changes in growth or production of range plants. For this system to be compatible with desert ecosystems, animals must be given the opportunity to graze plants of diverse growth habit and nutritional value. Often, a diversity of range types is necessary to provide year-long feed production. The main efforts of management are devoted to determining the number of animals consistent with rangeland carrying capacity, and distributing the animals equitably over the grazing area.

Seasonal grazing

This system recognizes that plants should be grazed during a favorable season. Grazing intensity must be kept within the limits where plants can sustain general use after their early critical growth stages or in their dormancy period. Recognition of the appropriate times for starting and stopping grazing is a critical management problem because of differences among plants in their early phenological development and because variations among years may cause range readiness to occur earlier one year than another.

Deferred grazing

This system delays the starting date of grazing to allow for seed production or achievement of a desired phenological stage. Some plants cannot survive early grazing year after year. Other plants may benefit from the absence of grazing in critical growth periods. Thus the problem of deferring grazing requires special arrangements for obtaining periods of non-use. Whether critical vegetation growth stages were recognized by primitive graziers is not known but present-day grazing managers have abundant, but not complete, information on desert species (Morton & Hull, 1973).

Deferred rotation

To provide a period of non-use and still allow grazing, animals are rotated among two or more grazing areas in successive seasons. An early season non-use period can encourage plant vigor by withholding defoliation until new growth is sufficient to support the photosynthate requirements of plants and restore their carbohydrate reserves (Trlica & Cook, 1972) or until seed production is assured (Sampson, 1951). Because early use is then imposed only once every two or more years, the productivity of all the range areas involved can be maintained. For example, Reynolds (1959) found an increase of desirable grasses from 30–60% of the composition on a desert shrub-grass range in Arizona where summer deferment was rotated among several grazing units.

Rotation grazing

A period of intensive use in rotation with a long period of non-grazing may be helpful in some vegetation types to maintain plant productivity and a balance in species composition. Various rotation time sequences for moving livestock to individual pastures or grazing areas have been devised (Hubbard, 1951; Acocks, 1966; Goodloe, 1969; Corbett, 1978) for areas where intensive management is possible. According to Heady (1975) a short period of intense grazing may prevent over-use and reduce species selectivity. However, improperly managed rotation grazing in desert regions could create problems because of the generally low productivity of deserts and lack of information about plant responses to defoliation at various phenological stages. When a season of high production is made possible by favorable weather, a short period of use may have less impact on the plant community than extended grazing over the entire period.

Intensive use over a short period followed by an extended period of non-use has been advocated as a means of reducing animal selectivity and increasing livestock use of certain rangeland types such as multiple camps in South Africa (Howell, 1977) or small pastures in Texas (Corbett, 1978). Wise management, both of range and animals, must seek to maintain a viable balance between producing animal gains and retaining plant vigor.

Rest-rotation grazing

This system of management involves the resting for one year of various areas or units in a rotation scheme (Hickey, 1966). The main principles of the grazing strategy were first described under western United States conditions by Smith (1899), who noted that the intermittent grazing and resting of land resulting from roving bands of buffalo (*Bison bison*) fostered continued rangeland productivity and sometimes improved the range plant community. Sampson (1913) observed that protection from grazing restored vigor, productivity and reproduction of overgrazed rangeland vegetation. Subsequent observations by Hormay (1956) led to the development of a system in which the grazing use of several range units followed a typical sequence of intense grazing for the first year or season, no grazing through the second season and the subsequent seed-producing period in the third year, grazing in mid-season of the third year to knock seeds from the plants and trample the plant seeds, and, in the fourth year, protection from early grazing to encourage seedling establishment followed by intense grazing to finish the sequence (Hormay, 1970). Flexibility in the order of events, timing of use and the number of livestock grazed are advantages described for the system to achieve the general goals of improving plant vigor and obtaining grazing for animals. Rest-rotation grazing systems have been imposed on many areas of desert rangeland in the western United States, but insufficient attention has been given to local conditions. Some of the difficulties encountered have been due to a lack of information concerning the time required for various range species to recover from intense grazing and a disparity in phenological stages of development for species making up the plant community. This lack of information almost precludes establishment of a definitive point for starting or stopping grazing in the middle of the grazing season. However, Herbel (1974) concluded that flexible grazing systems such as rest-rotation may provide a higher quality and quantity of forage for part of the year than season or year-long grazing use.

Manipulation of grazing use *versus* intensive rangeland improvement

Although the natural productivity of arid lands is inherently low, useful forage productivity levels can be maintained by applying appropriate methods. Past abuses from overgrazing can be overcome either by following a grazing system that provides an appropriate period of non-use and reduces selective grazing or by applying one or more intensive range improvement practices. Generally, any attempts to improve rangeland productivity must bring about one or more of the following: reduction of competition from undesirable plants; increased plant vigor and nutritional quality; a greater proportion of desirable forage plants; or an enhanced overall plant productivity. An extensive literature is discussed in a text by Valentine (1971) on methods and tools for range improvement. Only a brief overview of range improvement practices as they relate to desert lands will be attempted here.

Animals as tools

Primitive livestock herders could manage their forage resource only through their animals. Thus, animals were the management tools of primitive peoples and remain important to the present time (Stoddart *et al.*, 1975). Animals can exert a powerful influence on plant community productivity through the intensity, season, and duration of their grazing. Plant vigor, species composition, and community succession can be significantly shaped by animal use. A most important objective for grazing managers is to prevent animals from concentrating predominantly on palatable species at their critical phenological stages. Livestock also act as a dispersal medium in carrying seeds and other propagules to openings in plant communities where the trampling of animal feet may plant them.

How well people utilize the grazing animal as a management tool may largely determine whether the plant productivity of arid lands can be improved. For example, goats may be grazed on an area to use shrubs selectively (Provenza, 1978) and release other species from shrub competition. A high intensity–short duration grazing (Corbett, 1978) system may be used to increase the utilization of less-palatable species and thus equalize plant competition. A rest-rotation system may help to restore plant vigor and increase seed production and seedling establishment in areas where excessive grazing has debilitated the plant cover (Hormay & Talbot, 1961).

Range improvement methods

Fire

Pastoralists have traditionally used fire to reduce undesirable species, clear away accumulated litter and dead material, and to stimulate new growth of surviving species. Heady (1975) pointed out that fire has a greater effect on woody vegetation than on grasses. In arid lands of low productivity, insufficient understory fuel to carry a fire and wide spacing of shrubs precludes using fire as a tool except in years of good herbage production. Hence, burning may or may not be an effective practice (Pechanec, Plummer, Robertson & Hull, 1965). In so far as the main objective of burning is to reduce the species that have low forage value, those that are structurally adapted to survive and re-grow are of special concern because they may become more abundant rather than less. Fire may also scarify hard seeds and enhance germination. As a general rule, the first few years after burning are often good years for wildlife and livestock as they use the tender shoots of shrub re-growth in addition to the increased grass production.

Soil surface

Treatments of the soil surface with equipment designed to change its micro-relief patterns and retard surface run-off for greater infiltration of precipitation into the soil have been used to increase temporarily arid-land productivity. More effective plant use of limited precipitation is the principal basis for such increased productivity (Branson, Gifford & Owen, 1973). Water retained in pits, furrows, terraces or other surface indentations percolates into the soil and aids in establishment and growth of new plants in areas where the vegetation cover has been reduced, and accelerated run-off has deprived the site of water needed to sustain the vegetation. The indentations also trap seeds and litter, thus improving opportunities for seedling establishment in these safe sites (Eckert *et al.*, 1978).

Mechanical control

Undesirable species may be controlled by using rugged vehicles such as crawler tractors to pull out or push over trees and shrubs or by dragging implements such as rotary cutters, disc plows, or by using heavy chains to cut off, dig out, or pull over the target species and any others that are in the way, (Plummer, Christensen & Monsen, 1968). Some degree of selectivity is usually achieved only by choosing high productivity areas that have a great density of undesirable plants. Species escaping the treatments, and hence freed from competition, may drastically increase in numbers (Robertson & Pearse, 1945) unless kept in check by grazing, herbicide application, or by a follow-up seeding of species desired for forage or other values.

Mechanical control of undesirable species to open a plant community is often necessary for the subsequent seeding of desirable forage species. Many methods for mechanical control were developed in the 1950s and have been applied with varying success. Some of the biggest problems have been caused by seeding unadapted species, failure to observe natural topographic contours and concentrate on the most productive sites, removal of plant cover in years of inadequate rainfall, poor establishment of seeded species, re-growth of undesirable sprouting species and loss of habitat for resident wildlife. In the United States, mechanical control and seeding have worked best in plant communities dominated by *Artemisia tridentata* or *Juniperus osteosperma*. Generally, plant communities existing under less precipitation than 250 mm have not been successfully treated.

Chemical control

Another means of increasing arid-land forage production under certain conditions is by chemical control of undesirable species (Evans, Young & Eckert, 1978). Where a sufficient number of desirable species remain in the understory, reduction of competition for moisture and nutrients by controlling shrubby overstory species allows grasses in the understory to increase significantly. When forbs (broadleaf herbaceous species) are abundant in the understory, herbicide application may not be suitable because of the potential for damaging them. Valentine (1971) cited numerous examples where herbicide applications to sagebrush (*Artemisia tridentata*) and other shrubs resulted in an up to ten-fold increase in forage production.

Where understory species are lacking, chemical control of undesirable shrubs may be desirable to open the plant community for establishment of forage species. Another advantageous use of herbicides is to control shrub re-growth after fire. Vigorous re-growth of such genera as *Chrysothamnus*, *Quercus*, *Prosopis* and *Sarcobatus* may profitably be controlled by application of phenoxy herbicides.

In meeting the requirements of multiple users and effective ecosystem management, herbicide applications should be planned carefully and selectively to improve productivity of a broad range of species and increase wildlife habitat diversity.

Seeding

Replacement forage species may need to be seeded to restore the productivity of areas where the original vegetation has become depleted (Plummer, Hull, Stewart & Robertson, 1955). Seedling establishment in arid lands is difficult and seedings should only be undertaken after adequate planning and site preparation. Failure to use species adapted to harsh sites was the reported cause of failure in over 40 plantings of sites dominated by *Atriplex confertifolia* (Bleak, Frischknecht, Plummer, & Eckert, 1965). Hyder, Sneva & Cooper (1955) consider drought and improper seed cover as the two most serious problems affecting seedling establishment.

The controversy as to the value of native *versus* introduced species for seeding depleted rangelands that is currently being debated in the United States is easily solved by requiring that all species used must have a high degree of adaptability to local environmental conditions. Thus, in the western United States, where crested wheatgrass (*Agropyron desertorum*) has shown high adaptation to many arid sites, its continued use is justified on the basis of restoring production. The question of introducing an alien species into an ecosystem composed of natives is not easily solved. When seen as an alternative to the invasion of alien species from other ecosystems, the introduction of an adapted species gains considerable philosophical appeal. More research on native species of high forage value is needed so that they may receive greater use. This is especially true of palatable shrubs and forbs which have long been ignored in favor of grasses.

Integration

An essential requirement for proper management of arid lands is integration of improvement practices. Improper application of improvement methods or inadequate understanding of the anatomy physiology and phenology of a species have resulted in failure to control undesirable species (Robertson & Cords, 1957). Better results in vegetation management can be achieved only with adequate information on plant biology. Although some of the above practices may appear to be effective individually they have a greater chance of success when applied as parts of a logical sequence. For example, burning followed by chemical control of undesirable plant re-growth and proper grazing management will bring greater long-term benefits than will burning and little or no follow-up treatment. Whenever improvement practices are applied, livestock management must be in accord with the inherent carrying capacity of the land.

Summary

In contrast to agricultural systems that deal primarily with monocultures, range management must work with plant communities involving all of the interrelated biotic and abiotic influences that affect production of animal feed and habitat. Increasingly, the emphasis in range management is to adopt an ecosystem approach (Van Dyne, 1969) in considering the multiple influences/factors involved in research and management planning.

Whereas research in grasslands ecosystems has provided many answers to management problems, studies of desert ecosystems are fewer in number and cover somewhat more complex systems with severe constraints on productivity. Thus, management of desert rangelands must work on a more limited data base than used for grassland ecosystems. Fundamental information on biotic and abiotic components of the desert ecosystem often does not generally exist in sufficient depth to guide management practices. Further, information on particular ecosystem components too frequently lacks correlation with other components, thus making difficult the development of integrated management systems. The result has been trial and error in many range management and improvement practices.

A major thrust of the IBP research program has been to study the desert (rangeland) as a system and avoid a piecemeal approach to research. From this IBP effort, new ways to solve management problems have been formulated. Most important of all, however is a validation of the ecosystem approach to range management.

Management practices such as using grazing systems and manipulation of plant species, density and cover to increase useful forage production can be enhanced in their effectiveness with the availability of information and direction that has been developed from the integrated approach taken by IBP.

Future range research on desert ecosystems can and must be pursued with an orientation toward solving problems in relation to the whole system. In a like manner, management must consider the implications of actions taken in relation to the continued productivity of the biotic and abiotic components of the desert system.

References

Acocks, J. P. H. (1966). Non-selective grazing as a means of field reclamation. *Proceedings of the Grassland Society of South Africa*, **1**, 33–9.

Anderson, E. W. (1967). Rotation of deferred grazing. *Journal of Range Management*, **20**, 5–7.

Ayyad, M. A. & El-Ghareeb, R. (1972). Microvariations in edaphic factors and species distribution in a Mediterranean salt desert. *Oikos*, **23**, 125–31.

Bamberg, S. A., Kleinkopf, G. E., Wallace, A. & Vollmer, A. (1975). Comparative photosynthetic productions of Mojave desert shrubs. *Ecology*, **56**, 732–6.

Barbour, M. G., MacMahon, J. A., Bamberg, S. A. & Ludwig, J. A. (1977). The structure and distribution of *Larrea* communities. In: *Creosote bush: biology and chemistry of* Larrea *in New World deserts* (ed. T. J. Mabry, J. H. Hunziker & D. R. DiFeo), pp. 227–51. US/IBP Synthesis series, No. 6. Dowden, Hutchinson & Ross, Stroudsburg, Pennsylvania.

Beatley, J. C. (1967). Survival of winter annuals in the northern Mojave Desert. *Ecology*, **48**, 745–50.

Beatley, J. C. (1969). Biomass of desert winter annual populations in southern Nevada. *Oikos*, **20**, 261–73.

Blackburn, W. H. (1975). Factors influencing infiltration and sediment production of rangelands of Nevada. *Water Resources Research*, **11**, 929–37.

Bleak, A. T., Frischknecht, N. C., Plummer, A. P. & Eckert, R. E. (1965). Problems in artificial and natural revegetation of the arid shadscale vegetation zone of Utah and Nevada. *Journal of Range Management*, **18**, 59–65.

Branson, F. A., Gifford, G. F. & Owen, J. R. (1973). *Rangeland hydrology. Society for Range Management Range Science Series* 1. Denver, Colorado.

Cable, D. R. (1977). Seasonal use of soil water by mature velvet mesquite. *Journal of Range Management*, **30**, 4–11.

Caldwell, M. (1975). Primary production of grazing lands. In: *Photosynthesis and production in different environments* (ed. J. E. Cooper), pp. 41–73. Cambridge University Press.

Caldwell, M. M. & Camp, L. B. (1974). Below ground productivity of two cool desert communities. *Oecologia*, **17**, 123–30.

Caldwell, M. M., White, R. S., Moore, R. T. & Camp, L. B. (1977). Carbon balance, productivity and water use of cold-winter desert shrub communities dominated by C_3 and C_4 species. *Oecologia*, **29**, 275–300.

Charley, J. L. (1972). The role of shrubs in nutrient cycling. In: *Wildland shrubs – their biology and utilization* (ed. C. M. McKell, J. P. Blaisdell & J. R. Goodin), pp. 182–203. USDA Forest Service General Technical Report INT-1. Ogden, Utah.

Charley, J. L. & West, N. E. (1975). Plant-induced soil chemical patterns in some shrub-dominated semi-desert ecosystems of Utah. *Journal of Ecology*, **63**, 945–64.

Cook, C. W. & Sims, P. L. (1975). Drought and its relationship to dynamics of primary productivity and production of grazing animals. In: *Evaluation and mapping of tropical African rangeland*, pp. 163–8. Proceedings of a Seminar. International Livestock Centre for Africa, Addis Ababa.

Cook, C. W., Stoddart, L. A. & Harris, L. E. (1959). The chemical content in various portions of the current growth of salt-desert shrubs and grasses during winter. *Ecology*, **40**, 644–51.

Corbett, Q. (1978). Short duration grazing with steers – Texas style. *Rangeman's Journal*, **5**, 201–3.

Cottam, W. P. (1947). *Is Utah Sahara bound? Reynolds Lecture Series Bulletins*, vol. 37. University of Utah, Salt Lake City.
Dawson, N. M. & Ahern, C. R. (1973). Soils and landscapes of Mulga lands with special reference to south-western Queensland. *Tropical Grasslands*, **7**, 23–34.
Davies, H. (1962). *Tsetse flies in Northern Nigeria.* Gaskiya Corporation, Zaria, Nigeria.
Draz, O. (1969). *The 'Hema' system of range reserves in the Arabian Peninsula, its possibilities in range improvement and conservation projects in the East.* FAO/PL:PEC/13. Food & Agriculture Organization, Rome.
Eckert, R. E., Wood, M. K., Blackburn, W. H., Peterson, F. F., Stephens, J. L. & Meurisse, M. S. (1978). Effects of surface-soil morphology on improvement and management of some arid and semi-arid rangelands. *Proceedings of the First International Rangeland Congress*, pp. 299–302. Society for Range Management, Denver, Colorado.
Evans, R. A., Young, J. A. & Eckert, R. E. (1978). Use of herbicides as a management tool. In: *Sagebrush ecosystem: a symposium*, pp. 110–6. Utah State University, Logan.
Evenari, M., Shanan, L. & Tadmor, N. (1971). *The Negev: the challenge of a desert.* Harvard University Press, Cambridge, Massachusetts.
Garcia-Moya, E. & McKell, C. M. (1970). Contribution of shrubs to the nitrogen economy of a desert-wash plant community. *Ecology*, **51**, 81–8.
Goodin, J. R. & McKell, C. M. (1971). Shrub productivity – a reappraisal of arid lands. In: *Food fiber and the arid lands*, pp. 236–46. University of Arizona Press, Tucson.
Goodloe, S. (1969). Short duration grazing in Rhodesia. *Journal of Range Management*, **22**, 369–73.
Gutierrez-Garza, J. S. (1978). 'Potential for cattle grazing on sheep range in southwestern Utah'. Unpublished MS thesis, Utah State University, Logan.
Hadley, N. F. (ed.) (1975). *Environmental physiology of desert organisms.* Dowden, Hutchinson & Ross, Stroudsburg, Pennsylvania.
Halvorsen, W. L. & Patten, D. T. (1975). Productivity and flowering of winter ephemerals in relation to Sonoran Desert shrubs. *American Midland Naturalist*, **93**, 311–9.
Heady, H. F. (1975). *Rangeland Management.* McGraw-Hill, New York.
Herbel, C. H. (1974). A review of research related to development of grazing systems on native ranges of the western United States. In: *Plant morphogenesis as the basis for scientific management of range resources*, pp. 138–49. USDA Miscellaneous Publication 1271. Washington, DC.
Herbel, C. H. & Nelson, A. B. (1969). *Grazing management on semi-desert ranges in southern New Mexico.* US Department of Agriculture, Agricultural Research Service, Jornada Experimental Range Report No. 1, Las Cruces, New Mexico.
Hickey, W. C. (1966). *A discussion of grazing management systems and some pertinent literature (abstracts and excerpts) 1895–1966.* US Department of Agriculture Forest Service, Denver, Colorado.
Hormay, A. L. (1956). How livestock grazing habits and growth requirements of range plants determine sound grazing management. *Journal of Range Management*, **9**, 161–4.
Hormay, A. L. (1970). *Principles of rest-rotation grazing and multiple use management.* USDA Forest Service Training Text 4 (220). Washington DC.
Hormay, A. L. & Talbot, M. W. (1961). *Rest-rotation grazing – a new management system for perennial bunchgrass ranges.* US Department of Agriculture Production Research Report 51. Washington, DC.

Howell, L. H. (1977). Development of multi camp grazing systems in the southern Orange Free State, Republic of South Africa. *Journal of Range Management*, **31**, 459–65.

Hubbard, W. A. (1951). Rotational grazing studies in western Canada. *Journal of Range Management*, **4**, 25–9.

Hugie, V. K. & Passey, H. B. (1964). Soil surface patterns of some semi-arid soils in northern Utah, southern Idaho and northeastern Nevada. *Proceedings of the Soil Science Society of America*, **28**, 786–92.

Hutchings, S. S. & Stewart, G. (1953). *Increasing forage yields and sheep production on intermountain winter ranges.* USDA Circular No. 925. Washington, DC.

Hyder, D. L., Sneva, F. & Cooper, C. S. (1955). Methods for planting crested wheatgrass. In: *1955 Field Day report*, pp. 19–20. Squaw Butte – Harvey Range and Livestock Experiment Station Burns, Oregon.

IBP Secretariat. (1965). IBP News No. 2. Published in Rome under the auspices of the Special Committee for the International Biological Programme.

Johnson, D. L. (1969). *The nature of nomadism: a comparative study of pastoral migrations in southwestern Asia and Northern Africa.* Department of Geography, Research Paper No. 118, University of Chicago.

Kassas, M. & Imam, M. (1954). Habitat and plant communities in the Egyptian desert. III. The wadi bed ecosystem. *Journal of Ecology*, **42**, 424–41.

Le Houérou, H. N. (1973). *Peut-on lutter contre la désertisation?* Colloques International sur la Désertification Nouakchott Bulletin Institut Fondation Afrique Noire, Dakar.

Le Houérou, H. N. and Hoste, C. H. (1977). Rangeland production and annual rainfall relations in the Mediterranean Basin and in the African Sahelo-Sudanian zone. *Journal of Range Management*, **30**, 181–9.

Levitt, J. (1972). *Responses of plants to environmental stresses.* Academic Press, New York.

Lewis, J. K. (1969). Range management viewed in the ecosystem framework. In: *The ecosystem concept in natural resource management* (ed. G. M. Van Dyne), pp. 97–187. Academic Press, New York.

Lundholm, B. (1976). Domestic animals in arid ecosystems. *Ecological Bulletin (NFR)*, **24**, 29–42.

Lyford, F. P. & Qashu, H. K. (1969). Infiltration rates as affected by desert vegetation. *Water Resources Research*, **5**, 1373–6.

Mabry, T. J., Hunziker, J. H. & DiFeo, D. R. (eds.) (1977). *Creosote bush: biology and chemistry of* Larrea *in New World deserts.* US/IBP Synthesis Series, No. 6. Dowden, Hutchinson and Ross, Stroudsburg, Pennsylvania.

Martin, S. C. (1975). *Ecology and management of southwestern semi-desert grass-shrub ranges. The status of our knowledge.* USDA Forest Service Research Paper RM-156.

McGinnies, W. G. (1968). Vegetation of desert environments. In: *Deserts of the world* (ed. W. G. McGinnies, B. J. Goldman & P. Paylore), pp. 379–566. University of Arizona Press, Tucson.

McGinnies, W. G. & Arnold, J. F. (1939). *Relative water requirements of Arizona range plants.* Arizona Agricultural Experiment Station Technical Bulletin 80.

McGinnies, W. G., Goldman, B. J. & Paylore, Patricia (eds.) (1968). *Deserts of the world.* University of Arizona Press, Tucson.

McGinnies, W. G., Goldman, B. J. & Paylore, Patricia (eds.) (1971). *Food, fiber and the arid lands.* University of Arizona Press, Tucson.

Meigs, P. (1953). World distribution of arid and semi-arid homoclimates. *UNESCO Paris Arid Zone Programme*, **1**, 203–10.

Mooney, H. A., Ehleringer, J. & Berry, J. A. (1976). High photosynthetic capacity of a winter annual in Death Valley. *Science*, **194**, 322–4.

Morton, H. L. & Hull, H. M. (1973). Morphology and phenology of desert shrubs. In: *Arid shrublands. Proceedings of Third Workshop of the United States/Australia Rangelands Panel* (ed. D. M. Hyder), pp. 39–46. Society for Range Management, Denver, Colorado.

Mott, J. J. (1972). Germination studies on some annual species from an arid region of Western Australia. *Journal of Ecology*, **60**, 293–304.

Mott, J. J. (1973). Temporal and spatial distribution of an annual flora in an arid region of Western Australia. *Tropical Grasslands*, **7**, 89–97.

Muller, C. H. (1953). The association of desert annuals with shrubs. *American Journal of Botany*, **40**, 53–60.

Mulroy, T. W. & Rundel, P. W. (1977). Annual plants: adaptations to desert environments. *Bioscience*, **27**, 109–14.

Nicholson, S. E. (1978). Climatic variations in the Sahel and other African regions during the past five centuries. *Journal of Arid Environments*, **1**, 3–24.

Norton, B. E. (1974). IBP studies in the desert biome. *Bulletin of the Ecological Society of America*, **55**, 6–10.

Norton, B. E. (1978). The impact of sheep grazing on long-term successional trends in salt desert shrub vegetation of southwestern Utah. In: *Proceedings of the First International Rangeland Congress*, pp. 610–12. Society for Range Management, Denver, Colorado.

Novikoff, G. (1976). Traditional grazing practices and their adaptation to modern conditions in Tunisia and the Sahelian countries. *Ecological Bulletin* (*Stockholm*), **24**, 55–69.

Noy-Meir, I. (1973). Desert ecosystems: environment and producers. *Annual Review of Ecology and Systematics*, **4**, 25–51.

Obeid, M. & Mahmoud, A. (1971). Ecological studies in the vegetation of the Sudan. II. The ecological relationships of the vegetation of Khartoum Province. *Vegetatio*, **23**, 177–98.

Patten, D. T. (1978). Productivity and production efficiency of an Upper Sonoran Desert ephemeral community. *American Journal of Botany*, **65**, 891–5.

Pechanec, J. F., Plummer, A. P., Robertson, J. H. & Hull, A. C. (1965). *Sagebrush control on rangelands*. USDA Handbook 277.

Perry, R. A. (1970). Arid shrublands and grasslands. In: *Australian Grasslands* (ed. R. M. Moore), pp. 246–59. Australian National University Press, Canberra.

Petrov, M. P. (1976). *Deserts of the world*. Wiley, New York.

Phillips, T. A. (1971). Information concerning sagebrush stand density and its effect on sage grouse habitat. *Range Improvement Notes*, **17 (1)**, 3–9. USDA Forest Service, Ogden, Utah.

Plummer, A. P., Christensen, D. R. & Monsen, S. B. (1968). *Restoring big game range in Utah*. Utah Division of Fish and Game Publication 68-3. Salt Lake City, Utah.

Plummer, A. P., Hull, A. C., Stewart, G. & Robertson, J. H. (1955). *Seeding rangelands in Utah, Nevada, Southern Idaho and Western Wyoming*. USDA Handbook 71.

Provenza, F. D. (1978). Getting the most out of blackbrush. *Utah Science*, **39**, 144–6.

Raikes, R. L. (1969). Formation of deserts of the Near East and North Africa: climatic, tectonic, biotic and human factors. In: *Arid lands in perspective* (ed. W. G. McGinnies & B. J. Goldman), pp. 145–54. University of Arizona Press, Tucson.

Reitan, C. H. & Green, C. R. (1968). Weather and climate of desert environments.

In: *Deserts of the world* (ed. W. G. McGinnies, B. J. Goldman & P. Paylore), pp. 19–92. University of Arizona Press, Tucson.

Reynolds, H. G. (1959). *Managing grass–shrub cattle ranges in the southwest*. USDA Forest Service Agriculture Handbook 1962. Washington, DC.

Robertson, J. H. & Cords, H. P. (1957). Survival of rabbitbrush, *Chrysothomnus* spp., following chemical, burning and mechanical treatments. *Journal of Range Management*, **10**, 83–9.

Robertson, J. H. & Pearse, K. C. (1945). Range reseeding and the closed community. *Northwest Science*, **19**, 58–66.

Rodin, L. E. & Bazilevich, N. I. (1967). *Production and mineral cycling in terrestrial vegetation*. Oliver and Boyd, London.

Sampson, A. W. (1913). *Range improvement by deferred and rotation grazing*. USDA Bulletin No. 34. Washington, DC.

Sampson, A. W. (1951). *Range management principles and practices*. John Wiley, New York.

Schulman, E. (1956). *Dendroclimatic changes in semi-arid America*. University of Arizona Press, Tucson.

Sharon, D. (1972). The spottiness of rainfall in a desert area. *Journal of Hydrology*, **17**, 161–75.

Sheppard, W. O. (1968). *Report to the Government of the Sudan on range and pasture management*. FAO, Rome.

Shiflet, T. & Heady, H. F. (1971). *Specialized grazing systems: their place in range management*. US Department of Agriculture Soil Conservation Service.

Simpson, B. B. (ed.) (1977). *Mesquite: its biology in two desert shrub ecosystems*. US/IBP Synthesis Series, No. 4. Dowden, Hutchinson & Ross, Stroudsburg, Pennsylvania.

Slatyer, R. O. (1973). Structure and function of Australian arid shrublands. In: *Arid-shrublands. Proceedings of the Third Workshop of the US/Australia Rangelands Panel*, (ed. D. N. Hyder), pp. 63–73. Society for Range Management, Denver.

Smith, J. G. (1899). *Grazing problems in the Southwest and how to meet them*. USDA Division of Agrostology Bulletin 16. Washington, DC.

Stoddart, L. A., Smith, A. D. & Box, T. W. (1975). *Range management* 3rd edn. McGraw-Hill, New York.

Szarek, S. R. & Woodhouse, R. M. (1976). Ecophysiological studies of Sonoran desert plants. I. Diurnal photosynthesis patterns of *Ambrosia deltoidea* and *Olneya tesota*. *Oecologia*, **26**, 225–34.

Szarek, S. R. & Woodhouse, R. M. (1977). Ecophysiological studies of Sonoran desert plants. II. Seasonal photosynthesis patterns and primary production of *Ambrosia deltoidea* and *Olneya tesota*. *Oecologia*, **28**, 365–75.

Szarek, S. R. & Woodhouse, R. M. (1978). Ecophysiological studies of Sonoran desert plants. III. The daily course of photosynthesis for *Acacia greggii* and *Cercidium microphyllum*. *Oecologia*, **35**, 285–94.

Tiedemann, A. R. & Klemmedson, J. O. (1973). Effect of mesquite on physical and chemical properties of the soil. *Journal of Range Management*, **Z6**, 121–5.

Trlica, M. J. & Cook, C. W. (1972). Carbohydrate reserves of crested wheatgrass and Russian wildrye as influenced by development and defoliation. *Journal of Range Management*, **25**, 430–5.

Turner, F. B. (ed.) (1975). Rock Valley validation site report. US/IBP Desert Biome Research Memorandum 75-2. Utah State University, Logan.

Valentine, J. R. (1971). *Range developments and improvements*. Brigham Young University Press, Provo, Utah.

Van Dyne, G. (1969). *The ecosystem concept in natural resource management*. Academic Press, New York.
Wagner, F. H. (1979). Integrating and control mechanisms in arid and semi-arid ecosystems – considerations for impact assessment. In: *Proceedings of a Symposium on Biological Evaluation of Environmental Impact*. 27th Annual AIBS Meeting, New Orleans. Council of Environmental Quality, Washington, DC, in press.
Wallen, C. C. & Gwynne, M. D. (1978). Drought – a challenge to range management. In *Proceedings of the first International Rangeland Congress*, ed. D. N. Hyder, pp. 21–31. Society for Range Management, Denver, Colorado.
Webb, W., Szarek, S., Laurenroth, W., Kinerson, R. & Smith, M. (1978). Primary productivity and water use in native forest, grassland, and desert ecosystems. *Ecology*, **59**, 1239–47.
Went, F. W. (1942). The dependence of certain annual plants on shrubs in southern California deserts. *Bulletin of the Torrey Botanical Club*, **69**, 100–14.
Went, F. W. (1949). Ecology of desert plants. II. The effects of rain and temperature on germination and growth. *Ecology*, **30**, 1–13.
West, N. E. & Gasto, J. (1978). Phenology of the aerial portions of shadscale and winterfat in Curlew Valley, Utah. *Journal of Range Management*, **31**, 43–5.
West, N. E. & Skujins, J. J. (1978). *Nitrogen in desert ecosystems*. US/IBP Synthesis Series, No. 9. Dowden, Hutchinson & Ross, Stroudsburg, Pennsylvania.
Williams, R. E., Allred, B. W., Denio, R. M. & Paulson, H. A. (1968). Conservation, development and use of the world's rangelands. *Journal of Range Management*, **21**, 355–60.
Wilson, J. & Leigh, J. H. (1964). Vegetation patterns on an unusual gilgai soil in New South Wales. *Journal of Ecology*, **52**, 379–89.

Manuscript received by the editors August 1979.

21. Management of arid-land resources for dryland and irrigated crops

H. S. MANN

Introduction

The Indian or Thar Desert is spread over four Indian states and part of Pakistan. It is a vast undulating plain of sand interspersed by hillocks. In the recent past, this desert has sustained well developed civilizations, particularly in the Ghaggar and Indus river basins. In spite of the fact that it suffers from paucity of water, erratic rainfall, extremes of temperatures, and is probably the least productive land in the Indian subcontinent, it is the most thickly populated desert in the world. It has an average density of 46 persons per square kilometre as against about three in most other desert tracts.

The obsession with water scarcity, perhaps somewhat exaggerated and imagined, has obscured and deterred the development of the Indian arid region, as it has in other countries. Because of limited leaching, arid lands are rich in most essential plant nutrients, except nitrogen. High radiation levels; favourable temperature for plant growth for most of the year; fewer pests and diseases because of low humidity; and timely farm operations with infrequent disruptions by rain are some of the important assets of the arid regions. A healthy climate and, therefore, sturdy humans and livestock are additional and unique advantages for the scientific development of agriculture and animal husbandry in the region. In this chapter the scientific information and technology for development and management of arid region resources are discussed with emphasis on dryland and irrigated crops.

Climate

The average annual rainfall of the region varies from 150–500 mm with a coefficient of variation as high as 60–70%. The distribution of rainfall is erratic occurring mostly in the period July to September. During summer, the mean daily maximum temperature is generally 40 °C and 22–28 °C during the winter. The highest temperatures usually range from 44–45 °C during summer and 30 °C during winter. But the data on the highest extremes indicate that occasionally, the day temperature may reach 48–50 °C in summer and 38 °C during winter. The mean minimum temperature varies from 24–26 °C during summer and from 4–10 °C during winter. Occasionally the temperature drops

Table 21.1. *Seasonal means*[a] *of evaporation at selected centres* (mm day^{-1})

Centre	Winter	Summer	Monsoon	Post-monsoon	Annual
Bikaner	3.2	11.6	8.3	5.4	2647.4
Jodhpur	3.9	12.5	7.0	4.0	2615.2
Ahmedabad	5.7	11.6	5.4	6.3	2731.8

[a] Over 70 years.

to −5 °C. The mean diurnal temperature variation ranges from 12–17 °C. The mean relative humidity during summer varies from 36–59% and from 66–78% during the morning in the monsoon season. In the afternoon, however, humidity drops and the mean values range from 20–35% during summer and 48–60% in the monsoon season. The mean evaporation during summer exceeds 10 mm day^{-1} (Table 21.1).

The potential evapotranspiration values during summer (April to June) vary from 7–9 mm day^{-1}. The mean daily wind speed is highest during summer and monsoon season, 8–20 km h^{-1}. Data on selected climatic parameters are included in Table 21.2.

Land

The Indian desert is an undulating vast plain of sand interspersed by hillocks of the Aravalli series. Vast gravel plains form a part of its topography.

The Thar Desert has been subjected to marine transgressions during Jurassic, Cretaceous and Eocene periods. Some geologists believe that the sea receded during Miocene and Pliocene. Archaeologists think that this desert has a rather recent origin but some geomorphologists and biologists have put forward evidence to prove that it is actually much older.

Land use patterns in the Indian arid zone in 1970–71 and between 1951 and 1971 are shown in Tables 21.3 and 21.4. Millet and pulses, which have relatively low water requirements, are sown soon after the first effective rainfall in early July. The crop is harvested shortly after the end of the rainy season and the land is left exposed to high velocity winds. This annual disturbance of the soil not only depletes it of soil-binding vegetation but increases soil erosion, particularly on the semi-stabilized sand dunes. In fact, in certain areas of the Indian arid zone, rain-fed crops should be substituted by perennial grasses. However, as a consequence of high population growth, the sub-marginal and marginal lands have been put to agricultural use. In 1951, the net sown area in the arid zone of Rajasthan was 26.1%: it increased to 39.0% in 1961, and to 42.2% in 1971. This 58.6% increase has caused a reduction of about 25% in the pasture and rangelands and has led to consequent fodder shortages.

Table 21.2. *Climatic features of the Indian arid zone (1931–60)*. Source: Dr A. Krishnan.

Region	Season	Maximum temperature (°C)		Minimum temperature (°C)		Mean relative humidity (%)		Total rainfall (mm)	Mean potential evapotranspiration	Mean wind speed
		Highest	Mean	Mean	Lowest	Morning	Afternoon			
Bhuj	Winter	40	28	11	1	51	22	9	3.2	7
	Summer	48	38	25	14	65	35	31	7.7	17
	Monsoon	43	33	25	18	18	60	299	5.2	17
Barmer	Winter	39	27	12	−2	51	30	10	2.9	7
	Summer	49	40	26	12	57	28	33	7.7	13
	Monsoon	44	30	25	17	78	53	260	5.4	11
Jodhpur	Winter	38	26	11	−2	47	24	14	2.9	8
	Summer	49	40	26	9	45	20	39	7.9	15
	Monsoon	46	35	25	18	77	54	315	5.3	13
Bikaner	Winter	38	24	8	−3	60	34	15	2.0	5
	Summer	47	40	25	11	36	20	39	7.3	10
	Monsoon	46	37	26	19	69	48	236	6.5	11
Hissar	Winter	34	24	7	−4	72	41	38	1.9	6
	Summer	48	40	24	8	46	28	51	6.8	9
	Monsoon	47	36	26	16	73	57	317	5.6	8

Table 21.3. *Land-use patterns in the Indian arid zone (1970–71)*

Land-use category	Area (ha x 10^{-3})	Proportion (%)
Forests	338	1.28
Land not available for cultivation	6489	24.65
Sand dunes and culturable waste-lands (grasslands and pastures)	5295	20.12
Fallow land	3121	11.86
Net area sown	11083	42.09
		100.0
Total cropped area	12072	45.86
Area sown more than once	989	3.76
Total geographical area	26326	

Table 21.4. *Land-use pattern in arid zone of Rajasthan (excluding Ganganagar).* Source: Rajasthan Statistical Abstract, Jaipur

	Percentage of total geographical area of Western Rajasthan			Change (%)		
Land use	1951–52	1961–62	1970–71	1951–61	1961–71	1951–71
Forests	0.85	0.95	1.09	+12.13	+14.35	+28.22
Land put to non-agricultural use	1.99	2.79	3.02	+41.13	+8.36	+52.92
Barren waste permanent pasture+culturable+current fallow	71.03	57.26	53.67	−18.89	−6.18	−23.90
Net area sown	26.12	39.02	42.23	50.31	+8.34	+62.85
Area sown more than once	0.32	0.77	1.31	141.45	+70.03	+312.50

Human population

In most of the arid regions, the productivity of the land is rather poor, mainly due to a shortage of water, exteme temperatures and low soil fertility. In the Indian arid zone, 70% of the area is classified under Class IV–VI land types indicating a deteriorated condition. To compound the situation, the Indian desert is one of the most thickly populated in the world. The human population in the desert appears to be more prolific than that of the adjoining areas. During the period 1901–71, the growth rate of population in the arid zone of Rajasthan has been about 158% as compared with 125% increase in

Table 21.5. *Decinnial variation in population in Rajasthan Arid Zone, Rajasthan and India (1901–71)*. Source: Shri S. P. Malhotra

	Arid zone of Rajasthan (11 districts)		Rajasthan		India	
Year	Population ($\times 10^{-6}$	Growth rate (%)	Population ($\times 10^{-6}$)	Growth rate (%)	Population ($\times 10^{-6}$)	Growth rate (%)
1901	3.42	—	10.29	236.28	236.28	—
1911	3.67	+7.23	10.98	+6.70	252.12	+5.42
1921	3.41	−7.27	10.29	+6.29	231.35	−0.303
1931	3.94	+15.67	11.75	+14.14	279.02	+11.00
1941	4.74	+20.40	13.86	+18.01	318.70	+14.22
1951	5.53	+16.58	15.97	+15.20	361.13	+13.31
1961	6.95	+25.87	20.16	+26.20	439.24	+21.64
1971	8.84	+27.22	25.77	+27.83	547.37	+24.66
Overall increase	+5.42	+158.22	+15.47	+150.47	+311.08	+131.65

its five bordering administrative districts (Table 21.5). Within the desert, the percentage increase exhibits an inverse relationship with the amount of annual precipitation. In the regions receiving an annual rainfall of 300 mm, between 300 and 400 mm, and more than 400 mm; the percentage increase in population between 1931 and 1971 has been 143.8, 132.1 and 107.2, respectively. The daily needs of such a large population are the main cause of desertification. The fuel needs of desert-dwellers are never-ending; the requirement for housing material and thorny branches for protecting crops from wandering livestock, take such a large toll of the desert shrubs and trees that it is often difficult to find them close to the well scattered villages.

Livestock

In spite of the low productivity of the arid lands the Rajasthan Desert sustains a fairly high population of livestock. Paradoxically, along with the reduction in the grazing area during the last two decades, the livestock population has registered an alarming increase, from 9.4 million in 1951 to 18.1 million in 1971. The livestock density per 100 hectare rose from 72 in 1951 to 175 in 1971 in the desert districts (Table 21.6), but in the adjoining districts, the increase was not so spectacular (25% as against 293% in desert districts).

Besides livestock, wild animals, particularly rodents consume significant quantities of natural vegetation and they also damage crops and trees. The dietary demand of the rodent population, which fluctuates from 7.4 to 523 ha^{-1}, is so insatiable that it maintains an appreciable pressure on the already sparse vegetation and rangelands. The annual total food requirement of the

Table 21.6. *Livestock population* $\times 10^{-6}$ *and growth rate (%) in Rajasthan and arid districts from 1956 to 1971*

Animal	1956 Rajasthan	1956 Arid districts	1971 Rajasthan	1971 Arid districts
Cattle	12.1	3.8	12.5 (3.3)	5.8 (52.6)
Buffaloes	3.4	0.7	4.6 (35.3)	0.9 (28.6)
Sheep	7.4	4.6	8.6 (16.2)	5.1 (10.9)
Goat	8.7	3.6	12.2 (40.2)	5.7 (58.3)
Camel	0.4	0.3	0.7 (75.0)	0.5 (66.7)
Total	32.0	13.0	38.6 (20.6)	18.0 (38.5)
Density ($ha^{-1} \times 10^2$ grazing land)				
	160	104	238 (49)	175 (68)

rodent population (477 ha^{-1}) in the north-eastern Indian Desert during monsoon was found to be 1044 kg ha^{-1}. The annual production of edible (for sheep) grass species was estimated to average 865 kg ha^{-1} and the total forage production to be about 1100 kg ha^{-1}. These data reveal that in a rodent infested grassland, the rodents can consume the entire production of edible fodder, leaving virtually nothing for the livestock. Rodent activities expose the soil to the action of weathering agents and increase the erosion hazard. Rodent species exhibit a habitat specificity and are fairly abundant in the rocky, sandy habitats and in the rural environment, inclusive of the cropped fields.

Management technology

Multi-disciplinary integrated surveys

Detailed information and accurate data of the land resources is essential for scientific planning. Multi-disciplinary integrated surveys – using modern techniques like aerial photographs and field studies to assess soil, vegetation land capability, water resources, catchment characteristics, settlement patterns, etc. – have provided guidelines for the land to be used for dry-land crop production. Marginal lands now used for cropping have been identified and recommended for reversion to grassland and forest. A technique called composite mapping unit, namely the Major Land Resource Unit has been evolved for rational integrated planning.

To help further planning and management of these areas, the results of scientific survey are included in the Agricultural Atlas of Rajasthan (34 plates on $1:2 \times 10^6$ scale) published by the Indian Council of Agricultural Research. This atlas incorporates most of the information on the physical, biological and socio-economic environment. A geohydrological Atlas of the Central Luni basin ($1:5 \times 10^6$ scale) and a series of maps indicating ground water resources of Rajasthan ($1:2 \times 10^6$) have been prepared to depict ground water resources.

Sandy land, dunes and their production potential

Out of the total of 342000 km² in Rajasthan, about 214600 km² are sandy – mostly sand dunes concentrated in western Rajasthan. They occupy about 58% of the total area in the western Rajasthan. Some of the area is put under crops in favourable years. Studies on the dynamics of these sand dunes have indicated that they are of local origin and, if the source is controlled, the sand of western Rajasthan can be reclaimed. Properly managed, such areas could be more productive, particularly for grasses and shrubs.

The dunes are of five types: parabolic, longitudinal, transverse, barchan and shrub-coppice. In certain areas (Saila, Siwana, Balotra and Jalore) the longitudinal and transverse dunes are more predominant than parabolic, while coalesced parabolic dunes predominate in Bikaner (Kolayat and Lunkaransar tehsil), Barmer (Chohtan tehsil), Jodhpur (Shergarh and Phalodi tehsil) and north-western part of Jaisalmer district. These dunes, except transverse dunes, have been formed by the south-west prevailing winds. The transverse dunes occur tranverse to the prevailing winds. Studies have indicated that the dunes are of two systems – new and old.

The new system of dunes has formed mostly through faulty management of marginal lands which should be reverted to grasslands. At present, millets and low yielding pulses are cultivated. The soils are structureless with 3–4% moisture retention capacity. About 75% of the sand grains are 0.06–0.12 mm in diameter. Mean content of organic carbon is 0.09%, available phosphorus 5.0 kg ha^{-1} and potassium 170 kg ha^{-1}. Unstabilized dunes have considerable soil moisture at depths below 1 m even during the hot and dry summer months. More than 50% of the nitrogen is readily available as nitrates.

Techniques for stabilizing shifting dunes consist of (*a*) protection against all biotic interference; (*b*) effective micro-wind-breaks on the windward side of dune either in 5-m parallel strips or 5-m square chess board patterns; and (*c*) sowing grasses or transplanting trees and shrub species raised in sun dried earthen bricks on the leeward of the micro-windbreaks. Suitable species for afforestation of the shifting dunes are: trees such as *Prosopis cineraria, Prosopis juliflora, Acacia tortilis, Acacia senegal, Albizzia lebbek*; Shrubs such as *Calligonum polygonoides, Lycium barbarum, Acacia jacquemontii, Clerodendrum phlomoides* and *Zizyphus nummularia;* and grasses such as *Saccharum bengalensis, Lasiurus sindicus, Panicum antidotale, Panicum turgidum, Cenchrus ciliaris* and *Cenchrus setigerus*. In extremely arid conditions,

Acacia tortilis, Prosopis juliflora, Calligonum polygonoides and *Acacia jacquemontii* have been found suitable. An economic analysis of the dune stabilization technology indicated that the average annual cost of 760 rupees* ha^{-1} will be repaid after the 13th year.

Geohydrology of the desert

There is a general imbalance in the overall hydrological cycle of the Indian arid region. Along the ephemeral streams of the desert, a large number of wells have been dug and there is considerable scope for increasing their number. For better planning of ground-water exploitation or exploration, the areas surveyed are classified and mapped into over-exploited, sparsely-exploited, under-exploited and unexploited zones. Alluvial and blown sands are found to be the two most important potential water bearing geohydrological units in surveyed areas. Average ground-water reserves in the surveyed areas range from approximately 5000000 Ml–10000000 Ml of which 0.002–2.0% are utilized for irrigation and 0.05–0.10% for human and livestock consumption. It has been observed that stream flow affects both the water quality and water availability of the wells up to 1 km from the banks. About 65% of the surveyed area contains brackish ground waters. Total dissolved salts in ground waters ranged from 180 to 7000 p.p.m. Sodium hazard is medium to high. Methods are being developed to predict the salinity build-up in various soils with various types of available waters.

Arid agriculture (crops)

The foregoing discussion suspects that the arid region of India is an environment most suited to a silva-pastoral system, though crop production is widely practised for producing food grains for human consumption. It can be easily appreciated that for a community living near the subsistence level, crop production would be practised even though it leads to an eventual deterioration of the resources, particularly the soil. It is, therefore, important to determine the basic available water resources that can be put to most productive use in the arid region and also to develop technology for effective utilization of available water. Scientific studies have been in progress regarding the most efficient culture and irrigation methods to use in arid environments, and the adaptation of crops and other plant varieties including the best methods of using saline water whilst maintaining soil productivity.

Analysis of crop production conditions on dryland in arid regions reveals that an acute ecological imbalance of the various components of productivity is the sole, complex factor responsible for limiting the consistency of remunerative crop production.

* 8.3 Rupees are approximately equivalent to US $1.0.

The problems imposed by a low annual precipitation (366 mm) are accentuated by its erratic distributions from season to season, a high solar incidence of 450–500 cals cm^{-2} day^{-1}, and wind velocities of 10–20 km h^{-1} which result in a high potential evapotranspiration (6 mm day^{-1}) and a consequent high mean aridity index of 78%. The soils are sandy with low organic matter content (0.1%–0.45%) and have a very poor moisture-holding capacity (25–28%) and a high infiltration rate of 9 cm h^{-1}. Soil salinity and alkalinity, which extends to 45% of the unirrigated area, accentuates the situation.

Although research on dryland farming was initiated about 30 years ago, and certain useful practices for moisture and soil conservation evolved, no systematic attempt was made to initiate integrated research aimed at efficient management of soil, moisture and plant factors. The individual facts and practices developed by scientists working in different disciplines were not widely adopted because of their marginal impact on productivity and income. It is only when factors such as better tillage, improved soil conservation, water harvesting and storage, better varieties and better nutrition are made to interact that large yield and income increases per unit area are likely.

The conventional unirrigated crops, namely bajra (*Pennisetum typhoideum*), til (Sesame), kharif pulses, particularly moong (*Phaseolus radiatus*) and moth (*Phaseolus acontifolius*) are adapted to the existing conditions of extreme moisture stress. The average yield of bajra, jowar, moong and til are 263, 296, 168 and 99 kg ha^{-1}, respectively. It is only after the development of appropriate crop varieties both with high yield potential and the ability to complete their life cycle during the periods of moisture availability, that a firm base for remunerative dryland agriculture can be formulated. In other words, it is important that the flowering and grain development stages of crops should coincide with the periods of adequate soil moisture availability, which fall in the third or fourth week of August.

By adopting a flexible technology, it would be possible to maximize crop production in seasons of good rainfall and stabilize it under conditions of drought. This would be achieved by inter-cropping of appropriate crops coupled with water harvesting technology, inter-cropping of perennial forage grasses and annual grade legumes, and applying suitable mid-season corrections. Also, methods for improving soil fertility will have to be developed and the most efficient rates and methods of fertilizer application for principal dryland crops will need to be determined.

A suitable technology is required for increasing water intake rates in some problem soils, enhancing the moisture storage capacity and ultimately increasing the efficiency of stored moisture for crop production. Tillage methods, suitable for seed-bed preparation, moisture conservation and reduction of soil erosion under dryland conditions will have to be developed together with suitable implements and machinery. In order to overcome the challenging problem of soil crust formation and poor crop establishment, a suitable combination of agronomic and engineering techniques will have to evolve.

In view of the limited soil moisture availability for crop production, it is essential to develop suitable water-harvesting techniques for improving the moisture regime of soil and to provide for run-off re-cycling for supplementary irrigation. This approach enables the dryland farmers to harvest a crop even during periods of extreme drought; one irrigation from the harvested water would save the crop under such conditions.

Strategies to diversify crop production would involve unconventional crops, such as sunflower and castor as well as improved varieties of traditional crops, combined with the technology for water-harvesting, storage, moisture conservation and fertiliser use. Tailored to the prevailing variations in the weather, these would go a long way to provide an ecologic-economic basis for scientific management of the resources.

Water harvesting and run-off recycling

About 60% of the Indian arid zone is in western Rajasthan which receives scanty and erratic rainfall. Water is thus the main limiting factor for crop production in these areas. Here, water harvesting, holds promise. The Indian desert has numerous small and large local catchments scattered over areas with rocky and gravelly terrain. These form ideal foci for rainfall run-off storage and re-cycling.

Analysis of annual rainfall data at Jodhpur from 1901 to 1973 allows designation of four main rainfall distribution models representing: good (400 mm), normal (350–400 mm), average (250–350 mm), and below average (< 250 mm). Moisture losses as evapotranspiration, run-off and deep percolation have been computed with respect to each model. Run-off and deep percolation losses *in situ* (representing run-off potential) are estimated as 40, 35, 25 and 10% of the total rainfall received during good, normal, average and below-average rainfall years, respectively. The availability of water for each situation represented by these different models is shown in Table 21.7. The distribution of *kharif* rainfall to the total annual rainfall ranges between 84 and 90%. After accounting for run-off losses (10–40%), water availability for crop production ranges from 173 mm in an average rainfall year to 283 mm in a good rainfall year.

Interesting studies to establish a relationship between the size and the shape of micro-catchment and the yield of run-off were initiated during the monsoon of 1975. Run-off volume obtained from different catchments on an individual rainfall day (4 September, 1975) indicated that maximum run-off volume (33.7 mm) and relative amount (49%) was obtained from a rectangular catchment

Table 21.7. *Availability of water for crop production in the various rainfall distribution patterns.* Source: Dr R. P. Singh

Model description	Period (yr)	Annual rainfall (mm)	Kharif rainfall (mm)	Kharif rainfall as percentage of annual total	Run-off and deep percolation losses (mm)	Ratio percentage run-off: kharif rainfall	Availability of water (mm)
Good rainfall distribution (> 400 mm)	23	564.9	472.5	83.6	189.0	40	283.5
Normal rainfall distribution (350–400 mm)	12	366.0	329.1	89.9	115.2	35	213.9
Average rainfall distribution (250–350 mm)	17	292.8	229.6	78.4	57.4	25	172.7
Below-average rainfall distribution (< 250 mm)	21	275.5	138.0	78.6	13.8	10	124.2

Table 21.8. *Potential yields in good rainfall years*

Crop	Variety	Yield (kg ha^{-1})
Moong	Varieties M8 and S8	1400–1800
Cowpea	Varieties FS68 and K11	1500–1800
Guar	Varieties 2470 (12) and FS277	1700–2000
Sunflower	Varieties EC68414	1400–1600
Castor	Varieties Aruna, GH3 and Bhagaya (R63)	1500–2000

measuring 50 × 10 m. Further investigations are in progress to maximize the efficiency of the catchments and to develop efficient methods of storage and re-cycling of the harvested water.

Crop varieties for drylands

The life-cycle of crops must be matched to the rainfall pattern of the region. Suitable varieties of dryland crops like moong, cowpeas, guar, til, sunflower and castor with high yield potential and tolerance to drought conditions have been evolved. For instance, in good rainfall years, such as 1972 and 1973 the yields shown in Table 21.8 could be obtained provided an appropriate crop management technology was adopted.

Drought evasion

An alternative cropping strategy has been developed for when the onset of monsoon is delayed. It involves cultivation of grain legumes like moong, oilseeds like sunflower and castor instead of the conventional crops like bajra and til. Agronomic practices such as: deep sowing with a minimum of soil cover; planting of grain legumes and oilseed crops in a paired row system; split application of fertilizer at moderate levels with placement at sowing time; use of surface mulches to reduce excessive evaporation from the soil surface; and transplanting of 20–25-day old seedlings of bajra are some of the measures found to alleviate the effects of drought and stabilize crop production on drylands.

Increasing total productivity on drylands

In a good rainfall year the cropping intensity on drylands could be increased from the normal 50–100% to 200% by growing fodder crops like bajra or jowar or guar, followed by crops like cowpeas (FS68), moong (S8), sunflower (EC 68414), or castor (Aruna) which are suited to late sown conditions.

Besides increasing the cropping intensity, it is possible to obtain high productivity per unit area, per unit time and per unit of moisture used by planting cowpeas and moong as an inter-crop with sunflowers fertilized with 60 kg nitrogen^{-1} ha.

Inter-cropping of grain legumes in perennial grasses

A system of inter-cropping annual grain legumes like guar (*Cyamopsis psoraliodes*), moth (in good rainfall years), moong, and cowpeas (in average rainfall years) in established stands of perennial forage grasses such as *Cenchrus* spp. has been found promising. Moong varieties, 288-8, RS4, and T44 have been found suitable for inter-cropping with *Cenchrus* sp. The inter-cropping system, besides providing insurance against drought, ensures 20% higher productivity in terms of green forage yield if varieties like RS4 and T44 are inter-cropped with *Cenchrus setigerus*.

Fertilizer use and organic recycling

Optimum rates of nitrogen fertilizer, applied at the right time and in the right way result in higher yields and higher moisture use efficiency, particularly in the case of cereals and fodder crops. Irrespective of amount and distribution of rainfall, it is profitable to apply 20 kg nitrogen ha^{-1} to fodder and forage crops. Application of fertilizers to grain legumes and oilseeds, such as til, in years of low and erratic rainfall distribution, is unprofitable. Simple moisture conservation measures, such as the application of bajra husk at 4–6 t ha^{-1} increased the yield of bajra by 25% in drought years.

Because arid zone soils are deficient in organic matter, there is a need to devise an appropriate organic re-cycling technology that involves the use of the crop residues and organic wastes. An integrated crop nutrient management technology has to be perfected, embracing cultivation of legumes for nitrogen fixation and perennial forage grasses, such as *Cenchrus* sp. and *Lasiurus* sp. which contribute substantially to the soil organic matter. In this context, enhancing the efficiency of legumes for nitrogen fixation, by using suitable rhizobial strains and developing suitable plant types, assumes importance.

Crop production with limited irrigation

Because of the unique advantages available in arid regions for crop production, and social considerations there has been substantial investment in major and minor irrigation projects. The Ganga Canal built in 1920s in Rajasthan has made Sriganganagar (a district of Rajasthan) the 'granary of the State'. The Rajasthan Canal project which is more than half completed will have considerable influence on the arid-land productivity of western Rajasthan.

When completed, it will actually irrigate over a million hectares with a command area of 32 million hectares. The nature of the scientific problems and the management technology for optimum use of irrigation water for crop production involves an approach of a somewhat different nature and so far little work has been done on this aspect. Considerable work has, however, been done on optimal utilization of limited water supplies from tubewells, dugwells and ponds.

Trials with different crops have indicated that three irrigations for barley, two irrigations for raya and one irrigation for sunflower gave profitable yields in areas like Pali (420 mm rainfall). Improved irrigation methods such as sprinkler and drip (for vegetable crops) have been standardized for the optimum use of the limited water resource from the wells, dugwells and ponds.

Applying equal amounts of water via trickle irrigation to long-gourd, round-gourd, water-melon and potato, the yields were 55.8, 40.6, 82.3 and 30.4 t ha^{-1} respectively. Corresponding yields (excluding potato) were 38.6, 33.6, and 74.6 t ha^{-1} when these crops were irrigated by sprinkler, and 38.0, 29.5, 67.2 and 19.2 ha^{-1} by the furrow method. Water use efficiency was highest for trickle irrigation (1 kg of vegetable for 97, 105, 133 and 197 kg of water for water-melon, potato, long-gourd and round-gourd, respectively). The average amount of water required for irrigation by the sprinkler/furrow method was 113, 182, 219 and 255 kg for each kilogram of crop.

Trickle irrigation ensures about a 50% saving in water. In the potato crop, 160 mm of water via trickle irrigation produced 17.8 ha^{-1}, which was the same as obtained from 350 mm of irrigation (19.2 t ha^{-1}) in furrows. Thus 190 mm of water was saved by using trickle rather than furrow irrigation without adversely affecting production.

Further, indications were that trickle irrigation could be used for saline water, which would reduce production if applied by conventional irrigation. Even potato crops, known to be sensitive to saline water of 10000 mmhos cm^{-1}, gave yields of 14.4 t ha^{-1} under trickle irrigation, and the water use efficiency was equal to furrow irrigation with non-saline water. Contrary to popular belief, the plant root system under trickle irrigation adjusted to steady dripping and salinity accumulation.

Initial cost of trickle irrigation is 8000–12000 rupees ha^{-1}, which is rather high for farmers with limited resources. Therefore, a non-monetary crop management strategy has been developed. It has been found that 25-cm square planting of turnip and tomato, or 25-cm equilateral planting of cabbage reduced initial cost by 50%, in addition there was a 50% saving in water. Efforts are underway to bring down the cost by 75%. Extra returns will pay off initial cost in a season or two.

High yields, water economy and the possibility of using saline water give trickle irrigation definite potential, particularly for high value crops.

Salinity problem

Excess of salts in soil and ground water is a serious problem in the arid zone. Studies have shown that saline water up to 15 mmhos cm^{-1} can be used for irrigating wheat, (var Kharchia) without any impediment to nutrient uptake and the system is responsive to improved management conditions. The relative merits of wheat varieties, the conditions of fallow and input norms have been worked out for areas having saline ground water available for irrigation. Bajra collected from different areas has also helped to identify more tolerant strains.

The scientific work briefly reviewed here is indicative of the studies in progress. Potential for improving productivity of the land and available water resources has been indicated. Some important aspects like the effect of technical and management inputs on soil productivity are not adequately covered. Soil parameters, physical, chemical and other related aspects have undoubtedly been studied, but the data are still inadequate to permit fuller analysis or to reach even tentative conclusions. Work in this direction is in progress particularly with reference to the UNESCO's 'Man and the Biosphere' programme.

Acknowledgements

The work of a number of scientists at the Central Arid Zone Research Institute has been reviewed and is gratefully acknowledged.

The data on climatic factors was provided by Dr A. Krishnan, Head of Division of Wind Power and Solar Energy Utilization. Information and data on sociological aspects was provided by Shri S. P. Malhotra, Head of Division of Economics & Extension.

Dr R. P. Singh, Chief Scientist very kindly supplied information and data on dryland agriculture.

Manuscript received by the editors December 1975.

22. Recreation and tourism in arid lands

M. D. SUTTON

With all the popularity of the desert today as a habitat and haven for human beings, it is a little difficult to realize how recently the biome was regarded as dangerous, unwanted wasteland. It was scarcely considered part of the national heritage; there was something patriotic about 'making the desert bloom' with dams, cities, roads, and crops.

In many parts of the world, however, air conditioning has made travel and human habitation in deserts more comforable. Networks of highways and air traffic systems have made arid regions accessible. Accordingly, the natural regimen has become threatened by overpopulations of both migratory and resident human beings. As to what constitutes 'overpopulation', many factors come into play, but it should be kept in mind that the fragility of desert ecosystems and their slow recovery from damage renders them especially vulnerable to heavy use. Hence, a state of overpopulation may be reached with the entry of relatively few people.

Among the first responses to threatened desert habitats was the establishment of protected areas. Some of the larger and more significant desert parks or refuges are shown in Table 22.1. Of the approximately 300 units of the United States national park system, the state with the greatest number (24) is Arizona. The number of park units in the arid southwestern desert states far exceeds that in any other geographical region. This is not a mere accident, for the desert environment has shaped the scenery into unusual forms, guided the evolution of specially adapted forms of flora and fauna, and attracted prehistoric human immigrants whose ruined dwellings have been remarkably well preserved in the dry desert air.

The intended functions of such reserves included: (*a*) preservation of unique segments of the national heritage; (*b*) protection of flora and fauna; (*c*) provision of opportunities for outdoor recreation, scientific research, and public education; and (*d*) increase of economic benefits, including tourism.

The latter point has been a particular issue because the inherent values of desert biomes, mainly scenery and climate, are well suited to the establishment of tourism programs. It is little wonder that tourists come to the desert. They find pockets of air still fairly clear and clean, air with a magnification quality that enables them to see objects more than a hundred miles away. Sunshine is abundant. At Yuma, Arizona, for example, the average number of days each year with a cloud cover of three-tenths or less is approximately 275. In parts of Chilean and Australian deserts the figure may be even higher.

Table 22.1. *Some desert reserves in the United States*

Name and location	Area (ha)	Date of authorization or establishment	Number of visitors in 1973
Anza-Borrego Desert State Park, California	200000	1932	522666
Arches National Park, Utah	29356	1929	276000
Badlands National Monument, South Dakota	97321	1929	1399900
Big Bend National Park, Texas	283635	1935	341300
Cabeza Prieta National Wildlife Range, Arizona	344000	1939	800
Canyonlands National Park, Utah	135028	1964	62600
Capitol Reef National Park, Utah	96748	1937	311200
Carlsbad Caverns National Park, New Mexico	18702	1923	840100
Colorado National Monument, Colorado	7068	1911	552900
Death Valley National Monument, California/Nevada	827117	1933	606500
Desert National Wildlife Range, Nevada	635352	1936	20815
Dinosaur National Monument, Utah/Colorado	84420	1915	412700
Glen Canyon National Recreation area, Utah/Arizona	494752	1958	1209100
Grand Canyon National Park and National Monument, Arizona	350590	1908	2079000
Great Sand Dunes National Monument, Colorado	14667	1932	53200
Joshua Tree National Monument, California	224000	1936	592900
Lake Mead National Recreation Area, Arizona/Nevada	725342	1936	5534300
Organ Pipe Cactus National Monument, Arizona	132276	1937	89400
Petrified Forest National Park, Arizona	37676	1906	1072000
Saguaro National Monument, Arizona	31595	1933	390800
White Sands National Monument, New Mexico	58134	1933	663200

The weather is reliable, and visitors may reasonably expect to enjoy sunshine and clear skies during most of their stay. This makes the desert a retreat from less clement and predictable weather closer to the Poles, and provides an opportunity for outdoor recreation free of such accoutrements as parkas and snowmobiles. That, in turn, stimulates outdoor life–hiking treks (even Death Valley has become a challenge to endurance and competition walkers), collecting, photography, painting – amid plentiful sunshine and fresh air.

The consequent benefits to human health have been widely demonstrated and advertised, luring retirees and tourists in such numbers that agricultural and industrial pursuits have been overtaken in certain localities. Agriculturalists now have some pessimistic tinges in their outlook for agriculture in arid lands. They see both crops and grazing lands giving way to recreation or to land and wildlife preservation. As of 1975, under provisions of the United States Wilderness Act, more than five million hectares in all biomes have been removed from commerical exploitation. The setting aside of many more millions of hectares is pending in the national legislature. Moreover, the states have established parks, wildlife sanctuaries, and wilderness areas on their

Fig. 22.1. Saguaro cactus forest west of Tucson, Arizona, Saguaro National Monument.

own. One of the largest state parks in the United States is Anza-Borrego Desert State Park in southern California, measuring 200000 ha.

Comprehensive plans for regional development are taking into account more and more the needs for wildlife and scenic preservation. Examples are: Canaima National Park, Venezuela, the northern arid coastal parks of Colombia, the Argentine Patagonia, Galápagos Islands, Cordilleran volcanoes

Fig. 22.2. Monument Valley, Arizona.

of Chile, Bolivia, Ecuador, Panama, Costa Rica, and Guatemala – just to cite a few in the Western Hemisphere.

Another factor that favors recreational use of arid lands is the costliness of water desalinization and/or transport for residential, industrial, and agricultural purposes, providing one assumes that controlled tourist numbers would consume less water.

People look upon desert wild places as sites of less congestion than in the cities, another impetus for desert travel. Where the wild natural values remain intact, tourists refer to them as 'bizarre,' 'strange,' 'unreal.' Amateur botanists, ornithologists and other devotees of biological resources exert a real, if well-meaning, pressure on the land.

Tourists pay eager attention to traces of aboriginal cultures, such as outposts, ruined fortresses, cliff dwellings, and ancient irrigation systems. In Australia and the southwestern United States, for example, the modern descendents of aboriginal peoples sustain themselves in ways captivating to travelers on tourist routes. Obviously, the differing cultural aspects plus development of arts and crafts, are two major factors in the desert's fundamental appeal. According to Perry, however, (cited in Dregne, 1970) Australia's vast desert recreation potential remains virtually untapped. He

Fig. 22.3. Cactus forest Galapagos National Park, Ecuador.

Fig. 22.4. Desert forest, Santa Rosa National Park, Costa Rica.

explains this by citing the coastal concentration of human settlements which leads to sea-oriented recreation, and the long distances one must travel to reach desert recreation sites. Notwithstanding that view, the last five years have seen remarkable organization of park systems and management programs in such areas as the Northern Territory.

In the United States many elements that make the desert compelling receive interpretation in field museums, self-guiding trails and roads, and attractive books or booklets, which nurture an even more profound and undoubtedly wider interest in the region. One of the most effective and successful examples of this is the Arizona-Sonora Desert Museum, west of Tucson, Arizona, a focus of education dealing mostly with natural systems. Opened in 1952, the Museum is largely self-supporting. Its exhibits, many with live animals, include insects, reptiles, amphibians, fishes, birds, mammals, minerals, weather, soil, water, plants, marine life and geology. Of special interest are underwater and underground exhibits of living native animals in simulated natural habitats.

Progress in desert preserves

Although an economist may think of tourism as the major justification for establishing desert preserves (national and state parks, wildlife refuges, and recreation areas), the prevailing international philosophy is quite different. These preserves are set up to protect significant examples of a nation's natural heritage, to maintain at least remnants of the land's original identity, and from this act flow benefits, such as tourism.

The world's outstanding desert preserves have been officially set aside only within this century (Table 22.1). Their popularity, especially in the United States, has grown to unprecedented proportions, as the figures for 1973 visits show (Table 22.1).

The number of desert reserves established, or under consideration, in other countries is rising sharply. One commendable instance is the park and reserve system in the Patagonian province of Chubut, Argentina. It is unusual for being one of the few, if not the sole, park and refuge system administered by a government tourism agency. Under the direction of Antonio Torrejón, Director General of Tourism, the system has expanded to incorporate arid lands containing scenic red-rock valleys, petrified wood, and such life forms as Magellanic penguins, rheas, and guanacos.

Torrejón gained part of his expertise by studying in the United States and Canada. Returning to Chubut, he accomplished in less than two years some enormous tasks for which other persons might take a decade. He identified the principal reserves, got them set aside, hired rangers, initiated research programs, built museums, and inaugurated a training course – the annual Spanish-language Seminario Internacional Sobre Areas Naturales y Turismo (International Seminar on Natural Areas and Tourism).

On principle, a tourism agency would not ordinarily be deemed competent to administer such a comprehensive program, largely because tourism agencies do not employ public land-management experts. But Torrejón noted the decline of wildlife species for which Chubut was famous, and had little difficulty predicting what would happen to the provincial tourist industry if the fauna disappeared. He seized the initiative, employed the experts, and inaugurated a program of wildlife protection and public information. Within a short time the number of sea elephants on the arid coast of the Valdéz Peninsula multiplied from 1500 to 4000 and spread out along 40 km of shoreline. Sea lion colonies on the Golfo Nuevo similarly grew in population, from 50 to 300.

Torrejón may relinquish the program when a provincial organization is developed to administer the natural reserves of Chubut. In the meantime, he and his colleagues have recognized that tourism can save the wildlife and scenery, and that those assets are the life blood of an exhilarating tourism program for Patagonia.

Some of the best known deserts of the world lie in northern Chile, and are celebrated for their extreme dryness. The Chilean government, as part of a nationwide concentration on national park development, has established five parks north of Santiago: Fray Jorge, Talinay, Punta del Viento, Valle del Encanto, and Pichasca. These total about 10000 ha of desert landscape.

The Mexican government has also given recent emphasis to its national park program. Some of the first priorities, of course, include the semiarid uplands around Mexico City, Monterrey, and other population centers. But the more remote deserts are by no means being totally neglected. The government in 1975 set aside part of the Gran Desierto de Pinacate as a natural park – preserving some remarkable volcanic and biologic phenomena of the Sonoran Desert. This new area touches on the already established Organ Pipe Cactus National Monument and Cabeza Prieta National Wildlife Range. Altogether, this represents one of the world's largest desert complexes preserved for recreation and conservation purposes. The two United States areas alone comprise 476350 ha.

Mexico is also undertaking studies in the Sierra del Carmen across the Rio Grande from Big Bend National Park, Texas. This desert upland is being threatened by increasing human exploitation in a region newly opened by highways.

New highways are bringing visitors by the thousands into the fragile desert environments of the peninsula of Baja California. Such an invasion may seem very progressive to tourist developers, but responsible tourism and environmental preservation officials have misgivings about the changes that will surely be wrought on the unique Baja California landscape. Already, the beaches are becoming cluttered with ugly assortments of trailers, shacks, stands and other contributions to the tourist industry. So far, these adverse

Fig. 22.5. Big Bend National Park, Texas. Sierra del Carmen and Boquillas at dusk from hill near Boquillas. Photo: United States Department of the Interior, National Park Service photograph, by Roland H. Wauer.

developments remain near the major roads. But this type of use leads to pressures for more access roads, and offers an environment from which off-road vehicles can be launched into the desert wilderness. One can well imagine the dangers to vegetation and wild animals. Efforts are under way to ameliorate the situation and plan for wise use, but early reports from the desert suggest that the planning cannot be accomplished too soon.

The Canary Islands of Spain, much closer to Africa than to the Iberian Peninsula, exhibit many characteristics of North African deserts. Spanish officials of ICONA (the government agency for conservation of nature) have given particular attention to the preservation of unique endemic flora, although burgeoning urbanization makes that difficult. The volcanic island of Lanzarote is particularly arid, yet remains one of the most charming isles of the archipelago. At the time of writing, Timanfaya National Park is in the process of establishment, a procedure that would protect colorful lava landscapes, vents from which issue high-temperature air drafts, vast cinder fields, and 'gardens' of white lichens that impart to the terrain a frosted look. Tourism is already established in this area, and camel caravans take riders

Fig. 22.6. Santa Elena Canyon, Rio Grande (Rio Bravo). Big Bend National Park, Texas (right), proposed Sierra del Carmen National Park, Mexico (left).

up some of the cinder cones for panoramic views of the landscape. These trips are of short duration, perhaps an hour or so, but the effect on tourists is a lasting one, for they have utilized a local, historic form of transporation, and sampled the true Canary landscape. They also provided jobs for local residents.

Lanzarote officials have prepared land-use plans for the entire island, specifically setting aside about 25% of the land as natural reserves of one sort or another. Although such plans are ordinarily flexible, and in any country may be subject to political or industrial pressures, the very existence of such elaborations of policy – based, one hopes, on ecological and recreational research – represent a highly sophisticated advance in land-use planning.

The deserts of Australia have increasingly had portions reserved for the development of national parks and related areas. One reads or hears often of the famed Ayers Rock-Mount Olga National Park, but the Northern Territory government now has a complex system of reserves and a competent service of land managers to operate them and plan for the enlargement of the system. The New South Wales government possesses an exceptionally advanced park system that includes parks in arid regions. A unique value in Australian deserts is the existence of old aboriginal camps and early rock engravings, some of which are being conserved.

Desert regions of Africa and the Middle East still remain some of the least developed with regard to parks and reserves, but there is notable progress. In southern Africa, the original idea of Etosha National Park was to establish a park of 6.4 million hectares, combined with a surrounding buffer zone, which would equal a total protected area of nearly 16 million hectares. That would be nineteen times the size of Yellowstone National Park, larger than the entire national park system of the United States or Canada, and comparable in size to half of the state of Montana or two thirds of the British Isles. But the word is that administrative changes have substantially reduced the size of the area. It is indicative, nevertheless, of the immense size of arid zone parks of Africa.

There is a growing desire in northern Africa to undertake systematic efforts to create desert reserves, including sanctuaries in riparian forests along desert streams, and recreation areas along sea coasts. A brief survey of Mauritania was undertaken in 1974 by the government of that country, in cooperation with the United States National Park Service and Department of State. Much of the country consists of broad desert landscapes, sand dunes, rock-strewn plains, nomads herding sheep and goats, acacia trees, a little grass, and occasionally some shrubs. Color is often lacking, and many regions could be described as interesting but not of national park quality. There are genuine historic treasures. For example, an old adobe library at Chinguetti is said to house volumes dating back to the eighth century. The whole settlement, situated at the edge of the Sahara Desert, could be made into a 'living history' attraction with its timeless character and atmosphere of desert people, animals, and tents. Ouadane and other ancient towns merit special preservation also.

Over much of Mauritania the roads are unpaved or non-existent, and trails are few. But there are scenic canyons, oases, seashores, islands, and fishing villages that deserve improved access as the tourism program grows. A gazelle reserve exists near Nouadhibou, but the guard staff is in need of expansion and should have transportation and equipment. Some method needs to be arranged so that tourists can see gazelles, preferably in the wild, and in a manner not disruptive to the wildlife. Otherwise, tourists would find little attraction in such areas.

One principal problem in Mauritania, and it is virtually a world-wide problem, is the degradation of landscape by domestic animals such as goats, camels, sheep, and cattle. To the tourist these may be familiar and picturesque, but the thrill of seeing a wild jackal, or wart-hog, or monkey is a superior experience. In addition, from a sociological viewpoint the interest of native customs could be capitalized upon in Mauritania. Songs accompanied by beats on a drum, and the antics of pantomimists, provide memories as enduring as those offered to South Pacific tourists when they witness war chants in the Fiji Islands.

Less than 20 km from Ayoun, Mauritania, is a picturesque desert collection of eroded sandstone columns, arches, cliffs and tunnels worthy of national park status. Camel rides to scenic localities could be very popular, as they have been in the Canary Islands. The general conclusion of the Mauritania survey was that, for lack of suitable roads and lodging, the country was not yet ready for a massive tourism program, but that some day a good effort could be based on desert resources. Unfortunately, the Sahelian drought has upset plans; or perhaps pointed to the need for alternate, diversified uses of desert regions that are agriculturally marginal or submarginal.

In other parts of North Africa, vast expanses of the Sahara Desert are without preservation as parks or reserves. But in places tourism has become an industry seeking clients for the dry terrain and its attractions. Nowadays, one can enter the Sahara on a camel caravan to the Tassili N'Ajjer in Algeria's Touareg country and 'rough it' by hiking and camping under primitive conditions. Travelers may take off into the distant desert by four-wheel drive vehicles or sail down the Nile in a felucca.

In some cases, tours of desert areas may be diverted to river float trips because land access to points of tourist interest are restricted for reasons of military security. Some travel organizations book passengers on cruise vessels that have all the comforts, including swimming pools. To rugged individualists, that may not be true desert roaming, but it is at least an introduction to the desert and does save wear and tear on natural features.

Particularly active is Tunisia, which has established the 11625 ha Bou-Hedma State Park. This area has a relict population of gum-acacia and other xerophytic vegetation. Animal life includes gazelles, Barbary sheep, gundi, gerbils, jerboas, jirds, and houbara bustards.

Tunisian tourism officials have published colorful booklets on the country's historic sites and monuments, oases, seashores, and other attractions. By citing tourist values such as palm groves, white sand beaches, colorful mosques, and carpets from Gabes, the Tunisians encourage visitors to rent cars or hire drivers and go out into the hinterlands. The challenge of the desert is pointed out in brochure text: 'The land around [the huge salt lake], partly spongy and partly covered with a thin film of season's waters, has many surprises in store for one who dares to brave it.' A visitor is urged to observe the wealth of color at sunrise and sunset, experience the rapidly succeeding contrasts of heat, follow the fleeting mirages, and sense something of the lives of the poetic Djeridis, people who spend their lives in nearly barren deserts. And the healthful values of such environments are not ignored. The Tunisians point out, as Sallust, the Roman historian (86–34 B.C.), said, that 'most of the inhabitants of Africa die of old age and rarely succumb to illness'.

Remains of historic desert empires have had an air of mystery for modern desert travelers ever since John Lloyd Stephens visited the rose-red city of Petra, in Jordan, now a national park. In Tunisia, such ancient monuments form a significant lure in desert tourism appeals. The Tunisians promote Carthage, Dougga, the coliseum at El Jem, mosques, forts, mosaics, sculptures, and museums.

The desert tourist is also encouraged to hunt and, along the sea coasts, to swim and fish in the Mediterranean. Festivals of music and folk art are held. To make a visit relatively easy, there are air, sea, and land access, and all categories of hotels; the tourist may rent a car, charter a plane, or join a tour. Such activities, plus the fact that winter temperatures range around 11 °C, can be very alluring to Europeans tired of winter snow and ice, and unwilling to visit the desert in summer for fear of high temperatures. As a result, the desert no longer remains as quiet and untraveled as it used to be.

One nearly universal problem, in historic as well as natural areas, is the lack of good on-site interpretation – few signs if any, no self-guiding trails, no museums, and no uniformed technical personnel available to answer questions. While there are striking exceptions to this, because one does find a few interpretive devices here and there, the fact remains that the public presentation of much of the world's heritage leaves something to be desired. Visitors must often rely on multilingual tourist guides and/or booklets, and may have to depart from the scene without a real understanding and appreciation of the fundamental values of the area. In certain cases, the story, if not the landscape, may be debased by tourist guides, who create artificial geysers in volcanic parks, or by planners, who build commemorative monuments in places that are much too conspicuous.

In the Middle East, two preservation concepts have prevailed: retention of sites and ruins significant in the period of Roman occupation; and protection of scenically attractive landscapes. In Jordan, master plans have been developed for Petra, Jerash, Jericho, Qumran and other sites where outstanding ruins remain. Some work has also been done with natural areas such as Wadi Rum and Azraq, the latter site an oasis important to migratory birds. Walking, hiking, and camping trips are conducted into the Judean and Negev deserts of Israel and Egypt.

High on the arid Anatolian Plateau of central Turkey are the Asian steppes where human beings have fought nature for centuries, and where humans have fought humans. Before and after Alexander led his legions eastward across this plateau in the fourth century B.C. travelers used Turkey as a bridge of

Fig. 22.7. Turkey: Pamukkale National Park.

land between east and west. Some of the inns, or stopping places they used, now centuries or millenniums old, are still there. During the Middle Ages, these fortress-like inns, known as caravanserais, were established on major trade routes at intervals of a day's travel, that is, about 25 km apart. Many show the ravages of time and are crumbling into ruin, but others are being restored by the Turkish government, and the ornate facades give some indication of how stately and artistic they were.

The modern version of the caravanserai system is the Turkish network of Mocamps, which, in the opinion of some observers, is the world's finest system of outdoor camping facilities. Mocamps have been established near many historic sites – that at Kusadasi, for example, is not far from Ephesus National Park. Visitors either stay in a simple cabin if they have no bedding (a blessing for international tourists who come to the country by air), or pitch a tent and camp out. A central unit offers showers, rest room, laundry facilities, and electricity. Another unit nearby provides inexpensive meals, snacks, gasoline, oil and miscellaneous supplies. A special appeal of the Mocamps, apart from strategic location and convenience, is their cleanliness, maintained by an adequate summer staff that pays remarkable attention to neatness. The Mocamps, all privately operated, are consequently well used, and brochures provide full information about them.

Between 1967 and 1970, Turkey took steps to prepare comprehensive master plans for park development and management of some of its principal historic sites which have been designated national parks. These include Göreme, the refuge of early Christian monks; the Hittite capital of Hattusas in Boğasköy-Alacahöyük; the mountain fortress of Termessus; the seashore at Halicarnassus; the mineral springs and terraces of Pamukkale; and a bird sanctuary at Kuscenneti.

Even in Hungary, not generally thought of as a desert country, the flooding of the Tisza River has created a semi-desert plain of which 52000 ha have been designated Hortobagy National Park. The broad distribution of low-growing *Artemisia* spp. provides a desert aspect, as do, in a western US cowboy sense, the presence of original Hungarian cattle managed by horsemen in traditional costumes.

These are only a few samples of progress in preserving desert regions of high scenic quality and at least a minimum residual integrity. Of course, some desert preserves still exist only on paper, or are not yet government-owned, or have never had their boundaries delineated. No tourism facilities exist in such areas, of if so, they are badly located in the midst of prime scenic areas. But one by one, these 'abandoned' areas are coming under competent protection and administration, are being planned, and are being opened to the public. Perhaps most encouraging is the provision of informational and interpretive materials. Not only do we find the familiar tourist brochures, but self-guiding trail booklets, folders describing special historic values, museum leaflets, and sales publications on fauna, flora, geology, and related features. Where the demand exists, these materials are translated and printed in other languages.

Nothing is static about this process. Desert parks continue to be proposed, and publications continue to be issued or upgraded. But we can take only tentative and temporary comfort in this. The battle to sustain desert ecosystems undisturbed is a continuing one. Otherwise they get nibbled at administratively and politically until there is little left, or so vandalized for lack of protection that they have to be disestablished. By no means are all important deserts preserved; nor is there wholly adequate protection or interpretation, partly because of insufficient scientific data, as the North American example shows.

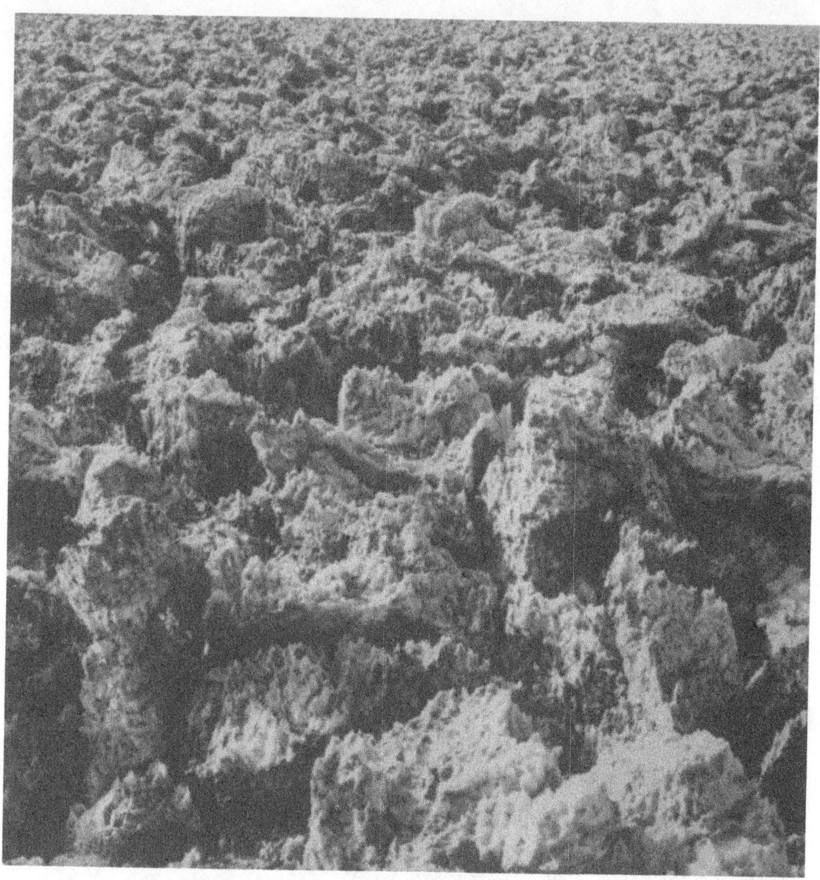

Fig. 22.8. Salt beds, Devil's Golf Course, Death Valley, California.

The North American situation: problems and pressures

The paucity of scientific data most assuredly applies to North American deserts. A case in point is the Chihuahuan Desert, in the states of Coahuila, Chihuahua, Texas, and New Mexico. In October, 1974, a symposium on the biological resources of that desert was sponsored by the United States National Park Service and the Texas Parks and Wildlife Department, and held at Alpine, Texas. While a commendable and even astonishing amount of information came from both Mexican and American sources, many contributors deplored the lack of fundamental data. In fact, some desert resources are disappearing before they can even be identified and analyzed, a lament heard not only in the Chihuahuan Desert but far beyond it.

Among the devastating effects are those resulting from human use of run-off

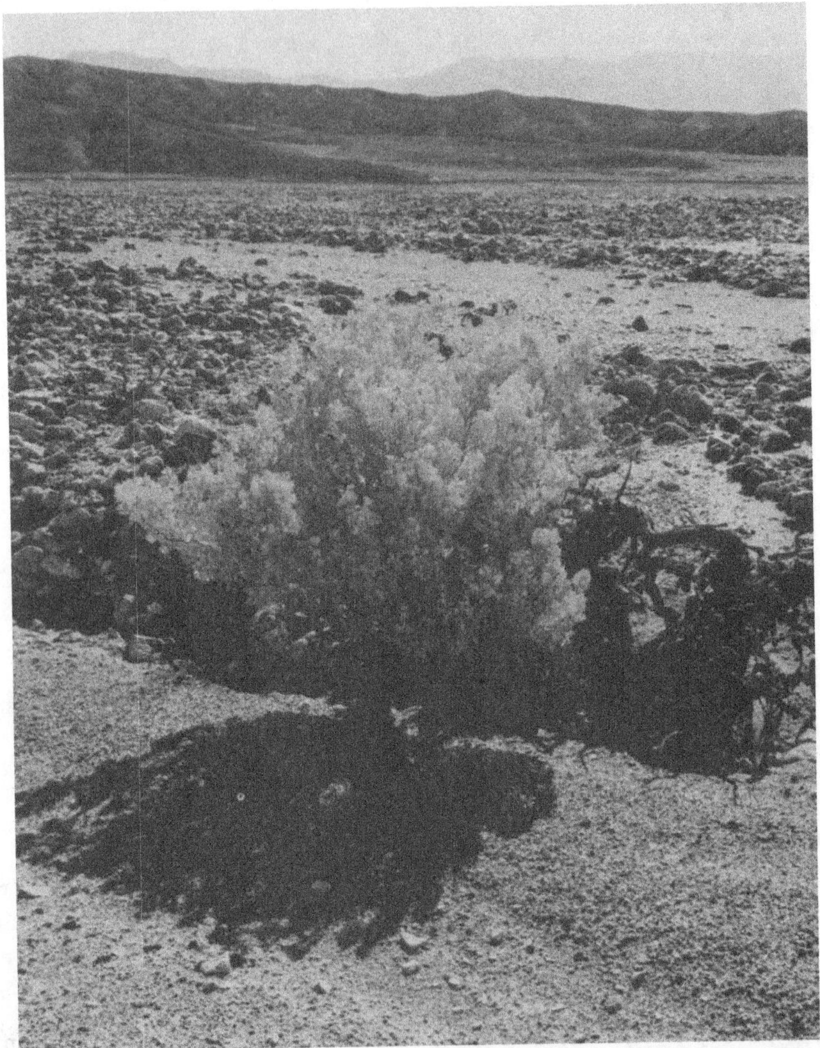

Fig. 22.9. Desert holly, Death Valley National Monument, California.

and 'fossil' water. Springs have dried up, ponds have been drained, water tables have dropped and marsh lands have diminished. Among the tragic losses are desert fish, not very plentiful even in the natural economy of desert biomes. Emergency measures have been taken to save the last of the desert pupfish near Death Valley National Monument, California/Nevada, and some such measures will be needed if Chihuahuan Desert fish are to survive. This is surely an economic matter as well as scientific, because an empty desert has little attraction for tourists.

Table 22.2. *Growth in numbers of visitors to selected desert parks in the United States*

Name and Location	1963	1973
Anza-Borrego Desert State Park, California	353936	522666
Big Bend National Park, Texas	114200	341300
Death Valley National Monument, California/Nevada	408100	606500
Joshua Tree National Monument, California	346200	592900
Lake Mead National Recreation Area, Arizona/Nevada	3349600	5534300
Petrified Forest National Park, Arizona	786000	1072000
Saguaro National Monument, Arizona	177000	390800
White Sands National Monument, New Mexico	413500	663200

In addition to the establishment of national and state parks to preserve significant portions of deserts, there have also been increases in the numbers of wildlife ranges, refuges, and wilderness areas. The results are encouraging. For example, the Desert National Wildlife Range, in southern Nevada, with an area of 635352 ha, has been crucial to the survival of desert bighorns. In the 40 years the Range has been in operation, the numbers of these animals have increased from a low of 300 to a high of 1700 in the late 1950s; the current population is estimated to be between 500 and 1000.

This is not principally a tourist-oriented area, but tourists are unquestionably interested in wildlife, and such reserves – almost 'fountains of life' – can provide stock naturally for depleted areas elsewhere. If these lands are kept free of excessive administrative manipulation, and free from pressures for potentially competing uses such as grazing, hunting, and mining, the tourism industry can grow and thrive without damage to the fragile natural resources.

Today more than four million hectares of public desert lands have been set aside in the United States (Table 22.1). Those accessible by road are heavily used, as the table shows and more remote wilderness areas, accessible by trail only, receive far fewer visitors. In general, there is an upward trend for use of all areas, however, even considering how obedient the tourist industry must be to economic and transportation conditions, and how vulnerable it is now to international political dislocations (see Table 22.2).

Martin, Gum & Smith (1974) have assessed the outdoor recreation demand in Arizona, a state with numerous upland and lowland desert resources. Comparing the tangible economic benefits of outdoor recreation with those of other uses of desert habitats is difficult. It has traditionally been done by measuring the gross expenditures for travel, accommodations, supplies, and such equipment as boats, motors and tackle. But this method has shortcomings.

Martin and his colleagues developed a system of evaluation based not on gross value but on the value added by a particular recreation opportunity.

'It is the net increase in the value of the resource produced by using the resource for recreation that is crucial,' they wrote; 'This represents a true net yield that can be compared with what the resource would yield if it were in an alternative use producing other services. It is the margin above the cost of taking advantage of the recreation opportunity which measures the real monetary value that would be lost if the recreation opportunity were not available. The gross expenditure figures in themselves do not measure this. They tell us the magnitude of the industry, in one sense, but they do not indicate the value of the losses that would be sustained if the particular recreation opportunity were to disappear, or the value of the net gain from an increase in a particular recreation opportunity.'

Martin and his associates estimated consumer benefits from the entire recreation experience, including not only licensed sportsmen but their families as well, without whom many trips might not have been made. Indeed, researchers suggested that the family is the decision-making unit, and that various recreation attractions beyond hunting and fishing need to be considered. On the basis of thousands of responses to questionnaires, it was determined that the consumer surplus value of general rural outdoor recreation in both high and low deserts far exceeded the consumer surplus value of hunting and fishing combined. Analysis revealed that in 1970 alone, 4905 384 household-trips were made by Arizona residents for purposes of rural outdoor recreation, generating $243 236 558 of consumer surplus value. This value represented the total net benefits of the state's natural resources in use for outdoor recreation by Arizona residents.

Such studies are valuable in developing land-use policies. They also help to set tourism and recreational use in proper perspective with agriculture, housing developments, and the industrial uses of water for farming and municipal purposes. The point has already been reached, in certain places, where desert resources are more beneficially utilized for tourist recreation than for more consumptive exploits.

Popularity has its problems. And, as we have seen, with the so-called blessings of modern technology, the desert is no longer safely remote: tourists can reach it, penetrate it, cross it, or go almost anywhere by means of that modern juggernaut, the off-road vehicle. A United States Department of the Interior Task Force Study in 1971 recognized the rapid rise in numbers of off-road recreation vehicles and documented the existence of five million of them. Expressing concern, but unable to provide detailed proof of the damage caused by such vehicles, the Task Force called for a unified government policy to regulate them, at least on public lands.

Subsequently, the President issued Executive Order No. 11644, which provided that the use of motorcycles, minibikes, trail bikes, snowmobiles, dune buggies, all-terrain vehicles, and others be limited to areas designated for their use. The principal object was to minimize damage to soils and

Fig. 22.10. View from Font's Point, Anza Borrego Desert State Park, California.

vegetation, reduce harassment of wildlife, and avoid destruction of esthetic and scenic values.

Certain states have also established regulations, and imposed penalties. In California's Anza-Borrego Desert State Park, for example, driving on closed roads, foot trails, cross country, or over vegetation is prohibited. All established routes of travel are marked. Persons operating motor vehicles of any kind are required to carry a valid operator's license for that class vehicle. All vehicles must be equipped with adequate mufflers, approved spark arrestors, and brakes meeting Vehicle Code requirements.

Establishing regulations, however, is one thing, enforcing them is another. Southern California alone has well over a million motorcycles registered, and sometimes hordes of these roar out onto the desert. Roger Luckenbach, of the Department of Geography at the University of California at Berkeley,

Fig. 22.11. View from Font's Point, Anza Borrego Desert State Park, California.

states that 'no other current form of recreation is capable of producing such a detrimental "residue". Although it may take only a split second for a dirt bike to cover a yard of terrain, the effects of its passing initiate far-reaching and diverse environmental degradation.' This is no idle statement: scientific researches have now confirmed such observations. Normally, the desert has a protective mantle of cobble or slightly indurated soil. When this is disturbed, the lighter subsurface soil is lost through deflation. Soil density increases, the upper layers become less permeable, and a new kind of 'pavement' forms.

Measurements taken through transects, quadrants, and bulk density analysis have shown that there was a statistically significant increase in the bulk density of soils disturbed by off-road vehicles. There is little enough porosity in desert soils anyway without human-caused compaction. The scarcity of organic matter means that soil particles are seldom formed into porous aggregates, and while the particles may be coarse, they are packed closely together. This adds even further to the delicate and sometimes marginal conditions under which plants (and thence animals) must survive, because the decrease in pore space means a decrease in retention and transmission of water in the soil. Hence the water runs off, there is less available for plant growth, and barren desert conditions ensue or increase.

The degree to which the desert is damaged depends, in part, upon the intensity and periodicity of use. Sometimes hundreds of motorcycles travel across the same route. A 'trail' for races may be 240 km long and 3 m wide. On desert slopes, vehicle ruts break up the soil cover and become new channels for the familiar rapid and violent erosion during and after summer cloudbursts. It may well be that the spreading effects from a few 'trails' eventually cause the removal of soil mantle from an entire hillside. Disturbing the soil also disturbs soil organisms, such as those involved in nitrogen fixation.

Nor is erosion the only result of off-road vehicle activity. Living plants have been directly damaged or killed, and seed beds destroyed. Riders in large concentrations utilize live and dead vegetation to fuel their campfires. The species diversity and numbers of individual perennial shrubs become markedly diminished. How long it takes for natural restoration of the vegetation depends on many factors; the process would probably not be very rapid.

Of course, plant species adaptive to disturbed soil will take root. Exotic ground cover, principally filaree (*Erodium cicutarium*) and an annual grass (*Schismus barbatus*) enter disturbed areas and flourish. But the damage to native herbaceous vegetation by off-road vehicles, such as the decrease of shrub density, has been scientifically documented and can be widespread. In using the same area repeatedly, motorcyclists tend to camp and to run their vehicles where the soil and vegetation have not yet been disturbed. Therefore, the circle of damage and destruction is enlarged. Parking and camping spots used for many years become almost completely devoid of vegetation. Moreover, auxiliary participants in races, such as family members, drive alongside the race routes, with consequent extension of the damage.

Invertebrate food sources may also suffer, which could adversely affect larger carnivores. The vehicular destruction of native plants, upon which many mammals depend for food, moisture, and shelter, has been shown to reduce the diversity of certain species. This obviously affects the capacity of land to support native wildlife. The existence of impoverished spots has been a warning that controls on vehicles and vehicle users are required. Or some place needs to be set aside specifically for off-road vehicle recreation use. However, such an area would in all likelihood be so heavily used that it would soon cease to be attractive to riders. The principal desire seems to be to roar across wild and undisturbed terrain.

Wagar (1964), in his pioneering study of wild land carrying capacity, observed that in every statement of carrying capcity there must be, at least implicitly, a statement of some management objective. He defined recreation carrying capacity as the level of use at which quality remains constant. Public land administrators need to develop methods of identifying and classifying quality, and assure that it remains high. This could be done by such means as regulating visitor use and behavior, sustaining adequate budgets for landscape management and maintenance, and basing administrative decisions on sound ecological research.

Apart from off-road vehicles and agricultural pursuits, the desert has other problems of encroachment by civilized society. New homes have so proliferated that once-wild desert valleys have become filled with urban sprawl that goes over the nearby mountains into adjacent valleys. The original desert is practically eliminated and another substituted. The result is an artificial transplanted desert around subdivision houses that Renée Dubos might call disposable cubicles for dispensable people. If we grant that desert living is highly suitable for retirement purposes, we need also observe that retirees make active tourists. That leads to increased automobile traffic and more desert paved over for ever larger parking lots to encourage more vehicle use.

The consequence is a creeping cloud of noxious fumes that literally turns the desert sun orange at midday and fills the air with eye-smarting, ill-smelling particulate matter.

In wilder areas, large numbers of recreationists entering a desert pose serious problems because other people may have to enter the same desert to supply their needs. It has been incumbent upon park planners and architects to hide or harmonize structures such as museums, restaurants or administrative buildings. Often the best solution is to locate support and administrative staffs completely outside the parks. Otherwise, one recurring problem is waste disposal. If the soil is thin, and the underlying rock a solid limestone or granite, park staffs simply cannot bury waste except by blasting a series of pits: they must haul refuse away. But neighbouring jurisdictions may have no desire to be dumping grounds for tourist-oriented agencies. One of the best ways to solve this matter has been to contract private waste disposal firms to haul the refuse far away and dispose of it in ways environmentally acceptable. Even that may not be simple when the nearest town of any size is 150 km distant. And waste disposal contractors may not be conveniently available. But it is a problem that should be dealt with in the early planning stages rather than saved for solution after people start to arrive.

Despite the difficulties, the onslaught of tourists and retirees shows little sign of abating. For the desert has been prescribed as an affective antidote to urban life (providing urban life hasn't also entered the desert). The recreational and educational values have obviously grown very rapidly in public recognition. Of course, the processes of artificially induced desertification are also painfully obvious, and will continue unless human beings modify their behavior or continue to establish large reserves. Otherwise the familiar process will play itself out: loss of productive vegetation cover, disturbance to the hydrological regime, and reduction in the water content of the soil.

Getting more people to appreciate the attributes of the desert biome is an endeavor to be highly commended. It is also becoming something of a commercial success. Certain trip-sponsoring organizations specialize in travels oriented toward natural history, nature photography and anthropology. One non-profit educational organization in the United States recently offered a journey among the Indian ruins of the American Southwest. The trip was quickly filled to capacity and was offered twice the following year. The organization sponsors trips to the Galápagos Islands and the islands of Baja California, excursions in four-wheel-drive vehicles to northern and eastern Africa, and ventures into the Utah Canyonlands country of the United States. It also sponsors short courses, seminars, symposia, and workshops in natural history, photography, and anthropology. College credit is offered in certain cases. These programs may well encourage a more profound appreciation of the natural environment than would general tourism programs, and should, as time goes on, convert large numbers to the cause of natural area protection.

Some companies offer desert tours on a more rugged scale. These include camping among the dunes, sharing trip chores, and following a flexible schedule. Transportation is sometimes by four-wheel-drive truck, which often has to go where roads do not. For accommodation, tents are more often used than hotels. Unpredictable weather and road conditions may mean on-the-spot modification of itineraries and schedules.

The routes for these tours take travelers across Asia from Istanbul to Bangkok. They traverse the searing center of Australia, from Adelaide via Alice Springs to Darwin. They are followed by minibuses over the Mexican Desert from Houston to Mexico City and points south. One safari trek leaves Nairobi for a 93-day tour through central and western Africa, north into the heart of the Sahara, and across the Atlas Mountains. Another lasts for 85 days and takes travelers up the Nile through Egypt and the Sudan, then southward to Zambia, Rhodesia and South Africa.

Trekkers on these tours investigate the lives and environments of nomadic inhabitants: they almost become nomads themselves. It is a way of life far different from the usual luxury tourist excursions, but it introduces travelers to wild and seldom visited desert lands and peoples.

A worldwide movement is afoot to save what flora and fauna remain, preferably in relatively undisturbed ecosystems. Keeping them reasonably 'undisturbed' while still admitting tourists is the problem, but it is a problem that has been solved in important places. We know how to admit millions of tourists without destroying fragile resources. The technique has been thoroughly tested. It is time that the fundamental knowledge accumulated be made more widely known so that it can be more widely used.

Management of arid lands

The economic values of desert tourism are reaching such proportions as to endanger the uncontrolled expansion of agriculture, grazing, and urbanization. It is even replacing these industries as the number one earner of foreign exchange. Yet the guiding principle, let us hope, will remain what it has been – that the desert forms part of a national identity and deserves to be protected for its own sake.

References

Department of the Interior Task Force (1971). *Off road recreation vehicles, a Department of the Interior Task Force study*. Washington, DC.

Dregne, H. E. (1970). *Arid lands in transition*. American Association for the Advancement of Science. Washington DC.

Martin, W. E., Gum, R. L. & Smith, A. H. (1974). *The demand for and value of hunting, fishing and general outdoor recreation in Arizona*. Agricultural Experiment Station, University of Arizona, Tucson.

Wagar, J. A. (1964). *The carrying capacity of wild lands for recreation*. Forest Science Monograph 7, Society of American Foresters, Washington, DC.

Other references

Although not appearing in the text, the following publications are listed for their general interest.

Berry, K. H. (1973) (ed.). *Preliminary studies on the effects of off road vehicles on the Northwestern Mojave Desert: a collection of papers*. Xerographed, Ridgecrest, California.

Featherstone, W. L. (1974). Mission to Mauritania. National Park Service Trip Report (manuscript), Washington, DC.

Green, R. & Sellers, W. D. (1964). *Arizona climate*. University of Arizona Press, Tucson.

Hodge, C. (1963) (ed.). *Aridity and man: the challenge of the arid lands in the United States*. Publication No. 74, American Association for the Advancement of Science, Washington, DC.

Wauer, R. H. & Riskind, D. H. (1975) (eds.). *Transactions of the Symposium on the Biological Resources of the Chihuahuan Desert region, US and Mexico*. National Park Service, Washington, DC.

Manuscript received by the editors May 1975.

23. Management of water resources in arid lands

J. L. THAMES & J. N. FISCHER

Introduction

Throughout the ages man has elected or been forced to settle in arid lands where fresh water supplies are often deficient, highly variable and of inferior quality. Traditionally, people in these regions have accepted the challenge of water management rather than live elsewhere. Only when water supplies have completely failed, or been made useless by salinization or massive siltation or when floods or wars swept away everything have they abandonded their efforts. Painful rebuilding frequently has been undertaken by later generations who, ironically, often re-invented the old methods of water management. In many cases these errors led to even greater failures as the resource base was progressively depleted. Most often, the reasons for failure lay more in the neglect of man to follow the rules of nature than in the capricities of nature itself.

The same process continues to a greater or lesser extent in most arid regions of the world today. Old techniques are being re-discovered or invented anew, and many of the old mistakes are being repeated.

Arid lands suffer a deficiency of water for man's desires. The simplistic approach to this problem has been the development of new water resources, the theory being that water is the only factor lacking to make the arid area productive. Unfortunately, this is not necessarily the case; there are other limiting factors such as stock carrying capacities, human population, etc. History amply shows that in the absence of adquate social controls, water resource development in a fragile arid system leads almost inevitably to an overextension of human activity, increasing the vulnerability of the system to unforeseen and unplanned for adversities. In arid lands, therefore, there is a thin line between what we have looked upon as underdevelopment and what is actually overdevelopment.

For example, overgrazing is perhaps the most serious misuse of arid lands, particularly among cultures that consider the number, not the quality, of animals as a criterion of wealth and status. With the development of a water resource on an arid range, even dead forage can be utilized but since water too distant from grazing areas is of little use, construction of more stock ponds has been undertaken. However, without effective grazing controls, each new watering source has historically led to increased herd size, overgrazing and a new center of expanding desert. Drought is blamed for the tragic situation in the sub-Sahara, for example, where both human and animal losses are counted in millions and where, paradoxically, outside aid has been developing water for years.

In order to be lastingly effective, water resource development must be preceded by comprehensive total land use and social planning with provisions for perpetual operational support. Without adequate controls on the use of limited water resources of arid lands, their overdevelopment can lead, in the long run, to regional or national underdevelopment.

Most arid lands have possibilities for the development and management of both surface water and sub-surface water. Surface water is a renewable resource in the sense that flow in ephemeral streams has short re-occurrence intervals that depend upon local climate. But its reliablility is only as certain as the rain, and rainfall in arid regions is not only slight but highly variable both within and between years.

Since the 1950s, considerable interest has been revived in developing more efficient use of arid-land surface water, largely prompted by accomplishments in Israel. There are six general practices associated with surface water management in arid areas:

(1) harvesting water by concentrating surface run-off from a larger catchment onto a smaller collection area to produce agricultural crops, forage or water for human and livestock use;
(2) utilizing the periodic flows of ephemeral streams for water spreading;
(3) maintaining water where it falls for on-site use by vegetation to improve the productivity of range lands;
(4) modifying the vegetation on watershed areas to improve water yields;
(5) erosion control; and
(6) evaporation control.

Groundwater may be replenished over relatively short time spans in some instances, but in most arid systems it is a finite resource established over thousands of years. Very careful management based upon a thorough understanding of the physical system is essential.

Water harvesting

Water harvesting is an ancient technique to provide water for small agricultural operations, stock use or domestic needs (Evenari, Shanan, Tadmor & Aharoni, 1961). It involves use of a catchment area, usually prepared in some manner to improve run-off efficiency and a collection area, in which either crops are grown or water stored. The United States and Australia, in particular, are active in the development of stock-watering systerms. In Israel, extensive research is being conducted on water harvesting systems for crop production. The potential of water-harvesting systems for providing domestic water supplies of small communities have yet to be fully developed, although the use of roof tops to collect water for household use has long been a world-wide practice.

Water harvesting techniques may be divided into four categories: (*a*) natural impervious surfaces; (*b*) land alteration; (*c*) chemical soil treatments, and (*d*) ground covers (Fraiser, 1975). Each method has associated costs, performance and maintenance requirements which influence its suitability for particular uses and sites.

If natural, relatively impermeable surfaces such as rock outcroppings are available, they may be used effectively as water-harvesting surfaces. A collection system must be added at the down-slope side of the outcropping to guide the water to a central storage point. Rock outcroppings are simple and durable water-harvesting systems of low cost. However, high run-off efficiencies should not be expected from all outcroppings; Burdass (1975) estimates that 45% efficiency is an average value for rock outcroppings. This value can be improved by the use of sealing compounds, such as asphalt, applied to the major fractures.

Simple earth smoothing and compaction is sometimes effective. The technique was used in early times in the middle east. Frith (1975) reported recent success using similar methods in Australia. Soil composition is an important factor with this technique. Success is generally higher on loam or clay-loam soils. Erosion may be a problem if land slope is too great and care must be taken to minimize slope and reach-length to reduce run-off velocity. In some cases, especially where the catchment area is large or where slopes cannot be economically reduced, it may be advisable to construct small berms (banks) across the smoothed land to reduce run-off velocities.

Chemicals are receiving increased attention as a means of improving the efficiency of run-off from areas that have undergone mechanical treatment. An example is the use of silicone water repellents to reduce infiltration. Myers & Fraiser (1969) reported a run-off efficiency of 90% after the first year of such a treatment but efficiency decreased to 60% after four years. Apparently, the silicone did not stabilize the soil, allowing the soil surface layer to be eroded.

Paraffin has also been used to increase water repellency. It stabilizes the soil to a greater extent than does silicone, but it frequently loses its effectiveness as a water repellent after freezing and thawing (Fink, Cooley & Fraiser, 1973). However, it has been shown that run-off efficiency can be restored if the soil surface is heated to 54 °C, a temperature frequency reached under natural insolation in many areas.

Research in Arizona indicates that small amounts of sodium applied to desert soils where the vegetation has been removed cause dispersion of the surface soil, reductions in infiltration rates and subsequent increases in run-off (Cluff, Dutt, Ogden & Stroehlein, 1971). Salt also retards weed growth which

Fig. 23.1. Concrete water-harvesting surface.

can be a serious problem on many catchment surfaces. To be effective, this type of treatment requires a minimum amount of expanding-type clay in the soil. However, even if sufficient quantities of clay are present initially, the clay may migrate downward in the horizon over time and reduce the effectiveness of the collection area. Compacting the soil surface helps reduce the migration.

Several methods of water harvesting have been developed using impermeable materials. Concrete has found widespread use but it is expensive. Cracks also develop and require maintenance but they can be repaired inexpensively, with bond strips of fibreglass applied with asphalt emulsion. Concrete surfaces have an expected life of about 20 years, which is excellent compared with other catchment surfaces (Fig. 23.1).

Air fields, such as the Grand Canyon Air Field in Arizona, have been used as water harvesting surfaces. An Arizona highway catchment system has also been in use for 15 years (Chiarella & Beck, 1975). Greater use has not been made of these large potential sources of impervious surface because of fear of pollution from vehicle oils. However, Chiarella & Beck have discovered no ill effects in livestock drinking the Arizona highway water. If highways were used as collection areas, great quantities of run-off water could be made available. For example, assuming an efficiency of 90% and an annual rainfall of 250 mm, each kilometer of two-lane roadway could produce about 2.8 Ml of water annually.

Sheet metal placed directly on the ground has been used with good success as a surface cover for catchment areas. A protective coating on the underside and on the exposed surface is necessary to prevent corrosion. In those cases where the metal has been placed on above ground structures, the frequency

Fig. 23.2. Fibreglass or polypropylene matting saturated with asphalt may be used effectively as a water-harvesting surface.

of failure has been high. Sheet metal treatment is expensive, but its average life is as long or longer than other surfaces.

Fibreglass or polypropylene matting saturated with asphalt have also been used successfully (Myers & Fraiser, 1974) (Fig. 23.2). It is particularly suited for use on rough terrain. The matting is used to form the surface of the catchment and the asphalt is the waterproofing agent. Asphalt discolors the run-off water but protective paints have been used to reduce the effect. Unevenness in the finished surface leads to only small losses in efficiency and a life of over five years may be expected.

Plastics or tar paper covered with a coating of gravel make very good water-harvesting surfaces. The method is not expensive and installation is not difficult, providing there is a source of clean, uniform gravel near the installation site. Gravel reduces deterioration of the membrane material, but it also traps considerable amounts of run-off water, particularly that from small rainfall events. If the gravel is too fine it it may also serve as a seed bed for weeds which will compete for run-off water, and if it is too large, the membrane material may be punctured during application.

Some of the most effective types of catchment surfaces are those constructed of artificial rubber membranes. They are easy to install providing the site has been smoothed carefully beforehand (Lauritzen & Thayer, 1966). In early studies, the lack of maintenance, improper installation or use of poor quality materials led to failures in many of these systems, and the method fell into

Table 23.1. *Relative cost and efficiency of catchments developed at the University of Arizona, December, 1972.* (After Cluff & Dutt, 1975)

Catchment methods	Approx. cost ha^{-1} ($US)	Efficiency (%)	Estimated life (yr)
Compacted earth[1]	400–600	30–60	Indefinite
Compacted earth			
Sodium treated[2]	720–1200	40–70	Indefinite
Graveled plastic[2]	2400–6000	60–80	20–25
Asphalt–plastic			
Asphalt-Chip-Coated[3]			
Polyethylene Reinforced	6000–7200	85–95	10–15
Polypropylene Reinforced	10800–12000	85–95	10–15

[1] Prices and efficiency of these catchments are dependent on soil type, cost of clearing and shaping. Maintenance consists of weed removal and recompaction as needed. Additional sodium chloride may be required periodically prior to recompaction for maintenance of the sodium-treated catchment.

[2] The variation in price of the catchment is primarily dependent on the cost of the gravel and to a lesser extent the cost of clearing and shaping. The cost of the 10 mil black polyethylene plastic to be used is relatively stable. Maintenance consists of adding gravel if necessary on exposed portions of the catchment.

[3] Prices of the systems are based on projection of small plots. Larger installations need to be made to firm up prices. Maintenance consists of recoating with asphalt and chips every 10–15 years.

disfavor. However, recent successes under proper installation and maintenance conditions have led to renewed interest in this type of system. A summary of efficiencies, lifetimes, and costs for various water-harvesting treatments is given in Table 23.1.

The geometric configuration of catchment areas depends upon the characteristics of the sealant used, limitations of site topography, and personal preference. A wide variety of shapes have been tested. In Australia, a common design is the roaded catchment (Fig. 23.3). The V-shaped catchment is popular in the United States. Simple sheet type catchments are used in Mexico.

Due to high variability of precipitation in arid areas, water harvesting can not always provide a completely reliable supply of water. For this reason, if harvested water is to be used for agricultural irrigation, drought resistant crops should be used, and generally, an alternate source of water should be available to insure a minimum water supply during drought conditions.

Water storage

A complete water-harvesting system includes a means to store the collected water. There are three basic storage systems: excavated pits; bags made of rubber or plastic; and tanks of steel or concrete. The most appropriate

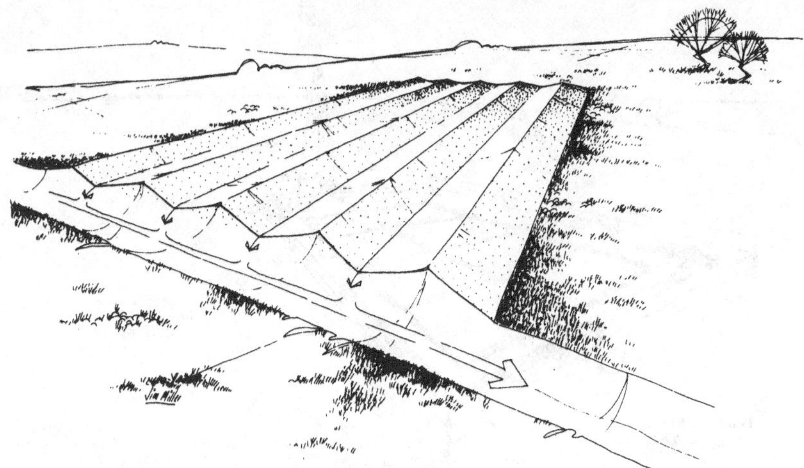

Fig. 23.3. Roaded catchments are a common water harvesting surface in Australia.

system for a particular site depends upon local conditions of soils, site accessability, available materials, cost of labor and material, and life requirements (Dedrick, 1975).

Excavated pits are the most common type of water storage system. They are easily constructed in flat areas with deep soils and are particularly convenient when special materials are not available. Caution should be exercised in areas where soils are shallow and where excavation may expose highly permeable sub-grade materials. Sealing is essential if this cannot be avoided and a wide variety of materials can be employed. Compacted earth is commonly used. Sodium bentonite has also been used successfully if mixed with soil to a depth of 15 cm. Mixing prevents cracking when the water level in the pit is lowered. Chemical additives such as sodium salts which act as dispersants have also been effective. Should soil properties not be amenable to compaction or chemical treatments, membranes, films or hard surface linings can be used. Asphalts, platics, synthetic rubber and concrete have all been installed with successful results (Dedrick, 1975).

Evaporation can be a serious source of water loss from excavated storage pits. Polystyrene rafts (Fig. 23.4) have been developed (Cluff, 1972) as aids in suppressing evaporation and have proven to be very effective. Other materials include butyl rubber or reinforced plastic, both of which have been shown to be effective on smaller ponds. In addition to reducing evaporation losses, physical covers prevent foreign material from entering storage areas and also reduce algal growth.

Storage bags of butyl-coated nylon (Fig. 23.5) offer completely closed systems which eliminate losses by evaporation and seepage. The bags are

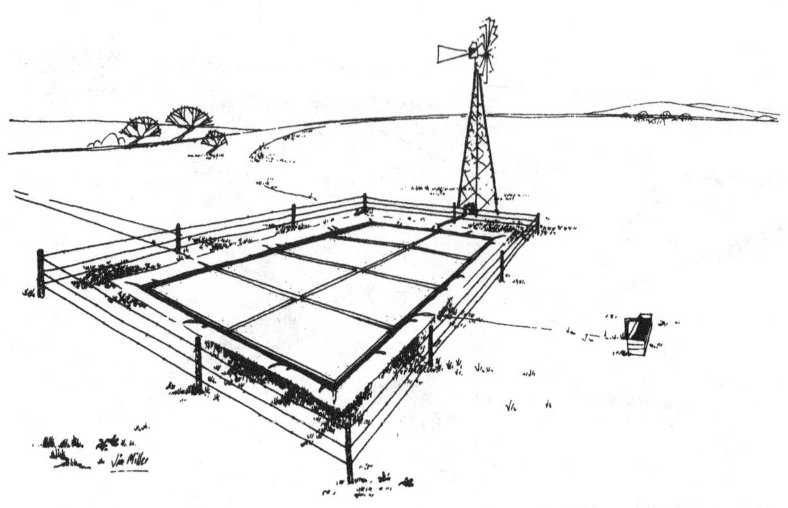

Fig. 23.4. Polystyrene rafts floated on water surfaces reduce evaporation from open storages.

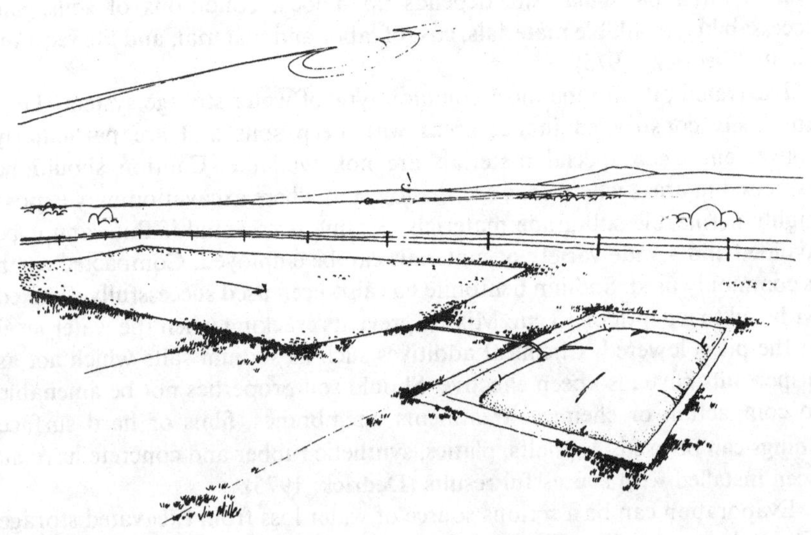

Fig. 23.5. Butyl-coated nylon water storage bag.

designed with water inlets, outlets and an overflow provision. They are susceptible to vandalism and vermin, but with proper management this danger can be minimized.

Tanks of metal or concrete are commonly used for storage of harvested water. They have a high initial cost, but greatly reduce seepage and evaporation and have a long life. A summary of advantages and disadvantages of various types of storage systems is presented in Table 23.2 (Fraiser, 1975).

Table 23.2. *Water costs for various water-harvesting treatments.* (After Frazier, 1975)

Treatment	Run-off (%)	Estimated life of treatment (yr)	Initial treatment cost ha^{-1} ($ US)	Annual[a] amortized cost ha^{-1} ($ US)	Water cost in a 500 mm rainfall zone ($ US Ml^{-1})
Rock outcropping	20–40	20–30	< 120	< 240	50–100
Land clearing	20–30	5–10	120–240	< 120	67–100
Soil smoothing	25–35	5–10	600–840	120–240	55–158
Sodium dispersant	40–70	3–5	840–1440	120–240	30–100
Silicone water repellents	50–80	3–5	1440–2160	240–480	50–158
Paraffin wax	60–90	5–8	3600–4800	600–1200	110–330
Concrete	60–80	20	24000–60000	2040–5280	420–1450
Gravel covered membranes	70–80	10–20	6000–8400	480–1200	100–282
Asphalt fiberglass	85–95	5–10	12000–24000	1680–5760	290–1110
Artificial rubber	90–100	10–15	24000–36000	2520–4920	415–890
Sheet metal	90–100	20	24000–36000	2040–3120	335–570

[1] Based on the life of the treatment at 6% interest.

Micro-catchments and strip farming

There are other water-harvesting systems which make use of run-off as it is collected, eliminating the storage requirements; among these are the related techniques of micro-catchments and strip farming. The use of micro-catchments involves the preparation of small catchment areas in which one to several plants are grown on the low side (Fig. 23.6). The collection area may range from 20 to 1000 m² depending upon the precipitation of the area and plant requirements. The micro-catchment procedure may be used in complex terrain where other water-harvesting techniques may be difficult to install.

Strip farming is a modification of the micro-catchment method. Berms are erected on the contour and the area between them prepared to serve as a precipitation collection area. Run-off occurring between the berms may then be concentrated above the down-slope berm to irrigate the crop or forage planted there.

Both micro-catchments and strip farming can be successful techniques in years of normal precipitation. However, in dry years most annual crops will fail. It is therefore advisable to select drought resistant plants for use with these systems.

Maintenance

All types of water-harvesting catchment and storage require some maintenance. Compacted and smoothed soil techniques, for example, are highly susceptible to weed invasions which can reduce catchment efficiency. Ground cover

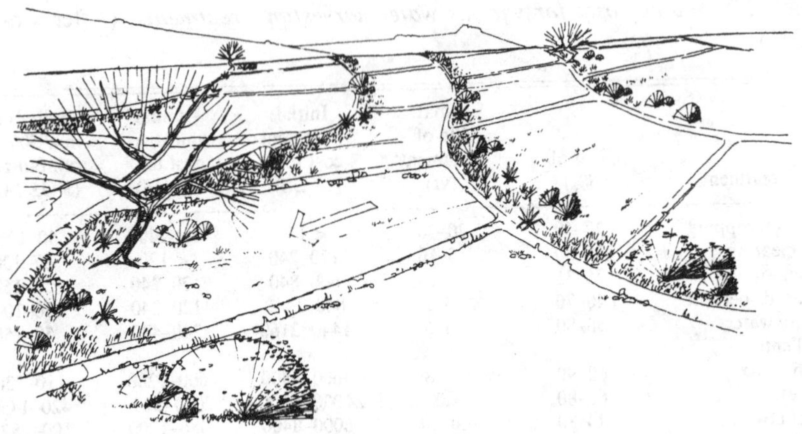

Fig. 23.6. Micro-catchment systems can be used to concentrate run-off from precipitation events.

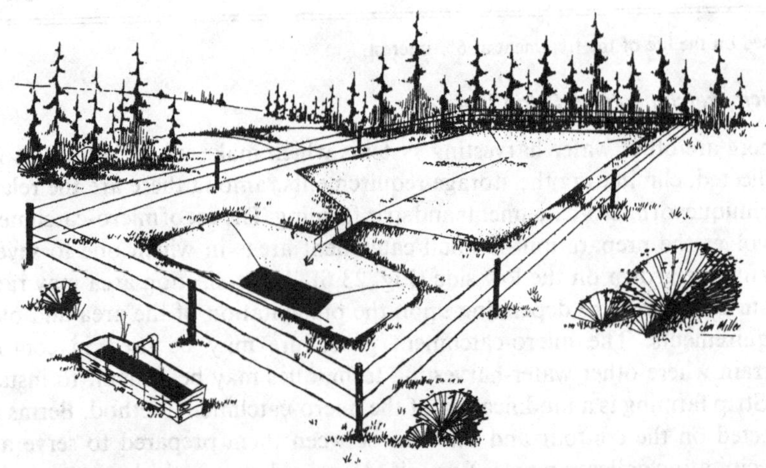

Fig. 23.7. Fencing is frequently necessary to reduce damage by livestock and wildlife to water harvesting systems.

treatments such as plastic membranes are easily damaged by livestock or rodents.

In order to avoid stock damage, catchments and storage areas are commonly fenced (Fig 23.7). Protection against rodents may be increased by installing a barrier of sheet metal or other appropriate material around the system area extending above and below the ground suface. This type of protection however, does not diminish the need for frequent visits to water harvesting sites to check for damage and conduct necessary maintenance. In areas which receive a limited number of run-off events per year, the loss of water from only one can be critical.

Water conservation on upland watersheds

Conservation of range lands is perhaps the most important aspect of water resource management on upland watersheds in arid lands. Maintenance and use of water where it falls (on site) for the production of vegetation and to control run-off is a first priority but must be followed by implementation of wise range management programs. Cultural treatments (mechanical manipulations) of the soil surface are often necessary before a range management system can be implemented. Their main purpose is to retain sufficient run-off for the establishment of a more permanent vegetative cover. They include a wide spectrum of structures such as pits, basins, contour furrows, contour trenches, diverters, check dams and water spreaders. The treatments have a life expectancy of only a few years, although maintenance can prolong this period. Some of the more common types of extensive cultural treatments follow.

Pitting

This is an old technique practised by the ancient Persians who dug pits at regular, closely spaced intervals on gently sloping lands and planted their crops in the depressions. In modern practice, pits about 30 × 60 cm and 15 cm in depth are dug with a modified farm disk harrow. Such a pitter can be used on slopes up to 30%, and can increase the surface retention storage of a watershed by 100 m^3 ha^{-1} (United States Forest Service, 1965). Although pitting has been reported to quadruple the production of perennial grasses on Sonoran desert ranges the first year, effectiveness was considerably less in subsequent years, attributed to excessive competition from annuals and forbs (Slaybach & Renney, 1972).

Basins

These provide considerably more retention storage space for run-off and endure longer than conventional pits. They generally measure 200 × 180 cm with a depth of 20 cm. They have been reported to improve range production five times more than conventional pits and nine times more than untreated areas (Slaybach & Renney, 1972). Neither pits nor basins require that the contour be followed exactly.

Fig. 23.8. Contour trenches concentrate overland flow for use by plants.

Contour furrows

These are small continuous ditches 20–30 cm deep which follow the contour. Many of the early trials with furrows failed, largely due to the difficulty of following the contour and lack of follow-up management. Normally, only every third or fourth contour is surveyed and cross bars are constructed within the furrow to act as baffles to lessen the effects of inaccuracy. On slopes greater than 5%, soil is thrown to the down-slope side. On slopes greater than 5% the soil can be thrown to either side. Studies with *Atriplex* spp. planted on furrowed slopes showed earlier spring growth and larger summer and fall growth and produced significantly higher seed crops than untreated areas (Weir & West, 1971).

Contour trenches

These are larger versions of contour furrows and can be used in gullied areas and on slopes up to 70% (Fig. 23.8). On slopes up to 30% trenches may be of the outside type (the excavated material holds the run-off) and on slopes up to 70% the inside type (the excavation holds the run-off) is used (United States Forest Service, 1965). On steep slopes closely spaced cross bars are necessary. Contour trenches are usually cut into deep soil horizons. If a clay layer is encountered the trench may seal. However, if the material is porous

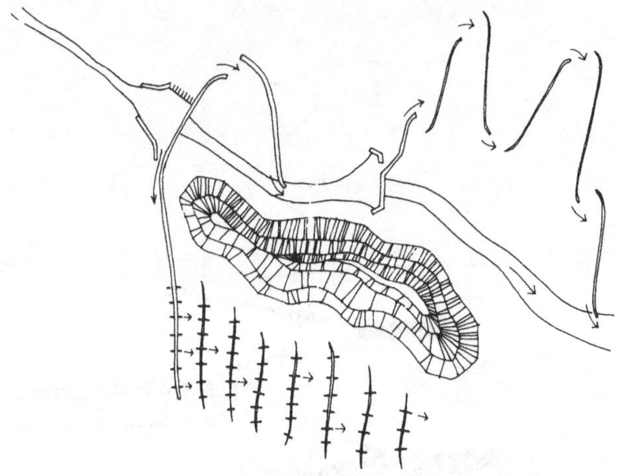

Fig. 23.9. General illustration of a water spreading system.

the soil may become supersaturated and lead to land slips and slides. During re-vegetation, the best initial stands occur in the trench bottoms with thinner stands on the excavated material and the thinnest on the excavation.

Water spreading

Water spreading is one of the more effective means of utilizing ephemeral stream flow for production of crops or forage. It is adaptable to gently sloping alluvial valleys with upstream watersheds that are subject to at least three or four run-off events annually. The simplest system (and one that probably predates irrigation and is still employed in many arid areas) is a series of low earth dams placed across a minor drainage to retain a portion of storm flow. Grain crops are usually planted directly in the wash behind the dams. If a natural levy is present on the drainage, the effective area of wetting can be increased by constructing larger dams which spill excess water around the ends into adjacent flood plains. If the drainage is sizeable, a more complex system can be used consisting of a diverter to conduct water away from the stream channel, a series of cascading dikes arranged in zigzag fashion down slope on the flood plain and a system to return large run-off events safely to the stream channel (Fig. 23.9). The diverter can conduct all or part of the stream flow directly into the dikes, which fill and spill into other dike areas at lower elevations, or direct a portion into storage for later release during dry periods.

Water spreaders are of two general types: (*a*) the ponding type, used on slopes up to 2% and on soils with low infiltration capacities; and (*b*) the flooding type used on slopes up to 5% and on soils of high infiltration

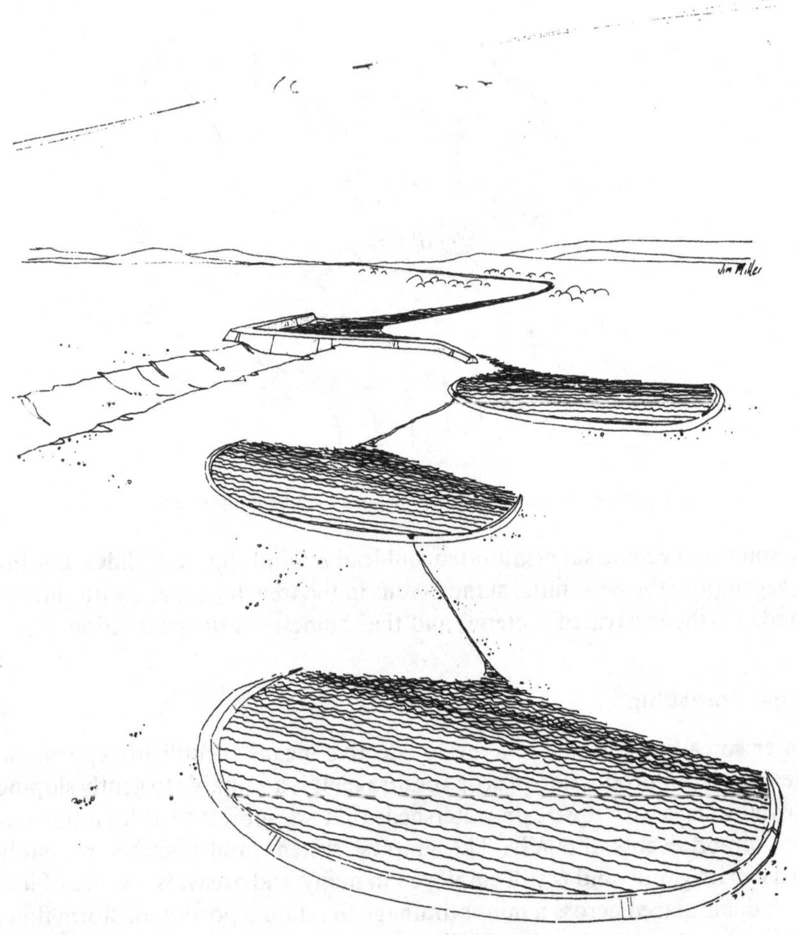

Fig. 23.10. Ponding type water spreading system.

capacities (Figs. 23.10 and 23.11). The ponding type is designed to allow water to spill from around the ends of the dike through broad spillways. The flooding type releases water from points along the face of the dike. Contour furrows may be used with either system to assure even spreading and prevent channelization.

The design of a water-spreading system requires basic data on watershed characteristics, precipitation and stream flow. Maintenance is of course, necessary and there is danger of adverse effects of sedimentation and salinization. Nevertheless, there are many potential sites in arid and semiarid lands suitable for water spreading.

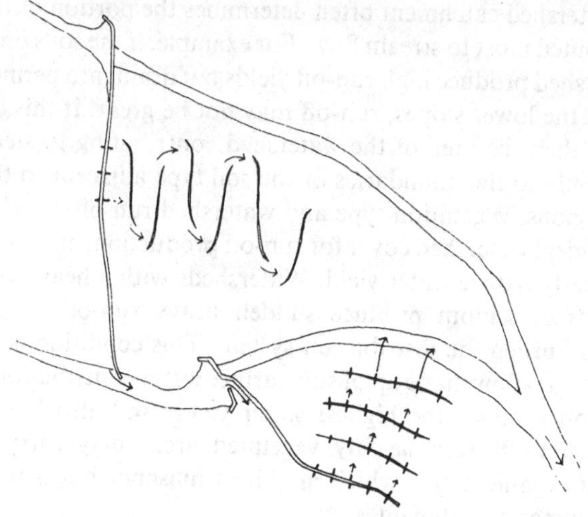

Fig. 23.11. Flooding type water spreading system.

Watershed characteristics

The parameters that characterize the water yield of a catchment basin, are highly variable and complexly interrelated. They include a great variety of geologic, topographic, soil and vegetative factors.

Geology is an important consideration in the design of water-spreading systems. If the rock layers or strata are inclined into the drainage area, conditions should be favorable for good run-off production. If the reverse is true, run-off may be very low. Level bedded formations often provide excellent collection areas.

Rainfall usually increases with elevation. Generally, therefore, the higher the basin, the greater the opportunity for sufficient run-off. Normally, north-facing slopes yield higher run-off (in the northern hemisphere). If the drainage area consists of many small, narrow valleys with steep slopes at right angles to the streams, the run-off will probably be rapid and total time of run-off short. With broad, flat valleys, the period of run-off will be much longer.

Stream grades also affect the period of run-off and peak flows. A long, narrow drainage will have a longer run-off period and a lower peak than one which is wide and relatively short. If stream gradients are steep, large amounts of bed load and suspended sediments may be carried during flow periods.

Soils greatly affect the run-off characteristics of watersheds. In general, shallow soils over clay or impervious rock yield high volume run-off and high peak flows. Those with high clay content behave similarily. Deep, permeable soils will absorb water rapidly and release it over a longer period of time. However, soils are rarely uniform over a catchment area. Their arrangement within a watershed catchment often determines the portion of the watershed that contributed most to stream flow. For example, if the soils on upper slopes of the watershed produce high run-off yields but drain into permeable alluvial material on the lower slopes, run-off may not be great. If this configuration is reversed, then the area of the watershed contributing to steam flow may be limited only to the boundaries of the soil type adjacent to the drainages.

In dry regions, vegetation type and watershed run-off are closely related. Grass is an ideal watershed cover for run-off production, and its improvement will not greatly reduce water yield. Watersheds with a heavy cover of grass, shrubs or trees seldom produce sudden heavy run-off or carry enough sediment to burden the distribution system. This condition is favorable for sustained stream flow made up of sub-surface water contributions. Areas bare of vegetation produce the highest water yields and also large volumes of sediment. Run-off from heavily vegetated areas may carry considerable amounts of organic debris which may be a nuisance but is much preferred to large volumes of sediment.

Water-spreading works best for deep-rooted crops which can tap stored water. Annuals require rain at the beginning of the growing season and usually at intervals thereafter. Plant varieties that can withstand intermittently dry and wet soils are most suitable. Generally, a minimum soil depth of 1.5 m is required for the crop growing area. Water storage facilities for supplemental irrigation are necessary for shallow soils.

Rainfall–run-off characteristics

The distribution of rainfall and the frequency of high intensity storms is important in water spreading. If the average annual precipitation is 200 mm and half of it occurs in June and July, it is desirable to know whether there are 15 days with rain during this period or only five in the average year. In general, to justify a water-spreading system, a minimum precipitation of 80 mm per rainy season is required if the season occurs during the cold period of the year. A higher amount is required if the season occurs during summer.

When run-off measurements are lacking, the slope–area formula is commonly used to compute peak discharge in channels. The computation is based on water marks and drift lines left by high flows. Sufficient high-water marks must be located along a reasonably straight, smooth portion of the stream channel to permit determination of the water-surface slope at the time of the peak. Cross sections of the channel are then determined, usually by leveling. With an estimate of the channel roughness, the peak rate of flow is computed by either the Chezy or the Manning formula. The volume of flow is difficult to estimate by this procedure, but if flow durations can be estimated from available run-off data, a simple triangular hydrograph can be used to estimate volume of flow.

Both volume and frequency of flow are important. While one flow per year may supply sufficient moisture to the soil to produce a satisfactory crop, frequent floodings will produce more yield. A small flow volume, especially if combined with infrequent flow, may not be worth the cost of diversion. A large volume may be too difficult to handle for the size of the available spreading area.

The desired rate of water application is also an important factor in system design and is dependent upon the infiltration characteristics of the soil, the tolerance of the plants to inundation and the normal season of available water. For agricultural crop production, the distribution of rainfall and run-off events during the growing season is crucial to a successful harvest, particularly for shallow-rooted crops. For forage production, where surface and internal drainage on the spreading area are good and the water supply abundant, applications up to 300 mm in depth over the area are not excessive, particularly if flows can be expected only once or twice during the growing season. If the spreading area is great and the water supply limited, the application of 75–150 mm of water, when available, may produce more forage than heavier applications on a smaller area.

If sediment of high silt content is deposited at low volumes per event, it may be beneficial. The spreading water tends to deposit the sediments in a gently sloping plane which fills the low spots and other irregularities. Silt deposited in such a manner actually improves the site for spreading purposes. It also adds to the soil depth and generally improves growing conditions. Unlike most types of sediment basins, a water-spreading system can function indefinitely without substantial loss of storage capacity. As sediment does accumulate, the height of the dikes can be increased and the storage capacity restored and enlarged. Rapid and heavy sediment accumulation, however, usually indicates serious soil erosion conditions in the drainage area upstream, a condition which should be remedied promptly to prevent rapid losses in storage capacity.

Design problems and maintenance

Few water-spreading systems are free from flaws when first constructed, and all require maintenance. Sedimentation is often a major problem. If sediment accumulation is great, there may be heavy annual maintenance work to raise the dikes and adjust dike design as needed. In some cases, channels behind diversion dikes may have to be cleared of sediment or the floor of the diversion raised to provide sufficient flow in channels. Very often maintenance costs can be reduced by improving vegetation on the upper watershed. Run-off delayed on the upper watershed may save dikes from being over-topped and breached and will reduce sedimentation.

High velocity concentrations of water along certain sections of a dike may

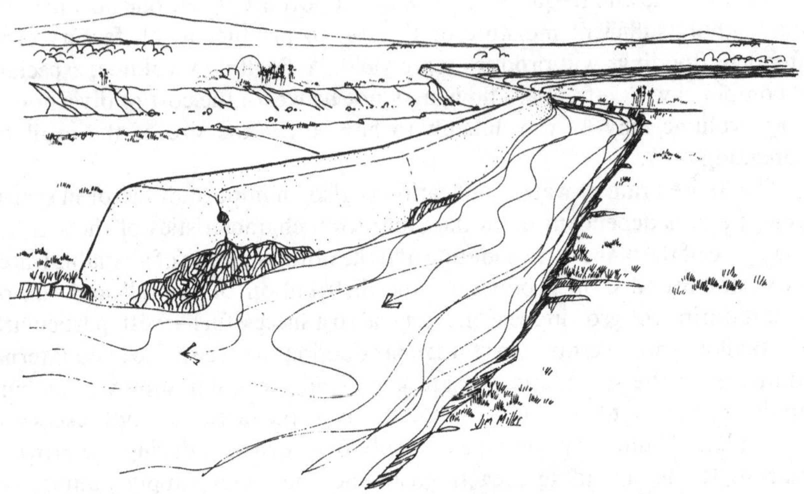

Fig. 23.12. Excessive flow velocity can cause damage to diversion dikes.

over-top or break through the dike (Fig. 23.12). This may require reducing the inflow at the head of the system, relocating or widening the spillway from the dike above, or constructing a diversion dike or ditch. The upper dike may have to be raised or reinforced. A concentrated flow may require additional small plugs or dikes in natural drainage ways, in old road ruts and trials, in low spots, in basin lips, in the borrow pits, etc., to turn out the water at frequent intervals. Often in wild flooding systems, only the primary dikes are constructed at first. Construction of secondary dikes is delayed until a flow of water has gone through the system to indicate where subsequent structures will be most effective.

Some areas may not receive sufficient directed water. A short ditch or dike directing water to the dry spot may solve the problem or a pipe may be installed to let water through the dike to the dry spot. Relocation of the spillway may be necessary. Irrigation of the entire spreader area should not be expected for every flow event, and if 65–85% of the area receives a good wetting from an average flow, the design is successful. The problem of too little or too much water coming through the openings on flood spreaders can be corrected by adjusting the size of the openings.

Improper diversion of water flowing into the borrow pit and that moving directly towards the next dike is a common problem that can be corrected with ditches or wing dikes. In straight channels which lie parallel to dikes, obstructions may deflect the current against the loose dike filling causing wash-outs or cave-ins.

Slight gullying at the spillways is common. If the gullies spread or entrench

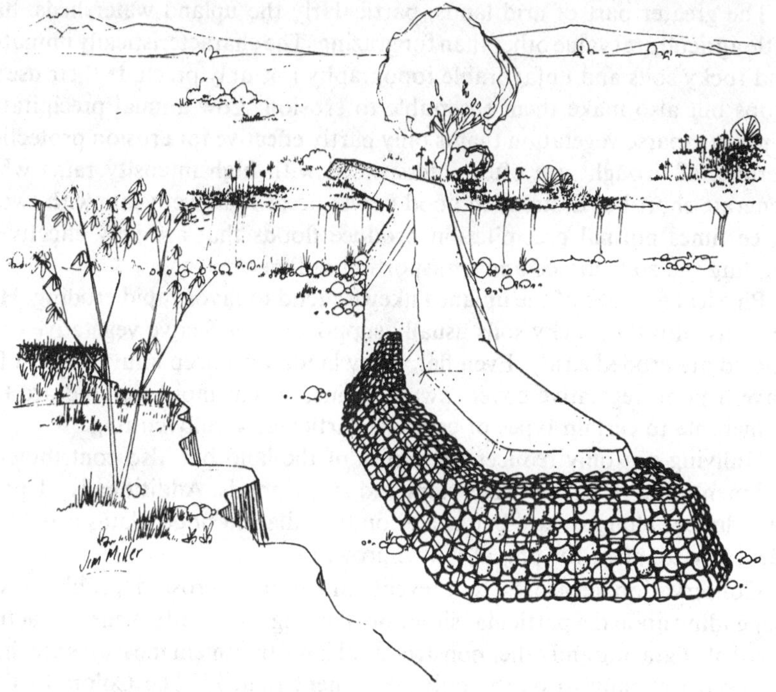

Fig. 23.13. Riprap is sometimes used to protect areas of dikes which are susceptible to erosion.

quickly, it may be necessary to widen the spillway by lowering the dike at the spillway end. Additional openings or pipes may also be necessary. Gullying in the return channel may also occur. If head cuts develop, the water may be diverted to more favorable locations. It is often impossible to avoid some erosion before an adequate vegetative cover can be developed to withstand erosive flows. Maintenance of the return channel may require protection from grazing.

Excessive widening or plugging of the spreader dike spillway may occur as a result of erosion at the end of the dike. A riprap face on the end of the dike usually prevents widening (Fig. 23.13). Sediment or debris deposits which plug the system may have to be removed or smoothed out from time to time.

Erosion and sedimentation control

Arid lands yield large quantities of sediment to streams causing serious erosion problems and resulting in losses of reservoir storage capacity and damage to valley bottomlands. Erosion control is almost always an integral part, and often the primary objective, of water resource management in arid regions.

Management of arid lands

The greater part of arid lands, particularly the upland watersheds, have little agricultural value other than for grazing. The characteristically immature and rocky soils and unfavorable topography not only preclude their use for crops but also make them suseptible to erosion. Low annual precipitation supports sparse vegetation that is only partly effective for erosion protection. Periods of drought are often interspersed with high intensity rains which generate short but destructive flood flows; occasional wet years with two or three times normal precipitation produce floods that are very effective in eroding the soil surface and transporting sediment.

Physical features of the uplands likewise tend to favor rapid eroding. Hilly terrains with thin rocky soils usually support an ineffective vegetative cover and so are eroded easily. Even flat valley lands with deep alluvial soils often have a poor vegetative cover, owing to insufficient moisture, and are thus vulnerable to certain types of erosion, particularly gully cutting.

Gullying not only reduces the utility of the land but also contributes to sedimentation in stream, reservoirs and stock ponds. Additionally, deposits of sediment along stream banks and on the adjacent flood plains provide an ideal environment for phreatophyte growth.

Conservation measures to prevent and correct erosion problems vary depending upon the particular situation. Among the land-treatment practices, control of grazing and other nonstructural land-treatment measures are firmly believed by many to be the only permanent remedy. The Colorado River Basin can be cited as one example. Beginning about 1942 there has been a gradual reduction in the ratio of suspended-sediment discharge to annual run-off. This reduction coincided with a reduction in livestock population and an active conservation program. Another dramatic example is the hydrologic change that occurred in the 1980s in the Sonoran Desert of southwestern United States. The history of severe overgrazing in the region during the previous two decades is well documented. Rivers that ran more or less consistently throughout the year became intermittent leaving the channels dry over much of their length (Hastings & Turner, 1972). Irrigation ditches were left high and dry and the valley floors became dissected as run-off cut steep-walled trenches in river channels and drainage ways.

The effectiveness of small structures such as rock and brush spreaders, gully plugs, and small diversion dams is easily evaluated since it is possible to observe how the structures act and the changes that occur in their immediate vicinity. An inherent weakness observed in many conservation structures is that it is often necessary to construct them without adequate hydrologic data. The inadequacy of many structures so designed is reflected in a shortening of their effective life by sediment and flood flows that arrive at a rate much higher than was anticipated.

In depleted arid lands, an urgent need in conservation at present is to develop methods for repairing the gullied channels that dissect once productive

Fig. 23.14. Check-dams serve to retard flows in gulleys and also allow sediment to settle. Eventual filling of the gulley will result.

valleys. All types of conservation measures must be evaluated on the basis of long-term benefits. The simple, inexpensive structure may prove in the long run to be quite expensive since its effectiveness may be only temporary. Larger structures, although more costly initially, frequently offer a more favorable economic return.

In many arid areas ancient man learned not only how to control gullies but how to use them for the collection of water and production of crops. A thousand years ago the Indians of Chihuahua and Sonora, Mexico, created field and garden plots in the sediment accumulated behind loose rock check-dams which they constructed across gullies and valleys. Similar and more sophisticated techniques have been practiced in other parts of the arid world. In some instances, substantial populations have been supported in this way.

In addition to growing plants, other aspects of gully treatment such as hydraulics, sedimentation, soils, and sometimes the logistics required for the management of the watershed must be considered. For instance, management may call for deposits of maximum possible depth at strategic locations to provide shallow gully crossings. The largest possible sediment accumulations will be required to achieve this objective. If sediment catch is a desirable objective, large dams should be built.

The most commonly applied engineering measure is the check-dam (Fig. 23.14). Forces acting on a check-dam depend on design and type of construction material. Non-porous dams with no weep holes, such as those built from concrete (Poncet, 1965; Heede, 1965; Kronfellner-Kraus, 1970), receive a strong impact from the dynamic and hydrostatic forces of the flow. These forces require strong anchoring of the dam into the gully banks. In contrast, porous dams release part of the flow through the structure, and thereby decrease the head of flow over the spillway and the dynamic and hydrostatic forces against the dam. Much less pressure is received at the banks than for non-porous ones. Once the catch basin of porous or non-porous dams is filled by sediment deposits, structural stability is less critical because the dam crest has become a new level of the upstream gully floor.

Loose rock can be used in different types of porous check-dams. They may be built of loose rock only, or the rock may be reinforced by wire mesh, steel posts, or other materials. They are readily constructed by manual labor (see Heede (1966) for details of design construction and maintenance of large rock dams).

Other types of filter dams feature vertical grids or grids installed at an angle to the the vertical. Such dams are described by Puglisi (1967), Kronfellner-Kraus (1970), and Fattorelli (1970).

Vegetation management

The conversion of vegetation from one type to another to increase run-off presents another aspect of management of arid-land resources for water. Normally, brush or tree species are replaced by grasses, the premise being that in this way both infiltration and transpiration will be reduced, leaving additional surface water available for run-off which may be utilized in drier valleys below.

Vegetation management for water is normally implemented on the mountain watersheds of arid zones for it is in these areas that most precipitation and run-off occur. Watershed vegetation types change with elevation, and the effect of manipulation on producing run-off varies with each vegetation zone (Ffolliott & Thorud, 1974).

At higher elevations, such as in Arizona's conifer forest vegetation zone, it has been found that patchcutting will yield run-off increases proportional to the area in cleared openings (Rich & Thompson, 1974). Most of the yield increase can be accounted for by decreases in transpiration and a subsequent decrease in soil moisture deficit, stimulating run-off. Another factor influencing the run-off increase is additional snow accumulation and melt rate in cut openings. Increases ranging from 25 to 45%, depending upon precipitation amount and duration, can be expected based upon a 33% patchcut of conifer vegetation replaced with grass (Fig. 23.15) (Rich & Thompson, 1974).

At slightly lower elevations water yield increases of 30–50 mm have been obtained by reducing tree density. Various harvesting patterns used include

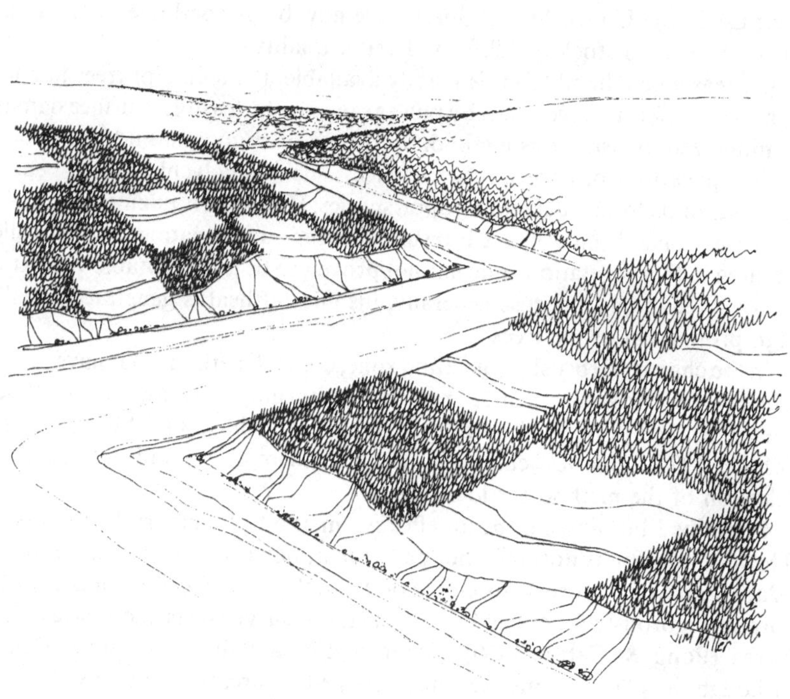

Fig. 23.15. Vegetation management practices, such as patchcuts, under proper conditions can result in increased water yields from watersheds.

strip cuts, thinning, clearcuts and shellerwood cuts. In an Arizona study of a *Pinus ponderosa* zone, increases in run-off varied from 16% in an area which was 32% clearcut in uniform strips to 103% in an area which was clearcut and thinned reducing basal area by 65% (Brown *et al.*, 1974).

Several vegetation removal techniques are used in management programs involving smaller coniferous species. One mechanical technique is cabling in which larger trees are uprooted by a heavy cable pulled between two bulldozers, smaller trees missed by the cable can be hand chopped. Seeding of grasses may follow. No significant increase in run-off from this zone has been reported following treatments by mechanical techniques (Clary *et al.*, 1974), due in part to the large depressions left by uprooted trees which intercept overland flow.

A second removal method in this zone involves the use of herbicides sprayed from a light plane or helicopter. An advantage of this method is that the ground surface is not disturbed. Increases in streamflow of 65% have been reported following application of 2.8 kg of picloram and 5.6 kg of 2,4-dichlorophenoxyacetic acid ha^{-1} (Clary, *et al.*, 1974). While herbicide use shows promise, environmental precautions have for the time being restricted their use in the United States. Studies are now being conducted to ascertain their effect upon stock, wildlife, and water quality.

In areas where hand labor is readily available, the felling of trees in zones of small coniferous vegetation by power saw may be feasible. Surface damage is minimized by using this method. Sprouting of stumps should be inhibited with applications of appropriate chemicals such as polychlorinated benzoic acid and/or picloram, depending upon species. However, no significant run-off increases have been reported resulting from this treatment. Controlled burning in this vegetation zone has not proven to be an acceptable vegetation removal technique, because generally, insufficient heat is generated to kill a high proportion of the trees.

In the chaparral (brush) vegetation zone, control methods that have proved effective in Arizona include prescribed burning, mechanical methods, chemicals, and prescribed burning followed by chemicals (Hibbert, Davis & Scholl, 1974). Species composition and seasonal rainfall influence the selection of the method for local use.

Prescribed burning can be an effective method of chaparral suppression. The modification is normally not permanent, however, due to the enormous adaptability of the brush species. For example, oak (*Quercus turbinella* G.) sprouts vigorously following fire. In one test it survived six successive annual burns (Pond & Cable, 1960). Other species are fire susceptible, that is, although the plants themselves may be killed by fire, the heat may stimulate their seeds to germinate. Characteristics such as these reduce treatment effectiveness and increase costs.

The primary mechanical brush suppression method is rootplowing. Rough terrain is the main obstacle to the use of this technique and its use on steep catchments is very limited Another disadvantage is that the soil surface is greatly disturbed increasing infiltration rates and reducing run-off.

As is the case in small coniferous vegetation zones, the most effective means of treating large areas of brush is by the use of chemicals. Of these the largest reliance has been placed on the phenoxy herbicides, primarily for economic reasons. Application is from the air. However, due to the resprouting capability of many of the chapparal species, five to six years of annual treatment may be necessary to achieve a high amount of kill. Use of herbicides should be carefully evaluated against possible environmental dangers.

Water increases resulting from conversion of the chaparral vegetation zone can be very significant. Hibbert *et al.* (1974) report increases ranging from

23 to 703% (6 to 68 mm) from treatment of brush watersheds. (Annual precipitation in the test area is 500 mm.) Normal treatment in the test cases was 100% brush removal; reductions of 50% gave a much smaller response.

The large scale management of desert grassland vegetation to increase run-off has not proved economical. Precipitation rates are too low and evaporation rates too high to produce sufficient water to justify treatment although water harvesting on a small scale has been successful.

In summary, vegetation modification can result in varying increases in run-off depending upon vegetation zone, treatment technique and treatment intensity. Given the essential nature of water and the demonstrated potential for increasing its availability through vegetation management, it is probable that decisions to manage vegetation in arid zones ultimately will be made.

Control of transpiration and evaporation

Transpiration losses

In the arid system, riparian species, particularly phreatophytes, are notorious consumers of water. In Arizona, in southwestern United States, it is estimated that phreatophytes consume 740 million m³ of water annually (Ffolliott & Thorud, 1974). Consequently, most of the research on transpiration control has been directed at phreatophyte removal.

Rootplowing is the most common mechanical eradication method. However, this technique also kills a large proportion of grass cover. When grasses form an appreciable part of the ground cover, therefore, rootplowing may not be satisfactory. In these cases a rotary type mower can be used to cut shrubs several inches above ground, but then sprouting becomes a problem requiring chemical follow-up. Also larger diameter trees cannot be felled by mowing, requiring the use of power saws to complete removal.

Chemical treatment of phreatophytes has not produced satisfactory results. Sprouting rapidly reduces treatment effects, and dangers of water pollution restrict wide useage. Antitranspirants have been used in stemming phreatophyte growth by reducing the rate of photosynthesis. While studies of this method show some success, the cost of treatment is high (Brooks & Thorud, 1971). Application of antitranspirants by aerial spraying has also presented problems since it does not reach the underside of the leaves. Manual application is time consuming and costly.

Only limited information is available on the actual change in the water balance after phreatophyte removal. While it is obvious that the water transpired by riparian vegetation will be saved by their removal, there may be an increase in other water losses such as transpiration by shallow-rooted species and evaporation from soil, stream banks, water table, and stream channel. The contention that phreatophyte removal will increase available water is therefore not unanimously accepted. There are also environmental objections to removal of phreatophytes such as the loss of wildlife habitat and recreational values. Furthermore, stream banks may be destabilized by the removal of vegetation leading to increased levels of sedimentation in streams and reservoirs.

Loss from bare soil

In some arid areas, particularly in the vicinity of man-made reservoirs and water transport and distribution systems, the local water tables may be close to the surface. If the water table is 1.5. m or less below the surface, evaporation loss from bare soil in these areas may be on the order of 50% of that from a free water surface. At water table depths of 1.5–3 m losses may drop to about 15–20%. With proper drainage management it is possible to maintain the water table at appropriate depths to minimize losses. Evaporation retardants such as bitumen emulsion have been applied to the soil surface but so far results have not warranted large scale operational programs.

Asphalt and plastic film barriers, laid 0.6 to 0.9 m below the surface to retard both percolation through the soil and capillary rise from shallow water tables are feasible in some situations. Asphalt has been found superior to plastic film in initial cost and durability.

Loss from free water surfaces

In deserts, water loss from free water surfaces may frequently exceed 10 mm day^{-1}. In addition to the loss of water, evaporation increases the salinity of remaining water. Early research of this problem was concerned primarily with the use of floating monomolecular films of long-chain alcohols. This procedure has not proven to be effective due to dispersive actions of wind. Also the water beneath the film increases in temperature due to insolation and where the film is removed by wind, evaporation may actually exceed control conditions.

Perlite ore distributed over pond surfaces may reduce evaporation by approximately 20% and, although subject to wind movement, will redistribute itself after the wind ceases (Cooley & Cluff, 1972). Other solid surfaces which have received attention as floating covers for ponds include styrofoam and foamed wax blocks. Evaporation reduction of over 50% may be achieved with these materials (Cooley, 1973). A key characteristic of both is their ability to reflect heat thereby preventing too great an increase in water temperature.

Groundwater

Despite the limited amounts of surface water normally available in arid areas, it is common to find substantial groundwater reservoirs beneath the same lands. In many cases this water is located at depths which permit economical extraction.

Groundwater may exist under either atmospheric or artesian pressure. In withdrawing water from aquifers under artesian pressure, it is important that extraction be managed to maintain artesian conditions for as long as possible. Loss of pressure increases pumping requirements for which power and equipment may not be available.

Recharge areas for arid land groundwater systems are usually located in mountain ranges. Precipitation in these areas is almost always higher than that occurring in the surrounding plains and valleys due to orographic effects. Water from recharge areas may travel great distances to reach the well where it is extracted. Because of the slow rates of groundwater movement, aquifer extraction rates frequently exceed recharge rates.

Groundwater bodies are not always formed by the accumulation of recharge. Many have been formed by rapid shifting of the land which covered river system and lakes. Recharge to such aquifers may be insiginificant.

The quality of arid-land groundwater is highly variable and depends upon its origin, travel path and the history of local surface activity. High salinity is common and frequently increases with depth. Quality may also vary at different areas within the same aquifer. In the Tucson Basin aquifer for example, water from pairs of wells located less than a few km apart vary in water salinity from less than 100 p.p.m. to more than 1000 p.p.m. (Dutt & McCreary, 1970).

Groundwater managers must maintain careful records of aquifer pumping volumes and water level changes. They are frequently faced with economic or social pressure to pump water in volumes far exceeding recharge and in such cases these records may be used to clearly demonstrate the consequences of heavy pumping. For the same reason, managers must have a thorough understanding of the aquifer physical system, such as, recharge areas, rates of recharge, and water quality variations within the aquifer. Without these management tools, errors are likely to occur, which may lead to aquifer depletion.

Acknowledgement

We would like to thank Mr James D. Miller for his skilled pen work in illustrating this chapter.

References

Brooks, K. W. & Thorud, D. B. (1971). Antitranspirant effects on the transpiration and physiology of tamarisk. *Water Resources Research*, **7**, 499–510.

Brown, H. E., Baker, M. D., Rogers, J. J., Clary, W. P., Kovner, J. L., Larson, F. R., Avery, C. C. & Campbell, R. E. (1974). *Opportunities for increasing water yields and other multiple use values on ponderosa pine forest lands*. United States Department of Agriculture Forest Service Research Paper RM-129, Forest and Range Experiment Station, Fort Collins, Colorado.

Burdass, W. J. (1975). Water harvesting for livestock in Western Australia. In: *Proceedings of Water Harvesting Symposium*, pp. 8–26, United States Department of Agriculture, Agricultural Research Service, ARS W-22 Phoenix, Arizona.

Chiarella, J. V. & Beck, W. H. (1975). Catchments on Indian lands in the southwest. In: *Proceedings of Water Harvesting Symposium*, pp. 104–14, United States Department of Agriculture, Agricultural Research Service, ARS W-22 Phoenix, Arizona.

Clary, W. P., Baker, M. B., O'Connell, P. F., Johnsen, T. N. & Campbell, R. W. (1974). *Effects of pinyon-juniper removal on natural resource products and uses in Arizona*. United States Department of Agriculture Forest Service Research Paper RM-128, Rocky Mountain Forest and Range Experiment Station, Fort Collins, Colorado.

Cluff, B. C. & Dutt, G. R. (1975). Economic water harvesting systems for increasing water supply in arid lands. In *Watershed management in arid zones*. (ed. J. L. Thames). Short course presented at Saltillo, Mexico by the Agency for International Development, March 1975. Asociacion el formento de la Sciencio, Mexico.

Cluff, C. B., Dutt, G. R., Ogden, P. R. & Stroehlein, J. L. (1971). *Development of economic water harvesting systems for increasing water supply*. Department of Water Resources Research Project No. B-005-AMZ, Completion Report. University of Arizona, Tucson.

Cluff, C. B. (1972). *Low-cost evaporation control to save previous stock water*. Arizona Farmer-Ranchman, Phoenix, Arizona.

Cooley, K. R. (1973). Evaporation reduction with reflection covers. *Journal of Irrigation and Drainage, Proceedings of American Society of Civil Engineering*, 99 (IR3), 353–63.

Cooley, K. R. & Cluff, C. B. (1972). Reducing pond evaporation with perlite ore. *Journal of Irrigation and Drainage, Proceedings of American Society of Civil Engineering*, **98** (IR2), 255–66.

Dedrick, A. A. (1975). Storage systems for harvested water. In: *Proceedings of Water Harvesting Symposium*, pp. 175–91. United States Department of Agriculture, Agricultural Research Service. ARS W-22. Phoenix, Arizona.

Dutt, G. R. & McCreary, T. W. (1970). *The quality of Arizona's domestic, agricultural and industrial waters*. Report No. 256, Agriculture Experiment Station, University of Arizona, Tucson.

Evenari, M. L., Shanan, L., Tadmor, N. & Aharoni, Y. (1961). Ancient agriculture in the Negev. *Science*, **133**, 979–96.

Fattorelli, S. (1970). Design of concrete check dams. *Monti e Boschi*, (1970), 3–20.

Ffolliott, P. F. & Thorud, D. B. (1974). *Vegetation management for increased water yield in Arizona*. Technical Bulletin 215. Agricultural Experiment Station, University of Arizona, Tucson.

Fink, D. H., Cooley, K. R. & Fraiser, G. W. (1973). Wax-treated soils for harvesting water. *Journal of Range Management*, **26**, 396–8.

Fraiser, G. W. (1975). Water harvesting: a source of livestock water. *Journal of Range Management*, **28**, 429–33.

Frith, J. L. (1975). Design and construction of roaded catchments. In: *Proceedings of Water Harvesting Symposium*, pp. 122–7. United States Department of Agriculture, Agricultural Research Service, ARS W-22. Phoenix, Arizona.

Hastings, J. R. & Turner, R. M. (1972). *The changing mile*. University of Arizona Press, Tucson.

Heede, B. H. (1965). *Multipurpose prefabricated concrete check dam*. United States Forest Service Research Paper RM-12. Rocky Mountain Forest and Range Experiment Station, Fort Collins, Colorado.

Heede, B. H. (1966). *Design, construction and cost of rock check dams*. United States Forest Service Research Paper RM-20. Rocky Mountain Forest and Range Experiment Station, Fort Collins, Colorado.

Hibbert, A. R., Davis, E. A. & Scholl, D. G. (1974). *Chaparral conversion potential in Arizona, part 1: water yield response and effects on other resources*. United States Department of Agriculture Forest Service Research Paper RM-126, Rocky Mountain Forest and Range Experiment Station, Fort Collins, Colorado.

Kronfellner-Kraus, G. (1970). Open torrent control dams. *Mitteilungen der Forstlichen Bundes Versuchsanstalt Wien*, **88**, 7–76.

Lauritzen, C. W. & Thayer, A. A. (1966). *Raintraps for intercepting and storing water for livestock*. United States Department of Agriculture, Agricultural Research Service Information Bulletin 307.

Myers, L. E. & Fraiser, G. W. (1969). Creating hydrophobic soil for water harvesting. *Proceedings of the American Society of Civil Engineers, Journal of Irrigation and Drainage*, 951, 43–54.

Myers, L. E. & Fraiser, G. W. (1974). Asphalt-fibreglass for precipitation catchments. *Journal of Range Management*, **27**, 12–4.

Poncet, A. (1965). Notes on erosion control and watershed management in the mountains north of the Mediterranean. *Revue Forestière, Francaise*, **10**, 637–61.

Pond, F. W. & Cable, D. R. (1960). Effect of heat treatment on sprout production of some shrubs of the chaparral in central Arizona. *Journal of Range Management*, **13**, 313–7.

Puglisi, S. (1967). Limpiego di dispositivi filtranti nella correzione dei torrenti. (The use of filter dams in torrent control.) *L'Italia Forestale e Montana*, **22** (1), 12–24.

Rich, L. R. & Thompson, J. R. (1974). *Watershed management in Arizona mixed conifer forests: the status of our knowledge*. United States Department of Agriculture Forest Service Research Paper RM-130. Forest and Range Experiment Station, Fort Collins, Colorado.

Slaybach, R. D. & Renney, C. W. (1972). Intermediate pits reduce gamble in range seeding in the southwest. *Journal of Range Management*, **25**, 224–7.

United States Forest Service (1965). *Watershed handbook, 2500*. Superintendent of Documents, Government Printing Office, Washington, DC.

Weir, R. W. & West, N. E. (1971). Seedling survival on erosion control treatments in a salt desert area. *Journal of Range Management*, **24**, 352–7.

Manuscript received by the editors March 1976.

24. Social aspects of managing arid ecosystems

M. D. YOUNG

Introduction

Most arid zone areas generally support an insignificant proportion of their countries' total population and hence their economies do not depend upon arid zone production.* Thus, in many parts of the world people living within the constraints set by arid ecosystems are isolated from the power structure which governs them.

This isolation has often meant that the needs and problems of people in the arid zone have been poorly understood which, in turn, has led to the introduction of policies that hinder, rather than benefit them.

In arid areas, effective rainfall is both unreliable and low, hence the way of life adopted by people living and working in arid areas throughout the world differs markedly from other areas. Most arid ecosystems are too unproductive to permit the economic establishment of improved pastures or the application of fertilizers, thus changes in production methods usually involve a change in a social system.

People in arid areas have adapted in different ways to live within the constraints set by their harsh environment. In the developing countries of the world, most people who live in the arid zones have evolved systems of pastoral nomadism without the influence of governments. More recently, some governments have sought to modify the social systems that govern nomadism and to directly influence the impact of pastoral land use. This is in direct contrast to the developed countries of the world whose governments have always had the ability to influence and manage pastoral land use by granting exclusive grazing rights over an area of land to specific individuals. The two different life-styles contrast vividly, one implies a fixed 'home' and the other a mobile 'home'. Both systems are limited by their inability to improve the resource they utilize.

In this chapter, the impact of irrigation schemes on the population in the arid zone is not considered in detail. The introduction of water to an arid ecosystem gives people greater control over their environment and releases them from many of the psychological pressures associated with arid zone communities. Recognizing this factor, many politicians have tried to introduce expensive irrigation schemes with limited success. Unfortunately people with

* In this chapter the significance of mining and petroleum industries within the arid zone is ignored.

little experience in managing arid ecosystems frequently underestimate their limited potential (Davidson, 1969, 1972; Young, 1979*a*).

Influencing pastoral land use

Pastoral nomadism has evolved to exploit the spatial and seasonal (temporal) variability of an arid ecosystem. In theory such managements permit greater rates of utilization than a sedentary grazing system. However, although higher rates of utilization are theoretically possible, their attainment requires astute resource management which is difficult to achieve when grazing rights are owned by the community at large. Often when an area is overstocked and/or overpopulated and the community is under pressure the short term returns to individual exploitation dominate. 'No rational pastoralist will reduce his breeding herd if he is grazing communal land unless he is guaranteed that his neighbours will do likewise' (Harrington, 1979).

It is probable that overgrazing and desertification begins when such communities are under pressure. This process continues in a cyclic fashion until either governments intervene or the vegetation becomes more resilient than the livestock and the people that attempt to live off it. In a subsistence pastoral economy, resilience is a physical concept while in a market economy it may be possible for economic factors to intervene before physical factors establish an equilibrium. In both developing countries and developed countries, livestock numbers frequently build up during wet periods, beyond drought grazing capacity of the areas, and with the onset of drought many of them eventually die. The difference between developed and developing countries is the linkage between the death of livestock and famine; in most developed countries famine is unknown, while in the Sahel and the Middle East it now appears that the nomadic population is regulated by the inability of livestock to obtain more nourishment from the vegetation (Tribe, 1977; Harrington, 1979). In developed countries, like Australia and the United States, some of these problems are avoided by maintaining a market economy and by granting exclusive grazing rights and responsibilities to individuals. This difference should not be interpreted as a recommendation for the replacement of pastoral nomadism with sedentary grazing, but rather a recommendation for the adjustment of existing social systems to permit resource management. There is an obvious need to develop methods of land use that enable individuals to reap the benefits of the investments they make.

Adjusting to changing conditions

Throughout history, people living in arid areas have constantly had to adjust to, and live within, the physical and social constraints of their ecosystem. It is understandable that such people are reluctant to leave their homeland even when it is overpopulated. People who have lived and worked in arid areas all their lives have generally done so out of preference or out of ignorance of other lifestyles. It is understandable that such people are reluctant to leave their homeland even when it is overpopulated. People under pressure often prefer to accept a lower standard of living than make a major social change. There are few biological and physical changes which can be made within the limits set by most arid ecosystems. These characteristics frequently make problems very difficult to rectify.

Social adjustments may be precipitated from within the community or alternatively they may result from the external influence of governments and development organizations. In developed countries, external influences often include transport subsidies, retraining programs and taxation concessions, while in developing countries they may include drought assistance, land reform and medical aid. When decisions are made from outside and imposed on a community, there is a greater likelihood that those living within the community will feel threatened and establish barriers to oppose change.

Arid zone social systems

While it may be possible to identify some of the prominent constraints to adjustment, it is very difficult to cause these adjustments to occur. Experience has shown that most attempts to impose adjustments on the structure of an arid zone society have been unsuccessful.

The ability of communities to resist change is rarely appreciated by administrators. In one part of Australia, Heathcote (1965) found that attempts by the government to establish small family farms managed by individual farmers were frequently frustrated by large established pastoral families who persistently bought back land which was taken from them by the government.

It is probable that changes are more easily achieved within developed countries, where the economic linkages to the outside world are stronger and the bonds to a sedentary pastoral lifestyle are weaker. In developed countries most people are accustomed to and frequently seek out government intervention in their affairs via the market place. A strong linkage to a market economy generally makes it easier for governments to influence land use by changing propery rights, by manipulating market prices and by supplying subsidies where necessary (M. D. Young, unpublished).

Most developed countries depend heavily on a strong market economy which supports a well established marketing and transport network. The importance of a reliable transport system cannot be overstressed as experience in Australia has shown that without a reliable transport system and a strong market economy, vegetation is frequently degraded (Heathcote, 1965; Bean, 1969).

Managing land use with property rights

Most arid areas of the developed world are managed under sedentary grazing systems which have only existed for relatively short periods. Such systems allocate exclusive grazing rights to individual pastoralists and allow a pastoralist to leave an area ungrazed with the knowledge that it will remain so. They depend heavily on the ability of governments to enforce and modify the grazing rights of individuals.

In Australia, five states have each developed different methods of manipulating arid-zone and land-use under a system of leasehold tenure which leaves exclusive grazing rights to an individual (Young, 1979*b*). One state, South Australia, has avoided interfering with market pressures, and relatively large stations (ranches or farms) have developed, managed by a pastoralist who employs several workers. In contrast, New South Wales has prevented its pastoralists from leasing large areas and has forced them to manage their stations unassisted. As a result of these policies, two different community structures have emerged: in South Australia, each station comprises several families living in close proximity to each other, while in New South Wales most families live in isolation. The result in New South Wales has been 'closer' uniform settlement with more towns whose sole function is to supply the recreational and supply needs of pastoralists. These towns are small and unless they serve other industries such as irrigation, mining or tourism, their fortunes depend heavily on an unpredictable grazing industry. The replacement of larger stations with smaller one-family properties has tended to create pastoral communities that are less able to adjust to changing physical and economic conditions (Young, 1979*a*).

Poverty, wealth and development

The recent famine in the Sahel and the desertification that ensued has stimulated many people to look for new ways to utilize arid zone resources. It is tempting for developing countries to look to Australia and America to identify new methods of land use and also to identify the mistakes which have been made by these countries in developing their administration systems. Experience in these countries has shown that there is a tendency for administrators to become over-optimistic in their planning and, under pressure from politicians, to implement development strategies that are beyond the potential of the limited arid resource. People often fail to realize the variable nature of arid zone production and that without water most arid ecosystems cannot be made more productive except by extensive periods of destocking. Further, without considerable subsidies and exclusive grazing rights, it is rarely possible to use fertilizers and conventional pasture re-seeding techniques to improve productivity.

Unfortunately, most successful forms of arid zone development involve achieving greater technical efficiency and adopting more conservative forms of arid-zone land use. The result is always a need for fewer and fewer people to live within arid areas. Thus many people in the arid zone are facing a prospect of relocation out of their homelands, into alien environments, providing new jobs and different lifestyles. As most arid lands are already over-populated there is a need to assist rather than hinder this process of reconstruction.

Alternative settlement systems

Within the developed world arid ecosystems are being used less and less. There is a tendency for people to adopt more conservative forms of land use and to use intensive forms of livestock management. In Australia, the number of livestock per man has been increasing and it appears likely that this trend will continue. Unless alternative occupations can be found it is likely that the region's population will continue to decline and its physical isolation and political insignificance will increase. Similar scenarios can be painted for other arid areas of the world. However, alternative forms of land use are developing. For example, more people are being attracted by the stark beauty of arid zone landscapes and regions such as America's Grand Canyon and Australia's Ayers Rock. In America, the relatively dry, unusually warm arid climate is also becoming popular amongst elderly people and many retirement communities are developing in its arid zone (Young, 1979*c*).

The impression in western countries is one of declining agricultural and pastoral use but growing recreational use, while developing countries are facing immense social upheaval. It appears unlikely that developed countries will be able to achieve a less exploitative use of the resources of arid ecosystems without the introduction of massive amounts of capital and astute government intervention.

References

Bean, C. E. W. (1969). *On the wool track.* Angus & Robertson, Sydney.

Davidson, B. R. (1969). *Australia, wet or dry? The physical and economic limits to the expansion of irrigation.* Melbourne University Press, Carlton.

Davidson, B. R. (1972). *The Northern Myth: limits to agricultural and pastoral development in tropical Australia*, 3rd edn. Melbourne University Press, Carlton.

Harrington, G. N. (1979). Grazing arid and semi-arid pastures. In: *Elsevier World Animal Science Series*, vol. 16 (ed. F. H. W. Morley). Elsevier, Amsterdam.

Heathcote, R. L. (1965). *Back of Bourke: a study of land appraisal and settlement in semi-arid Australia.* Melbourne University Press, Carlton.

Tribe, D. E. (1977). The conservation and improvement of resources: the grazing animal. *Philosophical Transactions of the Royal Society, Series B.* **278**, 565–82.

Young, M. D. (1979*a*). Influencing land use in pastoral Australia. *Journal of Arid Environments* (in press).

Young, M. D. (1979*b*). Differences between states in arid land administration. Land Resources Management Series No. 4.

Young, M. D. (1979*c*). Remote arid communities and the pastoral industry. In: *Present and future settlement systems in sparsely populated regions*. (ed. J. Holmes & E. Lonsdale), pp. Pergamon Press, Oxford.

Manuscript received by the editors June 1979.

25. Synthesis

M. EVENARI

Introduction

The International Biological Programme on arid lands is not the first biological–ecological study of desert regions. At the end of the last century and the beginning of this, ecologists, the first adepts of a comparatively new burgeoning science, began to be interested in studying desert plant and animal communities and the ecophysiological activity and behaviour of their components. I mention only a few of these investigators who could rightly be called the predecessors of IBP: G. E. Schweinfurth, G. Volkens, S. Passarge, S. v. Murbeck, W. A. Cannon, D. T. MacDougal, J. Walther, F. Shreve, O. Stocker, D. N. Kachkarov, E. P. Korovine, H. Walter, C. Killian & D. Feher, Th. Monod, and L. Emberger. Long before the term 'ecosystem' was born, these early authors, in many of their books and papers, realized that in order to understand the desert one has to consider all its biotic and abiotic elements.

Arid-land studies received a new momentum when UNESCO initiated its programme on arid lands and arranged a number of symposia on the various aspects of desert research. This enabled desert scientists of all disciplines involved to meet and discuss problems common to all deserts of the world. UNESCO published the results of these symposia and these publications constitute the foundations on which IBP was built. One of the most important aspects of these meetings was the long and sometimes heated discussions and the free interchange of opinions, something which was only possible because UNESCO kept the number of participants small. During these discussions it became clear that there were large gaps in our scientific knowledge of arid lands and that only an integrated approach could advance our understanding of the complex systems involved. It also became clear that the scientific community could not restrict itself to gathering knowledge about arid lands but must, in some way, get involved in the decision making process concerning the management of arid land resources and their management in general. At that time, long before anyone talked about 'desertification', we already knew that the existence of the arid lands and of their inhabitants was endangered. When IBP was established in 1964, three points became the aim of its arid-lands section (see Volume 1, p. xxv) and these two synthesis volumes are rightly called: 'Arid-land *ecosystems*; structure, functioning and *management*' (the italics are mine).

Only a limited number of countries set up national IBP committees which really functioned. The one in the USA was the most active, followed by

Australia and the USSR. But IBP's existence alone has stimulated the activity of many other countries in investigating arid-land ecosystems outside the official framework of IBP.

Most of the authors of these two volumes are either Americans or Australians and deal mainly with the arid lands of their own countries, which they naturally know best. This gives considerable scientific depth to their chapters but entails neglect of the contributions of non-Anglo-Saxon scientists to the research of arid-land ecosystems, their function, structure and management, much of which has been published in scientific journals and, *inter alia*, in two volumes on productivity (*Eco-physiological foundation of ecosystems productivity in arid zone*, ed. L. E. Rodin, Nauka, Leningrad, 1972, and *Photosynthesis and productivity in different environments*, ed. J. P. Cooper, Cambridge University Press, 1975). This restriction to American and Australian arid lands has also led to insufficient representation of our general knowledge of other large desert regions of the world such as the Sahara, the Middle Eastern deserts and the deserts of Central Asia. To cite only one example: Russian scientists from six permanent desert research stations, many universities and other research institutions have published many books (see for instance those edited by Rodin 1977, 1978 and Vosnessenski, 1977) and hundreds of papers on the arid lands of the USSR, their soils, climate, fauna and flora, ecosystems and productivity. Only a few are mentioned in the chapters of these two volumes. It is significant that most of the bibliographies attached to the various chapters contain, with few exceptions, only titles of papers written in English.

In these two volumes, three arid regions, all of which have very specific characters of their own, are not dealt with at all: South America, with the exception of the Monte region which some authors mention in connection with the question of convergence of deserts; the Chinese arid lands including the Gobi desert which is only mentioned in its Eastern part in Volume 1, Chapter 7; the Somali–Chalbi desert and the arid lands of south-west Madagascar, a very small but very peculiar area characterized by an endemic plant family (Didieraceae) which excels in strange morphological features.

IBP terminated officially in 1974 but this was not the end of its tremendous impact on arid-land research. Some of the loose ends were taken up by MAB of UNESCO, which has just published (Anon, 1979) a new, excellently detailed map of the world distribution of arid regions (*MAB Technical Notes* 7), a vast improvement over the previous UNESCO map prepared by Meigs (1953) which we have all been using until now. The joint IBP–MAB Tunisian Pre-Saharan project and similar projects still continue. IBP has stimulated arid-land research everywhere and in all fields as witnessed by the many relevant papers published in such scientific journals as *Oecologia, Ecology, Ecological monographs*; specific journals like the *Journal of Arid Environment* or the Indian *Annals of Arid Zone*; and by a number of books such as those edited by Hadley (1975), Orians & Solbrig (1977), Glanz (1977), and Mundlak & Singer (1977).

The purpose of this synopsis chapter, as stated by Perry (Volume 1, p. 2), is to try 'to pull together the contributions to both volumes and to indicate areas where further research could improve our knowledge and management of the arid lands'. In doing this I will try to take into account not only the research of IBP participants but also to include the results of research relevant to the basic theme of IBP conducted outside the official framework of IBP.

I will also review the various chapters critically, add some information based on my own experience and that of my colleagues in the Middle Eastern deserts.

Description and structure of arid ecosystems (Volume 1)

This section consists mainly of an up-to-date description of physiography, climate, soils, vegetation and fauna of the world's arid lands. These eight chapters lay the foundation for all the following chapters in summarizing our knowledge of the overall nature of the arid lands concerned.

From Chapter 2 it is clear that the North American deserts are better known today than most other deserts thanks to US IBP Desert Biome Program. This chapter contains a very instructive comparison of vegetation (characteristic species, life forms) and fauna (mammals, birds, reptiles, amphibians) of the four deserts (Sonoran, Mohave, Chihuahuan, Great Basin), which brings out the similarities and diversities of the species, genera and vegetation types. In spite of certain traits characteristic of each desert and in spite of differences in soil and climate (Great Basin is a 'cold' desert), these deserts basically form a continuum as far as species distribution is concerned. It is remarkable that at the generic level the flora of the Sonoran and Chihuahuan deserts show more similarity to the Monte region of Argentina than to the nearby North American Great basin desert. This is explained by the origin of the vegetation ('long jump dispersal from South America') which is discussed in Chapter 2 together with the history of fauna and man. The North American deserts date back at least to the Pliocene and man's 'desert culture' is quite old, dating from the Paleo-Indians (12000 B.P.) until the appearance of European man 'which was an ecological event of extreme consequence to the North American deserts' (p. 68), and 'only time will tell if they have been subjected to a new environmental factor outside of their tolerance range – modern man' (p. 69). This is equally true for the Australian arid lands. The chapters on North and South Africa, in addition to describing vegetation and fauna, deal with land use, management and their history. Both regions were already occupied by man hundreds of thousands of years ago (Pebble cultures, lower palaeolith), were the cradle of a number of stone age cultures and have an apparently uninterrupted history of use by man from early Pleistocene through prehistoric, protohistoric, historic times to the present, under climatic conditions different from those of today (see Chapter 49 for North Africa). The same is true of the Middle Eastern arid lands (for the Negev, see Marks, 1976, 1977).

Chapter 3 stresses the importance of run-off for understanding the hydrology, vegetation and early land use in North Africa where in Roman times run-off and water in general was totally controlled and used. This theme is largely neglected in most of the other chapters, although as far as early land use is concerned, run-off was the main water source for agriculture also in the arid lands of the Middle East, South Arabia and America (Evenari, Shanan & Tadmor, 1971).

A valuable feature of Chapter 4 (South Africa) concerns its bibliography which refers to many papers that were published in local journals not easily available. The most remarkable feature of the South African arid lands is that in the Namib, their most extremely arid part, fog is the main water source for plants and animals, as is the case also in the South American fog desert. The South African deserts abound in most specific and unique life forms (e.g. Welwitschia) and in plant and animal endemisms. The Namib alone contains about 200 endemic species of Tenebrionid beetles!

In Australia, as in North America, recent research has greatly increased our knowledge of arid lands, in this case of a continent which, with the exception of Antarctica, contains the largest arid area with about 80% of its surface classified as arid or semi-arid. The early separation of the Australian continent from the Gondwana landmass explains the many peculiarities of the continent and its arid lands in vegetation and fauna (richness of endemic families, genera and species, specific plant communities) and in man (aboriginees who remained at the level of stone age culture). Even the soils differ from those of other arid lands because of their very low levels of phosphorus and nitrogen. Although Australia was colonized by European man much later than America, the effect of his arrival and of the animals and plants which he deliberately or accidentally introduced to the Australian arid lands was, and still is, tremendous.

South-West Asia harbours the vast arid region of the Middle East (Sinai and the Negev, and the deserts of Syria, Jordan, Iraq and Iran), the Arabian peninsula and the Indian subcontinent (Thar desert). Our knowledge of the arid lands of this region is very varied. While the biotic and abiotic components, structure and function of the arid ecosystems of some sub-regions like Sinai and the Negev of Israel are known in considerable detail – a fact which is not made clear in Chapter 6 – other sub-regions, like the arid lands of Afghanistan, are practically *terra incognita* in this respect.

The main part of Chapter 7 (Central Asia) deals with production and biomass, a theme which will be discussed later.

In summarizing this descriptive section we may point out that in general our descriptive knowledge of arid-land ecosystems is far superior to our knowledge of their structure. But there are large gaps even in our descriptive knowledge of vast regions, gaps which are indicated in seven of the eight chapters of Part I. There is an urgent need to fill these gaps through further research since identification and description of the main ecosystems of all arid lands and their components is the basis for understanding their structure and function and for their management.

Besides plants and animals, soil micro-organisms are an important biotic component of all ecosystems. Although they probably only take second place functionally to the primary producers, our knowledge of them is meagre and for many arid-land regions non-existent. This gap in our knowledge is indicated by the fact that only Chapter 7 on Central Asia and Chapter 8 deal to any extent with soil micro-organisms. Future research of arid lands has to pay much more attention to this point.

Various biologists involved in the US IBP desert biome studies have investigated the convergent evolution in the deserts of the US and Argentina, that is their evolution, strategies, community patterns, resource utilization systems and degree of convergence (see Orians & Solbrig, 1977). It would be desirable to initiate a similar comparative study of all the world's deserts which would aim to find out whether there is a world-wide pattern of convergence in arid lands or whether certain deserts do not conform. If such convergences can be established, are there degrees of convergence which can be related to specific features of different deserts? What is the spectrum of convergences? Do they apply to strategies, life forms, species and resource diversity, etc.? There have been such attempts in the past (p. 310) but with the large amount of new information on all aspects of arid lands accumulated directly and indirectly by IBP, such an up-to-date study could bring out new facts and aspects of a basic biological problem. It should not only deal with convergence and the reasons why such convergence exists, but also with the question of why certain families (e.g. Chenopodiaceae), genera (e.g. *Acacia*) and species are found in all, or most deserts – apparently because of their 'constitution' (genetic make-up) – and others are restricted to specific deserts apparently because of their phylogenetic age.

Component processes (Volume 1)

Diversity and niche structure are basic concepts of modern ecological thought. They are difficult to define and quantify. Chapter 10 deals with the question of whether desert ecosystems can be characterized by a diversity and niche pattern specific to them. One general statement made by Pianka concerning species diversity is: 'Extreme aridity clearly must result in very low productivity and plant diversity' (p. 325). Very low production is certainly typical for extreme deserts but the statement of low species diversity in this generalized form may be questioned. It is true concerning the arido-active plant components of extreme deserts which are characterized by a very low species richness and production. Diversity of desert annuals seems to be determined by edaphic factors. Highly saline habitats are nearly mono-specific. Nearby leached habitats have a relatively high species diversity (e.g. the Dead Sea; Danin, 1976). Habitats with a high diversity of annuals in good years, have a low diversity in drought years. Different animal taxa, living in the same desert, differ in their species diversity. Consequently, a generalized statement of overall species diversity for all arid lands cannot today be made (see also Volume 2, Chapter 4, p. 59).

Diversity has more dimensions than just 'species diversity'. There is diversity of life forms, resource diversity (not necessarily identical with niche breadth and overlap), diversity of survival strategies, etc., which are all aspects of 'diversity' and which, including species diversity, are inter-related between themselves and with niche structure. A consideration of these diversities would probably bring out new aspects of the problem no less important than species diversity. But such an approach would need new definitions and formulations. If we take life form diversity of plants as an example, this would not include exclusively the traditional life forms of Raunkiaer, but also 'leaf tactics' (p. 325), life forms of Shreve (1942) (p. 312), life forms of Orshan (1953), Zohary (1954) and of Evenari, Shanan & Tadmor (1971). Animal life form diversity could be similarly treated by defining their life forms in various ways (e.g. 'geozonts' as counterparts of 'geophytes', 'zoophagi', 'carpophagi', 'detritophagi', 'phytoxylophagi', etc.*). In 'survival strategies' it should also be taken into account that many desert plants and animals have developed a number of alternative strategies (e.g. polycarpy). It could be that if these diversities could be quantified as has been done for species diversity it would emerge that the more extreme the environmental conditions, the higher the degree of these types of diversity.

It is one of the many merits of IBP to have drawn attention to the problem of diversity in deserts, and ecologists should devote much observation, experimentation, calculation and rethinking to this important problem.

Atmospheric processes (Volume 1)

This section deals with two specific climatic parameters (radiation, precipitation) in addition to the climatic data given in the foregoing descriptive part and accentuates the climatic heterogeneity of the desert. The section shows how much progress has been made in the last 10–15 years in quantifying atmospheric transport processes and microclimatic parameters near soil and vegetation surfaces of arid lands, two closely inter-related phenomena. A good example concerns the energy balance of leaves and canopies. The first desert

* 'Geozonts' is my proposition, the other terms were taken from Volume 2, Chapter 8.

ecologists, like Volkens (1887), already suggested that the structure of leaf surfaces, their cover by hairs and wax, leaf orientation, small size of leaves or lack of leaves, and open canopies are adaptive features of desert plants. They also knew that these adaptations were related to the 'energy balance' between organism and environment even if they did not yet use that term. Today their suggestions have been verified by experiment, calculation and conceptual models. The relevant chapters deal only with vegetation. Similar chapters, dealing with atmospheric transport processes around animals and energy balance between animal surfaces and environment, for which IBP and related investigations have provided much data, are sorely missing.

In relation to the coupling of water transport and carbon metabolism, Chapter 15 also touches upon the question of C_3, C_4 and CAM plants, a theme which will be discussed later.

The section on atmospheric processes is proof of how much IBP has contributed to the precise formulation, mathematical analysis and conceptual modelling of these processes, an important advance in understanding desert ecosystems which in its turn has stimulated new field research.

Soil processes (Volume 1)

The various soils that are typical of arid lands are described in Chapters 1–8, and their pedogenesis in Chapter 17. Peculiarities of pedogenesis of desert soils are ferruginization in automorphous soils, the formation of crusts of biological origin on auto- and hydromorphous soils and the large part which soil micro-organisms play in the cycling of organic matter and nitrogen in the upper 5–10 cm. These micro-organisms, which are heat and drought resistant, become active a few hours after the first rainfall of the season and remain active during the short periods when desert soils are wet. Of special interest is the specific case of a community of *Atriplex vesicaria* in which the relationship was studied between effective rains (i.e. rains triggering phytomass production), ineffective rains, mineralization caused by micro-organisms, the inorganic nitrogen pool in the soil and plant production. Since the activity of soil micro-organisms is stimulated by ineffective rains leading to the accumulation of mineral nitrogen it is suggested that phytomass production triggered by effective rain is related 'to the preceeding sequence of rainfall events, not just the rain to which the vegetation is currently responding' (p. 455). I stress this conclusion because most desert rains are 'ineffective' and therefore neglected by plant ecologists in their analysis of production processes although they affect microbial activity and are therefore also effective with regard to primary production.

Since most arid land soils are saline, Chapter 17 also deals with salt movement and accumulation in soils with special emphasis on the solontchak and takyr formation so intensively studied by Russian pedologists.

Plant processes (Volume 1)

The flowering of desert annuals is mainly controlled by an interaction of photo- and themoperiods. In most of the cases studied, the nature of the response is facultative, permitting the plants to flower and form seeds over a wide spectrum of photo- and thermoperiods, whereby the flowering of winter annuals is promoted by low temperatures and that of summer annuals by high temperatures. Under desert conditions this is advantageous since germination takes place only after a certain minimum of rainfall ('efficient' rain, cf. Chapter 19). These efficient rains, because of the unpredictable irregular rainfalls typical of all deserts, can occur any time over a number of months of the rainy season (in the Negev, for example, from October to April) causing the seedlings to develop under very varied photo-and thermoperiods according to the date of their germination. It is not certain if water stress has any effect on the flowering process of the annuals but the date of the first effective rain (early or late in the season) and availability of water (rain, or rain and run-off in special micro-habitats) affect the length of the growing season and the number of flowers formed and seeds produced. Desert annuals are therefore 'facultative ephemerals' with short (weeks) or long (months) life cycles, and many or few flowers and seeds according to the water status of the soil.

Besides winter and summer annuals there are in deserts also arido-active* biseasonal annuals, a very special group found in Middle Eastern deserts but more abundant in the deserts of Central Asia. They germinate in the Middle Eastern deserts with winter rain. Like winter annuals, after the first effective rain the seedlings develop a leaf rosette in winter and spring, at the beginning of summer the stem grows and the plants flower at the end of the dry summer and mature their seeds in autumn. Their flowering is controlled by a quantitative response to a combination of short days and high temperatures.

The flowering process of arido-active perennial desert plants is strictly controlled by water stress. Many of them do not flower at all in drought years. Flower bud formation is delayed in *Artemisia herba alba*, one of the foremost dominant plants of the Middle Eastern deserts, when water stress is eliminated by irrigation. In addition to water stress, photo- and thermoperiod play a part in flower and bud formation and *Artemisia* sp. responds to both in a facultative way. Unfortunately, few experiments have been conducted under controlled conditions on the flowering process of other arido-active desert perennials.

Arido-passive geophytes also do not flower in extreme drought years. In such years they remain dormant and do not sprout. They are able to stay alive and dormant for a number of years.

Desert plants have developed a variety of strategies for seed dispersal, synchronizing dispersal with environmental conditions favourable for germination and at the same time ensuring that the seeds of one plant do not all

* For the terms 'arido-active' and 'arido-passive', see Evenari *et al.* (1975*b*).

disperse at the same time. Thus the mother plant keeps a reserve of seeds for later dispersals, a good example is *Blepharis persica* (Gutterman, Witztum & Evenari, 1967). This behaviour counterbalances the dangers inherent in the erratic nature of desert rainfall where sufficient rain may fall to cause dispersal and germination but insufficient for the survival of seedlings. Most desert plants have developed mechanisms (heteroblasty, germination inhibiting structures or substances) to prevent simultaneous germination of all dispersed seeds, even when environmental conditions are propitious, and this has the same ecological effect: it guarantees that there is always a reserve of viable seeds either in or on the soils.

Amphicarpic annuals found in many deserts have developed an additional dispersal strategy. They carry many small aerial telechoric dispersal units which are fractionally dispersed over a wide area colonizing new sites for the species. Only in good rainy years do their seedlings have a good survival chance. The same plant also carries a few large subterranean atelechoric fruits which germinate out of the dead mother plant. The survival chances of these seedlings are high even in drought years. In drought years the plants form only atelechoric fruits.

There is a basic difference between germination and seedling survival of arido-passive winter annuals and arido-active perennials, according to our 15 years of observations in the Negev (Evenari & Gutterman, 1976). Germination of annuals occurred in all years. Even in the worst drought year (29.5 mm of rain) some seeds of the various annuals germinated and a certain percentage survived. In the worst year and in most other years the seeds of the perennials did not germinate and if there was any germination the seedlings did not survive. Only a certain percentage of seedlings from one specific year in which many perennial seeds germinated, actually survived and developed into mature plants. This is the reason why populations of perennial desert plants are divided into age groups with gaps between the groups as shown for *Zygophyllum dumosum* (Evenari, Shanan & Tadmor, 1971).

Knowledge of seed reserves in soils is one prerequisite for understanding population dynamics of vegetation. In this field the US IBP did the pioneering work and was followed by others (M. Evenari & Y. Gutterman in 1976 for the Negev, unpublished results). Most seeds in deserts are found on the soil surface and in the top 1–2 cm. In the Negev the larger numbers are concentrated below and around the perennial dwarf shrubs and in depressions, accumulated there by run-off and wind. These depressions serve as 'reservoir sites' of the population because run-off water collects in them and results in favourable conditions for germination even in drought years. The same applies, in some measure, to the area below the dwarf shrubs. From these localities, less favoured sites can be recolonized in more propitious rainy years. This fact is one of the reasons why the diversity of microtopography is of such vital ecological importance in deserts.

One of the most important contributions IBP has made to the understanding of desert ecosystems concerns plant processes controlling primary production (photosynthesis, respiration, stomatal movements, translocation, Chapters 20–23; Volume 2, Chapter 8). These contributions have been stimulated by the detection of the C_4 and CAM photosynthetic pathways and photorespiration, phenomena not known to the earlier plant ecologists. In turn this has stimulated intense activity by desert scientists in this field, as witnessed by a continuing flow of relevant papers.

The existence of desert plants is permanently endangered by the 'water-photosynthesis syndrome' (Stocker, 1976), a Scylla-and-Charybdis situation. The stomata have to be open to permit uptake of CO_2, but this may lead to excessive loss of water vapour. The regulation and control of stomatal movements is therefore vital for the survival of desert plants, especially those which are arido-active. In addition to the long known control of stomatal movement by soil and plant water potential and by light, we know today that leaf temperature and air humidity are no less important factors. All these parameters work in conjunction resulting in a high degree of sensitivity of the stomatal apparatus to internal and external parameters and their seasonal changes. A good example is the inter-relation between leaf temperature and plant water potential (Schulze *et al.*, 1973). At high water potentials, a temperature increase leads to opening of the stomata, at low water potential to their closing. The critical water potential is specific to each species.

Gas exchange of desert plants can also be regulated through residual resistance (mesophyll resistance, incipient drying (Livingston & Brown, 1912)). This is possibily the reason why certain desert plants are able to restrict transpiration without restricting CO_2 uptake and net photosynthesis (Schulze, Lange & Koch, 1972).

During the dry season the daily course of net photosynthesis of most arido-active perennials changes considerably. As long as soil water potentials are high the daily curves of net photosynthesis are one-peaked, with a maximum around noon. With increasing water stress they become two-peaked, with a pronounced noon depression, and at the end of summer the curves return to being one-peaked, but with a maximum in early morning followed by a sharp decrease leading to negative net photosynthesis when respiration rates (dark respiration in light or possibly photorespiration) are higher than rates of photosynthesis (Lange *et al.*, 1975). At this stage the total daily gain of net photosynthesis is for some plants still positive (e.g. *Hammada scoparia*) and for some negative (e.g. *Zygophyllum dumosum*), but even in the latter case they are still able, for some hours in the morning, to be photosynthetically active at extremely high water and osmotic stresses (e.g. *Artemisia herba alba*-xylem hydrostatic pressure of −163 bar, leaf osmotic potential −92 bar; Kappen *et al.*, 1972). Other parameters related to photosynthesis of desert plants show seasonal changes as well, for instance, the temperature optimum for photosynthesis and the upper temperature compensation point (Lange *et al.*, 1974).

Synthesis

Considerable progress has been made by IBP in understanding the role of dark mitochondrial respiration in the dark and light, peroxisomal photorespiration and the C_3, C_4 and CAM photosynthetic pathways. But there remain a number of questions. How exactly do environmental factors and the different pathways affect the two respiration mechanisms? Increase in temperature apparently leads to increase of both processes. But what about light intensity, water and osmotic potentials? There are indications in certain directions but no facts.

Many papers have discussed the ecological advantages of the C_4 and CAM pathways over the C_3. CAM plants 'decouple' transpiration and photosynthesis (p. 427) and have higher photosynthetic rates. C_4 plants have little or no photorespiration and high photosynthetic temperature optima. But in spite of these advantages a general statement that they are better adapted to desert conditions than C_3 plants cannot be made. In deserts like the Central Negev and Sinai, C_4 plants are rare and most of the dominant arido-active perennials are C_3 plants and by far the largest percentage of biomass is produced by C_3 plants (Winter & Thoughton, 1978). It may be true that CAM plants are better adapted to semi-arid regions and not so much to more extreme deserts and that C_4 plants to localities like the shores of the Dead Sea , with very high soil salinity, high temperatures but more soil water than is available in more arid deserts.

It is a shortcoming of Chapters 20 and 21 that they completely neglect water uptake, photosynthesis and respiration of poikilohydrous desert plants like algae and lichens although there is much data available, at least for lichens (see, for example, Kappen *et al.*, 1975, Lange, Geiger & Schulze, 1977).

A very serious gap in our knowledge is that despite an abundance of data on transpiration, photosynthesis and respiration of single plants, there is, to my knowledge, not a single integrated study of these processes concerning all the plant components of an ecosystem.

Our knowledge of translocation, partitioning and distribution to the various organs of assimilates in desert plants is meagre (Chapter 22). It may be safely assumed that in arido-passive desert geophytes a large percentage, and possibly the major part, of the dry matter produced during their active period is invested in the underground 'sinks' as storage material to be used when they restart growth of roots and later of shoots. The same is true for the annuals for which their sinks are their seeds.

In arido-active perennials most of the dry matter produced during the year and not used in respiration of the photosynthesizing organs, is invested in the roots. When the water stress during the dry season is removed or strongly decreased by artificial irrigation much less assimilate moves to the root sink (Evenari *et al.*, 1977). It seems therefore that in this group of desert plants water stress enhances assimilate transportation to the roots. By comparison with annuals, not much is invested in fruit and seeds, an investment which is totally suppressed in serious drought years.

In spite of the vital functions of roots (anchorage, uptake of water and inorganic ions, storage of assimilates and water, hormone synthesis, excretion of substances into the soil, mycorrhiza, rhizosphere) and their specific environment, which is so different from that of the above ground parts, our eco-physiological knowledge of roots in general, and roots of desert plants in particular, is most limited.

Roots of arido-active desert perennials are very flexible and change their growth and distribution pattern according to soil depth, texture and the distribution pattern of available water. Many of them are characterized by a combination of a vast network of lateral roots in the upper soil layer extending far beyond the crown of their above-ground parts and a system of vertical roots penetrating to considerable depths if water can be found there. Wherever the roots find water they develop a dense network of fine rootlets to exploit the water pockets which often exist, even in the dry season, below stones embedded in the soil. The vertical roots have a much greater capacity to penetrate rocks than is generally believed. We have found, for example, roots of *Atriplex halimus* growing in and through soft limestone rocks without seeing visible cracks through which they could have entered. It seems that these rocks are an important water source but there are no exact measurements. Some arido-active perennials develop only shallow, lateral, far-reaching roots like the cacti. These superficial roots must be highly drought-resistant since the upper soil layers dry out very fast and are subjected to high temperatures and great daily and seasonal temperature fluctuations. This is in sharp contrast to the deeper roots which live in a temperate environment because at a depth of 50–100 cm soil temperatures vary very little over the year. In the Negev, at a depth of 50 cm the temperature alternated during one year only from 16–24 °C and at 1 m depth the temperature was nearly constant at around 20 °C. The root system of arido-passive annuals is shallow and not strongly developed. The same is true for the arido-passive geophytes. One of their characteristics is the high speed with which they develop rootlets ('rain roots'). In *Carex pachystilis* we observed the first new rootlets less than 12 hours after the first rain of the season had wetted the upper 2 cm of soil and a few hours later their dense network enmeshed the upper soil layer (Evenari *et al.*, 1971). The same is true for the superficial root system of cacti (p. 581). The 'dogma' (Barbour, 1973) that the root:shoot ratio of desert plants is higher than that of plants of more mesic biota has been questioned, and it has been stated in a new 'dogma' (Volume 2, Chapter 3) that 'the xerophyte ratios turn out to be relatively low' (p. 35). In my opinion neither 'dogmata' are correct. Certainly not *all* desert plants have high root : shoot ratios and the root : shoot ratio is flexible even in the same species (see above). But more than 50% of the overall permanent phytomass of desert ecosystems is below ground if we do not take into account the phytomass of the annuals in very favourable rainy years. This is at least true for the Negev (Evenari *et al.*, 1976) and for similar true deserts like the ones of Central Asia (Volume 2, Chapter 8), and is probably not true for semi-deserts, especially those where cacti are dominant. Most of the phytomass in true deserts is found below ground; it is important since these parts, including roots, are not only energy sinks of assimilates but accumulate also the biologically active elements (potassium, calcium, phosphorus; Volume 2, Chapter 8, p. 185), and constitute 'a reserve of fertility in the soil'.

In two respects we must apparently change our conventional 'textbook opinion' of roots, at least for desert plants. Water and ion uptake is not restricted to the young rootlets, but takes place also in old roots which have already developed a periderm. Root hairs are not restricted to a short region near the apex of young roots and are not short-lived but extend in many desert plants over the whole length of young roots and persist until these roots develop periderms. We observed this in many desert plants of the Negev but did not test whether these persistent root hairs are still able to absorb water.

A problem which needs further research concerns mycorrhizas and rhizospheres which are present in many (or all?) desert plants. What is their exact function? Does the mycorrhiza participate in the water uptake process of desert plants? What is the relationship between the organisms living in the rhizosphere and the root, and what role do root excretions (organic substances, ions, p. 618) play in this regard? The hormonal activity and function of the roots is another problem needing clarification. Do hormones produced in roots (cytokinins, abscisic acid, giboerellins?) and transported to the shoots function as 'information carriers', informing the shoots on the water status of the soil and regulating their growth accordingly?

Modern plant physiologists have dedicated much research to ion transport in general and to ion transport through membranes in particular. But as far as desert plants are concerned one specific problem still remains: the 'luxury uptake' of so many desert plants (Chapter 24), which either dispose of part of the ions by excreting them in different ways (e.g. *Atriplex* spp., Reaumurea spp., Tamarix spp.) or store them (e.g. *Zygophyllum* spp., *Suaeda* spp.). Why is it that in some habitats some species exhibit luxury uptake and others the opposite, restricting uptake of the same ions to a minimum, and in some cases practically excluding them? What is the exact physiological mechanism of excretion of salts in quantities so large that they can salinize the soils below these plants?

The average litter production and 'necromass' (p. 648) of desert ecosystems is high when calculated in percentages of total phytomass, due to a combination of high mortality in the frequent drought years and continuous shedding of phytomass by the arido-active perennials during the dry season as the plant reduces its physiologically active surfaces. The amounts of litter produced in deserts vary largely from year to year as a function of rainfall, available water and production (Evenari *et al.*, 1975*a*, 1976).

Litter accumultion below dwarf shrubs and 'sub-trees' plays an important role in water and mineral cycling, and creates a specific micro-habitat for germination, seedling establishment (if it does not contain allelopathic substances) and for detritophagous animals and soil micro-organisms. We know very little about the speed of litter decomposition in deserts, which is supposedly slow, and practically nothing about the subterranean litter production and necromass.

On p. 661 it is stated that drought evading plants (arido-passive plants) 'comprise the majority of the desert flora'. This necessitates some comment. As far as species number is concerned this is true for the arido-passive annuals. But according to our experience in the Negev and Sinai, geophytes, and especially hemicryptophytes, make up only a small percentage of the total number of species. This may not be true for all deserts. With respect to the mean total permanent phytomass, the arido-active plants are the stable mainstay, whereas the above-ground phytomass of the arido-passive plants fluctuates from zero, in drought years, to more than 50% of the total, in good rainy years.

On p. 671 it is mentioned that some desert plants (four are named) are facultative annuals, biennials or perennials. In the Negev we have made the same observation concerning *Diplotaxis harra*, *Schismus arabicus* and *Blepharis persica*, which change their life cycle behaviour according to the environmental conditions. It may be asked whether this 'life-form-flexibility' is typical only in deserts or is as frequent in mesic biota.

Animal processes (Volume 1)

In most, if not all, arid lands most of the large native mammals which once populated these areas have by now either completely disappeared or have been severely reduced in numbers, and the wild grazing animals have been substituted by domesticated grazers. In the arid lands of the old world the introduction of man and of domestic livestock, has changed vegetation and environment. This is well documented for North Africa (Volume 2, Chapter 14). These changes began a long time ago and the new way of life of domestic livestock, together with the men that used them, became part of the environment which they changed but with which they created a new equilibrium, perhaps not too different from that which existed between the native grazers and their environment. For many centuries the nomad and his flocks co-existed with the native large mammals. But with the introduction of modern hunting weapons the large native fauna were murdered *en masse* and were substituted to a large degree by domesticated animals. This substitution, together with the modern means of fighting animal diseases and creating new sources of water, had a pronounced impact on the environment. In the new world and Australia the introduction of cattle, sheep, goats, camels and donkeys occurred in recent times with the same dire consequences. All this means that we have no knowledge of arid-land ecosystems in their primeval climax stage and can only speculate about it. It may be assumed that the larger wild mammalian herbivores existed in 'small mobile populations' (Volume 2, Chapter 5, p. 93). In Chapters 29–34 few data relating to domestic livestock will be found in spite of the fact that today they constitute most of the zoomass of arid lands. This fact must be kept in mind while reading the chapters on animals, discussing the few remaining large native mammals such as gazelles, deer, wild sheep and oryx, which make up only a small percentage of the total zoomass of deserts.

For most phytophagous and zoophagous desert animals, food is both a source of nutrients and water. The daily and seasonal changing availability of both sources regulate to a large extent the feeding patterns of animals. The night feeding habit of many phytophagous animals, for instance, is both an adaptation to the less stressing environmental conditions of the desert night and to the fact that at night more water is available in the eaten phytomass through dew formation and uptake of water vapour at high air humidities, as in the case of lichens and some higher plants (e.g. *Atriplex* spp., Chapter 41; *Prosopis tamarugo*, Went, 1975). Animals that eat seeds, containing little water (e.g. ants and rodents) increase the water content of the seeds 20–30% by storing them in their humid burrows. Many desert animals are food opportunists and omnivorous, which is fitting to the temporal and spatial variability patterns of available food. The coyote, for example, eats plants, rodents, rabbits, birds, grasshoppers and beetles.

The feeding habits, energetics and production of nematodes, protozoa and detritophagous animals, including soil-dwelling invertebrates, are scarcely dealt with in Chapters 29 and 30, apparently because we know little about them, although protozoa and detritophagi play an important part in litter turnover. In general our knowledge of the animal component of arid-land ecosystems decreases in descending the phylogenetic hierarchy from mammals, vertebrates, arthropods to protozoa, although it becomes more and more clear that the order of their ecosystemic importance runs in the opposite direction. It is relevant in this regard that poikilotherms have low rates of oxygen consumption and are more efficient in producing biomass than homeotherms.

Specific physiological mechanisms and behavioural patterns often working in conjunction, enable desert animals to cope with the two main dangers: overheating and dehydration. IBP has contributed much to our knowledge of both types of adaptations.

Most desert animals have a high heat tolerance. It is interesting to note that some can be supercooled, without damage, to temperatures below those which occur in their natural environment (p. 748). This is also true of desert lichens (Lange, 1965) and is apparently a constitutional characteristic of the respective genera or phyla. The normal body temperature of homeothermic birds is higher than that of mammals by 2–4 °C. Many desert birds and mammals possess mechanisms for facultative hypothermia. Body temperature can be regulated in many ways. White beetles and the white shells of desert snails like *Sphincterochyla zonata* (Yom-Tov, 1971) reflect a large percentage of the incoming radiation. Other animals possess structures insulating them against overheating. The black raven of the Negev (*Corax corvus ruficollis*), for instance, which is active during the hottest hours of a summer day, is insulated by air trapped in its plumage. This has been reported to create a temperature gradient of 34.7 °C between the surface of the black plumage (83.9 °C) and the skin (49.2 °C) as measured under intense solar radiation (Marder, 1973). Shaking its plumage aids the bird in dissipating heat.

Vasodilation, vasoconstriction and countercurrent heat-exchange are other means of regulating body temperature. Sweating, panting and gular fluttering serve the same purpose by evaporative cooling, but here, exactly as in plant transpiration, the cost of cooling may be very high because of the concomitant water loss. For the large majority of desert animals the main water source is water contained in their animal or plant food. Few desert animals drink water from the sparsely available springs, pools and water holes which persist in shady wadis and canyons even in the summer. Birds able to cover large distances are users of such water and even transport it in their plumage to their young (*Pterocles* spp.). The ibex (*Capra ibex nubiana*) and the gazelles (*Gazella dorcas*) in the Negev as well as the Bedouin goat contain large water reserves in their bodies, can survive a number of days without drinking and rehydrate rapidly during one short visit to a water source. This permits them to graze long distances from the water source (Shkolnik *et al.*, 1978). During the summer some large mammals dig holes into the wadi gravel with their hooves until they find the water stored there; this method is also used by the nomads. Toads and snails take up moisture from the soil. Some rodents and birds can drink highly saline water (even sea water) and eat succulent halophytes containing much salt because they have specialized renal functions and/or the presence of salt excreting glands. Beetles and isopods in the Negev (*Hemilepistus reaumuri*) drink dew drops, others, in the Namib, drink water condensed from fog on plants, stones and sand. Some beetles in the same area have developed 'fog basking' whereby they take up special positions enabling fog to condense on their own bodies. Others build fog trenches (Seely, 1979). For small animals metabolic water is the main and sometimes only water source. Some arthropods can take up water from a humid atmosphere.

Animals have developed various ways of avoiding or restricting dehydration: integumentary and epidermal structures limit water loss by evaporation; urine is highly concentrated, birds excrete nearly solid uric acid, faeces contain little water, GFR and UFR (p. 755) are reduced.

There are manifold temporal and spatial behavioural patterns to evade overheating and dehydration. In analogy to the arido-passive behaviour of plants one could call this 'spatial and temporal arido-passivity'. Diapause, aestivation, hibernation and large scale migration are temporal means of avoiding heat and water stress. But by far the most important strategy of desert animals consists of living in virtually air-conditioned burrows in which air humidity is high and temperature moderate. By analogy with 'geophytes' these animals should be called 'geozonts' (see p. 560). It is estimated that most of the animal species composing the major part of the desert zoomass are geozonts (ants, isopods, reptiles, rodents, etc.). Each has its own circadian and seasonal rhythmic pattern of above-ground activity for food and water collection, which is geared to the most favourable environmental conditions. Two good examples of such extreme geozontic behaviour in the Negev are the isopod *Hemilepustus reaumuri* (Shachak, Chapman & Steinberger, 1976; Dubinsky & Steinberger, 1979; Shachak, Steinberger & Orr, 1979) and the desert snail *Sphincterochyla zonata* (Yom-Tov, 1971*a*, *b*; Shachak, Orr & Steinberger, 1975; Shachak, Chapman & Orr, 1976; Shachak & Steinberger, 1980). *Hemilepistus* sp., which is a very important component of the desert ecosystem in the Negev Highlands is physiologically a 'non-desert' animal since it is easily dehydrated and its upper temperature limit is 35–36 °C. Only its behavioural pattern, through which it regulates energy inflow, enables it to thrive in the desert. The above-ground annual activity of *Sphincterochyla* sp. is limited to only 13–24 days.

Through their behavioural patterns, other non-geozontic desert animals exploit the diversity of the micro-environment, the many microniches so typical of the desert. The small, day-active Negev lizard, *Eremias guttulata* for instance, uses every shade-providing stone and every small rock fissure to avoid overheating and dehydration, rushing with swift movements in the heat of the summer day from microniche to microniche (Orr, Shachak & Steinberger, 1979).

One of the many analogies between desert plants and animals concerns reproduction. In many animals reproduction is supressed during drought years, as is also the case in arido-active perennial desert plants (p. 562), and is flexible in accordance with environmental conditions favourable for the survival of the new generation. Desert animals are therefore 'opportunistic breeders'. In general, rain and moderate temperatures coincide with reproductive peaks. In specific cases (certain lizards and rodents) reproduction is apparently related to rainfall pattern and intensity and/or to the availability of green forage which possibly contains substances not present in seeds, which are necessary for reproduction (p. 736).

Selection of appropriate nesting sites is another condition for survival. Some animal build their nests in nearly inaccessable or well-hidden localities, like the rock dove (*Columba livia*) in the Negev which nests in fissures and holes of steep cliffs. Many other animals lay their eggs in 'laying rooms' in the soil (for example the desert snails (Yom-Tov, 1971)) where they are well protected against dehydration and preying animals. Location, architecture and orientation of the nests of ground-nesting birds like the Negev desert lark (*Ammomanes deserti;* Orr, 1970) as arranged so as to avoid damaging temperatures and to guarantee ventilation.

Parental care is in no way restricted to birds and mammals. It was recently found in the Negev that male and female parents of the isopod *Hemilepistus reaumuri*, which leads a family life, bring food to their young (soil crusts and plant material in a fixed proportion) until they are able to forage freely outside their burrows (M. Shachak, personal communication).

Very little is known about patterns, rates and causes of mortality of desert animals. In *Hemilepistus* sp., for example, the main mortality occurs during the pairing phenophase when 55–75% of the population dies without our knowing the reasons (Shachak, Steinberger & Orr, 1979). But this is not the only gap in our knowledge of animal processes. Although many new and interesting data have been collected, the functional relationships between the various processes are not well known. Quantification of the impact of environmental factors on the various state variables is insufficiently known. What is the causal relationship between absolute amount, intensity and pattern of rain on the one hand, and spatial and temporal distribution, reproduction and development of desert animals on the other? What is the exact role of green phytomass and its components in fecundity and reproduction? What is the ecophysiology of detritophagi (including protozoa) which play such a decisive role in the overall energy flow of desert ecosystems?

Goodall (p. 845) mentions that the treatment of animal processes in these volumes may have been 'biased', 'reflecting the fact that most of the writers in this section are American'. I have said the same concerning other subjects as well but have to stress it again here because much material dealing with animal processes in deserts has been published, for example, by French authors on North Africa and by Russians on deserts of Central Asia, none of which is mentioned in this section on animal processes.

In conclusion to the sections on plant and animal processes, in Volume 1, I have summarised in Table 25.1 some analogies between the survival strategies of desert plants and those of desert animals.

Interactions (Volume 2)

Data on plant–plant interaction in deserts are contradictory. Some authors state that root competition (for water and minerals) is absent, others report observations and experiments showing that seedlings of certain arido-active perennials are absent or have high rates of mortality under the canopies of the same or other arido-active perennials (*Artemisia herba alba* seedlings near *Zygophyllum dumosum* or near adult plants of *Artemisia herba alba*; Friedman & Orshan, 1975). Mortality of annual seedlings is high in rainy years when

Table 25.1. *Comparison between the survival strategies of desert plants and animals*

Plants	Animals
A. *Evading behavioural strategies*	
Seasonal arido-passivity of whole plants (geophytes, hemicryptophytes)	Inactivity in time (diurnal and seasonal) and space (geozonts in burrows, other animals in above-ground shelters)
Anabiosis (poikilohydrous plants)	Reduced metabolic activity (Aestivation, hibernation)
Dormancy of seeds	Diapause
Growing in microniches (below shrubs, in run-on microsites, poikilohydrous plants below stones)	Habitation in and exploitation of microniches (below shrubs and stones, in caves, rock fissures, in soil, in litter, birds soaring in high air)
B. *Structural and eco-physiological strategies*	
1. Strategies to reduce waterstress	
Regulation of water loss (stomatal movements, residual resistance)	Restriction and regulation of water loss (concentrated urine, dry faeces, spiracular control of ventilation in beetles, reduction of GFR and UFR)
Xeromorphy	Structures reducing evaporation
High Water content (e.g. succulence in cacti)	Large water reserves (e.g. in snails)
2. Strategies to prevent death by overheating	
Transpirational cooling	Evaporative cooling
High heat tolerance (high temperature optimum and temperature compensation point of photosynthesis)	High heat tolerance (high body temperatures of homoiothermous animals, high lethal temperatures)
Mechanisms decreasing heat load (small leaves, orientation of leaves, stems instead of leaves as photosynthesizing organs)	Mechanisms decreasing heat load (high reflectance of incoming radiation, insolating structures)
3. Strategies to optimize water uptake	
Uptake of dew and water vapour (poikilohydrous plants, some phanerogams)	Uptake of dew and water vapour (arthropods, water enrichment of stored food)
Fast formation of rootlets after first rain	Fast drinking of large quantities of water when available (large mammals), uptake of water from wet soil (snails)
Uptake of highly saline water, high salt tolerance, salt excreting glands (halophytes)	Uptake of highly saline water, high salt tolerance, salt excreting glands (halozonts)
4. Strategies to synchronize reproduction and favourable environmental conditions	
'Waterclocks' of seed dispersal and germination	Sexual maturity, pairing and birth coincident with favourable conditions
Suppression of flowering (arido-active perennials) and sprouting (geophytes, hemicryptophytes) in extreme drought years	Sterility in extreme drought years

there is mass germination and high density, and low in drought or average years when only a few seeds germinate and density is low (Evenari, *et al.*, 1971; Evenari & Gutterman, 1976). It seems to me that this behaviour can only be explained by root competition or possibly allelopathy; but as noted in Volume 1, Chapter 3 there are no cases where allelopathy has been shown to exist under field conditions. The presence of crusts under these dwarf shrubs is not a likely reason for the absence of these seedlings since the same shrubs serve often as 'nurse plants' for annuals. It seems that there is root competition for seedlings but probably little or none for adult arido-active perennials.

Another controversial question concerns the distribution pattern of arido-active perennials. Is it regular, random or clumped? According to the site examined and to the method used to determine the pattern, one can make a case for any of the three patterns – and it seems as if each of the three patterns can be found in nature. Based on personal experience, I would say that the pattern will change according to the average amount of annual rainfall, its regularity or irregularity, average rainfall intensity, and the topography and soil type promoting or impeding run-off. In all pattern analyses, run-off (or run-on) has been neglected although it is an important factor involved in the distributional pattern of desert plants.

It is true that little is known of ecophysiology and interactions between host and epiphytes and parasites in deserts but one of the reasons is that the number of such plants in desert ecosystems is very limited.

The problem of animal–animal and animal–plant interaction is much more complicated than that of plant–plant interaction because the two are inter-related and there is obviously also a relationship between them, resource availability and abiotic parameters. Predators and the predator–prey relationship are of great importance in desert ecosystems and the percentage of predator species is high. It is suggested on p. 69 that this may be related to the 'pole-to-equator increase in the proportion of carnivorous species'. Could the reason be that animal prey is a rich source of water and that the predators can in this way survive in the desert without need of drinking water? Whether or not this is true, the impact of predator on prey and vice versa is a biotic parameter exerting constraints on the animal population.

How far do phytophagous animals compete for food and space and what is the nature of plant–animal interaction? Ants, rodents and to some extent birds compete for seeds and have a considerable impact on seed populations, which increase greatly when carpophagous animals are excluded. Over the ages a dynamic equilibrium has evolved in natural undisturbed ecosystems between these animals and seed production. The same is true for the porcupine (*Hystrix indica*) which feeds on bulbs, rhizomes and roots of a number of geophytes and hemicryptophytes (e.g. *Tulipa montana*, *Colchicum tunicatum*, *Urginea undulata*, *Bellevalia desertorum*, *Scorzonera pseudolanata*).

Recent research (Guterman, 1980) has shown that in spite of the large amounts of phytomass which the porcupine removes from the many digs (0.2–2.5 digs m^{-2}) it does not annihilate the species it eats because it digs out only the upper parts of their underground organs and leaves the lower parts from which most of the plants regenerate. The microhabitat created by the porcupine is quite favourable for the regrowth of plants since water and organic matter accumulate in the digs.

It is not clear how far herbivorous animals, eating above-ground phytomass, compete for food and how far food for them is a limiting factor. Wild grazing animals, before the introduction of domestic animals and phytophagous pests (rabbits in Australia) do not seem to have been constrained by food because of their mobility and small numbers, which were limited by availability of water. Sometimes grazing even seems to have a positive effect on plant production (Chapter 8, p. 194). For certain phytophagous groups like isopods and snails in the Negev and arthropods in general food is not a limiting factor. It may be supposed that the same is true for other herbivores, at least for those that are not food specialists ('generalized diets') and exhibit 'loose coupling ...to the food source' (Chapter 5, p. 97). In some deserts, as in Australia, termites occupy a 'pivotal position' (p. 90) because they eat very large amounts of plant material (more than 50% of the primary production (?) as stated in Chapter 11, p. 293), and together with the introduced domestic livestock they may destroy the natural grassland (Chapter 5, p. 90). It may be assumed that before the introduction of domestic animals, the termites were in equilibrium with their environment. In other deserts, termites are not such important consumers as in Australia.

Detritovores, about which we know little, are not limited by their food source. Their activity is enhanced in and around animal burrows and wherever litter and faeces accumulate.

An interesting point made in Chapter 4 is that 'Existing niche segregation and resource partitioning...can be considered a result of competition in the past, i.e. a kind of 'frozen competition'.

The animal–plant relationship is more than simply a consumer–food relationship since in certain cases green plant food is involved in reproductive animal processes (p. 94).

There are manifold interactions between the biotic and abiotic components of desert ecosystems and it seems that these interactions are more pronounced and have a greater impact on the structure and function of the whole system in deserts than in more mesic biota. Termites change soil structure and mineral distribution in soils. Anthills increase water intake and permeability of soils to water ('anthill effect', p. 375). Many shrubs accumulate soil (phytogenic hillocks, 'Nebka effect') in which water accumulates and its infiltration is increased. Temperature conditions there are favourable and attract animals to burrow, as happens with rodents (e.g. *Meriones crassus* below *Atriplex halimus* in the Negev), or snails to aestivate buried below these shrubs. It is also a preferred habitat for the germination of annuals. Litter accumulated below the shrubs and in soil depressions, brought there by wind and water, is the main habitat of detritovores and soil micro-organisms. These function as decomposers leading to enrichment of the soil with organic material and minerals, and therefore play a role in mineral cycling. In the Dead Sea valley, windblown loess is sedimented in the fissures of the thalli of bryophytes and lichens (Danin & Yaalon, 1980).

Soil underlying salt-excreting plants (*Tamarix* spp., *Reaumurea* spp., etc.) and plants accumulating salt in leaves and stems, are continuously enriched in salts by crystals that fall off or are washed off the leaves by rain, and by salt-containing litter. This changes soil qualities. Nitrogen fixing soil micro-organisms and some lichens growing on soil enrich the soil in nitrogen available to plant roots. In the Northern Negev bryophytes and Cyanophyceae stabilize sands (Danin, 1980).

Bacteria, fungi and Cyanophyceae growing on sandy surfaces make these soils unwettable, decrease infiltration and increase run-off (Savage, 1969; Rietveld, 1978). Stone cover and vegetation decrease run-off and increase infiltration (Evenari *et al.*, 1971; Shanan, 1975). This is why stones in deserts fulfill an important biological function. Plant cover, especially dwarf bushes, decrease deflation and erosion of soil. Slope affects run-off and infiltration processes considerably. In the Negev Highlands it was found for loessial slopes that the steeper the slope, the smaller the amount of water which runs off (Evenari *et al.*, 1971). The vegetation type changes as a function of the slope angle. In the Negev, as the slope increases, the predominance of *Artemisia herba alba* over *Zygophyllum dumosum* increases from 1.1 : 1 to 7.1 : 1 (Shanan, 1975). This in turn changes soil characteristics.

Endolithic lichens growing on hard limestone and dolomite in the Negev, Judaean desert and Northern Sinai, create pits and furrows on the rocks, changing runoff conditions and creating microniches (Danin, Orshan and Zohary 1975).

Physical and chemical properties of soils are changed drastically by large animals (porcupines, ground squirrels, etc.) and by small burrowing animals (ants, isopods, etc.). The amounts of soil turned over by these animals is very large. It has recently been shown in the Negev that the burrowing of porcupines, ants and isopods produces loose soil which significantly increases the silt load of run-off from slopes with burrows. Domestic grazing animals have their own specific impact on soil which they trample and compact.

Many cases of interactions between biotic and abiotic components of desert ecosystems are well documented and it is obvious that these interactions have far reaching consequences on the structure and function of the whole system. But only in few cases do the data permit quantification. Each interactive happening sets in motion a whole chain of events which has to be followed through in order to know the ecosystemic importance of the single original happening. In most cases this knowledge is missing or incomplete. In future, models should pay much more attention to biotic–abiotic interactions, some of which have been completely neglected, as for instance the impact of biotic and abiotic parameters on run-off and its effect on the biotic components of the system.

Population dynamics and production (Volume 2)

It is well known that the values of total peak phytomass and net annual primary production of arid lands are the lowest of all vegetation units known. It is less well known that their relative primary production (net annual production divided by peak standing phytomass) is high, much higher in fact than that of forests and similar to that of prairies and grasslands (Evenari *et al.*, 1976). Unfortunately, nobody, as far as I know, has yet calculated what I would like to call 'relative productivity', i.e. the annual phytomass produced per unit photosynthesizing surface. It could well be that these values would also be high in comparison to other biota.

Another typical feature of phytomass production in deserts is that the values of standing phytomass and production vary much from year to year. Years of measurements in the Negev have shown that for three dominant plant communities the above-ground annual production in a good rainy year is four to eight times larger than in drought years. In more extreme deserts this factor of maximum fluctuation is considerably higher. It could be shown that annual fluctuations of the total phytomass correspond not to the annual fluctuations of the rainfall but to the annual fluctuations of rainfall plus run-on (Evenari *et al.*, 1976). The fact that run-on water must be taken into consideration when trying to establish a correlation between primary production and available water has been largely overlooked. It has also been neglected that not only the absolute amounts of available water determine primary production but that a temporal factor is involved, that is the time of the first and last effective rains and the time interval between the effective rainfalls. In other words, the sequence of favourable and unfavourable moisture conditions, are important because photo- and thermoperiods and evaporation from soil become involved.

The large annual fluctuations of phytomass and primary production have a pronounced impact on the population dynamics of the vegetation since they cause changes in structure and composition of the plant component of the desert ecosystem. The greatest change concerns the arido-passive annuals. Their production strategy is 'Hurry before it is too late' whereas that of the arido-active plants is 'slow and steady wins' (Chapter 11, pp. 294–5). In drought years the arido-passive annuals are completely absent while in years with much available water and favourable temporal distribution of rainfall, permitting mass germination of annuals, their phytomass may account for 40–60% of the total peak phytomass.

The species density in such years is relatively large but the species composition changes from year to year and one could typify each such year according to the dominance of a specific species (the year of *Anthemis melampodina*, the year of *Matthiola livida*, etc.). The reason for this phenomenon is not known but it is possibly to be sought in the different germination requirements of each species.

The annual fluctuations of the above-ground and below-ground phytomass differ. One of the reasons is that in drought years the geophytes and hemicryptophytes do not produce any above-ground phytomass. During four consecutive years, for instance, the maximum fluctuation factor of the above ground phytomass of a *Hammada scoparia* community in the Negev was 4.43, and of the underground phytomass 2.75 (Evenari *et al.*, 1976). The below-ground phytomass which constitutes 'more than 50% of a desert community's biomass' (Chapter 12, p. 311) buffers or dampens the annual fluctuations and could be considered a permanent secure capital of desert ecosystems, their 'reserve of fertility'.

IBP has contributed much to our knowledge of secondary production and population dynamics of animals by using new methods of field observation and in-depth analysis of phenomena. The problem is more complex than that of primary production because it is a secondary process dependent on more variables, for instance the parameter of mobility ('nomadism') which is absent in primary production.

There are many similarities between the production patterns of plants and those of animals. Certain animals have zero production in drought years and could be compared to annual plants. The date of birth affects production as described for the germination of annual plants. Overall secondary production fluctuates from year to year and the various groups (mammals, reptiles, birds, rodents, etc.) differ considerably in their contribution to average production fluctuation. It seems that birds contribute little, mammals and reptiles more and arthropods perhaps most to the overall annual production. In each group a few species only are the main producers. Since in desert plants production and fluctuation patterns show a clear correlation with life forms and not with the systematic position of the plants* it can be asked if secondary production could not be correlated similarly to animal 'life forms'. This may bring out patterns that are not detectable when comparing production patterns of whole groups like 'mammals', 'rodents', etc.

The dissimilarities in production between plants and animals are most evident when the causal mechanisms responsible for the production fluctuations are considered. The annual fluctuations of animal populations are less

* The geophytic grass *Poa sinaica* has a production pattern different from that of the annual grass *Schismus barbatus*.

extreme than those of the primary producers. In both cases most of the fluctuations are coincident with rainfall variations but in some animal species the fluctuations are independent of rainfall. While in primary producers available water is the main causal agent, in animals other parameters are involved, such as quantity and quality of food, predation, competition, fertility, reproduction rates and mortality at certain phenophases, which are not linked to abiotic parameters (for instance high mortality of *Hemilepistus* spp. during the pairing phenophase, p. 572). All these factors are inter-related in a complex causal web in which biotic parameters play a much more important role than in plants. There are also cyclic oscillations of secondary production the like of which has, as far as I know, not been observed in plants. As in plants photo- and thermoperiods are also involved, especially in reproduction, where rain is the 'Zeitgeber' and photo- and thermoperiod the triggers. The Australian vertebrates are an exception where reproduction is apparently linked only to rainfall.

Many problems and questions remain to be answered by future research. It is stated in chapter 7, p. 161 'that food shortage appears to be a significant and pervasive limiting influence on desert animal populations'. In Chapter 9, p. 240, we read 'In terrestrial communities as much as 90% of the net primary production remains unharvested as living plant tissue and must be utilized as dead tissue by saprobes and soil animals'. All my personal experience tends to confirm the second statement and it seems to be certain that the bulk of the energy produced by the primary producers bypasses the secondary producers and reaches the decomposers. But how can the two seemingly contradictory statements be conciliated? Can it be that the first statement relates only to temporal food shortages and to green matter produced? Does it relate more to the quality of the food than to its quantity?

Chapter 7 discusses in depth the question of *r* and *K* selection. The author comes to an interesting conclusion which certainly will be questioned by many animal ecologists. He states 'In total, it appears that the major forces determining the demographic characteristics of desert animals are their phylogenetic momentum and whatever influences are associated with latitude. The space remaining for *r* and *K* selection, and the evidence for its existence cast serious doubt on its significance, *if not its reality*' (the italics are mine). I suppose that by 'phylogenetic momentum' the author relates to what others (Stocker, 1976 and earlier papers cited there) have called 'constitutional types' and by 'latitudinal influences' to photo- and thermoperiodic effects. Since the question of *r* and *K* selection has been mainly discussed by zoologists how do plant ecologists react to the author's statement?

Another question relates to the production efficiency of desert animals. Is it different from that of non-desert, secondary producers? We have considered the same question for primary producers on p. 577.

One of the main gaps in our knowledge of production and population dynamics concerns production and production fluctuations of decomposers of which we know very little. We can only state in the most general way (Chapter 11, p. 295) that their activity is 'moisture dependent and thus considerably variable'. This means very little in the absence of quantifiable data, which is unfortunate since the more we learn about desert ecosystems the more we realize the ecosystemic importance of decomposers. IBP has added much to our knowledge of production and production fluctuation of the various components of desert ecosystems. But we lack integrated knowledge of production fluctuation patterns of the whole system and its correlation with the rainfall patterns and patterns of water availability.

Mineral cycling (Volume 2)

After available water, the amount of organic matter and vital mineral elements is the second factor limiting production in desert ecosystems because they suffer from nutrient stress which in its turn affects water-use efficiency.

Detritus composed of litter, faeces, dead plants and animals is decomposed and mineralized by soil micro-organisms. Recent studies indicate that in deserts this process proceeds during wet periods faster than was believed earlier. Decomposing micro-organisms exist in all desert soils; even in regs and salinas, although they are poor in their respect. The number of microbes and microbial activity is greatest below perennial shrubs because of better water conditions, presence of mycorrhiza, rhizosphere effects and accumulation of plant litter. Accumulation and higher infiltration rates of water in these micro-habitats favour the downward movement and accumulation of minerals in deeper soil layers. Mineralization and mineral cycling in general have their annual, seasonal and daily cycles. Much is known about the annual cycles, very little about the seasonal and daily ones. These activity fluctuations are related to available water. Light rains, non-efficient for plant growth, initiate mineralization. The bulk of minerals is contained in the standing phytomass as is known from the many Russian papers on this subject. But animals play an important role in cycling and redistribution. Plant litter reaching the ground and decomposing there does not contain much nitrogen and phosphorus, because these two vital elements are, to a large extent, withdrawn by the plants from the parts to be shed, and stored in the remaining organs, especially when under water stress. But the absolute amount of nitrogen in litter is larger in deserts than in more mesic biota. Animal faeces which play a large part in mineral cycling, are rich in nitrogen, most of which is lost to the atmosphere. Controlled grazing has a beneficial effect on mineral cycling in contrast to over-grazing, which diminishes microbial activity. As far as nitrogen is concerned, detritus is not its only source for cycling. Some of it is added to the soil by nitrogen-fixing microorganisms, but the major part of the fixed nitrogen returns to the atmosphere. Mycorrhizal fungi apparently play a role in nitrogen redistribution and in the efficiency of mineral cycling in general.

'Biocycling' is only one aspect of mineral cycling, 'geochemical cycling' is the second aspect. Despite its apparently greater importance in deserts than in more mesic biota, it has not received the attention due to it. Chemical processes contribute to the decomposition of detritus. Physical processes increase the input into the cycling. Windblown dust and rainwater add minerals, including nitrogen. Run-on water distributes organic matter.

Leaching of plant surfaces by rain also adds minerals to the soil, especially excess salts excreted by the leaves. Leaching of soil redistributes minerals. Since this leaching is differential, abundant elements like calcium, sodium and chlorine accumulate and some of the vital ones may be lost through drainage to the ground water. Soil leaching may not be too important considering the scarcity of rain, but there are no measurements available.

IBP and allied research has stimulated modelling of the nitrogen and phosphorus cycles but we still lack an overall picture of mineral cycling in the desert as a whole.

Modelling (Volume 2)

One of the most important results of IBP, with a far-reaching impact on modern biology, concerns modelling. Biologists have always 'modelled', but theirs were word or diagram models demonstrating only certain relationships between different components of processes in the ecosystems. Only mathematical models, mainly developed by IBP, using analogue or digital computers have enabled the modeller to quantify these relationships and to predict, with a certain degree of probability, what would happen if a certain system variable were to change quantitatively. The value of a mathematical model for understanding the structure and function of an ecosystem or for prediction depends on the ingenuity and intuition of the modeller, on the quality of the mathematics used and on the quality of the input data. The IBP programme has succeeded in developing the necessary mathematical tools and has collected an enormous amount of facts on structure, function and processes of desert ecosystems used as input of the models. A number of important models of various natures have been built, which (and this is of the utmost importance for the degree of 'realism' of the models) could be validated at various sites. Basically the models are of two kinds: ad hoc (specific or process or subsystem) and general purpose (generalized or whole system) models.

Ad-hoc models have been made for such processes as photosynthesis, predation, competition, grazing, strategy of annual plants, cycling of various minerals such as nitrogen and phosphorus, primary and secondary production, water redistribution and its effect on primary production. While ad-hoc models try to answer specific questions, general purpose models should help us to understand whole ecosystems. By their nature these models are considerably more complex than ad-hoc models. With all their importance and the necessity for generalized models of desert ecosystems, their shortcomings lie in the fact that today's knowledge of their various components ('subroutines') is imperfect.

The main modelling activity concerning desert ecosystems has been concentrated in the US IBP and in Australia and to some extent in Israel. But there have been some attempts in this direction in the USSR, India, Egypt and Tunisia. It is one of the greatest contributions of IBP to have initiated and stimulated modelling of desert ecosystems, an activity which is continuing with increasing intensity, even after IBP has officially been terminated. Modelling of desert ecosystems has increased our understanding of these systems mainly by pointing out what, if any, effect a certain biotic or abiotic parameter has on the structure or function of the whole system or on its components or processes. This often leads to quite unexpected results, verified later by actual observations. It has helped to detect probable feedbacks and has enabled us to find basic gaps in our knowledge. Models also permit us to predict changes in the system, with a variable degree of probability, when certain parameters disappear or change quantitatively. They can thereby be a management tool. But I offer one word of caution, which may be obvious, but the obvious is sometimes forgotten or neglected. Models are tools and represent only a simulated reality and not reality. The nearness of simulation to reality, and therefore the quality and usefulness of the model, depends on the quality of the tools and the aptness of its user and on the quantity and quality of the input (i.e. the total evidence) collected and evaluated by the field ecologists. The latter is and remains the mainstay of ecology.

Management (Volume 2)

The purpose of the arid lands project of IBP was clearly stated in saying: 'The IBP aim is to further our understanding of the structure and functioning of these arid zone communities, so that we may be better able to predict the consequences of man's efforts to utilize the natural resources of these areas' (Volume 1, p. xxv). Clearly the last part of this sentence relates to management. Accordingly, the title of the two volumes in hand is: Arid land ecosystems: structure, functioning and management. But whereas 53 chapters are devoted to structure and function, only 5 deal directly with management, though the problems of management are touched upon casually in some of the other chapters. This disproportion is quite significant, indicating that within the framework of IBP the problem of how to optimize the use of the meagre natural resources of arid land ecosystems without damaging or destroying the system, has not been given the same attention as structure and function. Given the fact that it is mostly (and perhaps exclusively) biologists who have been involved in the arid lands IBP, this fact is not surprising.

When talking about 'management' some questions emerge: management for what – grazing, agriculture, recreation, etc.? Management for whom – autochtoneous nomads or semi-nomads, or allochtoneous introduced population? And at the back of our mind will be another question that will bother us even if we feel that it is not our problem: are we as scientists, as men who possess the knowledge upon which management decisions have to be based, to become involved as part of the decision making process or are we only to function as advisors? Should we help to turn the decisions into reality and if we feel we should, how can this be done? I am asking these questions not because I think that to answer them falls into the normal sphere of an ecologist's duty, but because the danger to the arid-land ecosystems is so great and their deterioration so rapid, as has been demonstrated in various chapters (14, 19 and others), that we should help to apply our accumulated knowledge to the practical solutions of management problems of arid lands. With all its unavoidable shortcomings, IBP has accumulated a large amount of knowledge of arid-land ecosystems. Part of this knowledge is contained in these two volumes and in the bibliography following each chapter. Part of it is found in the many ecological journals. In order to make the data easily accessible and retrievable they should be collected in an international computerized data bank which would be the basis for decision making and management practices.

The data collected by IBP, and related activities, which are important for the management of arid-land ecosystems cover a wide area. We know, first of all, the main constraints that the abiotic and biotic environment puts on primary and secondary production. We know the temporal and spatial dimensions of these constraints. We know that even slight changes, apparently affecting only one component or process of the system, have serious repercussions throughout the whole system because of the complex interactions between its components and its high degree of sensitivity. Modelling, simulation and systems analysis can predict, with a certain degree of probability, the effect of these changes.

We know, and this is of primary importance for any management, that in the overall productivity of the desert ecosystems there is a strong element of 'pulse-reserve' processes.

The first conclusion to be drawn from all this is that any successful management practice of desert ecosystems directed at a specific target cannot restrict itself to the management of the one resource concerned. The management planning has to take into account the whole ecosystem and the impact the introduced changes will have upon it.

The second conclusion is that the 'natural' resources of any desert ecosystem, biomass and water, are most limited, as has been mentioned time and again in these two IBP volumes. But some other resources most vital for modern man, like oil and certain ores, are – I may say unfortunately – concentrated in deserts in large quantities. The extraction, processing and consequent 'development' of these resources has a very serious and possibly catastrophic impact on the local desert ecosystems, on their flora, fauna and last but not least, on man and the social structure of the indigenous human population. This theme has not been dealt with in these IBP volumes. But if the ecologist has to deal with 'the consequences of man's effort to utilize the natural resources of these areas' we cannot neglect these specific consequences as we did in the past. It is appropriate to mention here another resource of arid lands which is unlimited: solar energy. If the various experimental efforts to use it directly (solar ponds, etc.) or indirectly (algal cultures, etc.) are successful, the desert ecologist will be confronted with another challenge.

Primary production of desert ecosystems has been utilized by man since time immemorial for grazing. Grazing by nomadic herds or flocks (including transhumance) is probably the most ancient management practice of desert ecosystems, which led to a dynamic state of equilibrium between plant, grazing animal and man, and was in itself an efficient way of utilizing the natural resources as long as the nomad did not learn to use new water sources excessively. Then overgrazing started, with all its known consequences. Today there are various methods of rational ecosystemic grazing and improvement of range land. There have been various attempts to arrive at optimization of grazing (Chapters 15 and 16) and of plant production in arid regions (Chapter 17), which should help to develop management models. But there seems still to be a large gap between knowledge and its practical large-scale application. As far as models are concerned one of the reasons may be that simulation results and the reality as established by field observations and experiments, are still too far apart.

Another lesson which IBP has helped to bring home concerns agriculture in arid lands. All the data on rainfall, production and desertification, as well as the practical experience, show that arid lands agriculture using rain where it falls is at best marginal in the infrequent rainy years, and is mostly highly destructive to the natural ecosystem. Agriculture is possible, however, even in areas with not more than about 100 mm of annual precipitation of winter rain, when using run-off water. Various methods of desert run-off agriculture have been developed over the last 3000 years and are being developed today. Modern techniques can be applied in order to increase run-off artificially (Chapter 23). Irrigation with water brought in from the outside, by river diversion, etc., is and can be used to make arid lands agriculturally productive, but only if this water resource is managed correctly, taking into account what IBP has taught us about soils (Volume 1, Chapter 17), soil–plant–atmosphere interaction (Chapter 2), water management and potentials of primary production (Chapter 17). If the management if faulty, disasters occur and there are only too many cases in which this has happened.

Recreation, tourism and permanent modern urban settlements are 'resources' of arid lands which are becoming increasingly popular (Chapter 22), leading to a human mass invasion of certain deserts. Thus 'the original desert is practically eliminated and another substituted' (Chapter 22). This new human activity, a kind of return of modern man to the desert, has created new problems like the accumulated 'residue' which people leave and the impact of 'off-road vehicles' on soil, vegetation and the whole ecosystem. One of the worst problems is the modern desert town and settlement which, with very few exceptions, were and are being set up without any ecological planning. There are enough examples of this in the US and Israel. If this trend to overpopulate arid lands continues it may be necessary to devise new, revolutionary methods using the one plentiful resource of deserts, solar energy, to drive closed system agriculture and self-contained large cities (Revel, 1977).

The penultimate chapter of Volume 2 deals briefly with the social aspects of arid-land management. Social aspects of arid-land management are mostly neglected in spite of the fact that any such management involves tremendous social upheavals. This is to some degree true even for developed countries but much more so for developing countries. The outstanding examples are the nomads and semi-nomads. As pointed out above, up to our times their way of life was in equilibrium with their surroundings. For various reasons, discussed in a number of publications not contained in our two IBP volumes on arid lands, the nomadic way of life is doomed and rapidly disappearing. But what is going to happen to the nomads? Can they be forced to become sedentary? Do we want to keep them in reservations like the American Indians? Can they be integrated into urban societies? Up to now no practical solution to this problem has been found. This is only one example concerning the oldest inhabitants of deserts. The social and environmental problems posed by modern desert dwellers are no less acute. Since man is part of the arid-land ecosystem and affects it drastically, we ecologists have a duty to help solve the problems involved, something which we have not done so far.

Conclusion

IBP and related research on arid lands have produced a great number of new facts, opened new vistas, and new approaches and formulated new concepts, contributing much to our knowledge of arid lands and their ecosystems. But 'knowledge' and 'understanding' are not synonymous terms and it must be asked if an integration of all this knowledge enables us to arrive at a better understanding of arid lands in general and their ecosystems in particular. Do we know the qualities and processes bestowing on arid lands their specific nature, making them different from more mesic areas? Can we quantify these differences? Can we do the same for arid-land ecosystems concerning their structure and function?

All arid lands have much in common and at the same time differ from each other. The main common features are:

(1) rainfall is erratic and unpredictable with great annual fluctuations around the mean;

(2) mean annual potential evapo–transpiration is much larger than mean annual precipitation;

(3) the wind-erosional cycle determines land-forms to a large degree in conjunction with the water-erosional cycle;

(4) similarity of main land-forms and soils;

(5) large amounts of precipitation do not infiltrate but run-off is of great biological importance;

(6) availability of water is the main limiting factor for most (but not all) arid-land organisms, hence sparse vegetation cover with poor vertical stratification;

(7) similarity of survival strategies of organisms; and

(8) extreme importance of heterogeneity of macro- and micro-topography for survival of organisms.

The main dissimilarities between different arid regions are:

(1) dissimilarity of mean annual precipitation from near zero to 600 mm;

(2) large spectrum of degree of aridity as measured by various aridity indices ranging from extreme arid to semi-arid;

(3) dissimilarity in pattern of precipitation – winter rain, summer rain, winter and summer rain – with ensuing dissimilarity in efficiency of rainfall;

(4) dissimilarity in mean annual temperatures – hot, temperate and cold deserts;

(5) dissimilarity in mean annual humidity of air;

(6) dissimilarity in main sources of water-rainfall versus fog and high air humidity; and

(7) dissimilarity in plant and animal taxa due to differences in history and geography with ensuing differences in dominating life forms.

The similarities and dissimilarities are well quantified by all kinds of indices and IBP has done much to increase the precision of quantification. But there are whole regions which have been neglected and for which basic data are missing or insufficient (e.g. Somali-Chalbi, Central Gobi). Other regions have not been dealt with adequately in these two volumes (e.g. Central Asian cold deserts). This is true also for the dissimilarities of arid lands which should have been more accentuated because they necessitate different management practices.

The arid-land ecosystems are as complex or even more complex than their more mesic counterparts. They are in no way 'simple'. Their water regime is characterized by its great quantitative and temporal heterogeneity which determines its 'pulse' nature. One single pulse event and not the absolute yearly or mean amount of rainfall may be decisive for one year's behaviour of the system's biotic components. As the consequence of this rainfall pattern, arid-land ecosystems are pulse–reserve systems with ensuing large biomass fluctuations. The fluctuations of the phytomass are larger than those of the zoomass which buffer the overall biomass fluctuations. The permanent underground phytomass of the perennial plants fulfills the same function. Underground plant organs, seeds and food and water stored in, and by, animals serve also as sink reservoirs and as 'memory of the system between pulses'. The fluctuations are the cause of the system's lability but their fast reaction to the pulses of the environment and the existence of the 'memory' endow the system with resilience which in its turn gives it long-term stability.

The basic survival strategies of the various biotic components of the system are to a large extent analogous. The system's spatial biotic and abiotic heterogeneity which creates a large diversity of macro- and microniches is of vital importance for the survival of its biotic components. The more extreme the environmental conditions, the greater becomes the importance of the environmental heterogeneity.

The food and water webs of the system are intrinsicly interwoven since for most animals food is also a source of water. The main energy flow of the system by-passes the secondary producers and goes directly to the decomposers. Because of its lability, the system is highly sensitive to interference.

Many questions remain unanswered. What is the importance of overall species diversity for resilience and stability of the system? Can life-form diversity and diversity and flexibility of survival strategies be measured, and if so, are the arid-land ecosystems characterized by specific features of these diversities? How far is the quantity and quality of primary production and phytomass an overall limiting factor for secondary producers? It is true, as some authors suggest (e.g. Noy-Meir, 1979), that because of the large dependence of the biotic components upon the abiotic components arid-land ecosystems are characterized by a weak coupling between species ('independent population concept') and that this gives the system stability? If this is so, it would mean that MacMahon's opinion (Volume 1, p. 30) 'the essential nature of communities is that they are the sum of a series of species' ('individualistic' approach) is right at least for arid-land ecosystems. This would also mean that the desert ecosystems as a whole do not possess specific qualities which would be more than the integral of all qualities of their components. This problem is worthy of discussion in depth.

Acknowledgment

Sincere appreciation is extended to Dr David Goodall and Dr Immanuel Noy-Meir for review of the manuscript and to my colleagues Dr Avinoam Danin, Hebrew University, Jerusalem, Dr Yitzhak Guttermann and Dr Moshe Shachak, Desert Research Institute, Ben Gurion University of the Negev for their permission to use unpublished data.

References

Anonymous (1979). Map of the world distribution of arid regions. *MAB technical notes* 7. UNESCO, Paris.

Barbour, M. G. (1973). Desert dogma re-examined: root:shoot ratio and plant spacings. *American Midland Naturalist*, **89**, 41–57.

Danin, A. (1976). Plant species diversity under desert conditions. *Oecologia*, **22**, 251–9.

Danin, A. (1980). *Desert vegetation of Israel and Sinai.* Sifrait Poalim, in press.

Danin, A., Orshan, G. & Zohary, M. (1975). Vegetation of the Northern Negev and the Judean Desert of Israel. *Israel Journal of Botany*, **24**, 118–72.

Danin, A. & Yaalon, D. H. (1980). Silt fixation by lichens and mosses in the Dead Sea valley, Israel (manuscript).

Dubinsky, Z. & Steinberger, V. (1979). The survival of the desert isopod *Hemilepistus reaumuri* (Audouin) in relation to temperature (*Isopoda, Oniscoidea*). *Crustaceana*, **36**(2), 147–54.

Evenari, M., Shanan, L. & Tadmor, N. (1971). *The Negev: the Challenge of a Desert.* Harvard University Press, Cambridge, Massachusetts.

Evenari, M., Bamberg, S., Schulze, E. D., Kappen, L., Lange, O. L. & Buschbom, U. (1975*a*). The biomass production of some higher plants in Near Eastern and American deserts. In: *Photosynthesis and productivity in different environments*, ed. J. P. Cooper, pp. 121–7. Cambridge University Press.

Evenari, M., Schulze, E. D., Kappen, L., Buschbom, U. & Lange, O. L. (1975*b*). Adaptive mechanisms in desert plants. In: *Physiological adaptation to the environment*, ed. F. J. Vernberg, pp. 111–29. Intext Educational, New York.

Evenari, M. & Guttermann, Y. (1976). Observations on the secondary succession of the plant communities in the Negev desert, Israel. I. *Artemisiatum herbae albae.* Etudes de biologie végétale, ed. R. Jacques, pp. 57–86. Paris CNRS, Gif-sur-Yvette.

Evenari, M., Schulze, E. D., Lange, O. L., Kappen, L. & Buschbom, U. (1976). Plant production in arid and semi-arid areas. In: *Water and plant life*. Ecological studies 19, ed. O. L. Lange, L. Kappen & E. D. Schulze, pp. 439–51. Springer Verlag, Berlin.

Evenari, M., Lange, O. L., Shulze, E. D., Kappen, L. Buschbom, U. (1977). Net photosynthesis, dry matter production and phenological development of Apricot trees (*Prunus armeniaca* L.) cultivated in the Negev Highlands (Israel). *Flora*, **66**, 383–414.

Friedman, J. & Orshan, G. (1975). The distribution, emergence and survival of seedlings of *Artemisia herba alba* in the Negev desert of Israel in relation to distance from the adult plants. *Journal of Ecology*, **63**, 627–32.

Glanz, M. (ed.) (1977). *Desertification*. Westview special studies in natural Resources and Energy management. Westview Press, Colorado.

Gutterman, Y. (1980). The relation between the porcupine and the desert geophytes and hemicryptophytes of the Negev Highlands. Israel *Journal of Botany*, in press.

Gutterman, Y., Witztum, A. & Evenari, M. (1976). Seed dispersal and germination in *Blepharis persica* (Burm.) Kuntze. Israel *Journal of Botany*, **16**, 213–34.

Hadley, N. F. (ed.) (1975). *Environmental physiology of desert organisms*. Dowden, Hutchinson & Ross, Stroudsburg, Pennsylvania.

Kappen, L., Lange, O. L., Schulze, E. D., Evenari, M. & Buschbom, U. (1972). Extreme water stress and photosynthetic activity of the desert plant *Artemisia herba alba Asso*. *Oecologia*, **10**, 177–82.

Kappen, L., Lange, O. L., Schulze, E. D., Evenari, M. & Buschbom, U. (1975). Primary production of lower plants (lichens) in the desert and its physiological basis. In: *Photosynthesis and productivity in different environments*, ed. J. P. Cooper, pp. 133–43. Cambridge University Press.

Lange, O. L. (1965). Der CO_2 Geswechsel von Flèchten bei tilefen Temperaturen. *Planta*, **64**, 1–19.

Lange, O. L., Schulze, E. D., Evenari, M., Kappen, L. & Buschbom, U. (1974). The temperature-related photosynthetic capacity of plants under desert conditions. *Oecologia*, **17**, 97–110.

Lange, O. L., Schulze, E. D., Kappen, L., Buschbom, U. & Evenari, M. (1975). Photosynthesis of desert plants as influenced by internal and external factors. In: *Perspectives of biophysical ecology*. Ecological studies 12 (ed. D. M. Gates & R. B. Schmerl). Springer Verlag, Berlin.

Lange, O. L., Geiger, J. L. & Schultz, E. D. (1977). Ecophysiological investigations on lichens of the Negev desert V. A. model to simulate net photosynthesis and respiration of *Ramalina maciformis*. *Oecologia*, **28**, 247–59.

Livingston, B. E. & Brown, W. H. (1912). Relation of the daily march of transpiration to variations in the water content of foliage leaves. *Botanical Gazette*, **53**, 309–30.

Marder, J. (1973). Body temperature regulation in the brown-necked raven (*Corvus corax ruficollis*) I and II. *Comparative Biochemistry and Physiology*, 45A: 421–30, 431–40.

Marks, A. E. (ed.) (1976, 1977). *Prehistory and Paleo-environments in the Central Negev, Israel*, 2 Vol. SMU Press, Dallas.

Meigs, P. (1953). World distribution of arid semi-arid homoclimates. *UNESCO Arid Zone Research*, **1**, 203–9.

Mundlak, Y. & Singer, S. F. (ed.) (1977). *Arid zone development*. Ballinger, Cambridge, Massachusetts.

Netschaeva, N. T., Antonova, K. G., Karshenas, S. D., Movkhammedov, G. & Noorberdie, M. (1979). *Productivity of the Vegetation of the Central Karakum in connection with different regimes of utilization*. Nauka, Moskva, pp. 254.

Noy-Meir, I. (1978). Structure and function of desert ecosystems. *Abstracts of the Second International Congress of Ecology (INTECOL)* Vol. 1, p. 265.

Orians, G. H. & Solbrig, O. T. (eds.) (1977). *Convergent evolution in warm deserts*. Dowden, Hutchinson & Ross, Stroudsburg, Pennsylvania.

Orr, Y. (1970). Temperature measurements at the nest of the desert lark (*Ammomanes deserti deserti*). *Condor*, **72**, 476–8.

Orr, Y., Shachak, M. & Steinberger, Y. (1979). Ecology of the small spotted lizard (*Eremias guttulata guttulata*) in the Negev desert (Israel). *Journal of the Arid Environment*, **2**, 151–61.

Orshan, G. (1953). Note on the application of Raunkiaer's system of life forms in arid regions. *Palestine Journal of Botany*, **6**, 120–2.

Ravel, R. (1977). Let the water bring forth abundantly. In: *Arid zone development*, ed. Y. Mundlak & S. F. Singer, pp. 191–200. Ballinger, Cambridge, Massachusetts.

Rietveld, J. (1978). *Soil non-wettability and its relevance to surface runoff on sandy soils in Mali*, p. 179. Wiegenmyen Report Production Primaire au Sahel.

Rodin, L. E. (ed.) (1977). *Productivity of vegetation in arid zone of Asia*. Nauka, Leningrad (in Russian).

Rodin, L. E. (ed.) (1978). *Management of vegetation in arid zone of Asia*. Nauka, Leningrad (in Russian).

Savage, S. M. (1969). Contribution of some soil fungi to water repellancy in soil materials. *Proceedings of a Symposium on Water repellant soils, Riverside, California*, pp. 241–257.

Schulze, E. D., Lange, O. L. & Koch, W. (1972). Oekophysiologische Untersuchungen an Wild- und Kulturpflanzen der Negev-Wüste. II. Die Wirkung der Aussenfaktoren auf CO_2 Gaswechsel und Transpiration am Ende der Trockenzeit. *Oecologia*, **8**, 334–55.

Schulze, E. D., Lange, O. L., Kappen, L., Buschbom, U. & Evenari, M. (1973). Stomatal response to changes in temperature at increasing water stress. *Planta*, **110**, 29–42.

Seely, M. K. (1979). Irregular fog as a water source for desert beetles. *Oecologia*, **42**, 213–27.

Shachak, M., Orr, Y. & Steinberger, Y. (1975). Field observations on the natural history of *Sphincterochyla (S.) zonata* (Bourguignat 1853) (= *S. boissieri Charpentier 1847*). *Argamon, Israel J. Malac.* **5**, 20–46.

Shachak, M., Chapman, E. A. & Steinberger, Y. (1976). Feeding, energy flow and soil turnover in the desert isopod *Hemilepistus reaumuri*. *Oecologia*, **24**, 57–69.

Shachak, M., Chapman, E. A. & Orr, Y. (1976). Some aspects of the ecology of the desert snail *Sphincterochyla boissieri* in relation to water and energy flow. *Israel Journal of Medical Science*, **12**, 887–91.

Shachak, M., Steinberger, Y. & Orr, Y. (1979). Phenology, activity and regulation of radiation load in the desert isopod *Hemilepistus reaumuri*. *Oecologia*, **40**, 133–40.

Shachak, M. & Steinberger, Y. (1980). The bioenergetics of the desert snail *Sphincterochyla (S.) zonata* (Bouguignat, 1835) = *S. boissieri* (Charpentier, 1847) (manuscript).

Shanan, L. (1975). Rainfall and runoff relationships in small watersheds in the Negev Highlands. PhD Thesis, Hebrew University of Jerusalem.

Shkolnik, A., Chosniak, I. & Maltz, E. (1978). Water economy and exploitation of desert pastures by Beduin goat, ibexes and gazelles. *Abstracts of the Second International Congress of Ecology (INTECOL)*, Vol. I, p. 350.

Shreve, F. (1942). The desert vegetation of North America. *Botanical Review*, **8**, 195–246.

Stocker, O. (1976). The water-photosynthesis syndrome and the geographical plant distribution in the Saharan deserts. In: *Water and plant life*, Ecological studies 19 (ed. O. L. Lange, L. Kappen & E. D. Schulze), pp. 506–21. Springer Verlag, Berlin.

Volkens, G. (1887). *Die Flora der aegyptisch-arabischen Wüste*. Bornträger, Berlin.

Vosnessenski, V. L. (1977). *Photosynthesis of desert plants*. Nauka, Leningrad (in Russian).

Went, F. (1975). Water vapour absorption in Prosopis. In: *Physiological adaptation to the environment* (ed. F. J. Vernberg), pp. 67–75. Intext Educational, New York.

Winter, K. & Thoughton, J. H. (1978). Photosynthetic pathways in plants of coastal and inland habitats of Israel and Sinai. *Flora*, **167**, 1–34.
Yom-Tov, Y. (1971*a*). Body temperature and light reflectance in two desert snails. *Proc. Malac. Soc. London*, **39**, 319–26.
Yom-Tov, Y. (1971*b*). The biology of two desert snails *Trochoidea* (*Xerocrassa*) *Seetzeni* and *Sphincterochyla boissieri*. *Israel Journal of Zoology*, **20**, 231–48.
Zohary, M. (1954). Hydr.-economical types in the vegetation of Near East deserts. In: *Biology of deserts*, pp. 56–67. Institute of Biology, London.

Manuscript received by the editors February 1980

Index

Note: **References in bold type refer to chapters; references in italic type refer to tables, figures and plates**

Abies spp., 378
abiotic components, effects of biotic components on *see* biotic components
Acacia amentacea, 14
Acacia aneura (mulga): and animal interactions, 89–90; and nutrient cycling, 308, 336
Acacia cyanophylla, 326, 348, *349*, 375
Acacia decussata, 110
Acacia ehrenbergiana, 39
Acacia greggii, *36*, *339*
Acacia gummifera, 364
Acacia jacquemontii, 14, 485, 486
Acacia ligutata, 375
Acacia raddiana, 364
Acacia senegal, 14, 485
Acacia spp., 25, 34
Acacia tortilis, 485, 486
Acamptopappus shockleyi, *36*
Acantholepis frauenfeldi, 336
Acantholimon renustum, *36*
Acrididae, 132
Achromobacter spp., 341
Actinomycetes, 330
Acytonema spp., 344
Acynonix jubatus, 378
Addax nasomaculatus, 94
Allenia subaphylla, 171, *172*, *177*
afforestation, of shifting dunes, 485–6
Africa, desert reserves of, 504–6
agonistic behaviour, of animal spp., 72–4, 76
agriculture (crop production), 486–93; crop varieties, 490; drought evasion, 490; fertilizer use, 491; inter-cropping, 491; and limited irrigation, 491–2; and salinity problem, 493; water harvesting and run-off recycling, 488–90; *see also* Indian Desert
Agriophyllum latifolium, 172
Agropyron desertorum, *19*, 471
Agropyron fragilis, *186–7*, 193
Agropyron inerme, 114
Agropyron spicatum, 35
Ailopus spp., *92*
Aimophila carpalis, 156, *212*
Albizzia lebbek, 485
Alcephalus babalus, 378
Aleppo pine forests, plant communities of, 359–60, *365*; *see also Pinus halepensis*
alfa grass steppes, plant communities of, 360–1, *365*
algae, 330; algal crusts and reduction of soil erosion, 111; and nitrogen fixation, 304, 344
Algeria, 23–4
Alhagi camelorum, *36*
Alhagi mororum, 20
Alhagi persarum, 172
Allactaga elator, 73
allelopathy, in desert plants, 38–9
Allenrolfea occidentalis, 16
Allium mongaolicum, *182*
Allium polyrrhizum, *182*
allogenic succession, 371–3; erosion, 371–2; sedimentation, 373
alluvial fans, 272–4, *284*, *288*
alluvial flats, 276, *284*
Alternaria, spp., 331
aluminium, cycling of, in plants, *184*
Alyssum desertorum, 172, *177*
Ambrosia dumosa, *36*, 281, *282*, 326
amensalism, *see* allelopathy
America, North, deserts of, 557; birds, habitat preferences of, 54; and plant–plant interactions, **33–46**; plant spp. diversity, 60–1; reserves of, 495, *496*, 497, *498*, *502–3*, 509–18; secondary production in, 200–1, 202–40; shrub patterns in, 40–3
Amitermes abruptus, *89*
Amitermes agrilus, *89*
Amitermes laurensis, *118*
Amitermes obtusidens, *89*
Amitermes perarmatus, *89*
Amitermes ritiosus, *121*
Ammodendron argenteum, 185, *186–7*, 188, 193
Ammodendron conollyi, *170*, 171, *172*, 173; gas exchange, *178*
Ammomones deserti, 572
ammonification, 340
Ammophila arenaria, 17
Ammospermophilus harrisii, production by, *203*, 205–7, *243*, *250*
Ammothamnus lehmannii, 171, *172*
Amphispiza belli, 245
Amphispiza bilineata, 211, *212*; production by, 238–40, 245
Anabaena spp., 344
Anabasidetum articulatae arenarium, *23*
Anabasis aphylla, 177, 372; mineral cycling in, 183
Anabasis aretioides, 326, *327*, 344, 346
Anabasis oropediorum, 362, 375
Anabasis salsa, 173, 185, *186–7*, 188
Anabasis spp., 25, *37*
Anacyclus clavatus, 361, 376
Anacyclus cyrtolepidioides, 363
analysis of data *see* data analysis
Andropogon scoparius, 14

Index

animal–animal interactions, **51–76**; agonistic behaviour, 72–4; interspecific competition, 53–62; modelling of 52–3, 63; predation, 62–72, 76
animals: animal–animal interactions *see* plant–animal interactions; energy flow and water dynamics, 285–92, 371, 378; and grazing, 468–9; life-history patterns *see* life-history; and mineral dynamics, **334**, 335–8; modelling, of processes, 398–9; population variations, 127–35, 159–62; processes, 398–9, 568–72; and soil effects, 112–21, 328; survival strategies, *573*; temperature regulation of, 570; *see also particular groups and species*
annual plants, model of life history, 389–91
annual variations, in vegetation, 21–3
Antechinomys spp., 88
Anthemis melampodina, 578
Anthemis pedunculata, 363
ants: anthill effect, 375; and denudation of soil surface, 114–15, 116; and plant interactions, 86–7, 98–100; seed-harvesting spp., and energy flow, 288, 290; species separation of, 55–6, 58, 60; *see also particular groups and species*
Antilocapra americana, 94, 138
Anza-Borrego Desert State Park, USA, 513–14
Aphanopleura leptoclada, 172
Apicotermes spp., 117
Apodemus flavicollis, *243*
arable lands, effects of rainfall fluctuations on, 366–9
Aral Sea, 22
Argania spinosa, 364
Argentina: reserves in, 500–1; shrub patterns and available moisture, 43–4, 72
Argyrolobium uniflorum, 363
ARIDCROP model, 404, 439–43
Aristida acutiflora, 371
Aristida ciliata, 371
Aristida karelinii, *170*, 172; gas exchange, *176*
Aristida obtusa, 363
Aristida pennata, 172
Aristida plumosa, 371
Aristida pungens, 374
Aristida spp., *96*, 335
Arizona, USA; secondary production in, 202–14; *see also* Sonoran Desert
Aroga websteri, 68
arroyos (water courses), 15, 274–5
Artemisia campestris, 361, 363, *368*, 374
Artemisia frigida, *182*
Artemisia herba-alba, 15, *23*, *24*, 562, 572; dynamics of, 360, 361, 362; erosion effects on, 372; and plant–plant interactions, 34–5; and rainfall fluctuations, 367–9; and steppization, 376; water and energy flow, 278, *279*
Artemisia kemrudica, 171, *172*
Artemisia monosperma, 14, 21, 328, 331
Artemisia pauciflora, 19
Artemisia spp., *37*, 185, 378
Artemisia terrae-albae, 173, *177;* mineral element cycling in, 183, *184*, 185, *186–7*, 188
Artemisia tomentalla, 193
Artemisia tridentata, 15, 16, *36*; herbicide control of, 470; and nutrient cycling, 308, 311, 336, 347; root systems of, 333
Artemisia turanica, 177
Artemisia vachanica, 22
Artemisietum, 19
Arthrocnemon glaucum, 16, 363
Arthrocnemetum-Limoniastretum monopetalae, 16
arthropods: population dynamics of, 132, 289–90, 291–2, 295; predator–prey ratios, *69*; species diversity, 61; species separation of, 54–5; *see also* microarthropods *and particular groups and species*
Arundo plinii, 364
Asclepiadaceae, 337
Asilidae, 69
Aspergillus spp., 330, 331, 332, 341
Asphodelus microcarpus, 20, 370
Aster xylorhiza, 14
Astragalus ammodendron, 193
Astragalus chivensis, 172
Astragalus filicaulis, *177*
Astragalus gombiformis, 374
Astragalus gombo, 374
Astragalus longipetiolatum, 171, *172*
Astragalus micropterus, *36*
Astragalus paucijugas, *170*, *176*
atmospheric processes, in deserts, 560–1; *see also* evaporation, precipitation
Atraphaxis serarschanica, 177
Astragalus spp., 316
Atractylis candida, 363
Atractylis humilis, 376
Atractylis serratuloides, 360, 361, 376
Atriplex canescens, 105
Atriplex confertifolia, 14, *36*, 45, 105, 106, 471
Atriplex corrugata, 14
Atriplex dimorphostegia, 180
Atriplex glauca, 363
Atriplex halimus, *24*, 347, 363, 364, 374, 566, 576
Atriplex lentiformis, *36*
Atriplex leucoclada, *36*
Atriplex malvana, 363
Atriplex mollis, 363
Atriplex nummularia, 105, 109, 333
Atriplex nuttallii, 14, *115*
Atriplex paludosa, 105
Atriplex spp., 108, 310, 315, 341, 530, 569; salt accumulation by, 105; water uptake, 109, 110
Atriplex tridentata, 106, *349*
Atriplex vesicaria, 39, 86–7, 105, 328, *334*, 336, 338, 339, 345, 346–7; nutrient cycling, *302*, 303, 308, 311, *342*; root systems of, 333
Atta vollenweideri, 116
Auriparus flaviceps, *212*
Austracis spp., 92
Australia, 39, 558; animal populations of, 131; desert reserves of, 503–4; ecosystem modelling in, 404–7; kangaroo species segregation, 54; and plant–animal interactions *see* plant–animal interactions; predator–prey systems, 67, 68; termite activity in, 113–21; vegetation, spatial variations in, 15; water dynamics in, 277; water uptake by plants in, 109–10

Index

Austroicetes cruciata, 52
autogenic succession, 373–5; anthill effect, 375; dune stabilization, 374–5; effect of plant cover, 373–4; 'nebka effect', 374
Avedat Desert, Palestine, 15
Avena sativa, 435, *436*, *439*
Avena sterilis, 18, *436*
Avicennia marina, 16
Avra Valley, Arizona; shrub patterns and available moisture, 40, *41*
Azotobacter chroococum, 343
Azotobacter spp., 314, 331, *342*, 343–4

Bacillus spp., 345
bacteria, 576; nitrogen fixation, 343–4; *see also* micro-organisms
BACROS simulation, of transpiration coefficients, 437–9
bajadas *see* alluvial fans
basins *see* swales, watersheds
Bassia obliquicuspis, 110
Bassia spp., *93*, 105
beetles, tenebrionid: life-history patterns of, 144; population dynamics, 133, 137; populations, regulation of, 159; species segregation, 75
Bellevalia desertorum, 575
Berberis trifoliata, 14
Bettongia spp., 94
biomass data, 286–8, 290–2; *see also* production, productivity
biotic components: and effect on abiotic components, **105–21**, 575–6; animal effects on soil, 112–21; dynamic behaviour of, and water and energy flow, **271–95**; plant effects on soil, 105–21
birds: agonistic behaviour of, 74; breeding season effects, 154, 155, 156; clutch size, 147–8, 149–50, 151, 153, 155–6; energy flow, 289, 295; habitat preference of, 54, 571–2; and plant interactions, 89; population variations, 127–8, *130*, 131, 136–7; predatory influence, 70, 71–2; production by, 211–14, 238–40, 244–5; species diversity of, 61; *see also particular groups and species*
Bison bison, 94, 467
Blepharis persica, 563, 568
blowouts, 11
Boerharia spp., *93*
Bombyliidae, 69
Bootettix punctatus, production of, 215
boron, cycling of, 316
Bos ibericus, 378
Bos primigenius, 378
Botrytis pyramidalis, 330
Brachanthemum gobicum, *174*, *175*, *182*
Brassica tournefortii, 363
breeding of animals, 147–56; *see also* reproduction
Brickellia incana, *36*
Bromus tectorum, 35, 114, 346
Boutelona eriopoda, 22
burning *see* fire
burrowing, effects on soils, 115–17

Cadaba rotundifolia, 40
calcium, in soil: and cycling of, in plants, *184*; and effect of termites on, 119–21
California, 21, *510*, *511*, 513–4; root systems of perennials in, 34; *see also* Mohave Desert
Calipepla squamata, *130*, 147
Calligonum aphyllum, 177
Calligonum arich, 364, 375
Calligonum azel, 364, 375
Calligonum caput-medusae, *170*, *176*
Calligonum comosum, 59, 337, 364
Calligonum polygonoides, 485, 486
Calligonum setosum, 172
Calligonum spp., 14, 171, *172*, 173, 326
Calotropis procera, 337
Camelus bactrianus, 94
Camelus dromedarius, 94, 378
Camelus thomasi, 378
Campylorhynchus brunneicapillus, 74, *130*, 131, *212*, 213, 214
Canis anthus, 378
Canis familiaris dingo, 67
Canis latrans (coyote), 52, 54; and predator-prey system, 64, 65–7, 71; and population dynamics, *130*, 131; and species separation, 56–7, 63, *65*
Capparis decidua, 39
Capparis spinosa, 14
Capra ibex nubiana, 570
carbon, organic, soil content of: and effect of plants on, 107–8, 303–4; and effect of termites on, 119–21
Cardiosis spp., 59
Carduus gaetulus, 363
Carex pachystylis, 362, 566
Carex physodes, *170*, *172*, 185, 188–9, 193
Carlina involucrata, 370
Carnegiea gigantea, 213
carnivores, species separation of, 56–7
Carpodacus mexicanus, *212*
Carthamus lanatus, 370
Cassia armata, *36*, 208, 338, 339
Cassia spp., 378
catchments *see* watersheds
cattle, and compaction of soils, 112–13, *see also* domestic stock
Cedrus spp., 378
Celtis pallida, 14
Cenchrus ciliaris, 371, 485
Cenchrus setigerus, 485, 491
Cenchrus spp., *96*, 491
Centaurea dimorpha, 363
Centaurea nicaeensis, 376
Centrioptera muricata, 136, 140
Centurus uropygialis, *212*, 213
Cerastes cerastes, 292
Cercidium floridum, 39
Cercidium spp., 213
Cereus giganteus, 44
Ceratonia siliqua, 378
Chaenactis spp., 45
chamaephytic steppes, plant communities of, 362–3, *365*, 371
Chamerops humilis, 378

Index

check-dam, and control of erosion and sedimentation, 539–40
Chelaner spp., 86, 87, 288
chemical effects, on soils: by plants, 105–12; by termites, 112–21
Chenopodiaceae, 109–10, 173, 179, 378
Chenopodium murale, 180
Chihuahan Desert, USA, 54, 56, 60, 129, 131–2, 447–8; and animal biomass, 286–8, 290–2; litter production, 292–3; primary production, 283–5; water and energy flow in, 271, 272–7
Chile, desert reserves of, 501
Chilopsis linearis, 37
chloride concentrations, in soil under plants, 105–6
chlorine, cycling of, 184, 315
Chloris spp., *93*
Choriotis spp., 147
Choristoneura fumiferana, 133
Chortoicetes terminifera, 52; life cycle of, 92; and plant interactions, 91–3, 99; population dynamics, *130*, 133, *134*, 135–6
Chroococaceae, 330
Chrozophora gracilis, 172
Chrysanthemum coronarium, 361, 376
Chrysothamnus spp., 471
Cistus libanotis, 360, 361
Cistus villosus, 375
Citellus beecheyi, 116
Citrullus colocynthis, *36*
Cleistogenes songorica, *182*
Cleome arabica, 363, 370
Clerodendrum phlomoides, 485
Clethrionomys glareolus, *243*
Clethrionomys rutilis, *243*
Climacoptera lanata, 172
climate *see* atmospheric processes, Indian (Thar) Desert
climatic fluctuations: over long periods, 378–9; rainfall effects, 366–71; and variations in vegetation, 21–5, 365–71; *see also* precipitation
Clostridium spp., 314, 331, *342*, 343
clutch size, of birds, 147–8, 149–50, 151, 155–6, 238–9
concrete, for water harvesting, 522
Cnemidophorus spp., 54, 57
Cnemidophorus tigris, 57, *130*, 149; production by, 201, 228–34, 238
Colaptes chrysoides, *212*
Colchicum tunicatum, 574–5
Coldenia canescens, 15
Coleogyne ramosissima, *36*
Colinus virginianus, 147
Collema coccophorus, 109, 344
Colorado Desert, USA, 58, 72
Columba livia, 571
Columbiformes, 147
commensalism, of plant communities, 44–6
Commiphora spp., 14
compaction of soil, and effects of animals on, 112–13, 114–15
competition: root, 34–8; interspecific *see* interspecific competition; *see also* plant–plant interactions
component processes, of deserts, 559–60
Compositae, 378; *see also individual species*
Condalia oborata, 14
Condalia obtusifolia, 14
Corax corvus ruficollis, 570
Coronilla scorpiodes, 376
coyote *see Canis latrans*
Cressa cretica, 16
Cricetinae, 73, 129
crops *see* agriculture
Crotalaria burhia, 336
Crotalus spp., 292
Crotaphytus wislizenii, 68, 73, *130*, 244
Crucianella filifolia, 172
Crucianella maritima, 17
Ctenodactylus gundi, 378
Cubitermes spp., 117
Cucurbita palmata, *36*
cultigene vegetation (weeds), 361, 363, *365*, 368, 380
cultivation *see* agriculture
Cupressus dupresziana, *327*, 330, 338
Cupressus spp., 378
Curlew Valley, Utah USA: and spatial variations of vegetation, 15, 65, 67
Cutandia dichotoma, 363
Cutandia divaricata, 363
Cyamopsis psoraliodes, 491
Cyanophyceae, 576

Dactylis hispanica, 371
Dactyloctenium radulans, 92
Danthonia caespitosa, *334*, 336
data analysis, of shrub patterns and moisture availability, 39–44
Death Valley, California, USA, 21, *510*, *511*
decomposition, microbial: of litter, 292–3; model of, 401; and nutrient cycling, 304–5, 314; *see also* litter, dynamics of
Delphinium camptocarpum, 172
demography *see* population
Dendroctomus ponderosae, 133
Dendroctomus refipennis, 133
denitrification, 345
denudation of soil surface, by animals and insects, 114–15
desert reserves *see* reserves
desertization, 377
Diarthon vesiculosum, 172
Dichanthium annulatum, 336
Dichondra spp., *93*
Digitaria nodosa, 371
Diplotaxis harra, 568
Diplotaxis muralis, 374
Dipodidae, 73, 129, 146
Dipodomys merriami, 54, 58, *130*, 154; production by, 202, 203–7, *243*, 249–50
Dipodomys microps, *130*
Dipodomys ordii, 54
Dipodomys spp., 242, 265
dispersal of plants, 426–7
Distichlis stricta, 16
diversity, and interspecific animal competition, 59–62, 559–60; and precipitation, 59–62

domestic stock, 568–9; energy flow in, 290, 295; in Indian Desert, 483–4; management of arid-land resources for, **455–72**; and manipulation of grazing, 468–9; and plant interactions, 94–7; and nutrient cycling, 308, 336
Drepanotermes perniger, 90, 113
Drepanotermes rubriceps, 113, *118*, *120*
Drepanotermes spp., *89*, 114
dunes, 11, 59; afforestation of, 485–6; production potential of, 485–6; stabilization of, and spatial variations of vegetation, 16–17, *18*, 374–5
dynamics, long-term, of aridland vegetation, **357–82**, 448–9; and climatic fluctuations changes, 365–71; ecological changes over long periods, 378–9; methodology, 357–9; present-day vegetation, 359–65; time-scales, *358*

earthworms, and effects on soil by burrowing, 116
Echinops spinosissimus, 17
Echiochilon fruticosum, 363, 374
Eclipta spp., *93*
ecosystem dynamics, modelling of, **385–407**; animal processes, 398–9; ARIDCROP model, 404, 439–43; in Australia, 404–7; GAME, 406–7; grazing, 391–4; in Israel, 402–4; life cycle of annual plants, 389–91; NEGEV model, 402–4; plant processes, 396–7, **433–44**; *see also* simulations; soil processes, 399–402; in United States, 388–402; WATBAL, 404
Egypt: spatial variations in vegetation, 13–14, 15, 16, 17, 20; root systems of perennials in, 34
Eleagnus spp., *37*
Elephas atlanticus, 378
Eliomys melanurus, 68
Elizaldia violacea, 63
Elyonurus tripsacoides, 14
Emex spinosus, 374
Encelia farinosa, 38, 326
energy flow: and litter dynamics, 292–3; in predators, 290–2; *see also* water and energy flow
Ephedra lomatolepis, *186–7*, 188
Ephedra nevadensis, *36*
Ephedra slata, 364
Ephedra spp., 179
Ephedra trifurca, *37*
Ephedra strobilacea, 171, *172*; gas exchange, *176*; photosynthesis, 179
ephemerals, effects of precipitation and temperature on germination, 21–2
epiphytism, 45
Equus mauritanicus, 378
Eragrostis cilianensis, 92
Eragrostis eriopoda, 406
Eragrostis papposa, 363
Eremias guttulata, 571
Eremophila gilesii, 311
Erica scoparia, 378
Eriogonum fasciculatum, *36*
Erodium circutarium, 515
Erodium glaucophyllum, 14
erosion, 109–11, 371–3, 539–40
Erpodium hirtum, 278, 279, *280*, *284*
Erucaria vesicaria, 367, 375, 376
Eryngium spp., 370
Erythropygia spp., 147
Escalante Desert, Utah, USA, 106
Ethiopian Desert, 40
Eucalyptus corynocalyx, 110
Eucalyptus spp., 375
Eudrilus eugeniae, 116
Euphorbia cheirolepis, 172
Euphorbia densa, 172
Euphorbia guyoniana, 374
Euphorbia terracina, 363
euro *see Osphranter robustus*
Eurotia eversmanniana, *177*
Eurotia lanata, *36*, 45
Eurotia spp., 15
evaporation, control of, 544

Fagonia arabica, 326, 333, 340
Fagonia kaherica, 13
Fagonia mollis, 14
Fagonia spp., 348
Fagus silvatica, 378
Fallugia paradoxa, *37*, 139
Felis lynx, 378
Ferocactus wislizenii, 34
Ferula assa-foetida, *177*
fibreglass matting, for water harvesting, 523
Filago spp., 25
fire, use of, for rangeland improvement, 469, 542
Flourensia cernua, *37*, 276
forage production: ecology of, 456; increased by herbicides, 470–1; *see also* grazing
forests *see* afforestation; Aleppo pine
Franseria dumosa, *36*, 38, 40, 45, 326
Fraxinus spp., 378
Fumana ericoides, 375
Fumana thymifolia, 360, 375
fungi, 576; and mineral cycling, 330–1, 332; mycorrhizal, 304, 331, 581; *see also* micro-organisms
Fusarium spp., 331

GAME model, 406–7
garrigue (degraded forest), 360, *365*
gas exchange of plants, *176*, *177*; *see also* photosynthesis and respiration
Gazella cuvieri, 378
Gazella dorcas, 378, 570
Gazella leptoceros, 129
Gazella spp., 94
Genista microcephala, 361
Genista saharae, 364
geohydrology of Indian Desert, 486
Geococcyx californianus, 148, 155–6; production by, *212*, 213
geological formations, and spatial variations in vegetation, 17–18
Gerbillinae, 73, 129, 146
Gerbillus dasyurus, 68
Gerbillus pyramidium, 129–31; breeding season, and rainfall effects, 155

Index

Globularia alypum, 360, 375
Gobi Desert: photosynthesis, 181–3; water relations of plants, 173–5
Gopherus agassizi, 137, 150
Gorgon taurinus prognu, 378
grasshoppers, 60; and denudation of soil surface, 114; and plant interactions, 90–3, 99; production by, 215
Grayia spinosa, *36*, 346
grazing, and grazing systems, 580, 584; continuous, 466; deferred, 466; evolution of, 463–4; and growth forms of plants, 20–1; manipulation of, 468–72; model of, 391–4; overgrazing, 370, 519, 550; and rainfall fluctuations, 370–1; regulated, 465–6; rest-rotation, 467–8; rotation, 467; seasonal, 466
Great Basin Desert, USA, 401; animal populations, *65*, 68, *130*; 'nurse plants' of, 45; population dynamics, 138; predator–prey ratios, 69; root competition between plants, 35; variations in primary production, *128*
Great Salt Lake, Utah, USA; salinity and spatial variations of vegetation, 16
groundwater, extraction of, 544–5
Gymnocarpos decander, 376

habitat: preferences in birds, 54, 571–2; spatial effects, 428
Halimodendron spp., *37*
Halocnemon strobilaceum, 16, 363
Halocnemum spp., 348
halophilous crassulescent steppes, plant communities of, 363–4
halophytes, 315
Haloxylon ammodendron: gas exchange, *176*; mathematical model of population, 188–92; mineral cycling, 185, *186–7*, 188; photosynthesis, 180, 183, *182*; water relations of, *170*, 173, *175*
Haloxylon articulatum, 14, *36*
Haloxylon persicum, *170*, 171, *172*, 173; gas exchange, *176*; mineral cycling in, 185; photosynthesis, 179, 180
Haloxylon saliconricum, 14, 21
Haloxylon scoparium, *335*
Haloxylon spp., *37*
Hammada schmittiana, 363, 370
Hammada scoparia, 15, 278, 361, 370, 372, 376, 564, 578; production by, 283, *284*, *285*
Hammada spp., *37*
Haplophyllum pedicellatum, *172*
Helianthemum kahericum, 15, 360, 361, 376
Helianthemum lippii, 363, 374
Heliotropium arguzioides, *170*, *176*
Hemerodromus africanus, 156
Hemilepistus reamuri, 570, 571, 572
herbicides, 541–2; and increase of forage production, 470–1
herbivores, and plant interactions in desert ecosystems *see* plant–animal interactions
Herniaria fontanesii, 360
Heteromyidae, 72, 73, 120, 146, 153, 286–7
Heteromys desmarestianus, 72, 144
Hilaria jamesii, 14
Hilaria mutica, 276, 292
Himantopus himantopus, 137
Hipparhenia hirta, 371
Hippocrepis bicontorta, 363
Hippocrepis scabra, 361
Hodotermitinae, 114
Hordeum murinum, *436*
Hordeum sativum, 436, *439*
human interference *see* man; management
Hyaena striata, 378
Hymenoclea salsola, *36*
Hymenoptera, 289
Hypecoum pendulum, *172*
Hyperiodrilus africanus, 116
Hyparrhenia hirta, 18
Hystrix cristata, 378
Hystrix indica, 574

Idria spp., 45
Ifloga spicata, 363
Indian (Thar) Desert, management of, for crops, 278, **479–92**; arid agriculture, 486–93; climate, 479–80; geohydrology of, 486; human population of, 482–3; land 480–2; livestock of, 483–4; production potential of, 485–6; surveys of land resources, 484–5
insects: and denudation of soil surface, 114–15; diversity of species, 61; populations dynamics, 133, 139; predator–prey ratios, 69; *see also particular groups and species*
interactions, in desert ecosystems, 9–25, 572–7; animal–animal *see* animal–animal interactions; nature of, 9–13; plant–animal *see* plant–animal interactions; plant–plant, *see* plant–plant interactions; spatial variations of vegetation, 13–21
interspecific competition between animals, 53–62; and diversity, 59–62, 559–60; empirical evidence for, 54–8; hypotheses of, 53
invertebrates: burrowing effects of, on soil, 116–17; population dynamics, 132, 137; predator–prey ratios, 69; *see also particular groups and species*
Iphiona mucronata, 14
Iridomyrmex spp., 87
Iris songarica, 172
iron, cycling of, in plants, *184*
irrigation, 584; and crop production, 491–2
Israel, ecosystem modelling in, 402–4
Israeli desert system *see* Negev Desert

jackrabbit, blacktailed *see Lepus californicus*
Jaculus jaculus, 129–30, 378; reproduction of, and rainfall effects, 155
Jaculus orientalis, 73
Jaculus spp., 287
jerboas, territorial pattern of, 73
Joshua Tree National Monument, California, USA, 21; mammalian production in, 208–10; *see also* Mohave Desert
Juniperus phoenicea, 360, 361, 375
Juniperus spp., 378

K-selection of animal species, 143–54, 158, 266–7, 579
Kalahari Desert, 59
kangaroos, species segregation of, 54
Karakum Desert: gas exchange in plants, *176*; photosynthesis, 178–83; water relations of plants in, 69–75
Kochia prostrata, 177, 193
Koeleria salzmanni, 363
Koelpinia linearis 172
Krameria grayi, *36*
Krameria parvifolia, 281, *282*; production by, 283, *284*
Krameria parviflora, 15, *36*
Kyzylkum Desert, USSR, 73; gas exchange of plants, *177*; photosynthesis, 178–81

Lagomorpha, 146
Lagorchestes spp., 94
Lamium amplexicalue, *177*
land use, and ecological changes, 379–80; *see also* management
Lappula caspia, 172
Larrea cuneifolia: distribution of, and available moisture, 43–4
Larrea divaricata, 15; pattern analyses and available moisture, 39, 40, *41–2*; and plant–plant interactions, 34, 46; root:shoot weight ratios, *36*, *37*
Larrea spp., 21, 45; leaf extracts of, and allelopathy, 38–9
Larrea tridentata, 208, 215, 272–3, 274, 326; litter production, 292; and nitrogen in soils, 338–9; production of, 283, *284*, *285*, 290, 456; root systems of, 333
Lasiurus hirsutus, 14, 20
Lasiurus sindicus, 336, 485
Lasiurus spp., 491
Launaea arborescens, 336
Launaea resedifolia, 363
Launaea spp., 25
Laurus nobilis, 378
leaf extracts, and allelopathy, 38–9
Lecidea crystalifers, 109
Lepidium spp., *93*
Lepidoptera, 133, 289
Lepus americanus, 131
Lepus californicus (black-tailed jackrabbit), 52; life-history patterns of, 146–7, 155; population trends, 64, 65–8, 71, *130*, 131, 143
lichens, nitrogen fixation and, 344
life-cycles, modelling of, 389–91
life-history patterns, of animals, 92, 143–56, 266–7; *r* and *K* selection, 143–54
life-spans of perennial plants, 86
Limoniastrum quyonianum, 374
Limonium pruinosum, 14, 16
Liomys salvini, 144
Liomys saliana, 72
Liriodendron spp., 126
litter, dynamics of, 292–3, 311, 568
livestock *see* domestic stock
lizards: agonistic behaviour, 72, 73–4; demographic characteristics, 148–9, 150, 153; energy flow in, 291–2; niche separations of, 54, 55, 57, 60, 62; population dynamics, *130*, 137, 138, 139, 156; predator–prey systems and, 68; production by, 201, 210, 228–38, 244, 255; *see also particular species*
locusts: interaction with plants, 90–3, 99
Locusta migratoria, 52
Locustana pardalina, 52, 70
longevity of plants, 86
Lophortyx californicus, 147
Lophortyx gamebelii, 70, *130*, 131, 135, *137*, 140–1, 147, 152; production of, *212*, 213
Lotus pusillus, 363
Lycieto-Limoniastretum halimi, 16
Lycium andersonii, *36*
Lycium arabicum, 20
Lycium barbarum, 485
Lycium pallidum, *36*
Lycium spp., 25
Lygaeidae, 57
Lygeum spartum, *24*, 361
Lynx rufus, 54

Macropus fuliginosus, 52
Macropus spp., 94
Macrotermes subhyalinus, 113
Macrotermitinae, 114, 119
magnesium content, of soils: cycling of, in plants, *184*; and effect of termites on, 119–21
Maireana aphylla, 110, 328
Maireana astrotricha, 105
Maireana pyramidata, 39
Maireana restita, 336
Maireana sedifolia, 109–10
Maireana spp., 109
Malacothrix spp., 45
Malcolmia africana, *177*
Malcolmia grandiflora, 172
Malva parviflora, 370, 374
Malye Barsuki: gas exchange in plants, *177*; mineral cycling, 185, *186–7*; photosynthesis, 178–81
mammals: and plant interactions, 93–7, 99, *334*; population dynamics of, *130*; production by, 202–10, 217–28, 241, 243, 245–50, 252–4; *see also* domestic stock; *particular groups and species*
man: and climatic fluctuations, 371; interference by, 194, 196; *see also* management
management of desert ecosystems, for livestock forage, **455–72**, 584; chemical control, 470–1; continuous grazing, 466; deferred grazing, 466; deferred rotation, 466–7; ecology of forage production, 456; evolution of grazing systems, 463–4; grazing manipulation, 468–72; improvement of, 461–2; by fire, 469, 542; mechanical control, 470; natural plant production, 456–61; nomadism, 464; rangeland improvement, 468–72; regulated grazing systems, 465–6; rest-rotation grazing, 467–8; rotation grazing, 467; seeding, 471; soil surface treatment, 469–70; strategies of, 462–8; transhumance, 464–5; *see also* grazing; Indian Desert

Index

management, of water resources, **519–45**; conservation on upland watersheds, 529–31; erosion and sedimentation control, 537–40; groundwater, 544–5; micro-catchments and strip farming, 527, *528*; vegetation management, 540–3; storage of water, 524–6; transpiration and evaporation control, 543–4; water harvesting, 488–90, 520–9, *527*; water spreading systems, 531–7
manganese, cycling of, in plants, *184*
Marrubium alysson, 370
Mastotermes darwiniensis, *89*
Mastotermitidae, *89*
Matthiola livida, 578
Mausolea eriocarpa, 171, *172*
Medicago hispida, *436*
Medicago polymorpha, *334*, 336
Medicago spp., *93*, 338
Medicago truncatula, 375
Megaleia rufa, 52, 89, 94–7, 99; population fluctuations, 129, 136
Megleia spp., 94
Membracidae, 133, 290
Meridae, 290
Meriones crassis, 576
Meriones libycus, 131
Meriones shawi, 378
Meriones unguiculatus, 131
Meranoplus spp., 86, 87
Mesembrianthemum cristallinum, 374
Mesembryanthemum forskalie, 14
Messor spp., 86, 288
meteorology *see* atmospheric processes
Mexico, desert reserves in, 501
microarthropods, 57, 293
microbial decomposition *see* decomposition
micro-catchments, 527, *528*
Microcerotermes distinctus, *89*
Microcerotermes serratus, *89*
Microcoleus spp., 330
micro-organisms, in soil, 328–33, 341–3, 559, 561; nitrogen fixation, 343–4
Micropallus whitneyi, *212*
Microtus ochrogaster, *243*
Migda site, Negev Desert, water and energy flow in, 279–81
migration, 427
Mimus polyglottos, *212*, 245
mineral cycling, 183–8, 193–4, 316, 580–1; animals and, *334*, 335–8; micro-organisms in, 328–33; nitrogen dynamics, 338–45; organic matter and, 326–8; phosphorus dynamics, 345–6; potassium dynamics, 346–7; and primary productivity, 183–8, 194; root systems and, 333–5; short-term dynamics of, **325–50**; sodium chloride, 347–50; *see also* nutrients, cycling of
models and modelling, **385–407**, 434–43, 448–9, 581–2; of animal–animal interactions, 52–3, 63; of arid ecosystem dynamics *see* ecosystem dynamics; of *Haloxylon ammodendron*, 188–92; of secondary production, 251–5; spatial effects *see* spatial effects; of water distribution, 422–4; *see also* simulation
Mohave Desert, USA, 25, 58, 61, 68, 72; animal populations, *130*, 131, 149, 155; variations in primary production, *128*, 129, 283–5; productivity of, 19, 458, *459*; shrub patterns and available moisture, 40–4; water and energy flow dynamics, 281–2; *see also* Joshua Tree National Monument; Rock Valley
moisture availability: for crop production, 488–90; moisture retention, 194; and pattern analysis of shrubs, 39–44; root competition for, 34–8, 46; and spatial distribution of vegetation, 13–21; *see also* water and energy flow
Molothrus ater, *212*
Moltakea callosa, 331, *342*, 344
Monomorium spp., 86
Mucor spp., 330
Mucorales, 332
mulga *see Acacia aneura*
Mus musculus, 55, 131
Myiarchus cinerascens, *212*
Myiarchus tyrannulus, *212*
Mycobacterium spp., 341
mycorrhizae, 304, 331, 581

Nama hispidum, 15
Namib Desert, 55, 59, 70
Nasutitermes spp., *89*
Nasutitermes triodiae, *120*, *121*
Nasutitermitinae, 114
'nebka effect' in succession, 374
Negev Desert, Israel, 348; modelling of ecosystem, 402–4, 434–43; plant–plant interactions in, 35; predation, 68; productivity, 19, 283–5; spatial variations of vegetation, 14, 15, 18; water and energy dynamics, 278–81
nematodes, soil, production by, 216–17
Neotoma albigula, 139
Neotoma lepida, 74
Neotoma spp., 129, 161
Nerium deander, 364
Nevada, 22; secondary production in, 210, 215, 234–40; *see also* Mohave Desert; Rock Valley
New Mexico, 22; *see also* Chihuahuan Desert
New South Wales, Australia, 39
niche separation, 54, 55, 57, 60, 62
Nitraria retusa, 16, 20, 364, 374
Nitraria sibirica, *175*, *182*
Nitraria spp., 378
nitrification, 340–3
Nitrobacter spp., 341
nitrogen in soils: ammonification, 340; cycle of, *313*; cycling of, in plants, *184*, 185, 311; deficiency of, 313–14, 317; denitrification, 345; dynamics of, 338–45; fixation, 304, 314, 342–4; nitrification, 340–3; effect of plants on, 107–9, *184*, 303–4, 308, 460; proteolysis, 340; effects of termites on, 119–21, 304
Nitrosomonas spp., 341
Noaea mucronata, *361*
Nodulana spp., 330
Noea spinosissima, *36*
Nolletia chrysocomoides, 363, 374
nomadism, 427, 464
Nostoc spp., 344

Notomys spp., 131, 287
Novomessor spp., 288
'nurse plants', 44–5
nutrients cycling of, 184, **301–17**, 336, 342, 346–7; accumulation of, 307; deficient elements, 313–15; and domestic stock, 308, 336; excessive elements, 315–16; influence of precipitation, 301–2; intra-seasonal variation, 309; structural variability, 303–4; temporal variability, 304–7; turnover rates, 309–12; *see also* mineral cycling, nitrogen in soils

Odocoileus spp., 94
Oenanthe lugens, 147
Oenanthe spp., 148
Olea europea, 378
Onopordon arenarium, 363, 374
Onychomys leucogaster, 54
Onychomys torridus, 54; production by, *203*, 205–7, 241, 243, 250
Operophtera brumata, 63, 158
Opuntia bigelovii, 39–40
Opuntia lindheimeri, 14
Opuntia occidentalis, 74
Opuntia app., 45, 213, 288
organic matter, in arid soils, 326–8
Orthoptera, *289*
Oryctolagus cuniculus, 66, 141
Oryx leucoryx, 94
Oscillatoria spp., 330
Osphranter robustus, 94–7, 99
Osphranter spp., 95
overgrazing, 370, 519, 550
Ovis camadensis, 94
Ovis tragelaphus, 378
Oxalis spp., *93*

Panaxia dominula, 133
Panicum antidotale, 485
Panicum obtusum, 277, 292
Panicum turgidum, 14, 20, 21, 331, 332, 485
Panthera leo, 378
Panthera pardus, 378
Parmelia conspersa, 109
Parthenium incanum, *37*, 38
PASTOR model, of grazing, 391–4
patterns of plant distribution, 39–44; *see also* spatial effects; vegetation, spatial variations
Peganum harmala, 172, 341, 370, 374
Penicillium spp., 330, 331, 332
Pennisetum dichotomum, 348
Pennisetum typhoideum, 487
perennial plants: and animal–plant interactions, 87, 88; and chemical effects on soils, 105–9; life-spans of, 86; pattern analysis and available moisture, 39–44; root systems of, and plant–plant interactions, 34–8
Periploca loevigata, 364
perlite ore, for evaporation reduction, 544
Perognathus formosus, 72, *130*, 159; production by, 202, 217–28, 242–3, 245–7, 248–9, *250*, 252–4
Perognathus longimembris, *130*
Perognathus parvus, *130* 135, 140
Perognathus spp., 242, 245
Peromyscus crinitis, 73
Peromyscus eremicus, 73; production by, 202, *203*, 206–7, 249, *250*
Peromyscus spp., 286–7
Peruvian Desert, 242
Phainopepla nitens, 148
Phalaris minor, *436*
Phaseolus acontifolius, 487
Phaseolus radiatus, 487
Pheidole spp., 86, 87, 288
Phillyrea media, 375, 378
Philolithus densicollis, *130*, 133, 136, 137, 140
Phoenix dactylifera, 364
phosphorus, cycling of, in plants, *184*, 311; deficiency of, 314–15; dynamics of, 345–6; relation to plants, 107–9, 303–4, 308, *460*
photoperiod, and effects on breeding season, 155, 156
photosynthesis and respiration, *176–7*, 178–83, 192–3, 564–5
Phragmites communis, 364
Phragmitetum communis, 16
Phrynosoma cornutum, 73
Phrynosoma platyrhinos, 234–8
phylogenetic momentum, 151
physical effects, on soils *see* soils
Phyllotis griseoflavus, 72
Pinus halepensis, 360, 378; *see also* Aleppo Pine
Pinus ponderosa, 541
piosphere effect, 428–30
Pipilo fuscus, 156, *212*
Pistacia atlantica, 364
Pistacia lentiscus, 378
Pittosporum spp., 378
Plantago albicans, 361, 363
plant–animal interactions, **85–100**, 574–5; ants, 86–7, 98, 99–100; birds, 89; grasshoppers and locusts, 90–3, 99; large mammals, 93–7, 99, *334*; rodents, 88, 98, 195; termites, 89–90, 98, 99–100; *see also* grazing; piosphere effect
plant–plant interactions, **33–46**; allelopathy, 38–9; pattern analyses and available moisture, 39–44; positive associations, 44–5; root systems of perennials, 34–8, 333, 566
plants: allelopathy, 38–9, 46; commensalism, 44–5, 46; community composition, variations in, 13–19; growth forms, variations in, 20–1; model of life-history of annuals, 389–91; plant–animal interactions *see* plant–animal interactions; plant–plant interactions *see* plant–plant interactions; plant processes model, 396–7, **433–44**; processes, 562–8; production of, 456–61; and soil effects, 105–12; survival strategies of, *573;* water relations of, 169–75; *see also* vegetation; *particular groups and species*
playas (sinks), 16, 277, *288*
Poa bulbosa, 278, *279*, 362
Poa sandbergii, 336
Poa sinaica, 15
Poa spp., 22, *37*
pocket-mice, 217–28, 286–7; *see also* particular species
Pogonomyrmex occidentalis, 114–15

Index

Pogonomyrmex rugosus, 288
Pogonomyrex spp., 86, 288
Polioptila minuta, 212
Polygonum equisetiforme, 363
polystyrene rafts, for water storage, 525–6
population dynamics, **125–62**, 289–92, 295, 563, 577–80; animal life-history patterns, 143–56; demographic mechanisms, 136–8, 143–54; environmental mechanisms, 138–9; mathematical model, 188–92; magnitudes of variation, 126–35; precipitation effects, 135, 139, 371; regulation and limitation, 156–62; short-term fluctuations, 125–43; temporal patterns, 139–43; *see also Lepus californicus*
Populus alba, 378
Populus spp., *37*
potassium, cycling of, in plants, *184*; dynamics of, 346–7; and effect of termites on, 119–21
Poterium spinosum, 18, *23*
precipitation, 42, 264, 586; and variations in animal populations, 135, 139, 371; annual variations, 21–2; and breeding season timing and intensity, 154–6; and diversity of species, 59–62; and germination, 21–2; and grazing effects, 370–1; long-term variation in, 366–71; magnitude of variation in, 126–7; nutrients affected by, 301–2; primary production variations, 125, *126*, *128*, 366–7; and water dynamics, 272–5, 278, *280*, *281*, *287*, 290–1, 294–5
predation, 62–72, 76; empirical evidence of, 65–72; energy flow in predators, 290–2; predator–prey systems, 65–71; predatory influence, 70–2; theoretical considerations, 62–4
primary production, 240–1, 283, *284–5*, 290, 456–61, 577–80; magnitude of variation in, 127, *128*, 129, 283–5; and variation in precipitation, *126*, 127, *128*, 366–7; *see also* productivity, primary
production, by desert animals, **199–256**, 578; birds, 211–14, 238–40, 244–5; mammals, 202–10, 218–28, 241, 243, 245–50, 252–4; and contrast with other environments, 241–3; efficiency of, 247–50; grasshoppers, 215; intraspecific variations, 243–7; lizards, 201–2, 210, 228–38, 244, *250*, 251, 253–5; models of, 251–5; and procedures of investigation of, 200–2; termites, 210, 238, 241
production, potential of Indian Desert, 485–6
production, primary *see* primary production
productivity, primary, 19, **169–97**, 458–9; comparisons of producers, 283–5; cycles of mineral elements, 183–8, 193, 194; mathematical models of, 188–92; photosynthesis and respiration, *176–7*, 178–83, 192–3, 564–5; variations of, 18–19, 21–5; water input and transpiration, 169–75; *see also* primary production
Prosopis cineraria, 485
Prosopis glandulosa, 274–5, 290
Prosopis glandulosa var. *torreyana*, *37*
Prosopis juliflora, 14, 39, 485, 486
Prosopis spp., 14, 34, 471
Prosopis tamarugo, 569
proteolysis, 340
Psammomys obesus, 371
psammophilous steppes, plant communities of, 363
Psammomtermes spp., 337
Pseudomonas spp., 345
Pseudomys spp., 131
pseudo steppes, plant communities of, 364
Pseudotsuga mensiezi, 61
Psoralea lanceolata, 326
Psyllidae, 290
Pterocles senegallus, 129
Pterocles spp., 147, 570
Pulicaria crispa, 346
Purshia tridentata, 45
Pycnostyctus spp., 92
Pyrethrum achilleifolium, *19*
Pyrrhuloxia sinuata, *212*

Quercus coccifera, 378
Quercus faginea, 378
Quercus ilex, 378
Quercus spp., 471
Quercus suber, 378
Quercus turbinella, 542

r-selection of animal species, 143–54, 158, 266–7, 579
radiotelemetry, and predation mortality rate, 63, 65
rainfall *see* precipitation
Rajasthan Desert, India: spatial variations in vegetation, 14; *see also* Indian Desert
Randonia africana, 347
rangeland management *see* management
Rattus spp., 131
Rattus villosissimus, 68
Reamuria negevensis, 15
Reaumuria soongorica, 173, *174*, *175*, *182*
Reaumuria spp., 25
Reboudia pinnata, *436*
recreation and tourism, **495–518**, 584–5; desert reserves in USA, 495, *496*, 509–18; progress in desert preserves, 500–8
Recurvirostra avosetta, 137
Red Sea, salinity and spatial variations in vegetation, 16
regulation of populations, 156–62
reproduction of animals, 147–56, 238–9
reserves, **495–518**; progress in, 500–8
resources, of arid-lands, management of *see* management
respiration *see* photosynthesis and respiration
Retama retam, 14, 364, 374
Retama spp., 25
Rhagodia spinescens, 110
Rhamnus lycioides, 375
Rhantherium suaveoleus, 363
Rhazya stricta, 328, 331, 346
Rheum tataricum, 173, 185
Rheum turkestanicum, 172
Rhinoceros merki, 378
Rhinoceros simus, 378
Rhinotermitidae, *89*

Rhizopus spp., 330
Rhombomys opimus, 141
Rhus pentaphyllum, 364
Rhus tripartitum, 364
Ricinus communis, 375
Rock Valley, Nevada, USA, 242; lizards in, 228–34; pocket mice in, 217–28, 286–7; secondary production in, 216–17; *see also* Mohave Desert; Nevada Desert
rodents: agonistic behaviour of, 72–4; breeding season and effects of rainfall on, 155; life-history patterns of, 144–5, 146–7, 151–2; and plant interactions, 88, 98, 195; population dynamics of, 129–31; production of, 202, *203*, 205–7, 217–28, 241–3; species separation of, 54, 55, 58, 60; water and energy flow, 286–8; *see also particular groups and species*
rootplowing, 542, 543
root: shoot ratio, *35–7*, 46
root systems, of perennials, 34–8, 333, 566; and mineral cycling, 333–5; and plant–plant interactions, 34–8
Rosmarinus officinalis, 360, 361, 375
Rubus ulmifolius, 378
Rumex tingitanus, 374

Saccharum bengalensis, 485
Saccharum spontaneum, 375
Sahara Desert: animal population variations, 129; climatic fluctuations, 366; reserves in, 505
Salazaria mexicana, *36*
Salicornia rubra, 16
Salicornia utahensis, 16
Salicornietum fruticosae, 16
Salicornietum herbaceae, 16
salinity, and spatial variations of vegetation, 15–16, *17*
Salix canariensis, 378
Salsola arbuscula, 171, *172*, 173
Salsola gemmascens, 172
Salsola orientalis, 172, *177*
Salsola passerina, 173, *174*, *182*
Salsola richteri, *170*, 171, *172*, 173; gas exchange, *176*; photosynthesis, 180
Salsola rubens, 172
Salsola sclerantha, 172
Salsola spp., 25, *37*
Salsola tentranda, 363
Salsola tetragona, 363
Salsola vermiculata, *36*, 362, 363, 375
Salsola zygophylla, 363
Salsoletum tetrandrae, 16
salt concentration, of soils: effects of plants on, 105–9; and problems of crop production, 493; *see also* sodium chloride
Salvadora oleoides, 14
sand dunes *see* dunes
Sapindus spp., 378
Sarayia site, Negev Desert: water and energy flow in, 278, *279*
Sarcobatus, spp., 471
Sarcobatus vermiculatus, 106, 347, 348, *349*
Sarcopoterium spinosum, 278, *279*
Sceloporus graciosus, 149–50
Sceloporus undulatus, 149
Schedorhinotermes actuosus, *89*
Schismus barbatus, 515
Schismus calycinus, 363, 568
Schistocerca gregaria, 52, 91; population variations, 127, *134*, 142
Schumannia karelinii, 172
Scincus scincus, *334*, 337
Scorophularia saharae, 374
Scorzonera pseudolanata, 575
Sde-Boqer site, Negev Desert: primary production, 283–5; water and energy flow, 278–9, *280*
seasonal variations, in vegetation, 23–5
secondary production *see* production, by desert animals
sedimentation, 373, 539–40
seed consumers, energy flow, 286–9, 295
seeding, for replacement of forage species, 471
selection systems *see* *K*-selection; life-history patterns; *r*-selection
selenium, cycling of, 316
Senecio gallicus, 363
Senecio subdentatus, 172
sheep, and plant interaction, 95–7; *see also* domestic stock
Silene arenaroides, 374
Silene colorata, 363
silicon, cycling of, in plants, *184*
simulations, of plant production, **433–44**; ARIDCROP model, 404, 439–43; BACROS model, 437–9; transpiration coefficient, 434–9; *see also* models
sinks *see* playas
Sisymbrium irio, 374
Smilax aspera, 378
Smithopsis spp., 88
Smirnovia turkestana, *170*
social aspects, of management, 549–53, 585
sodium, cycling of, in plants, *184*, 315
sodium chloride, dynamics of, 347–50
sodium monofluoroacetate (1080), 56
soils: animal effects on, 112–21; chemical effects on, (by plants) 105–12, (by animals) 112–21, 576; compaction of, 112–15; denudation of, 114–15; desertization of, 377; management of, 469–72; micro-organisms in *see* micro-organisms; and mineral cycling *see* mineral cycling; model of processes, 399–403; nitrogen in *see* nitrogen in soils; organic matter in, 107–8, 303–4, 336–8, 424–6; plant effects on, 13, 17–18, *19*, 105–12; processes in, 399–402, 561; physical effects on, (by plants) 109–13, (by animals), 117–19, 576; and steppization of, 375–6; and trampling effects, 114; water dynamics, *273–7*, *280–1*; *see also* erosion
Solanum spp., *93*
Sonoran Desert, USA, 21, 22, 24, 34; animal populations of, *130*, 146–7, 153, 155; animal species separation, 56, 60; predatory species, 69; shrub patterns and available moisture, 40–4
Sorex spp., 250
Spain, desert reserves of, 502–3
Spartium junceum, 378

Index

spatial effects, in modelling, **411–30**; animal habitats, 428; heterogeneity of sites, 411–15; nomadic migrations, 427; organic matter and nutrients, redistribution of, 424–6; piosphere effect, 428–30; plant survival and dispersal, 426–7; variation in vegetation, 13–21; water input, redistribution of, 415–24; *see also* patterns of plant distribution
species diversity *see* diversity
species segregation, 54, 75
species separation, 54–8, 60
Sphincterochyla zonata, 570, 571
Sphingidae, 133
Spizella breweri, 245
stabilization of dunes, 16–17, 18, 324–5
Stemphylium ilicis, 332
steppes, plant communities of, 362–3, *365*; alfa grass, 360–2; chaemaephytic, 362–3; halophilous crassulescent, 363–4; psammophilus, 363; pseudo, 364
steppization, 375–6
Sterculia diversifolia, 110
Stipa gobica, *182*
Stipa lagascae, 371
Stipa parviflora, 371
Stipa tenacissima, 360, 361, 362, 367–9
Streptomyces spp., 346
strip farming, 527
Struthio camelus, 129, 378
Suaeda asphaltica, 347
Suaeda erecta, *16*
Suaeda fruticosa, 16, 348, 363
Suaeda mollis, 364
Suaeda pruinosa, *24*, 363
Suaeda torreyana, 16
Suaeda vermiculata, 16
succession, 267–8; allogenic, 371–3; autogenic, 373–5
Sudan: spatial variations in vegetation, 15; pattern analyses, of vegetation and available moisture, 39
sulphur, cycling of, in plants, *184*
Sus crofa, 378
Swainsona sweinsonioides, 314
swales (basins), 276
Sylvia nana, 129
Sylvilagus audubonii, 146–7
Sylvilagus audubonii cedrophilus, 146–7
Sylvilagus spp., 153

Tamarix africana, 364
Tamarix aphylla, 364
Tamarix boreana, 364
Tamarix gallica, 364
Tamarix mannifera, 16
Tamarix spp., *37*, 326, *327*, 336, 337, 348, 374, 576
Tamiasciurus hudsonicus, *243*
Tassili n'Ajjer, 330, 337, 338, 343
Tatera spp., 287
Taterillas spp., 287
Taurotragus derbyanus, 378
Taxidea taxus, 56, *57*
temperature: annual variations, 21–2; and effects on energy flow, 272–82, 294–5; and germination, 21–2
temporal variations: in nutrient cycling, 304–7; in insect species, 61–2; and population dynamics, 139–43; in vegetation, 21–5
Tenebrionidae *see* beetles, tenebrionid
Tephrosia purpurea, 336
termites, 58, 575; and plant interactions, 89–90, 98–100; production of, 201–2, 210, 238, 241; and effects on soils, 112–21, 304; *see also particular groups and species*
Termitidae, *89*
territoriality, 73
Tetraclinis spp., 360
Tetracme recurvata, 172
Texas, spatial variations in vegetation, 14–15
Thamnosma montana, 38, 45
Thapsia garganica, 370
Thar Desert *see* Indian Desert
Thymelaea hirsuta, 14, 17, 370
Thymelaea microphylla, 361, 363
Thymelaea nitida, 361
Thymelea spp., 378
Thymelaea tartonraira, 361
Thymus capitatus, 375
Thymus hirtus, 360, 361, 375
Tilia spp., 378
tortoise *see Gopherus agassizi*
tourism *see* recreation
Tournefortia sogdiana, *172*
toxins, water-soluble, of desert perennials, 38
Toxostoma curvirostre, *212*, 213
Toxostoma lecontei, 129; productivity by, 238–40, 245
trampling effects, on soils, by animals, 114
transcinnamic acid, 38
transhumance, 464–5
transpiration, in plants, 169–75; coefficient, modelling of, 434–9; control of, 542–4; intensity of, 171–2
Triodia irritans, 39
Triodia pungens, 94, 95, *96*
Triticum sativum, *436*, *439*
Trochoidea seetzeni, 68
Troglodytes aedon, 148
Tulipa montana, 574–5
Tumulitermes bastilis, *121*
Tumulitermes pastinator, *120*
Tumulitermes spp., *89*
Tunisia, Pre-Saharan zone; productivity, variations in, 18–19; reserves of, 505–6; water and energy flow, 282
Turkey, desert reserves of, 506–8
Typheto-Scirpetum littoralis, 16
Tyto alba, 67

Ulmus scabra, 378
Ulmus spp., 378
Uma notata, 72
United States of America, desert reserves in, 495, 496, 497, *498*, 501, 502–3, 509–18; ecosystem modelling in, 388–402; *see also* America, North
Urginea undulata, 575
Urnisa guttulosa, 92
Urocyon cinereoargenteus, 54

Uromastix acanthinurus, 292, 336, 337
Urosaurusornatus, 139
Ursus arctos, 378
Uta stansburiana, 68, 72, 73, *130*, 137, 149, 159; production by, 201–2, 210, 238, 244, *250*, 251, 253–4
Utah, spatial variations in vegetation, 14, 22; *see also* Curlew Valley; Escalante Desert; Great Basin Desert
Uzbekistan, 22

vegetation: and climatic fluctuations, changes in, 21–5, 365–71; composition of, 13–19; management of, to increase water yield, 540–3; maps, *359*; present-day plant communities, 359–65; spatial variations, 13–22, 65, 67; temporal interactions, 21–5; variations in form, 20–1; *see also* dynamics, long-term, of arid-land vegetation; plants
Veromessor spp., 288
Veronica biloba, 172
Veronica campylopoda, *177*
vertebrates: population fluctuations, 136–7; soil effects of burrowing, 115–16; *see also particular groups and species*
Vitamin A shortage, and population dynamics, 138
Vulpes fulva, *57*
Vulpes macrotis, 71
Vulpes vulpes, 67
Vulpia myuros, *334*, 336

wadis, 10, 11–12
WATBAL model, 404
water: availability of *see* moisture availability; and erosion of soils, 111; modelling of, 422–4; redistribution of, 415–24; relations of plants to 169–75; *see also* precipitation; water and energy flow
water and energy flow, **271–95**, 558; and animal growth and behaviour, 285–92; and consumer factors, 286–92; dynamics of water, 272–82; resources, management of *see* management; spreading systems, 531–7; temperature effects, 272–82, 294–5
water-courses *see* arroyos
water resources *see* management of water
watersheds, upland, conservation of, 529–31; characteristics of, 533–4
water uptake, 109–10; *see also* transpiration
weeds *see* cultigene vegetation
Willow Creek Basin, Montana, USA: total soil moisture stress, *17*
wind erosion, of soil, and role of plants, 109–10

Xanthium spinosum, 373
xerophytes, root:shoot ratio, 35–7, 46

Yucca elata, *37*, 274
Yucca schidigera, *36*
Yucca spp., 287

Zanthoxylon fagara, 14
Zeiraphera griseana, 133
Zenaida asiatica, *212*
Zenaida macroura, *212*
Zilla macroptera, 347
Zilla spinosa, 14, 20, 21, 336, 346
Zitelletum spinosae, 326
Ziziphus lotus, 364, 374
Zizyphus nummularia, 485
Zygophyllaceae, 344
Zygophylletum albae, 16
Zygophylletum dumosi, *23*
Zygophylletum, 19
Zygophyllum album, 326, 333, 335, 344, 347
Zygophyllum coccineum, 14, 344
Zygophyllum dumosum, 14, 15, 35, 347, 563, 572
Zygophyllum fabago, *36*
Zygophyllum simplex, 335, 348
Zygophyllum thoxylon, *174*
Zygophyllum waterlotii, 347
Zygophyllum xanthoxylon, 173–5, *182*